Comparative Vertebrate Anatomy

Comparative Vertebrate Anatomy

seventh edition

A Laboratory Dissection Guide

Kenneth V. Kardong

Washington State University

Edward J. Zalisko

Blackburn College

with major new artistic contributions by

Kathleen M. Bodley

COMPARATIVE VERTEBRATE ANATOMY: A LABORATORY DISSECTION GUIDE, SEVENTH EDITION

ISBN 978-0-07-765705-5
MHID 0-07-765705-5

Senior Vice President, Products & Markets: *Kurt L. Strand*
Vice President, General Manager, Products & Markets: *Marty Lange*
Vice President, Content Production & Technology Services: *Kimberly Meriwether David*
Managing Director: *Michael Hackett*
Marketing Manager: *Patrick Reidy*
Director, Content Production: *Terri Schiesl*
Senior Content Project Manager: *Melissa Leick*
Buyer: *Susan K. Culbertson*
Compositor: *Laserwords Private Limited*
Typeface: *10/12 Times New Roman*
Printer: *R.R. Donnelley*

All credits appearing on page or at the end of the book are considered to be an extension of the copyright page.

Library of Congress Cataloging-in-Publication Data

Kardong, Kenneth V.
    Comparative vertebrate anatomy : a laboratory dissection guide / Kenneth V. Kardong, Washington State University,
    Edward J. Zalisko, Blackburn College ; with major new artistic contributions by Kathleen M. Bodley. – Seventh edition.
        pages cm
Includes index.
    ISBN 978-0-07-765705-5 (alk. paper)
    1. Vertebrates—Dissection—Laboratory manuals.   2. Anatomy, Comparative—Laboratory manuals. I. Zalisko,
Edward J. II. Title.

    QL812.5.K37 2015
    596.078--dc23

                                                                    2013045422

The Internet addresses listed in the text were accurate at the time of publication. The inclusion of a website does not indicate an endorsement by the authors or McGraw-Hill Education, and McGraw-Hill Education does not guarantee the accuracy of the information presented at these sites.

www.mhhe.com

*"For Willemina, Kyle, and Jason,
and for Amy, Ben, Sarah, and Jennifer"*

Contents

Preface viii

1 Introduction 1

Classification and Comparison 1

Defining the Chordates 2
Chordate Characteristics 2

Studying Advice 5
Laboratory Strategy 5

Designing for Students 7

2 Protochordates 8

Protochordates 8
Hemichordata 9
Cephalochordata 11
Urochordata 15

3 Agnathans—Examination of a Primitive Vertebrate: The Lamprey 18

Agnathans 18
Myxini—Hagfishes 18
Petromyzoniformes—Lampreys 18
The Origin and Evolution of Vertebrates: Shifts in Feeding Mechanisms 18

Adult Lamprey Anatomy 22
External Anatomy 22
Sagittal and Cross Sections 22

Anatomy of the Lamprey (Ammocoete) Larva 27
Whole Mount 27
Cross Section through the Pharynx 28
Cross Section through the Intestine 28

4 Vertebrate Integuments 29

Introduction 29
Dermis 29
Epidermis 29

Examination of Vertebrate Integuments 31
Fishes 31
Tetrapods 33

Specializations of the Integument 37
Nails, Claws, and Hooves 37
Horns and Antlers 37
Baleen 38

5 Skeletal Systems 40

Tissues of the Skeletal System 40
Cartilage 40
Bone 41

Divisions of the Skeletal System 42

Postcranial Skeleton 43
Vertebrae and Ribs 43
Girdles and Limbs 53

Skull 66
Introduction 66
The Composite Skull 66

Cranial Skeleton 68
Elasmobranchii—Shark 68
Osteichthyes—*Amia* (Bowfin) 71
Amphibia—*Necturus* 71
Reptilia 72
Aves 78
Dorsal and Lateral Views 79
Ventral and Posterior Views 79
Sclerotic Ring and Hyoid 79
Mammalia—Cat 79

Teeth 83
Tooth Anatomy 83
Sharks—Homodont and Polyphyodont Teeth 84
Amphibia/Salamander—Homodont and Polyphyodont Teeth 84
Reptiles/Alligator—Homodont and Polyphyodont Teeth 84
Mammal—Heterodont and Diphyodont Teeth 84
Pattern of Evolutionary Change 85

6 Muscular Systems and External Anatomy 88

Introduction 88
Terminology 88
Actions 88
General Muscle Groups 89
Dissection 89
The Process 89
General Safety Precautions 89
External Anatomy 89

Shark Dissection 89
External Anatomy 89
Skinning the Shark 91
Musculature 91

Necturus   99
   External Anatomy   99
   Dissection   100
   Musculature   100

Cat   105
   External Anatomy   105
   Approach   105
   Musculature   105

7   Digestive Systems   131

Introduction   131

Shark   131
   Pleuroperitoneal Cavity   131
   Pericardial Cavity   133

Necturus   136
   Pleuroperitoneal Cavity   136
   Pericardial Cavity   138

Cat   138
   Salivary Glands   138
   Oral Cavity   138
   Structures of the Thoracic Region   141
   Abdomen   143
   Patterns & Connections   145

8   Circulatory and Respiratory Systems   146

Introduction   146
   Respiratory System   146
   Circulatory System   146

Shark   146
   Pericardial Cavity   146
   Respiratory System and Branchial Arteries   146
   Pleuroperitoneal Cavity   150

Necturus   154
   The Heart, Branches of the Ventral Aorta, and
      Gills   154
   Dorsal Aorta and Lungs   156
   Hepatic Portal System   160
   Systemic Veins   160

Cat   160
   Heart and Its Major Vessels   161
   Respiratory System   166

   Precava   167
   Dorsal Aorta   168
   Postcava   170
   Hepatic Portal System   170

9   Urogenital Systems   172

Shark   172
   Male Urogenital System   172
   Female Urogenital System   175

Necturus   177
   Male Urogenital System   177
   Female Urogenital System   179

Cat   179
   Male Urogenital System   179
   Female Urogenital System   182

10   Nervous Systems   185

Introduction   185
   Nerve Function   185
   Spinal Nerves   185
   Cranial Nerves   186

Shark   189
   Dissection Procedure   189
   Brain (Dorsal Aspect)   190
   Cranial Nerves   192
   Brain (Ventral Aspect)   194
   Brain (Sagittal Aspect)   194

Sheep Brain   196
   Overview—Whole Brain   196
   Cranial Nerves—Whole Brains
      (Figure 10.11)   197
   Brain Regions—Sagittal Sections
      (Figure 10.12)   200
   Brain Stem—Parts (Figure 10.13)   202
   Spinal Nerves   203

Glossary   204
Credits   210
Index   211

Student Note Book   1–148

Preface

Hands-on Engagement

One of the most effective ways to master a subject, and make it one's own, is to meet it directly by hands-on engagement. A laboratory accompanying a course in comparative vertebrate anatomy provides this direct, practical experience. Students are encouraged to become active participants in their own mastery of vertebrate design. Although there is no single "vertebrate morphology" and no single way to run a laboratory, a course in vertebrate anatomy includes modern expectations in a discipline that should be a centerpiece of a university curriculum in organismal biology. First and foremost, the basic animal architecture—the anatomy—should be understood in a clear and concise way. Second, the significance of the discovered anatomy must be considered, beyond the business of putting names on parts.

The anatomy raises questions about how the systems work and holds clues to the remarkable story of how vertebrates evolved. An understanding of these form-and-function relationships and the evolutionary history gives meaning and significance to the structure that is studied. With this philosophy in mind, we have prepared a laboratory dissection guide that sets the foundations and the tone for instructors who wish to provide a challenging laboratory course.

Learning and Instructional Aids

We have prepared a richly illustrated laboratory manual that carefully guides students through dissections. Throughout the dissections, we pause strategically to bring the students' attention to the significance of the material they have just covered. Further, we have organized this laboratory guide to make clear to students how the text is used, and to connect the dissections found here to related material in the textbook

Vertebrates: Comparative Anatomy, Function, Evolution, Seventh edition, by Kenneth Kardong

For the Instructor

- **Flexibility of Use**—In most chapters, instructors will find a depth of coverage that is sufficient for a traditional course, yet structured to allow less detailed coverage where desired, thereby providing the flexibility you may need. You can mix and match. Chapters are written to permit flexibility in the course sequence. Each chapter can be studied as a distinct unit, which gives instructors the flexibility to build a laboratory that suits their needs. However, we believe the sequence of chapters in this laboratory guide is a reasonable compromise of practical and conceptual goals. For example, the earlier chapter on protochordates introduces the historical background to the vertebrates. We have placed the examination of external anatomy just before the muscular systems, since most laboratories begin with "dry" material, then move later to "wet" preserved specimens. Dragging out wet, preserved cat, shark, Necturus, or other vertebrates during the first days of a course is much less convenient. Instead, bringing these preserved animals out just prior to the muscle dissections is an economical use of the instructor's time and makes a nice transition from dry to wet dissections for students.

- **Integration**—This laboratory manual, a self-contained guide to dissection of representative vertebrates, is designed to stand alone. However, many classrooms will be using the textbook *Vertebrates: Comparative Anatomy, Function, Evolution,* Seventh edition, by Kenneth Kardong. Users of the Kardong textbook may take note of these icons ▯, which are cross-references from the dissection guide to supporting pages in the textbook. Their purpose is to integrate the laboratory with the lecture, and the hands-on identification of anatomical parts with the functional and evolutionary significance of the discovered morphology. This is intended to reinforce our overall philosophy of getting students to think critically about the material, thereby increasing the usefulness of the textbook and the dissection guide.

- **Companion Website (www.mhhe.com/kardong7e)**—This dissection manual remains a centerpiece of the study of vertebrate architecture, addressing our first goal of providing a clear and concise guide to the examination of vertebrate anatomy. But students benefit from other activities that engage them further in thinking about the function and evolution of vertebrate systems. On our website we provide functional labs, laboratory exercises that you can use as supplements or to extend the study of functional and evolutionary aspects of the anatomical material addressed in the laboratories. We have tested these activities in our courses and found students engaged in and appreciative of the connections between the laboratories and lectures. These functional and evolutionary laboratory exercises are organized with current ideas of instructional pedagogy in mind. Each exercise steps the student through an inquiry system, building step by step, and concluding with insight into the functions and/or evolution of the anatomy examined in lab. Instructors facilitate and students engage in these critical thinking supplements.

In addition to the supplemental labs, you will find images from the seventh edition of Kenneth Kardong's *Vertebrates: Comparative Anatomy, Function, Evolution* on the website, ready for use in your course.

For the Student

For you as a student, perhaps coming to this subject for the first time, we have provided a variety of special features designed to support your efforts. If you use these resources and let them work on your behalf, it will greatly aid not only your work in the laboratory but also in the lecture part of the course. Here they are:

- **Integrated Illustrations**—The written descriptions that guide you through the dissections are accompanied by high quality and informative figures. You will work through the written text to discover the anatomical parts for yourself, but the integrated figures allow you to do so in an efficient and economical manner. We are especially pleased with the usually difficult soft-tissue systems that appear throughout the dissection guide. The animals are posed in more helpful orientations that clarify different points in the dissection.

- **Student Art Notebook**—This icon beside a figure in the laboratory manual indicates that the figure is included in the Student Art Notebook, which is at the end of the manual. To facilitate the review of anatomical structures, the labels of these figures have been replaced with numbers referenced to a key, which is located on the same page as the art or on the facing page. The art notebook figures can also be removed in order to place homologous systems of various animals next to each other for easy comparison. This invites you to think in terms of comparison from group to group and to think of the evolutionary changes these representative animals help to illustrate.

- **Boxed Essays**—"Patterns & Connections" and "Form & Function" boxes invite consideration of the larger implications of the anatomy, moving from dissection to the significance. What is the functional significance of the positions, patterns, shapes, and interrelationships of the anatomy under dissection? What might this suggest about the evolution of

characters that define derived groups of vertebrates? These questions help correlate the laboratory exercises with the lectures.

- **Design and Organization**—The text is organized to be accessible and to provide clear direction about what you are to do.

- **General Introduction**—To understand the context in which the dissections are undertaken, each chapter and some chapter sections begin with a general introduction to the anatomy. With this introduction, the text then carefully guides you through the material, using precise anatomical terminology to verify identifications.

- **Dissection Directions**—To distinguish between the background text and specific dissection directions, the dissection directions are set off with an icon 🦖 and indentation from the textual narrative.

- **Keyed Terminology**—Anatomical terms and nomenclature are differentiated so that the student may easily refer to them. The anatomical structures that are feasible to locate are in **boldface;** important conceptual terms are in *italics*. A **Glossary** at the end of the dissection guide is a reference for quick review of some of these terms.

- **Synonyms**—Common synonyms are indicated in parentheses.

New to This Edition

This latest edition of our laboratory manual incorporates valuable suggestions of expert reviewers and our students who use and edit the manual as we use it in our laboratories. For greatest clarity and accuracy, we have reformatted some of the illustrations, edited labeling, and rewritten anatomical descriptions. In addition, we have made the following changes.

- A new table of anatomical terminology.
- New "Patterns & Connections" boxed essay, the microbiota of vertebrate digestive tracts.
- Revised many new labels of existing figures to improve recognition of structures described in the laboratory text.
- Made extensive changes in the text to improve the consistent use of terminology and anatomical directions.
- The laboratory manual has been updated to direct students to related sections of the new seventh edition of the companion textbook *Vertebrates: Comparative Anatomy, Function, Evolution*, by Kenneth Kardong.

Acknowledgments

In addition to support we have received from family, friends, and colleagues, we also express our appreciation for the constructive advice offered by our Blackburn College and Washington State University comparative vertebrate anatomy students who used the prior editions, and the reviewers who helped shape previous editions:

William Anyonge
 University of California,
 Los Angeles
David Bardack
 University of Illinois,
 Chicago
Michael Cassiliano
 University of Wyoming
Joseph Clark
 Blackburn College
Terry D. Jones
 Stephen F. Austin State
 University
E. Dale Kennedy
 Albion College
Gavin R. Lawson
 Bridgewater College
Kevin Lumney
 The Ohio State
 University
Martin M. Matute
 University of Arkansas,
 Pine Bluff
Earl H. Meseth
 Elmhurst College

Virginia L. Naples
 Northern Illinois
 University
Maureen Scott
 Norfolk State University
David Shomay
 University of Illinois,
 Chicago
Amanda Starnes
 Emory University
Randall L. Tracy
 Worcester State
 College
Ralph G. Turingan
 Florida Institute
 of Technology,
 Melbourne
Gene K. Wong
 Quinnipiac University
Mark L. Wygoda
 McNeese State
 University
Christopher J. Yahnke
 University of Wisconsin
 Stevens Point

In addition, we thank the following reviewers who provided thoughtful and constructive suggestions for this seventh edition:

Robert Aldridge
 Saint Louis University
Deborah Anderson
 St. Norbert College
Clare Chatot
 Ball State University
George Cline
 Jacksonville State
 University
Claude Gagna
 New York Institute of
 Technology
DeLoris Hesse
 University of Georgia

Mason B. Meers
 University of Tampa
Sue Ann Miller
 Hamilton College
Chris Murdock
 Jacksonville State
 University
Susan Murphy
 Our Lady of the Lake
 University
Maureen Scott
 Norfolk State University
Charles M. Watson
 NcNeese State University

We also appreciate the enduring support of many associates at McGraw-Hill: Lynn Briethaupt, Lori Bradshaw, and Melissa Leick and the production staff. Finally, we would like to thank Gavin Lawson, Faye Prevedell, Julie Brown, Hazen Audel, Laszlo Meszoly, Emily Green, and especially Kathleen M. Bodley for their artistic contributions. Special recognition goes to Tamara L. Smith whose personal and professional support has been so important and much appreciated during revisions of this laboratory manual.

Introduction

Classification and Comparison

Comparative vertebrate anatomy examines the structure of vertebrates classified in the phylum Chordata, subphylum Vertebrata. To best understand the history of vertebrate evolution, we need to examine representative forms, such as those indicated in figure 1.1. Unfortunately, detailed anatomy, especially of key fossils, is not always known or easily studied. Instead, we must focus on convenient, representative animals that exhibit ancestral and derived traits. For example, this laboratory manual provides for a detailed dissection of a shark, a salamander, and a cat. These three animals generally represent a cartilaginous fish, an amphibian, and a mammal, respectively. However, these modern

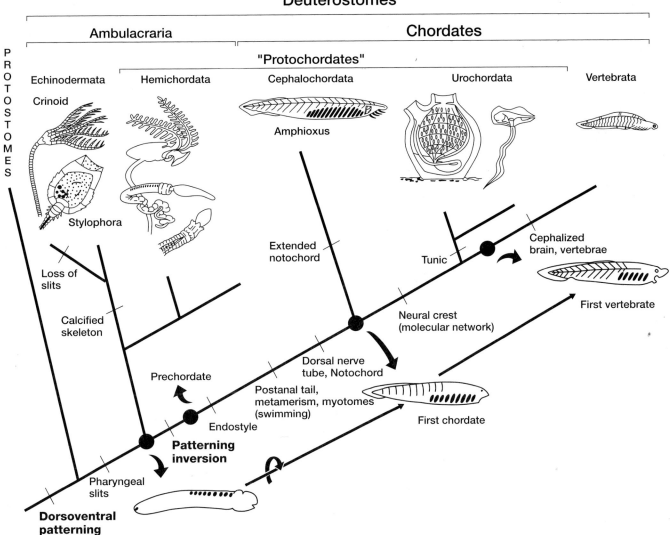

FIGURE 1.1 Chordates. One possible evolutionary sequence is shown using traditional names. An understanding of these groups reveals how vertebrates changed through time. But because many of these early groups do not survive or are not easily accessible, we study selected living vertebrates, which we hope provide a reasonable introduction into these evolutionary events. Students must remember that these modern examples may have changed significantly since their groups first evolved.

animals have evolved, sometimes considerably, since their ancestors first appeared hundreds of millions of years ago. Therefore, as we examine these living representatives, we must remember that they are also specialized for current lifestyles.

Although only three animals will be dissected in detail, many more animal groups will be examined in the protochordate, integument, and skeletal system chapters (chapters 2–5). Thus, you should learn a chordate classification scheme (table 1.1) and be able to use it. We suggest that for each animal you study, you know its Linnaean classification from the phylum down to the class (when available). In addition, you should be able to completely classify the three animals that are dissected in detail (table 1.2).

Defining the Chordates

Chordates are neither the most diverse nor the largest of the animal phyla, although in terms of the number of species, they come in a respectable fourth behind arthropods, nematodes, and mollusks. Chordates have a fluid-filled internal body cavity termed a *coelom*. Chordates, along with other animals possessing such a coelom, are grouped together as the *coelomates*. Among these coelomate animals, two apparently distinct and independent evolutionary lines are present. One line of coelomates, called the *protostomes*, includes mollusks, annelids, and arthropods together with assorted smaller phyla (figure 1.2). The other line, the *deuterostomes*, includes several other small phyla, plus echinoderms, hemichordates, and chordates. The distinction between protostomes and deuterostomes rests upon certain embryological characteristics (table 1.3; figure 1.3). Chordates are classified as deuterostomes: The mouth forms opposite to the blastopore; generally, cleavage is radial; the coelom is an enterocoelom; and the skeleton arises from mesodermal layers of the embryo.

We should be clear from the beginning, however, about the character of the chordate phylum itself. It is easy to forget that two of the three chordate subphyla are technically *invertebrates*. The Chordata traditionally divides into three subtaxa of unequal size: the Cephalochordata, commonly represented by amphioxus; the Urochordata, commonly known as sea squirts or tunicates; and the largest group, the Vertebrata, or vertebrates. Strictly speaking, the "invertebrates" include all animals except members of this last group, the "vertebrates."

The earliest chordate fossils appear in the Cambrian period. Although later chordates would evolve hard bones and teeth that preserved well, ancestors to the first chordates likely had soft bodies. Early chordates left little fossil trace of the evolutionary pathway taken from prechordate to chordate. Thus, to decipher chordate origins we use evidence from anatomical clues carried in the bodies of living forms. To be able to evaluate the success of these attempts at tracing chordate origins, we first need to identify the features that define the chordates. We will then attempt to discover the animal group that is the most likely evolutionary source for chordates.

Chordate Characteristics

At first glance, the differences among the three chordate subtaxa are more apparent than the similarities that unite them. Most vertebrates have an endoskeleton, a system of rigid internal elements, of bone or cartilage, beneath the skin. The endoskeleton participates in locomotion, support, and protection of delicate organs. Some chordates are terrestrial, and most use jaws to feed. But cephalochordates and urochordates are all marine animals, none are terrestrial, and all lack a bony or cartilaginous skeleton, although their support system may involve

---

TABLE 1.1 Traditional Classification

Phylum Chordata

Subphylum Cephalochordata (amphioxus/*Branchiostoma*)

Subphylum Urochordata (tunicates)

Subphylum Vertebrata (extant groups)

    Class Myxini (hagfish)

    Class Petromyzoniformes (lampreys)

    Class Chondrichthyes

        Subclass Elasmobranchii (sharks and rays)

        Subclass Holocephali (chimaeras)

    Class Osteichthyes

        Subclass Actinopterygii (paddlefish, sturgeon, gar, bowfin, bass, tuna, etc.)

        Subclass Sarcopterygii (lungfish and lobe-finned fishes)

    Class Amphibia (frogs, salamanders, and caecilians)

    Class Reptilia

        Subclass Parareptilia (turtles)

        Subclass Eureptilia (snakes, lizards, tuatara, and crocodilians)

    Class Aves

    Class Mammalia

        Subclass Prototheria (monotremes: duckbill platypus and echidna)

        Subclass Theria

            Infraclass Metatheria (marsupials: kangaroo, opossum, etc.)

            Infraclass Eutheria (placentals: rodents, apes, cats, etc.)

TABLE 1.2 Linnaean Classification of Representative Vertebrates

Common Name	Dogfish Shark	Mudpuppy	Domestic Cat
Subphylum	Vertebrata	Vertebrata	Vertebrata
Class	Chondrichthyes	Amphibia	Mammalia
Subclass	Elasmobranchii	Lissamphibia	Theria
Order	Selachimorpha	Urodela	Carnivora
Family	Squalidae	Proteidae	Felidae
Scientific Name	*Squalus acanthias*	*Necturus maculosus*	*Felis domestica*

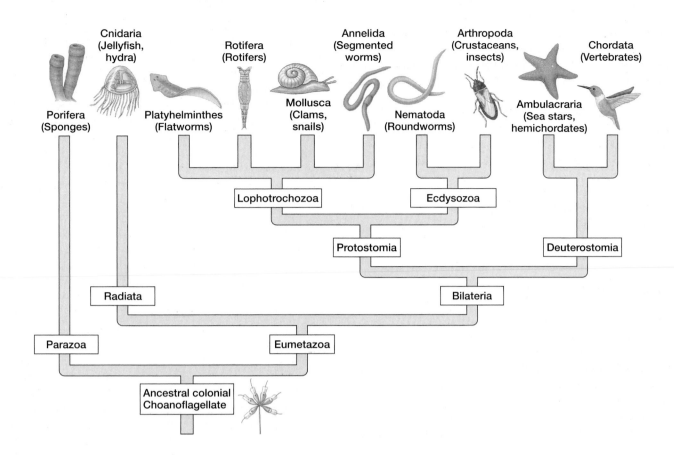

FIGURE 1.2 Diagram of the relationships of the major metazoan groups. The coelomate body plan and deuterostome form of development separate the echinoderms, hemichordates, and chordates from all other phyla.

TABLE 1.3	Protostomes Versus Deuterostomes
Protostomes	**Deuterostomes**
Blastopore (mouth)	Blastopore (anus)
Spiral cleavage	Radial cleavage
Schizocoelic coelom	Enterocoelic coelom
Ectodermal skeleton	Mesodermal skeleton

rods of collagenous material. Cephalochordates and urochordates are suspension feeders, using a sticky sheet of mucus to collect food from streams of water passing over a filtering apparatus. All three subtaxa, despite these superficial differences, share four chordate synapomorphies: 1) a notochord, 2) an endostyle or thyroid gland, 3) a dorsal and tubular nerve cord, and 4) a postanal tail. These join pharyngeal slits arising earlier in Hemichordates. Together these five characteristics help diagnose the chordates.

Pharyngeal Slits

The *pharynx* is a part of the digestive tract located immediately posterior to the mouth. During some point in the lifetime of all chordates, the walls of the pharynx are pierced, or nearly pierced, by a longitudinal series of openings, the *pharyngeal slits.* The term "gill slits" is often used in place of pharyngeal slits for each of these openings in the wall of the pharynx, but a "gill" proper is a specialized structure composed of tiny plates or folds that harbor capillary beds for respiration in water. In many primitive chordates, these openings serve primarily in feeding, but in embryos they play no respiratory role. Therefore, the term "gill slits" is misleading.

Pharyngeal slits may appear early in embryonic development and persist into the adult stage, or pharyngeal slits may be overgrown and disappear before the young chordate is born or hatched. Whatever their eventual embryonic or adult fate, all chordates show evidence of pharyngeal slits at some time in their lives.

When slits first arose, they likely aided feeding. As openings in the pharynx, they allowed the one-way flow of a water current: in at the mouth and out through the pharyngeal slits (figure 1.4). Secondarily, when the walls defining the slits became associated with gills, the passing stream of water also participated in respiratory exchange with the blood circulating in the capillary beds of these gills. The water entering the mouth could bring suspended food and oxygen to the animal. As it exited through the slits and across the gills, carbon dioxide was given up to the departing water and carried away. The current of water can thus simultaneously support feeding and respiratory activities.

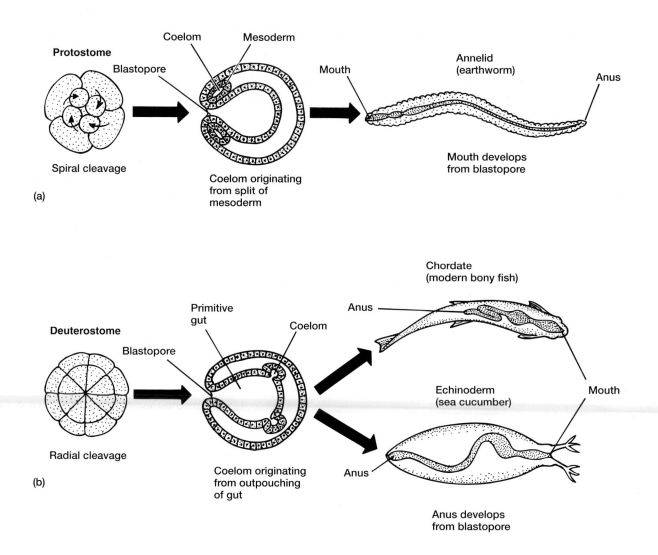

FIGURE 1.3 Protostomes and deuterostomes. The Bilateria are divided into two major groups on the basis of embryonic characteristics. (a) Protostomes show spiral cleavage, coelom formation by splitting of mesoderm, and derivation of the mouth from the blastopore. (b) Deuterostomes show radial cleavage, coelom formation by outpocketing, and derivation of the anus from or in the vicinity of the blastopore.

In primitive chordates, the pharynx itself is often expanded into a pharyngeal or branchial basket, and the slits on its walls are multiplied in number, increasing the surface area exposed to the passing current of water. Sticky mucus lining the pharynx adheres to food particles from suspension. Sets of cilia, also lining the pharynx, produce the water current. Other cilia gather the food-laden mucus and pass it into the esophagus. This mucus-and-cilia system is especially efficient in small organisms that are *suspension feeders,* those that extract food floating in water. Such a ciliary mucus feeding system is prevalent in primitive chordates and in groups that preceded them.

Notochord

The notochord is a slender rod that lies dorsal to the coelom but beneath and parallel with the central nervous system. The taxon takes its name, Chordata, from this structure. The notochord is typically composed of a core of cells and fluid that are encased in a tough sheath of fibrous tissue. The notochord is a hydrostatic organ with elastic properties that resist axial compression. It lies along the long axis of the body, allowing lateral flexion and preventing longitudinal collapse of the body during locomotion.

Endostyle or Thyroid Gland

The thyroid is involved in iodine metabolism, and so is the endostyle, leading some to suggest that one (endostyle) is the predecessor of the other (thyroid). If so accepted, then all chordates (and possibly hemichordates) have endostyles (urochordates, cephalochordates, larval lamprey) or thyroids (adult lamprey, all other vertebrates).

Dorsal and Tubular Nerve Cord

The nerve cord in chordates lies above the gut and is hollow along its entire length (figure 1.4a,e). The major nerve cord in most invertebrates is ventral in position, below the gut,

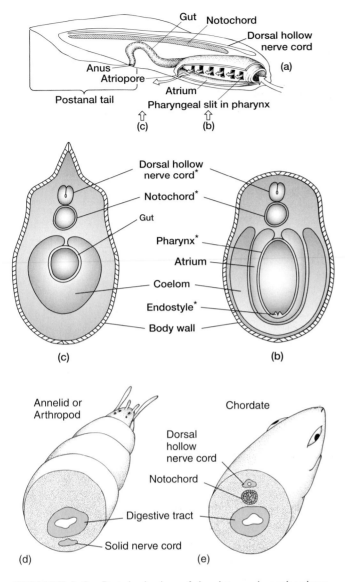

FIGURE 1.4 Basic body plans of chordates and nonchordates. (a) Generalized chordate. A single stream of water enters the mouth into the pharynx, then exits through several pharyngeal slits. In many lower chordates, water exiting through the slits enters a common enclosing chamber, the atrium, before returning to the environment via the single atriopore. (b) Cross section through the pharynx showing the tube (pharynx) within a tube (body wall) organization. (c) Cross section through region posterior to the pharynx. Asterisks indicate chordate synapomorphic characters. (d) Basic body plan of an annelid or arthropod. In such animals, a definitive nerve cord, when present, is ventral in position, solid, and below the gut. (e) Basic chordate body plan. The nerve cord of chordates lies in a dorsal position above the gut and notochord. Its core is "hollow," or more correctly, a fluid-filled central canal.

and solid (figure 1.4d). The advantage, if any, of a tubular rather than a solid nerve cord is not known, but this distinctive feature is found only among chordates.

Postanal Tail

Chordates possess a postanal tail that represents a posterior elongation of the body beyond the anus. The tail is primarily an extension of the chordate locomotor apparatus, which is composed of segmental musculature and a notochord. The tail provides an increased surface area important for thrust and maneuverability in an aquatic environment.

Chordate Body Plan

What is common, then, to all chordates are four synapomorphic traits: 1) notochord, 2) endostyle/thyroid, 3) dorsal and tubular nerve cord, and 4) postanal tail. Pharyngeal slits, a primitive trait, are also a prominent feature of all chordates. These may be present only briefly during embryonic development, or they may persist into the adult stage. But all chordates exhibit these distinctive characteristics at some point during their lifetimes. Taken together, they are a suite of characters found only among chordates. Chordates are also bilaterally symmetrical (figure 1.5b) and show segmentation. Blocks of muscle, or *myomeres,* are arranged sequentially along the body and tail as part of the outer body wall.

Now that we have described the basic and secondary characteristics of chordates, we turn our attention to the ancestry of this group. Biologists interested in such questions often consult an assortment of primitive animals that in some ways seem in their structure and design to have preceded chordates phylogenetically and in other ways seem to represent chordates at their earliest stages. These animals are the protochordates, and they will be addressed next in chapter 2.

Studying Advice

Laboratory Strategy

Association, correspondence between parts, and facility with descriptive terminology are as important to learn as the names of structures. In reading through the written descriptions, you are encouraged to learn the parts more by their association with one another than just as a picture atlas. For example, the certainty in identifying a blood vessel comes from confirming the region it serves, a muscle from noting its attachments, a bone from its shape and position, and so on. These features are less obvious or less likely to be noted if you rely on pictures rather than working through, and hence actually thinking through, a specimen described in words.

Uncertainty, impatience, and a lust for speed in completing an assignment will tempt you to rely more on picture-books than upon your own thinking resources. Yes, the material can be learned by brute force, but such knowledge vanishes quickly and your skill in becoming conversant and comfortable with descriptive anatomy will remain rudimentary. We thus urge you to work with and, hence, think through the written descriptions.

Approaches

Two approaches to descriptive laboratory anatomy are presently in vogue. One is "regional anatomy," wherein a region of the body, for example the forelimb, is dissected and the bones, muscles, blood vessels, nerves, and other structures are examined simultaneously. The advantages are that you are introduced to everything on one dissecting pass and, philosophically, the integrated nature of structures is stressed. The second approach is a "systems anatomy." Each system (skeletal, muscular, digestive, etc.) is examined separately. Our text takes this approach. Systems are treated as entities rather than as integrated with one another. That is a disadvantage. Special responsibility

thus lies with each student to add systems together functionally as each new one is learned. Knowledge and understanding should accumulate, not displace each other. The advantage to a systems approach is that it facilitates and focuses comparisons between representative animals. Making comparisons and understanding the adaptive reasons for changes so revealed are the major goals of the course.

In the laboratory manual, "left," "right," "dorsal," "ventral," and other directional words refer *to the animal's* left, right, and so on, not to yours as you peer down on the specimen. The planes of the body are illustrated in figure 1.5.

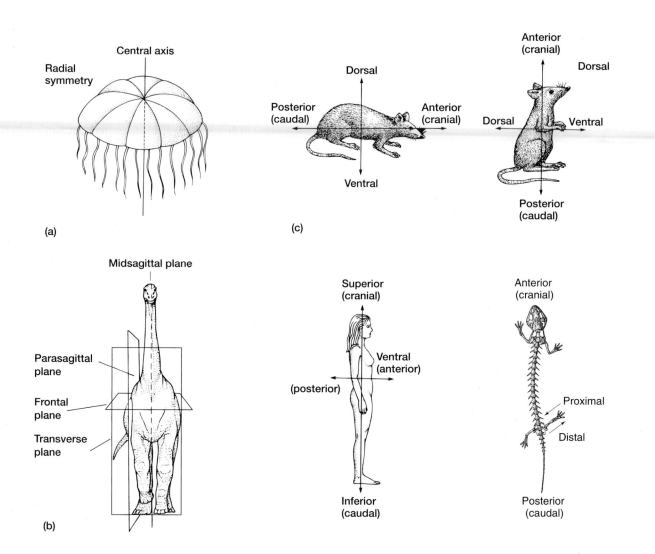

FIGURE 1.5 Body symmetry. Radial and bilateral are the two most common body symmetries. (a) Radially symmetrical bodies are laid out regularly around a central axis. (b) Bilaterally symmetrical bodies can be divided into mirror images only through the midsagittal plane. Lateral and medial refer to direction from a midsagittal plane. (c) Dorsal and ventral refer to back and belly, respectively, and anterior and posterior to cranial and caudal ends, respectively. In animals that move in an upright position (e.g., humans and birds), superior and inferior apply to cranial and caudal ends, and ventral and dorsal apply to anterior and posterior sides. (d) Lateral and medial refer to direction from a midsagittal plane. Proximal (close to) and distal (away from) refer to the main mass of the body.

TABLE 1.4 Anatomical Terminology

Terms	Definitions/Descriptions
Left/Right	The animal's left and right, independent of our perspective
Anterior/Cranial	Towards the head end of the body
Posterior/Caudal	Towards the tail end of the body
Dorsal	Towards the back
Ventral	Towards the belly
Medial	Towards the midline (mid-sagittal plane) of the body
Lateral	Away from the midline (mid-sagittal plane) of the body
Proximal	Towards the point of attachment of an appendage
Distal	Away from the point of attachment of an appendage
Superficial	Towards the surface
Deep	Away from the surface

Study Tips

Read the assigned section of the laboratory manual before each laboratory. This will give you an overview, help organize your work, and introduce you to the structures to be located.

Make a few, simple drawings after completing a section. The drawing should be uncomplicated. Delicate shading and subtle coloring should be avoided. The drawing should be an outline of the structure: forelimb bones, jaw muscles, branches of the dorsal aorta, and so on. These drawings are a personal reference for you and a personal check to see if you have identified structures in a way that makes sense to you. To make a simple, but accurate drawing, you must have done a good dissecting job and intellectually have some idea of the topography discovered. Don't overdo it. Don't make lots of drawings; just a few simple, quick, but accurate ones for your use and benefit.

The most effective studying occurs frequently and for relatively short periods of time (an hour or less). Just as athletes practice their sports on a regular schedule, and not just the day before a game, reviewing material and mastering terminology comes not with a sudden burst of effort just prior to an exam. Instead, regular and steady progress, with frequent breaks to rest the mind, yields the greatest depth of retention and recall.

Designing for Students

As a student, the style of your participation in a laboratory differs from how you approach the lecture. But keep in mind that the subject is the same. Lab and lecture are simply different methods to the same goal, namely to becoming conversant with the descriptive terminology and to understanding vertebrate form, function, and evolution. To help set the conceptual context in which you undertake dissections, each chapter and some chapter sections begin with a general introduction. The actual examination of specimens and identification of parts then follows. To distinguish between background material and specific dissection directions, dinosaur icons and indented text are used. This is intended to make it clear as to what you are to read as background and what you are to do as part of the actual dissection. The structures feasible to locate on the specimens are in boldface type; important terms are italic. Take a moment to scan a few of the later chapters and notice how this design feature of the laboratory manual is used.

2 Protochordates

Protochordates

Between the vertebrates and other invertebrates lie the "protochordates," which usually include three groups: the hemichordates, cephalochordates, and urochordates (figure 2.1). Although they are an informal assemblage of animals and are not a proper taxonomic group, protochordates share some or all five features of the fundamental chordate body plan. Living protochordates are themselves products of a long evolutionary history, independent from other taxa. Their anatomy is relatively simple, and their phylogenetic position ancient. The hope is that within living members of the protochordates are traces of the evolutionary steps from prechordate to chordate. A closer look at these protochordates will indicate why they provide tantalizing clues to the origin of the chordates.

All protochordates are marine and feed by means of cilia and mucus. But, as young larvae they often live quite different lives than they do as adults. As larvae, they may be *pelagic,* living in open water between the surface and the bottom. Although unattached, most free-floating larvae have

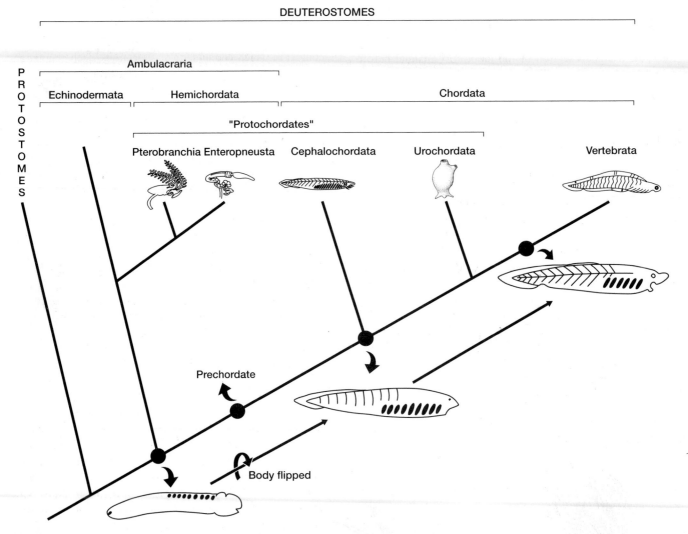

FIGURE 2.1 Phylogenetic relationships within the deuterostomes. Note that between the Ambulacraria (Echinodermata + Hemichordata) and Cephalochordata a body inversion occurs, reversing the dorsoventral axis. Other major changes in character states are shown along the way.

Based on Mallatt, 2007.

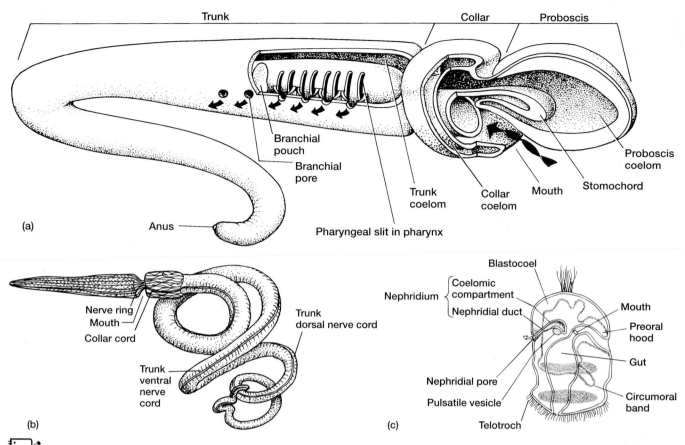

Trunk   Collar   Proboscis

Branchial pouch

Branchial pore

(a)

Anus

Trunk coelom

Collar coelom

Mouth

Proboscis coelom

Stomochord

Pharyngeal slit in pharynx

Nerve ring
Mouth
Collar cord

Trunk dorsal nerve cord

Trunk ventral nerve cord

(b)

Blastocoel

Nephridium { Coelomic compartment

Nephridial duct

Mouth

Preoral hood

Gut

Nephridial pore

Pulsatile vesicle

Circumoral band

Telotroch

(c)

FIGURE 2.2 Hemichordata, Enteropneusta. (a) Generalized acorn worm. Proboscis, collar, and trunk regions are shown in partial cutaway view, revealing the coelom in each region and the associated internal anatomy. Within the proboscis is the stomochord, an extension of the gut. The food-laden cord of mucus (spiral arrow) enters the mouth together with water. The food is directed through the pharynx into the gut. Excess water exits via the pharyngeal slits. Several slits open into a common compartment, the branchial pouch, that in turn opens to the environment by a branchial pore. (b) General structure of the acorn worm, *Saccoglossus*. (c) Hemichordate, generalized tornaria larva.

(b) Source: G. Stiasny, 1910, "Zur kenntnis der lebenweise von Balanoglossus clavigerus," Zoolisches Anzeiger 35, Gustav Fischer Verlag.

limited locomotor capability and are therefore *planktonic,* being moved from place to place primarily by currents and tides rather than by their own efforts. But as adults, they are usually *benthic,* living on or within a bottom marine substrate. Some *burrow* into the substrate or are attached to it—that is, they are *sessile.* Some adults are *solitary,* living alone. Others are *colonial* and live together in associated groups. Some are *dioecious* (literally, "two houses"), with male and female gonads in separate individuals; others are *monoecious* ("one house") with both male and female gonads in one individual.

Hemichordata

Members of this taxon are marine "worms" with some links to chordates and other links to echinoderms. They share with chordates unmistakable pharyngeal slits (figure 2.2a) and similar embryonic origin of the dorsal nerve cord. Although typically solid, parts of the dorsal nerve cord (figure 2.2b) may be tubular in some species. However, hemichordates lack a notochord and postanal tail, hence the name "hemi-" or half-chordates. As larvae, some pass through a small, planktonic stage, the *tornaria larva* (figure 2.2c). This planktonic larva is equipped with ciliated bands on its surface and a simple gut.

The ciliated structure, simple digestive system, and planktonic lifestyle of the tornaria larva resemble the larva of echinoderms. In fact, some scientists argue that the similarities testify to a phylogenetic link between hemichordates and echinoderms.

Further, hemichordates, like both echinoderms and chordates, are deuterostomes. The mouth forms opposite to the embryonic blastopore, and hemichordates exhibit the characteristic deuterostome patterns of embryonic cleavage and coelom formation. The similarities of hemichordates to the larval design of echinoderms on the one hand, and adult chordates on the other, are tantalizing. Perhaps hemichordates stand close to the evolutionary route taken by early chordates and still hold clues to the evolutionary origin of the chordate body plan. But, it must be remembered that living hemichordates are themselves millions of years departed from the actual ancestors they might share with early prechordates. Their own evolution has dealt them specialized structures serving sedentary habits.

Within the hemichordates are two taxonomic groups: the *enteropneusts,* burrowing forms examined next, and the *pterobranchs* (figure 2.3a–b), usually sessile forms not discussed further.

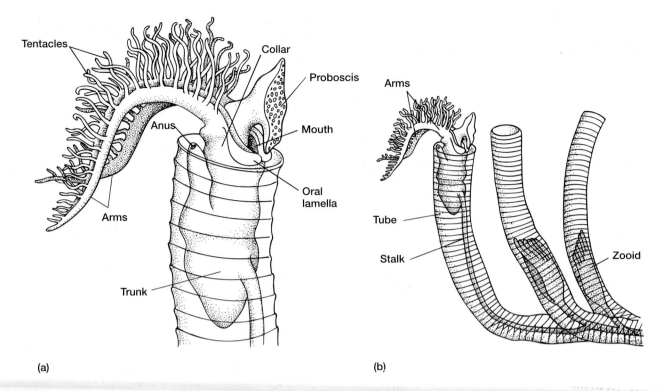

FIGURE 2.3 Hemichordata, Class Pterobranchia. (a) The sessile pterobranch, *Rhabdopleura*. Note that it is built upon the same body plan as acorn worms—proboscis, collar, trunk—but these are modified, and the whole animal lives in a tube. (b) Pterobranchs in tubes. When disturbed, the stalk shortens to pull them inside to safety. When merged into a colony, each contributing individual is often termed a zooid.

After Dawydoff.

Enteropneusta—Examination of an "Acorn Worm"

🦖 Examine the anatomy of an adult acorn worm, usually embedded in a plastic block. The enteropneusts, or acorn worms, are marine animals of both deep and shallow waters. Some species reach over a meter in length, but most are shorter. They live in burrows lined by mucus and have a body with three regions—**proboscis, collar, and trunk**—each with its own coelom (figure 2.2a). Identify these three regions of the body.

The proboscis, used in both locomotion and feeding, includes a muscular outer wall that encloses a fluid-filled coelomic space. Muscular control over the shape of the proboscis gives the animal a useful probe to shape a tunnel or inflate against the walls of the burrow to anchor the body in place. Tucked away in their burrows, many species ingest loosened sediment and extract the organic material. After passing through their simple gut, the remaining material is expelled out the anus as a casting, deposited on the surface of the substrate. Acorn worms are fragile, so when collected they commonly break. If this specimen is complete, the *anus* (figure 2.2b) will be located at the terminus of the trunk.

Other species are suspension feeders, extracting tiny bits of organic material and plankton directly from the water. In these forms, the synchronous beating of cilia on the outer surface of the proboscis sets up water currents that flow across their mucous surface (figure 2.4). Suspended

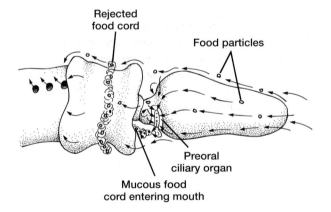

FIGURE 2.4 Ciliary mucous feeding. The direction of movement of food and mucus is indicated by arrows. Food material, carried along in the current produced by surface cilia, travels across the proboscis and is directed to the mouth where it is captured in mucus and swallowed. Rejected food material collects in a band around the collar and is shed.

After Burdon-Jones.

materials adhering to the mucus on the proboscis are swept along ciliary tracks to the mouth. The muscular lip of the collar can be drawn over the mouth to reject or sort larger food particles. The excess water that enters the mouth exits through the pharyngeal slits (figure 2.4).

The *stomochord* (figure 2.2a) arises embryologically as an outpocketing from the roof of the embryonic gut anterior to the pharynx. In the adult, it retains a narrow connection to what becomes the buccal cavity but usually enlarges as it projects forward as a preoral diverticulum into the cavity of the proboscis. The surface of the stomochord is associated with components of the vascular and excretory systems. Its walls consist of epithelial cells, like those of the buccal cavity, as well as ciliated and glandular cells. Its interior is hollow and communicates with the space of the buccal cavity. Excretion probably occurs partly through the skin, but acorn worms also possess a dense network of capillaries termed the *glomerulus,* within the proboscis, the presumed excretory organ.

🐦 A dissection scope may be necessary to see the following detail. Numerous slits are located along the lateral walls of the pharynx. Sets of adjacent **pharyngeal slits** open into a common chamber, the **branchial pouch,** which in turn pierces the outer body wall to form an undivided exit opening, the **branchial pore** (figure 2.2a). Excess water departing from the pharynx thus passes first through a slit, then through one of several branchial pouches, and finally through a branchial pore to reach the outside.

Other Systems The circulatory system is represented by two principal vessels: a *dorsal* and a *ventral blood vessel* not easily distinguished in whole specimens. The blood, which contains few cells and lacks pigment, is propelled by muscular pulsations in the major blood vessels. The nervous system consists mainly of a diffuse network of nerve fibers at the base of the epidermis (figure 2.2b). Dorsally and ventrally, the *nerve network* is consolidated into longitudinal nerve cords joined by nervous interconnections. Often this *collar nerve cord* is partially tubular. The method of embryonic origin from dorsal ectoderm and retention of a tubular structure in a few species suggest homology to the dorsal and tubular nerve cord of chordates. The gonads are housed in the trunk; the sexes are separate, and thus dioecious; fertilization is external.

Cephalochordata

Living cephalochordates are clearly chordates, built upon the characteristic pattern that includes pharyngeal slits, dorsal and tubular nerve cord, notochord, endostyle or thyroid gland, and postanal tail (figure 2.5a). Yet they are anatomically simple, with an approach to food gathering we have seen in other protochordates, namely suspension feeding based upon a pharyngeal filtering apparatus involving an atrium. Amphioxus,[1] or the lancelet, the most commonly studied cephalochordate, has a thin and elongate body extending only a few

centimeters long. It prefers coarse sediments in coastal waters and lagoons, well aerated by tides but not churned by heavy wave action. Adults can swim but usually live buried with the front of their head protruding into the water.

Cephalochordate Systems

The general structure of cephalochordate body systems will first be described. Then you will be directed to identify specific structures in whole mounts and representative cross sections. Often, a single slide will contain a composite of representative cross sections.

Filter Feeding The *oral hood* encloses the anterior entrance to the pharynx and supports an assortment of food-processing equipment (figure 2.5b,c). Projecting from the free edge of the oral hood are *buccal cirri* (figure 2.5b) that prevent entrance of large particles. The inside walls of the oral hood hold ciliated tracts that sweep food particles into the mouth. The coordinated motion of these cilia gives the impression of rotation and inspired the name *wheel organ* for these tracts (figure 2.5b,c). The posterior wall of the oral hood is defined by the *velum,* a partial diaphragm, that supports short, sensory *velar tentacles* (figure 2.5b). The velum and velar tentacles are both involved in sampling and sorting food carried in on the entering stream.

Major ciliated food corridors line the pharynx (figure 2.5d). The ventral channel is the *endostyle* (figure 2.6), the dorsal channel is the *epibranchial groove* (figure 2.6), and the inside edges of the primary and secondary pharyngeal bars carry *ciliary tracts* (figure 2.6). One of these dorsal tracts, usually located below the right side of the notochord, bears a ciliated invagination known as *Hatschek's pit* or *Hatschek's groove* (figure 2.5b,c), which secretes mucus to help collect food particles. It is an invagination in the roof of the oral cavity, a similarity shared with the vertebrate *pituitary gland,* part of which forms similarly by invagination from the roof of the buccal cavity. This has led some to propose that Hatschek's groove may have an endocrine function like that of the pituitary gland.

Mucus, secreted by the endostyle and secretory cells of the pharyngeal bars, is driven up the walls of the pharynx by cilia. Food particles adhere to the mucus, which is collectively swept dorsally to form a thread in the epibranchial groove and then conveyed to the gut. Slits in the walls of the expanded pharynx allow exit of a one-way feeding current driven by cilia. Excess water passes out these *pharyngeal slits* (figure 2.5b) to the atrium and finally departs posteriorly via the single *atriopore* (figure 2.5a).

Digestive System Parts of the cephalochordate digestive system may be precursors to vertebrate organs. For instance, the *endostyle* of amphioxus collects iodine, as does the pharyngeal endocrine gland of vertebrates, the thyroid gland. The midgut cecum, a forward extension of the gut, is thought by some to be a forerunner of both the liver (because of its position and blood supply) and of the pancreas (because cells in its walls secrete digestive enzymes). Whatever their phylogenetic fate, these and other parts of

---

[1]The term *Branchiostoma* wins claim to the official scientific generic name for most members of this subphylum, because this name was bestowed, alas, two years before the now better known and more widely used term Amphioxus. We thus use *Branchiostoma* as the Latin name of the genus, but in a spirit of charity will use "amphioxus" (without italics, or capital) as the common name.

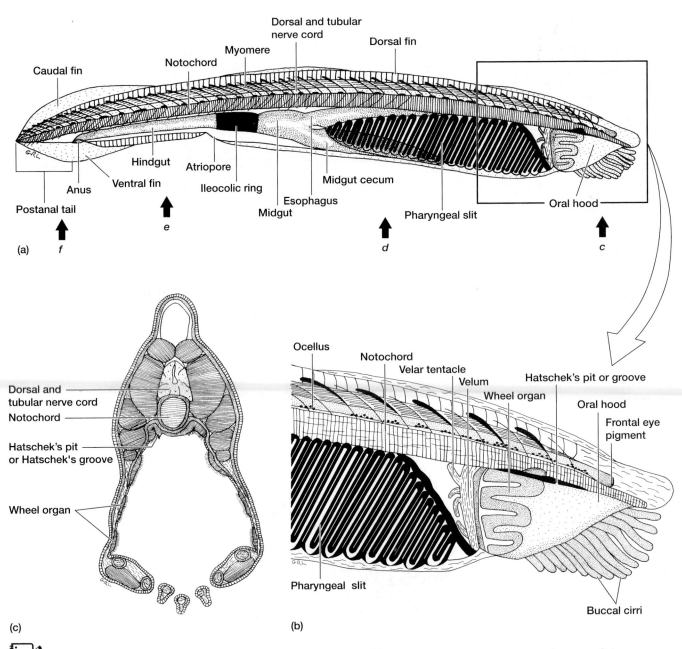

FIGURE 2.5 Cephalochordate, amphioxus. (a) Lateral view. (b) Enlargement of oral and pharyngeal regions. Selective cross sections, (c–f), are indicated in (a), and shown in (c) oral hood; (d) pharynx; (e) hindgut; (f) postanal tail.

amphioxus are structurally distinct from vertebrate organs, a reflection of the specialized demands of suspension feeding.

Nervous System and Muscles   The tubular nerve cord of amphioxus does not enlarge anteriorly into a differentiated brain. The segmental *myomeres,* V-shaped blocks of muscle along the body (figure 2.5a), are used for swimming and burrowing. They make contact with the spinal cord in a unique fashion. Most vertebrates have motor nerves that reach out peripherally from the spinal cord to the muscles. But in amphioxus, thin processes of the muscles reach centrally to the surface of the spinal cord.

Unique Notochord   Unlike notochords of other chordates, the notochord of amphioxus consists of a transversely arranged series of striated muscle cells. The developmental origin of the cephalochordate notochord is from the roof of the embryonic gastrocoel, as in most other chordates. But the muscle cells of the cephalochordate notochord are unique and set it apart from all other protochordates and vertebrates. When these muscle cells of the notochord contract, the tough notochordal sheath prevents ballooning, so internal pressure rises, and the notochord stiffens. Stiffening may strengthen burrowing, or it could increase the intrinsic vibration rate of the animal to aid fast swimming.

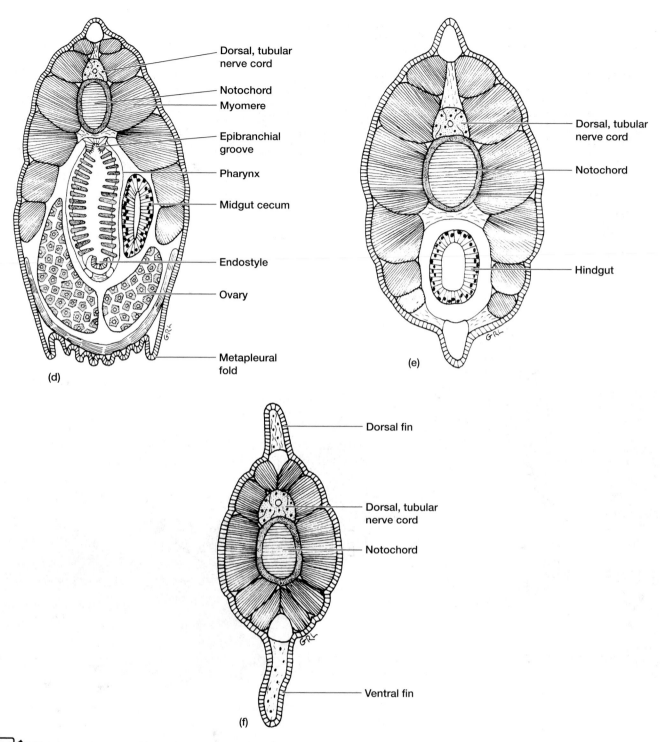

(d)

Dorsal, tubular nerve cord
Notochord
Myomere
Epibranchial groove
Pharynx
Midgut cecum
Endostyle
Ovary
Metapleural fold

(e)

Dorsal, tubular nerve cord
Notochord
Hindgut

(f)

Dorsal fin
Dorsal, tubular nerve cord
Notochord
Ventral fin

FIGURE 2.5 *Continued.* (d) pharynx; (e) hindgut; (f) postanal tail.

Circulatory System Blood flow and circulation in amphioxus occurs in the same general pattern as in vertebrates. However, in two major respects, circulation in amphioxus departs from that of vertebrates. First, there is no heart. A swelling at the confluence of the returning veins is, because of its location in the circulation, generally termed a *sinus venosus* (a vertebrate heart chamber), but it lacks any pulsations. Instead, the *hepatic vein, ventral aorta*

(endostylar artery), and others pump blood using specialized *myoepithelial cells* in their walls. The second difference from the vertebrate circulatory system occurs in the pharynx. Here, several parallel vessels travel in each pharyngeal arch from the ventral to the dorsal aorta, rather than the single aorta typical of vertebrates (figure 2.6). Furthermore, each primary pharyngeal bar contains a narrow extension of the body cavity, the coelomic space (figure 2.6).

Examination of Amphioxus Structure

🦎 *External Anatomy (Whole Mounts) (figure 2.5a)* In a whole mounted specimen, note the head with **oral hood** ending in projecting **buccal cirri,** the tail with expanded **caudal fin,** and the body with a single **dorsal fin** running lengthwise along its dorsal surface. Also notice repeated V-shaped blocks of muscle, or **myomeres,** that are arranged sequentially along the body and tail as part of the outer body wall. Myomeres are separated by connective tissue called *myosepta.* Myomeres are also connected to the notochord.

🦎 *Internal Anatomy (Whole Mounts) (figure 2.5a,b)* Also on this whole mount, note the long **notochord** beginning well into the head and continuing to the base of the caudal fin. Above the notochord and parallel with it is the **dorsal** and **tubular nerve cord.** Within the oral hood can be seen fingerlike projections of the **wheel organ** actually applied to the inner surface of the oral hood. The **velum,** bearing **short velar tentacles,** is an incomplete partition at the back of the oral hood, but from this lateral view is seen as a line defining the posterior boundary of the oral hood. Behind the velum begins an extensive region, the **pharynx,** pierced by **pharyngeal slits.** Midventrally, the pharynx bears an **endostyle** and middorsally, an **epibranchial groove,** both seen to best advantage in cross section (figure 2.6).

Posteriorly the pharynx constricts into an **esophagus** that continues into the gut itself divisible into a forward projecting **midgut cecum,** and caudally into the **hindgut,** which empties through the **anus** near the base of the caudal fin. About halfway between pharynx and anus is the **atriopore,** which allows escape of water that streams over branchial arches, into the **atrium** (figure 2.6), and exits via the atriopore.

Specimens used to prepare whole mounts are usually small and sexually immature, making the identification of the gonads difficult. The gonads of sexually mature specimens are found along the ventral edge of the body (figure 2.6).

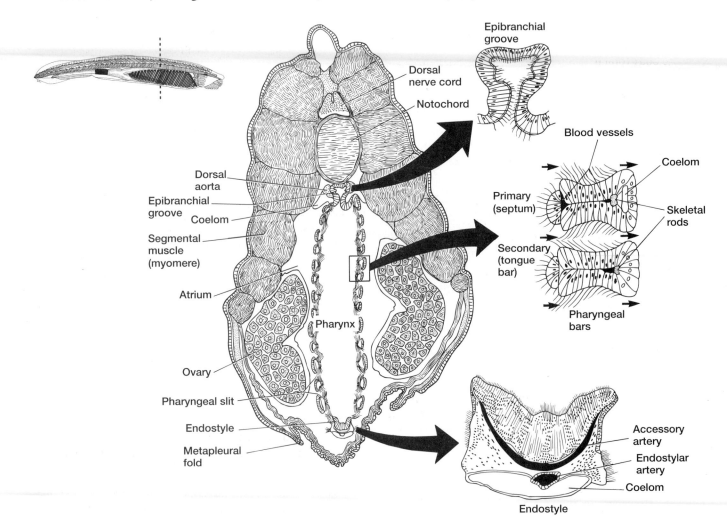

FIGURE 2.6 Amphioxus in cross section. The slanted pharyngeal bars encircle the pharynx. Enlarged at right are individual pharyngeal bars cut transversely at right angles to their long axis. The coelom continues into some of the branchial bars. The cross section is drawn at about the point indicated in the insert, upper left.

After Smith; Moller and Philpott; Baskin and Detmers.

Larval and adult protochordates typically have different structures and lifestyles. For example, although protochordates possess some (hemichordates) or all (urochordates, cephalochordates) five characteristics that define a chordate—notochord, pharyngeal slits, dorsal and tubular nerve cord, endostyle or thyroid gland, and postanal tail—these features often occur in only the larva or adult. In addition, adults are usually benthic, living on the ocean bottom, while their larval stages are planktonic, floating unattached in water. Yet despite these differences in structure and lifestyle, most protochordates share a common mode of feeding: extracting suspended particles from a stream of water moved by cilia. These food particles are collected on sheets of mucus and directed to the gut. Excess water drawn in with the food is diverted to the outside of the body through lateral pharyngeal slits.

The structures associated with filter feeding often occupy a significant portion of the body, up to about half the total body surface in cephalochordates. The composition of these filtering devices is typically complex and delicate. It is therefore no surprise that the larger protochordates (adult tunicates and adult cephalochordates) typically have a protective sheath or wall that guards the filtering region from damage by the environment.

So why do these various animals share a common feeding system? One way to look at this would be from a functional point of view. Given the relatively small size of the larvae and adults, suspended food particles (composed of microscopic organisms and pieces of larger organisms) may be the most abundant food source. Another way would be to consider their evolutionary history. Each of these protochordates has inherited a system of mucus production and cilia that are well suited for moving currents of water and extracting suspended food. Finally, other feeding systems that chew food typically require hard skeletal systems or tough plates that have not evolved in these groups.

Representative Cross Sections (figures 2.5c–f and 2.6) The smallest cross section is through the anterior end of the body (oral hood region, figure 2.5c). Other cross sections are from selected sites more posteriorly. Examine at low magnification the cross section that passes through the pharynx (figures 2.5d and 2.6). The **notochord** lies in about the center of this section; note its structure, outer sheath of fibrous tissue enclosing irregular material (cells), and spaces (fluid filled in life). Again notice the notochord below the **nerve cord.** Below the notochord lies the **pharynx;** slanting **branchial arches** are cut in cross section and so define the pharynx with an oval series of represented arches. In cross section, the **epibranchial groove** forms the middorsal, and the **endostyle** the midventral channel of the pharynx.

Examine two adjacent branchial arches at higher magnification. In each, the core of pink tissue is the **pharyngeal bar,** the space between arches the **pharyngeal slit.** Slightly darken the field by closing the substage diaphragm. This should help show the surface **cilia** on each arch that project into the pharyngeal slit. (A magnification near 400× may be required, but check with your instructor before using oil immersion, probably the highest magnification on your microscope.)

Open the substage diaphragm, return to low magnification, and move to the last cross section taken from farther posteriorly in the amphioxus body. Reidentify structures already mentioned: notochord, dorsal and tubular nerve cord, and pharynx (branchial arches, epibranchial groove, endostyle). Next to the pharynx lies the tubular **midgut cecum** (figure 2.5a,d), and next to it, but still within the body cavity, the **gonad** (testis or ovary depending upon the sex of the individual).

Finally, identify in cross section the **myomeres, metapleural folds, ventral fin,** and the **dorsal fin** (figures 2.5d–f and 2.6).

Urochordata

At some point in their life histories, urochordates generally show all five shared, derived chordate characteristics: *notochord, pharyngeal slits, tubular nerve cord, endostyle* or *thyroid gland,* and *postanal tail.* Consequently, they are proper chordates traditionally found within the Chordata. Urochordates are specialists at feeding on suspended matter, especially very tiny particulate plankton. In some, the pharynx is expanded into a complex straining apparatus, the *branchial basket.* In other species, the filtering apparatus includes mucus that is secreted by the epidermis and surrounds the animal. There are about 2,000 species, all of which are marine.

Urochordates are divided into three taxonomic groups. Larvae of the Ascidiacea are free swimming until they become attached to firm rock surfaces, piles, or floats and metamorphose into sessile adults. Members of the Larvacea and Thaliacea are permanently pelagic. Urochordate literally means "tail back-string," a reference to the notochord and postanal tail. The more common name, *tunicates,* is inspired by the flexible outer body cover, the *tunic,* secreted primarily by the underlying epidermis. This tunic (sometimes called a *test*) characterizes members of the largest group, Ascidiacea, the only group we examine further.

Ascidiacea—"Sea Squirts"

Ascidians, or sea squirts, are marine animals that are often brightly colored. Some species are solitary; others colonial. Larvae are planktonic, but adults are sessile.

Larva The larva, sometimes called the *ascidian tadpole,* does not feed during its short sojourn of a few days as a free-living member of the plankton. Instead, the larva disperses to and selects the site at which it will undergo a dramatic metamorphosis and take up a permanent residence as an adult. *Only the larval stage exhibits all five chordate characteristics simultaneously* (figure 2.7a).

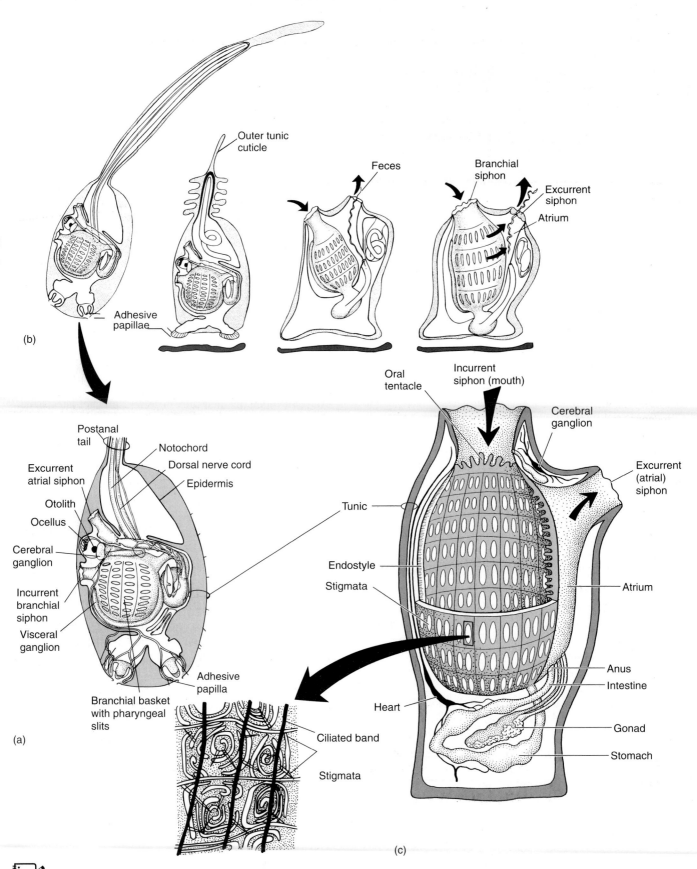

FIGURE 2.7   Urochordate, Class Ascidiacea. (a) Larval head of the ascidian *Distaplia occidentalis*. (b) Metamorphosis of ascidian larva *Distaplia*. Left to right, the planktonic, nonfeeding larva settles and attaches to a substrate. Adhesive papillae hold the larva, the tail contracts, and the outer cuticle of the tunic is shed. (c) Generalized adult urochordate. The flow of water is shown as it enters the incurrent siphon and passes through modified pharyngeal slits (stigmata) into the atrium and out the excurrent siphon. At the right, several of the highly subdivided pharyngeal slits are shown.

(a) Based on the research of R. A. Cloney.

Under a compound microscope, examine the mounted larval specimen to identify four of the five main chordate characteristics. Note the swollen **body** and thin **tail** (1). The **tubular nerve cord** (2), and **notochord** (3), which supports it, both extend from the body to the tail (figure 2.7a). The **branchial basket,** an enlargement of the pharynx, is recognized by the presence of tiny, regular rows of perforations, the **pharyngeal slits** (4) (figure 2.7a). Water is drawn in through an **incurrent siphon** and expelled through the **excurrent siphon.** The *anus* empties internally into the excurrent siphon and thus does not clearly mark the point beyond which the tail continues.

Other features can be identified. Several **adhesive papillae** on the anterior end of the body are involved in eventual attachment of the larva when it settles down (figure 2.7a). Usually two very black spots are present: the tiny, marble-shaped spot, the **otolith,** is a sensor of gravity; the irregular black spot, the **ocellus,** is a light-sensitive eyespot (figure 2.7a).

Other organs are present but difficult to identify on most slides.

Metamorphosis of Larva to the Adult   Most of the chordate features that debuted in the larva, namely the *notochord, tail,* and *dorsal* and *tubular nerve cord,* disappear in the forming adult (figure 2.7b). Although the pharynx persists and even expands, it becomes highly modified. The slits in its walls proliferate and each subdivides repeatedly, producing smaller openings, termed *stigmata.* This radically remodeled pharynx forms the barrel-shaped *branchial basket* (expanded pharynx plus numerous stigmata) of the adult sea squirt (figure 2.7c).

Adult   The tunic, composed of a unique protein, *tunicin,* a polysaccharide similar to plant cellulose, forms the body wall. The tunic and epidermis attach the base of the animal to a secure substrate and enclose the branchial basket, the viscera, and a large atrial cavity (figure 2.7c). *Incurrent* (branchial) and *excurrent* (atrial) *siphons* form entrance and exit portals for the stream of water that circulates through the body of the tunicate (figure 2.7c). Particulate matter is extracted from the passing stream of water by a netlike sheet of mucus lining the branchial basket. Rows of cilia forming a *ciliated band* deliver the food-laden mucus to the *stomach,* which in turn conveys it to the *intestine* (figure 2.7c). The current of water sieves through the *stigmata* and passes out of the branchial basket to the space between basket and tunic,

the *atrium* (figure 2.7c). From here it exits the animal via the excurrent siphon.

On adult specimens that are mounted in plastic and not cleared and stained, it is only practical to see the two **siphons** (one incurrent, the other excurrent) and the tough external covering, the **tunic.**

The nervous system consists of a *cerebral ganglion* located between the siphons (figure 2.7c). From each end of the ganglion, nerves arise that pass to the siphons, gills, and visceral organs. Contractions of bands of smooth muscle running the length of the body and encircling the siphons bring about changes in the shape and size of the adult.

Blood in a tunicate is propelled to the organs and tunic by contractions of a tubular heart. After a few minutes, the flow reverses to return blood along the same vessels to the heart. No specialized excretory organ has been found in tunicates.

All ascidians are hermaphrodites; both sexes occur in the same individual (monoecious), although self-fertilization is rare. Solitary ascidians reproduce only sexually, while colonial ascidians reproduce sexually and asexually. Asexual reproduction involves *budding.* The rootlike *stolons* of the base may fragment into pieces that produce more individuals, or buds may arise along blood vessels or viscera.

Overview

Although unsettled and controversial in its specifics, the origin of chordates lies somewhere among the invertebrates, a transition occurring in remote Proterozoic times. Within the chordates arose the vertebrates, a group of vast diversity that includes some of the most remarkable species of animals ever to grace the air, waters, and surface of Earth. Within the early chordates the basic body plan was established. Feeding depended upon the separation of suspended food particles from the water and involved a specialized area of the gut with walls lined by cilia to conduct the flow of food-bearing water. Pharyngeal slits allowed a one-way flow of water. Locomotor equipment included a notochord and segmentally arranged muscles extending from the body into a postanal tail.

Feeding and locomotion were activities that favored these novel and specialized structures in chordates. Subsequent evolutionary modifications would center around feeding and locomotion and continue to characterize the wealth of adaptations found within the later vertebrates, the subject of chapters 4–10.

3 Agnathans—Examination of a Primitive Vertebrate: The Lamprey

Chapters 3 through 10 focus on members of the Vertebrata (= Craniata). As a first sampler of vertebrates, to see some primitive structures, and to examine a possible early phylogenetic link to protochordates, a lamprey and its larva are examined first.

Agnathans

The vertebrate story begins with agnathans. A mouth is of course present, but these "jawless" fishes lack a biting apparatus derived from branchial arches. Vertebrates have a deep past, debuting within the early Cambrian explosion of other animal types half a billion years ago. However, vertebrate history has yielded slowly to view. The remarkable fossil impressions of soft-bodied vertebrates come from the very dawn of vertebrate origins—*Haikouella* and *Haikouichthyes* (figure 3.1). Another early and curious group is the conodonts, small vertebrates lasting into the age of dinosaurs. Further, during the Cambrian, bony shards of carapace attest to the presence of still other early agnathan vertebrates that possessed a bony body. These were the **ostracoderms** (meaning "bone" and "skin"), ancient early vertebrate fishes encased in bony armor. Hagfishes and lampreys carry this history of jawless vertebrates into the present. Together, these two living groups are known as **"cyclostomes,"** (meaning "round" and "mouth"). They are often treated as proxies for the most primitive of vertebrates. But they are highly modified, adapted to specialized lifestyles, and therefore depart in many ways from the general ancestral state.

Myxini—Hagfishes

The hagfishes are a very primitive vertebrate with no true vertebrae, no teeth, and reduced eyes with limited vision. They are eel-like scavengers that feed on the insides of dead or dying animals, for example invertebrates and other fishes (figure 3.2a–b). They use teethlike processes on their muscular tongues to rasp flesh from prey. Slime glands beneath the skin release mucus through surface pores. This mucus, or "slime" as it is called, may serve to release them from the grip of a predator or clog its gills.

Ovaries and testes occur in the same individual, but only one is functional, so hagfishes are not practicing hermaphrodites. Eggs, but no larval stages, have been found, so development is thought to be direct, without metamorphosis.

Like lampreys, hagfishes lack bone or surface scales, which was once taken as evidence of close affinity between them. However, lack of bone is now thought to be a secondary loss, occurring independently in both hagfishes and lampreys.

Petromyzoniformes—Lampreys

A lamprey uses its oval mouth to grasp a stone and hold its position in a water current. In parasitic forms, the mouth clings to live prey so that the rough tongue can rasp away flesh or skin, allowing the fish to open blood vessels below and drink of the body fluids. The marine forms often migrate long distances to reach spawning grounds where fertilized eggs are deposited in a prepared nest of loose pebbles. A lamprey larva, termed the *ammocoete larva,* hatches from an egg. Unlike its parents, the ammocoete is a suspension feeder that lies buried in loose sediment with only its mouth protruding up from the bottom. Upon metamorphosis, the ammocoete transforms into a parasitic adult. In some species, the larval stage may last up to seven years, at which time metamorphosis yields a nonfeeding adult that reproduces and soon dies.

Lampreys have single medial fins (figure 3.2c) but lack paired fins or lobes. Individual blocks of cartilage ride atop a lamprey's prominent notochord, forming a relatively simple vertebral column (figure 3.3). Like their ostracoderm relatives, lampreys are jawless. But unlike ostracoderms, they lack the bony exoskeleton. In fact, lampreys lack bone entirely. Despite these differences, the single medial nasal opening and strikingly similar brain and cranial nerves are sufficient to classify lampreys with similar ostracoderms.

The Origin and Evolution of Vertebrates: Shifts in Feeding Mechanisms

In chapter 2, protochordates were examined and discussed as links between invertebrates and vertebrates. But what became of the first prevertebrates? Some evolved into modern cephalochordates and urochordates, while others evolved into agnathans and other vertebrates (figure 3.1). It is this agnathan story that we explore next.

The evolution of early vertebrates was characterized by increasingly active lifestyles, hypothesized to proceed in three major steps. These steps primarily involve changes in feeding strategies and the related evolution of new pharyngeal structures.

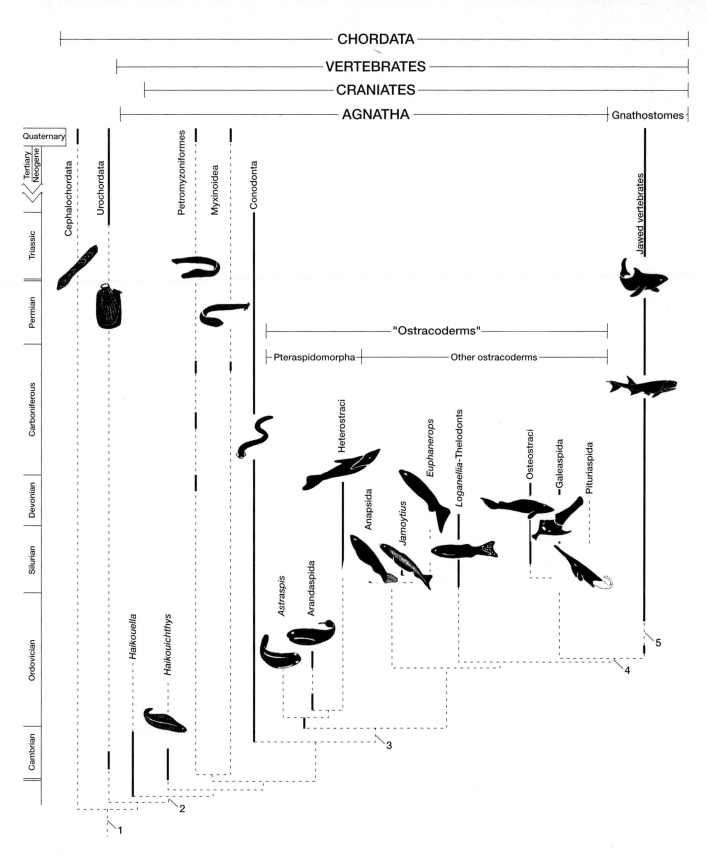

FIGURE 3.1 **Agnathan Phylogenetic relationships.** Dotted lines indicate the likely phylogenetic relationships and inferred geologic range. Solid lines show stratigraphic ranges. Dermal bone fragments from the Late Cambrian imply early presence of ostracoderms, probably an unnamed member of the Pteraspidomorpha. Major synapomorphy nodes: 1. Notochord, tubular and dorsal nerve cord, pharyngeal slits, postanal tail, endostyle (thyroid); 2. Cephalized brain, vertebrae; 3. Extensive dermal skeleton, lateral line system in grooves; 4. Pectoral fins; 5. Jaws, pelvic fins. "Ostracoderms" in quotes to remind that it is a paraphyletic group.

Modified from Donoghue, Fore, and Aldridge, with additions based on Janvier and on Mallatt.

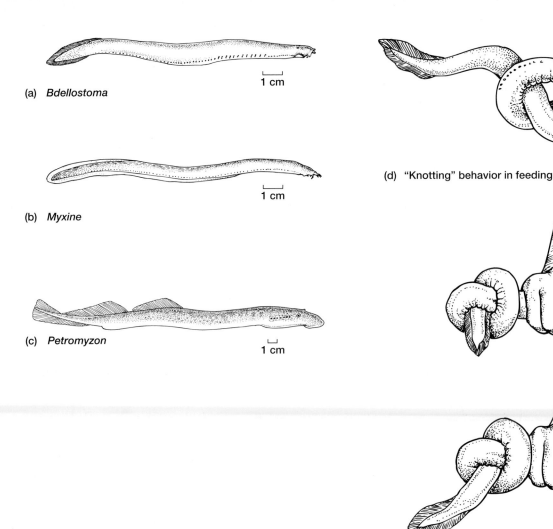

(a) *Bdellostoma*

1 cm

(b) *Myxine*

1 cm

(c) *Petromyzon*

1 cm

(d) "Knotting" behavior in feeding

(e) "Knotting" behavior in escape

FIGURE 3.2 Cyclostomes. (a) The slime hag *Bdellostoma*. (b) The hagfish *Myxine*. (c) Lamprey, *Petromyzon*. (d, e) "Knotting" behavior in a hagfish.

After Romer and Parsons, 1986, The Vertebrate Body, Saunders College Publishing, from Dean.

Step 1: Prevertebrates

As noted in chapter 2, hemichordates, cephalochordates, and urochordates rely upon suspension feeding using ciliary pumps to propel water past a mucus-covered surface. This system was shared by the first prevertebrate. But as the prevertebrates evolved, two changes occurred in the structure of the pharynx that produced a muscular instead of ciliary pump. First, the pharynx developed an encircling band of muscle, which then contracted and forced water out of the pharyngeal slits. Second, strong and spongy cartilage replaced the collagen of pharyngeal bars. Thus, when the muscular band contracted, these cartilage bars were bent, and water was forced through the pharynx. But when the muscular band relaxed, the flexed cartilaginous supports sprang back to expand the pharynx, restore its original shape, and draw in new water. Initially, this new muscular pump probably supplemented the prevertebrate ciliary pumps in moving water through the pharynx. But as animals increased

in body size, the muscular pumps became more significant as the surface ciliary pumps became less effective. The appearance of the new muscular pump thus removed the size limits imposed by the ciliary pump. Along with its contribution to more efficient feeding, the muscular pump also addressed the demands of another evolving key vertebrate innovation, gills. Protochordates have pharyngeal slits, but not gills. **Gills** are complex, folded respiratory organs in the pharynx, constructed of highly specialized capillary beds that rim these slits. Gills are bathed by water, laden with suspended food and high in oxygen content, circulating through the pharynx. Placed in this current of water, gills necessarily increase the resistance to fluid flow through the pharynx. Therefore, besides serving feeding, the muscular pump also helped push water across the newly evolving gills, thereby supporting the increased respiratory demands in this active prechordate. The first agnathans, much larger than the prevertebrates, appeared.

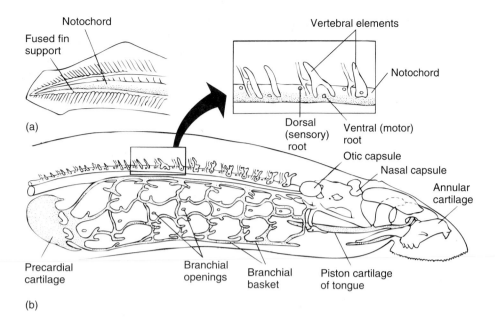

FIGURE 3.3 Lamprey skeleton. (a) Enlargement of caudal body section of the lamprey. (b) Anterior end of the lamprey with enlargement of the axial skeleton illustrating the prominent notochord. Note that only a few cartilaginous vertebral elements are present.
After M. J. Wake (ed.), 1979, Hyman's Comparative Vertebrate Anatomy, *University of Chicago Press, p. 203.*

Step 2: Agnathans

The evolution of the new muscular pump and variations on this design resulted in the ensuing diversification of these jawless fishes. At this point, most agnathans were envisioned to be suspension feeders on an unusually thick soup of particles, or deposit feeders, mud grubbers that pushed their mouths into loose organic or silty mud and drew in sediment, rich in organic particles and microorganisms. Although cilia and mucus of the branchial basket still served to collect these passing particles from suspension and transport them into the esophagus, the new muscularized pharynx, not cilia, forced the stream of rich organic material through the mouth.

Step 3: Gnathostomes—The Jawed Vertebrates

Prevertebrates, with their ciliary pump, and probably early agnathan vertebrates with their muscular pump, fed on suspended particles. Food-bearing currents carried enough organic material past mucus, where some collided with it and was gathered up. But the transition from agnathan to gnathostome involved a switch away from this feeding method. Transitional species plucked individual food particles selectively from suspension or off of surfaces. These food items may have been escaping zooplankton or heavier pieces of food that required a more powerful effort to be ingested. Thus, new designs evolved that allowed for the forceful expansion of the pharyngeal pump followed by secure mouth closure to prevent escape of captured food. These new designs led to the advent of jaws powered by quick muscle action and the evolution of active predation.

Lampreys and hagfishes, the only agnathans alive today, represent the second step in the evolution of jawed vertebrates. Both lack bone (and jaws) entirely and are specialized for parasitic or scavenging lives that depend on a rasping tongue to scrape up tissue for a meal. We therefore examine in detail a lamprey as an intermediate between prevertebrates and the gnathostomes.

Patterns & Connections

The Great Lakes of North America generally flow from west to east, draining over Niagara Falls into Lake Ontario before flowing out to the Atlantic Ocean. In the 1920s and 1930s, canals and locks were added that bypassed the great falls, permitting the movement of barges and ships between the Great Lakes and the Atlantic. These canals also allowed sea lampreys, a parasitic species previously limited to Lake Ontario, to enter and spread throughout the upper great lakes, devastating the fish communities. Once attached to prey, the parasitic adult lamprey uses its rasping tongue (figure 3.4b) to erode the side of its prey and consume its body fluids. Lamprey larvae develop in streams that flow into the great lakes. Current methods to control lampreys include the application of chemicals that interfere with natural pheromones, or the spreading of toxins into streams where larvae lampreys develop. Consult the U.S. Government Web site to learn more about the hundreds of introduced species disrupting natural ecosystems at www.invasivespeciesinfo.gov/.

Adult Lamprey Anatomy

Specific directions for finding structures in sections are indicated at the end of each system section. The boldfaced terms indicate structures to find in the sections. Italic terms are key structures in the text descriptions that are not to be identified at that point in the laboratory manual.

External Anatomy

🦎 Examine a whole preserved lamprey (figure 3.4a) and note the basic parts: the **head** includes a circular mouth, properly termed a **buccal funnel,** that is lined by soft **papillae** and hard, horny **teeth** (figure 3.4b). Near the center of the buccal funnel is the **tongue** with hard teeth at the tip. Just above the tongue is the **mouth opening,** which leads to the internal **buccal cavity** (best seen in the sagittal section). On the sides of the head, behind the mouth, are a pair of **eyes.** Dorsally, between the eyes on the top of the head, lies a single median **nasohypophyseal opening.** Just behind this opening is a lightly pigmented spot, below which is the **pineal gland,** or third eye. Behind the eyes on each side of the body run a series of seven paired, round openings, the **external gill slits.**

Follow the body of the lamprey posteriorly until coming upon the midventral opening of the caudal end of the digestive tract; the slightly raised **urogenital papilla** lies within this opening. The part of the body between the last gill slit and this papilla is the **trunk.** The region posterior to this papilla is the **tail.** Two unpaired **dorsal fins** run along the middorsal side of the body; the tip of the tail is fringed by the **caudal fin.** Hold the lamprey up to the light to back-illuminate the numerous **fin rays** (figure 3.4a), slender cartilages that internally support both dorsal and caudal fins. Return the lamprey to the dissection tray and note the impressions in the skin of underlying **myomeres,** muscle segments arranged along the body. Myomeres are best seen in the tail region.

Sagittal and Cross Sections

🦎 Examine a lamprey in which a middorsal sagittal plane extends from the front of the head to a few inches behind the last gill opening (figure 3.4c). Also examine a lamprey cut in cross section at the following levels through: (1) the pineal gland and eyes, (2) the third or fourth gill slit about midlength along the branchial region, (3) the liver and intestine, and (4) the tail (see figure 3.6d).

Respiratory System

415 Vertebrate gills are designed for gas exchange with water. Specifically, gills are dense capillary beds supported by skeletal elements in the branchial region. Ventilation in lampreys involves the muscular pump of the *buccal cavity* actively driving water across the internal gills.

In lamprey species with a prolonged adult stage, the adult lamprey feeds by attaching its circular mouth to the sides of living prey. In such species, the mouth grips the prey, making it unavailable for entry of water through the *pharyngeal slits.* Instead, water exits *and* enters through the external pharyngeal slits (figure 3.5). Muscle compression and relaxation of the branchial apparatus drive this water, which moves tidally in and out of the branchial pouch via the associated slits. A partition divides the *pharynx* into a dorsal *esophagus* and a ventral *branchial tube* (respiratory tube) that serves as a water channel (figure 3.4c). The *velar tentacles* at the tip of the branchial tube selectively close this water channel during feeding to ensure that food moves from the buccal cavity to the esophagus.

🦎 **Sagittal Section** Identify the path of water that moves from the **buccal cavity,** through the **pharynx,** past the **velar tentacles,** through the **branchial tube,** and out the **internal gill slits** to the gills.

Cross Section Examine a cross section through a gill slit in the branchial region (figure 3.6b). Trace the path of water drawn in through the mouth as it next flows laterally from the central **branchial tube,** through the **gill lamellae,** into the **gill pouch,** and then out of the body through the **external gill slits.**

Digestive System

In lampreys, the alimentary canal is a straight tube leading from mouth to anus without coils, folds, or major bends. The ciliated *esophagus* runs directly from the pharynx to the *intestine.* No distinct stomach is present, and lampreys lack a spleen. Diet includes small particulate matter, blood and tissue rasped from prey, and detritus. Storage in an expanded stomach before entering the intestine would be of little value, so food passes directly from the esophagus into the intestine.

521

The adult intestine is lined with an epithelium that contains numerous gland cells dispersed along internal *folds* that form a modest *spiral valve* (figure 3.7). Digestive enzymes are released into the anterior intestine, and mucus is secreted into the posterior intestine. In parasitic lampreys, the anterior intestine is especially important in absorption of fats. In addition, this region of the intestine of marine forms holds swallowed salt water and is important in osmoregulation. The posterior section of the intestine is important in protein absorption and elimination of biliverdin, a pigment of bile produced by the *liver* and released to the intestines. The liver, located just caudal to the heart, and the *intestine* lie within the *pleuroperitoneal cavity,* a portion of the *coelom.*

🦎 **Sagittal Section** Return to the **velar tentacles** located at the junction of the **esophagus** and **branchial tube.** Trace the esophagus caudally to its position above the heart. The gut continues beyond the heart as the **intestine,** extending ventrally between the **gonad** and **liver.** The liver, gonad, and intestine reside within the **pleuroperitoneal cavity.**

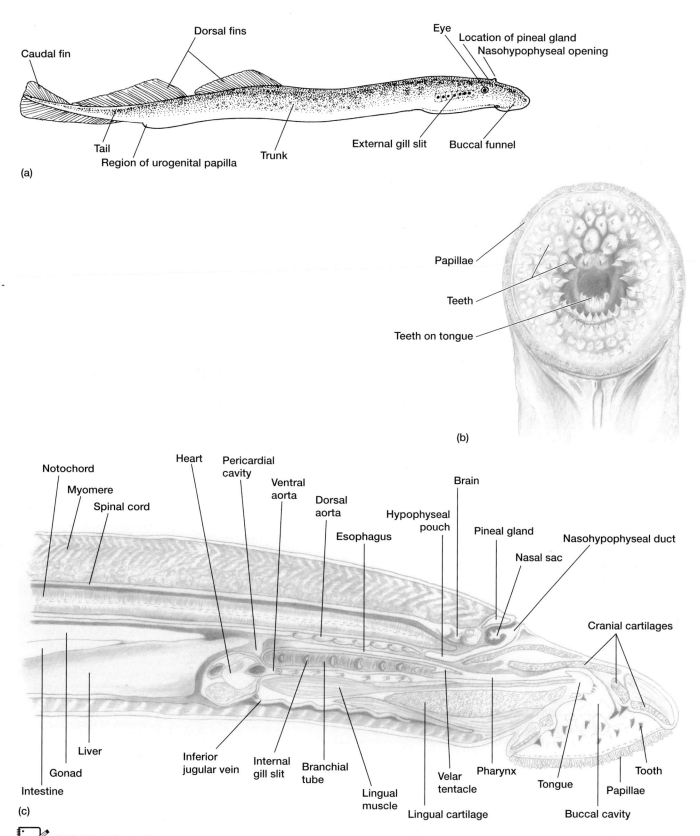

FIGURE 3.4 The adult lamprey, *Petromyzon*. (a) External morphology of the adult lamprey. (b) Ventral view of the structure of the mouth. (c) Midsagittal section of the anterior half of the body.

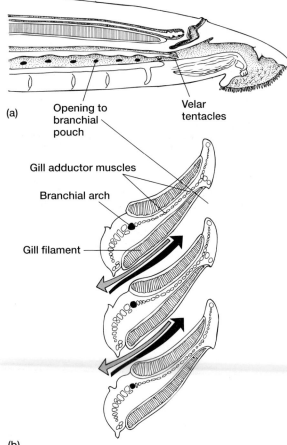

(a)

Opening to branchial pouch

Velar tentacles

Gill adductor muscles

Branchial arch

Gill filament

(b)

FIGURE 3.5 Ventilation in the adult lamprey. (a) Longitudinal section. Because the adult lamprey's mouth often is attached to prey, water must alternatively enter as well as exit via pharyngeal slits. Thus, unlike most fishes, gill ventilation in the lamprey is tidal. (b) Frontal section of the left three gill arches. Arrows indicate tidal flow of water; black inflow, gray outflow.

After Mallatt.

Cross Section Examine a cross section through the liver and intestine (figure 3.6c). Identify the **liver** on the ventral side and above it, the **intestine,** both surrounded by the **pleuroperitoneal cavity.** The walls of the intestine are folded, forming part of the **spiral valve.**

Circulatory System

The lamprey heart (branchial heart) is a typical fish heart. It resides in another portion of the coelom, the *pericardial cavity.* Blood flows sequentially through the four heart compartments: the *sinus venosus, atrium, ventricle,* and muscular *conus arteriosus.* One-way valves are present between compartments. The luminal walls of the conus arteriosus are thrown into folds collectively forming the *semilunar valves,* which prevent reverse blood flow. After leaving the conus arteriosus, blood is distributed to the *ventral aorta,* the *afferent branchial arteries,* and finally to the delicate *gill capillaries.* Blood collected from the gills is distributed to the

head through the *carotid artery* and to the trunk and tail via the *dorsal aorta* and *caudal artery* (an extension of the dorsal aorta in the tail region).

The *caudal vein* drains the tail into the paired *posterior cardinal veins.* The paired *anterior cardinal veins* drain the head. The left anterior and left posterior cardinal veins fuse with their right counterparts to form the right *common cardinal vein,* which enters the *sinus venosus.* Veins draining the gut enter the liver, forming a true *hepatic portal system* common in the vertebrates.

➤ **Sagittal Section** A series of blood vessels can be identified cranial to the **heart** in a midsagittal section (figure 3.4c). (The blood vessels in your specimen may be injected with colored latex, making the identifications easier.) Just ventral to the notochord is the unpaired **dorsal aorta.** The **ventral aorta** originates from the cranial side of the heart and extends below the branchial tube. Numerous **afferent branchial arteries** extend laterally from the ventral aorta. Their points of departure may appear as holes along the length of an uninjected ventral aorta. The **inferior jugular vein,** draining the cranial end of the body into the **sinus venosus,** runs along the ventral edge of the body. Finally, identify the **pericardial cavity** surrounding the heart.

Cross Section Examine a cross section through the eyes (figure 3.6a). The paired **carotid artery** is located below the brain and medial to the eyes. Next examine a cross section through the branchial region (figure 3.6b). The paired **anterior cardinal veins** are located on the ventral sides of the notochord. The single **dorsal aorta,** which distributes blood to the trunk and tail regions, is found just ventral to the notochord. Identify the **ventral aorta** just below the branchial tube. Afferent branchial arteries depart the ventral aorta laterally and extend into the gills, but identification relies upon an unlikely fortuitous section.

Examine a cross section through the liver (figure 3.6c). Just ventral to the notochord, the **dorsal aorta** is found with the paired **posterior cardinal veins** on either side. The hepatic portal system extends through the liver but is impractical to distinguish.

Examine a cross section through the tail (figure 3.6d). Just ventral to the notochord is the **caudal artery,** and just below it the larger **caudal vein.**

Muscular System

In fishes, the axial musculature arises directly from the embryonic and segmental myotomes. Once fully differentiated into the adult musculature, the blocks of axial musculature retain their segmentation, but they are termed *myomeres* to distinguish them from the formative embryonic myotomes from which they arose. Successive myomeres are separated from each other by connective tissue sheets, the *myosepta.* Myosepta extend inward, become attached to the axial column (primarily the notochord), and join successive myomeres into muscle masses. The horizontal septum

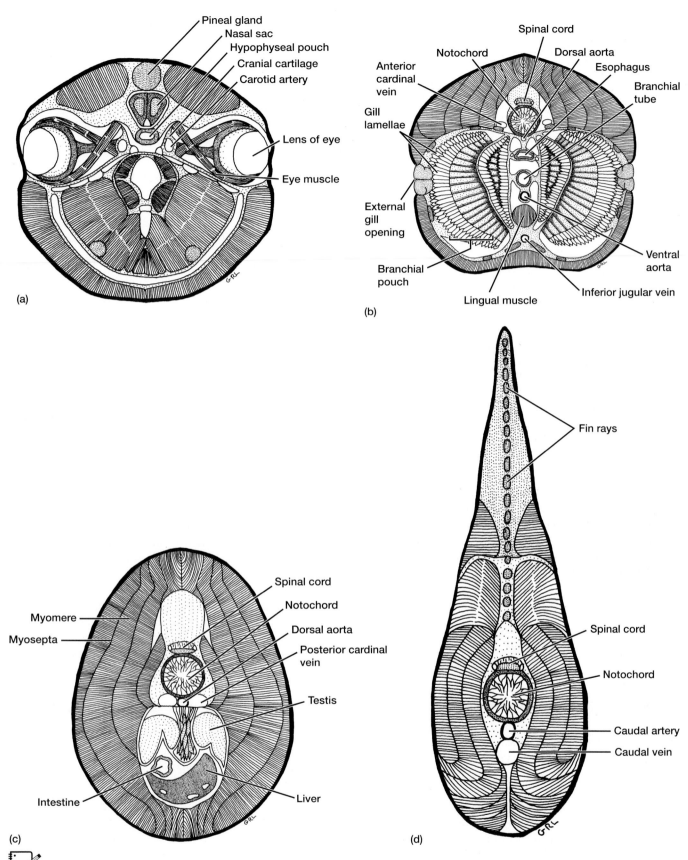

FIGURE 3.6 Cross sections of the adult lamprey. (a) Section through the pineal gland and eyes. (b) Section through the 3rd or 4th branchial slit. Two sets of gill lamellae are seen. The section passes through the most lateral lamellae in the region of an external gill slit. The medial lamellae are slanted laterally and caudally to communicate with a more posterior gill slit. (c) Section through the liver and intestine. (d) Section through the tail region.

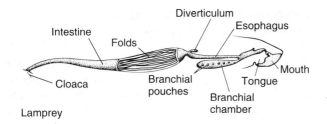

Intestine
Folds
Diverticulum
Esophagus
Cloaca
Branchial pouches
Branchial chamber
Tongue
Mouth
Lamprey

FIGURE 3.7 Digestive tract of the adult lamprey.
From Dean.

of the skeleton is not present in lampreys but is present in all gnathostome fishes, where it divides the myomeres into epaxial (dorsal) and hypaxial (ventral) muscle masses.

The axial musculature of lampreys supplies the major propulsive forces for locomotion, and, not surprisingly, constitutes the bulk of the body's musculature. Viewed from the lateral surface, the myomeres are folded into W-shaped, zigzag blocks. A contraction spreading within the axial musculature alternates from side to side, developing characteristic waves of lateral undulation. These powerful bends produced by the axial musculature are responsible for developing the body's lateral thrusts against the water and driving the lamprey forward.

🦖 **Sagittal Section** Identify the **myomeres** dorsal to the spinal cord (figure 3.3c). The **lingual muscles** and associated **lingual cartilages** extend ventrally and caudally from the tongue that they control.

    Cross Section Examine a cross section through the liver (figure 3.6c). The lateral body walls are formed by several layers of muscle, each a separate **myomere** separated from other myomeres by connective tissue sheets called **myosepta.**

Skeletal System

Most of the skeletal components listed here are not easily seen in intact whole specimens. Limit your identification of these skeletal components to those shown in figure 3.3, unless you have been provided with a specially prepared lamprey skeleton.

The skeleton of lampreys is composed of cartilage and tough connective tissues, with no trace of bone (figure 3.3). The cranial endoskeleton consists of several prominent cartilages forming the *olfactory* and *otic capsules* and the *piston cartilage,* which helps to manipulate the tongue. In addition, a series of pharyngeal cartilages form an unjointed *branchial basket,* which supports the roof of the pharynx and the lateral pharyngeal slits, and forms a shield around the heart (*precardial cartilage*). A lamprey's mouth is neither defined nor supported by jaws. Instead, an *annular cartilage* that supports the buccal area is itself supported by the cartilaginous braincase (figure 3.3). The *notochord* provides most of the axial support for the body, extending

forward from the tail to a point just ventral to the brain. As muscles apply their forces, the notochord must act as a compression girder, resisting telescoping of the body that might otherwise result.

🦖 **Sagittal Section** Identify the **notochord, lingual cartilage,** and portions of the **cranial cartilages** cut in sagittal section and identified in figure 3.4c. Note how little of the skull is formed by the skeletal elements.

    Cross Section Identify the cranial cartilages in a cross section through the eyes (figure 3.6a). The **notochord** can be seen in the sections through the branchial region, the liver, and the tail (figure 3.6b–d).

Reproductive and Urinary Systems

Adult lampreys are *dioecious*—either male or female. Gametes, either sperm or ova, are produced by an unpaired *gonad* (either *testis* or *ovary*) located above and behind the liver. The gametes are released into the coelom of the body during mating season and collected into short channels at the caudal end of paired kidney ducts. These ducts extend through the *urogenital papilla* to release the gametes out of the body. Although fertilization is external, the male possesses a penis-like tube that can be closely applied to the female's cloaca to bring the eggs and sperm into close proximity during mating.

The sex differences in cloacal structure are most pronounced during the mating season. Therefore, unless your specimen was collected at this particular time, distinguishing the sexes based upon cloacal structure will be difficult. You will identify a gonad later in the cross sections.

Lampreys possess *opisthonephric kidneys* (see glossary for definition) generally located just dorsal to the caudal half of the gonad. The paired kidneys are each drained by an *archinephric duct,* which unite caudally into a urogenital sinus drained through the tip of the *urogenital papilla.*

🦖 **Sagittal Section** The cranial end of the **gonad,** immediately dorsal and caudal to the liver, is the only reproductive or urinary structure found in a sagittal section of the anterior end of the body (figure 3.4c). The many follicles of **ovaries** make them appear to be more granular than the texture of the **testes.** Compare the gonad in your lamprey to gonads in other lampreys to speculate on the sex of your specimen.

    Cross Section Examine a cross section through the liver (figure 3.6c). The **gonad** is located just dorsal to the liver and intestine. The many developing follicles make the **ovary** appear more granular than the texture of the **testis.** A cross section caudal to the liver is required to see the **opisthonephric kidneys,** positioned just dorsal to the caudal half of the gonad.

Nervous System and Special Senses

The *spinal cord* is located just dorsal to the notochord, consistent with the basic chordate body plan. The lamprey brain exhibits the basic vertebrate organization, with an immense concentration of neurons in the head. However, the brain is disproportionately smaller than that of most other vertebrates. The forebrain region, associated with smell, and the midbrain region, associated with vision, are well developed. However, the cerebellum and the hindbrain, associated with taste and acousticolateral systems, are both small. The medulla oblongata is one of the largest brain regions as it is associated with the trigeminal nerve, which controls the extensive musculature associated with sucking.

Lampreys possess ten pairs of cranial nerves that provide sensory information to the brain (e.g., olfactory and optic nerves associated with smell and vision, respectively) and send motor signals out to muscles of the head and pharynx (e.g., oculomotor and trochlear nerves controlling eye muscles). The nerves of lampreys are unique in two ways: (1) the dorsal and ventral roots of the spinal nerves do not join lateral to the spinal cord and (2) all of the nerves are unmyelinated.

The *pineal gland,* identified previously on the middorsal side of the head, functions as a light-sensing third eye, and is particularly well developed in lampreys. The gland is actually formed of two sacs of unequal size, in which the right sac is predominant. The gland appears to control fluctuations in body color between night and day. In addition, it may sense light rhythms associated with seasonal activities such as metamorphosis and reproductive cycles.

Lampreys possess only inner ears associated with paired sets of two semicircular canals that provide sensory information about the position of the body. All gnathostome vertebrates possess a pair of three semicircular canals.

The *nasal canal* extends ventrally toward the mouth where it reaches the expanded *olfactory sac.* Near the beginning of the olfactory sac, another duct departs ventrally and expands to form the *hypophyseal pouch.* As the shape of the pharynx changes, this pouch is expanded and compressed, moving water in and out of the nosohypophyseal opening. This ventilation of water past the olfactory sac assists the sense of smell by moving more molecules past the sensory surface.

➤ **Sagittal Section**   Trace the **spinal cord** cranially until it expands to form the **brain,** just beyond the cranial end of the notochord (figure 3.4c). Immediately above the brain is the lightly colored **pineal gland,** located just behind the **nasohypophyseal opening,** identified previously when examining the external anatomy. Water is pumped in through the nasohypophyseal opening, down the **nasohypophyseal duct,** and past the dark **nasal sac** to the expanded blind-ended sac called the **hypophyseal pouch.**

Cross Section   Examine a cross section through the eyes (figure 3.6a). Identify the middorsal **pineal gland.** The **hypophyseal pouch** is positioned just ventral to the **olfactory sac,** which is in turn just ventral to the **pineal gland.**

Anatomy of the Lamprey (Ammocoete) Larva

The larval lamprey looks so much like the cephalochordate amphioxus that it was originally described in the nineteenth century as an adult cephalochordate and given the name *ammocoete.* Although we now recognize the ammocoete as a larval lamprey, the name has stuck.

Whole Mount

Like all of the preceding protochordates examined, this larva is a filter feeder. However, the fine organic matter extracted from the stream of water is pumped through the *pharynx,* not by ciliary action, but by the muscular *velum.* Fine food particles are selectively captured in the mucus, lining the parabranchial pouches next to the pharyngeal arches using an endostyle like that in cephalochordates. In addition, the endostyle in amphioxus and the larval ammocoetes is similarly involved in thyroid hormone production. In fact, as the ammocoete larva undergoes metamorphosis, the endostyle is converted into a thyroid gland that releases hormones directly into the circulatory system. Depending upon the species, the ammocoete larva may persist for several years, before transforming into the sexually mature adult.

➤ Obtain a whole mount of a small ammocoete larva and study it using the low power magnification on your microscope. Refer to figure 3.8a for help in finding the boldfaced structures. Examine the gross morphology of the larva, noting the **pharynx, oral hood, oral tentacles, dorsal** and **caudal fins,** and **segmented myotomes,** all similar in structure and function to that seen in amphioxus. The muscles of whole mount specimens have been chemically treated, making them translucent and difficult to identify. Focus carefully up and down and look for these repeated muscle units.

Distinguish between the **notochord** and the **spinal cord** just dorsal to it. The spinal cord is confirmed by tracing it forward to the head, where it expands forming the **brain.** You may identify the three divisions of the brain separated by constrictions defining the **prosencephalon** (most anterior), **mesencephalon** (middle), and **rhombencephalon** (most posterior). Just lateral to the brain, note a dark **eye spot** that is only slightly sensitive to light.

When feeding, the muscular pharynx expands drawing water through the **mouth** into the **oral cavity** surrounded by the **velum.** Gas exchange between the blood and filtered water occurs as the water passes over the **gill filaments** in the **gillpouch** and out the **external gill slits.** Food particles captured in the mucus pass down the pharynx into the **esophagus,** then the **intestine,** and finally out of the **anus** located near the base of the tail. The **liver, heart,** and perhaps the **gallbladder** can be seen on the ventrolateral side of the cranial end of the esophagus. A few tubules of the small **pronephric kidney** may be seen just dorsal to the heart.

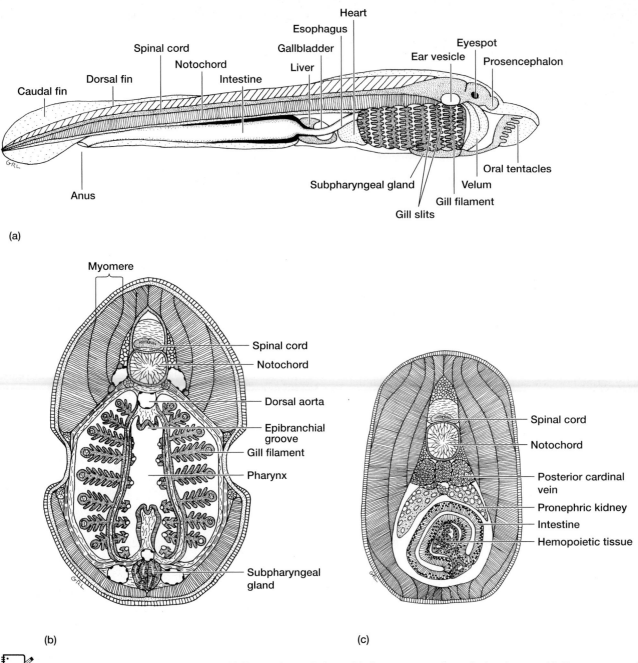

FIGURE 3.8 Ammocoetes larva. (a) External morphology. (b) Cross section through the pharynx. (c) Cross section through the intestine.

Cross Section through the Pharynx

🦖 Use figure 3.8b and a cross section of the pharynx to help you identify the following structures: **myomeres, spinal cord, notochord, dorsal aorta, epibranchial groove, pharynx, gill filaments,** and **subpharyngeal gland.**

Cross Section through the Intestine

🦖 Use figure 3.8c and a cross section through the intestine to help you identify the following structures: **spinal cord, notochord, posterior cardinal vein, pronephric kidney, intestine,** and **hemopoietic tissue.**

4 Vertebrate Integuments

Introduction

The vertebrate integument is one of the largest organs of the body and forms some of the most varied structures found within vertebrates. For example, the epidermis produces hair, feathers, baleen, claws, nails, horns, beaks, and some types of scale, and the dermis gives rise to dermal bones and the osteoderms of reptiles. Collectively, the epidermis and dermis form teeth, denticles, and the scales of fish. As the critical border between a vertebrate and its environment, the integument, or skin, has a variety of specialized functions, including:

- forming part of the exoskeleton
- thickening to resist mechanical injury
- establishing a barrier to the entrance of pathogens
- holding the shape of the organism
- participating in regulation of osmotic control
- presiding over the movement of gases and ions to and from the circulation
- absorbing needed heat or radiating the excess
- housing sensory receptors
- contributing feathers for locomotion, hair for insulation, and horns for defense
- containing pigments to block harmful sunlight or to display colors

The integument is a composite organ. The superficial layer, in contact with the outside environment, is the *epidermis,* below it the *dermis,* and between them the *basement membrane* (figure 4.1). The epidermis derives from the ectoderm, and the dermis derives from mesoderm and mesenchyme. Each contributes to a portion of the basement membrane, a layer unique to vertebrates. Below the dermis, between the integument and deep body musculature, is a transitional region made up of very loose connective tissue and adipose termed the *hypodermis* in microscopic examination, or termed the *superficial fascia* in gross anatomical dissection (figure 4.1).

Dermis

The most conspicuous component of the dermis is the fibrous connective tissue, which is composed mostly of collagen fibers. The collagen fibers may be woven into distinct layers, termed *plies,* which are best at resisting forces in specific directions. In addition, the dermis of many vertebrates directly produces plates of bone through intramembranous ossification. Because of their embryonic origin and initial

FIGURE 4.1 General structure of vertebrate integument. The epidermis differentiates into a stratified layer often with a mucous coat, or mucous cuticle, on the surface; within the dermis, the collagen often forms distinctive plies, or layers, that taken together constitute the stratum compactum. The basement membrane lies between epidermis and dermis. Between the dermis and the deeper layer of musculature is the hypodermis, a collection of loose connective tissue such as adipose tissue.

position within the dermis, these bones are termed *dermal bones.* They are prominent in ostracoderm fishes but appear secondarily even in derived groups, such as in some species of mammals.

Epidermis

The epidermis of many fishes and amphibians produces mucus to moisten the surface of the skin. The mucus seems to afford some protection from bacterial infection and helps ensure the laminar flow of water across the surface. In amphibians, the mucus also keeps the skin from drying when the animals are on land.

In terrestrial vertebrates, the epidermis covering the body often forms an outer *keratinized* or *cornified* layer, the *stratum corneum.* Rapidly dividing cells at the base of the epithelium push more superficial cells toward the surface. These superficial cells tend to undergo an orderly self-destruction, accumulating protein products during their demise. Collectively, these various protein products are termed *keratin,* and this orderly process of their formation is *keratinization.* The resulting superficial stratum corneum is a nonliving layer that serves to reduce water loss through the skin in dry, terrestrial environments.

Keratinization and formation of a stratum corneum also occur where friction or direct mechanical abrasion damage the epithelium. For example, the epidermis in the oral cavity of aquatic and terrestrial vertebrates often exhibits a keratinized layer, especially if the food eaten is unusually sharp or abrasive. In areas of the body where friction is common, such as the soles of the feet or palms of the hands, the cornified layer may form a thick protective layer, or *callus,* against mechanical damage (figure 4.2).

In addition, scales form within the integument of many aquatic and terrestrial vertebrates. Scales are basically folds in the integument (figure 4.3). Where dermal contributions predominate, especially in the form of ossified dermal bone, the fold is termed a *dermal scale;* where the epidermal contribution predominates, especially in the form of a thickened keratinized layer, the fold is termed an *epidermal scale.*

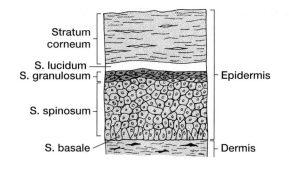

FIGURE 4.2 Keratinization. Where mechanical friction increases, the integument responds by increased production of a protective keratinized callus, and the stratum corneum becomes thickened as a result.

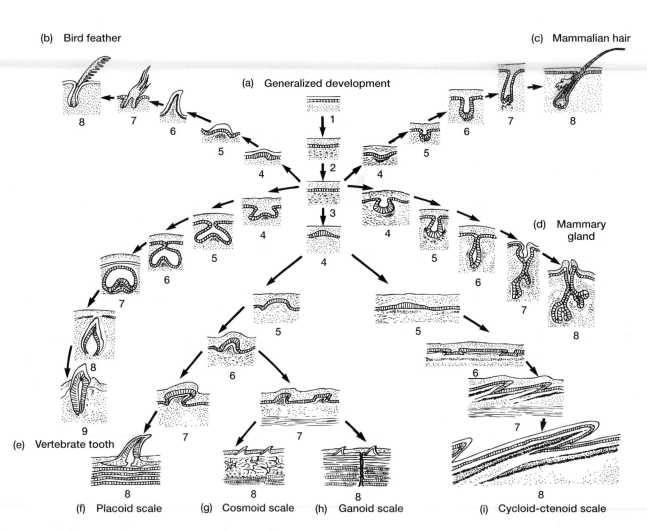

FIGURE 4.3 Skin derivatives. (a) Out of the simple arrangement of epidermis and dermis, with a basement membrane between them, a great variety of vertebrate integuments develop. The interaction of epidermis and dermis gives rise to feathers in birds (b), hair and mammary glands in mammals (c and d), teeth in vertebrates (e), placoid scales in chondrichthyans (f), and the cosmoid, ganoid, and cycloid-ctenoid scales in bony fishes (g–i).

Based on the research of Richard J. Krejsa.

Examination of Vertebrate Integuments

The following figures are artist composites that likely do not reflect what will be found on a single microscope slide. Examine several representative sections to best identify structures.

Fishes

Epidermis

With few exceptions, the epidermis of most living fishes is alive and active, with a nonkeratinized surface covered by mucus. The mucous layer is formed from various individual cells in the epidermis with contributions from multicellular glands.

Dermis

Collagen within the dermis is regularly organized into plies that spiral around the body of the fish, allowing the skin to bend without wrinkling. In some fishes, the dermis has elastic properties. When the swimming fish bends its body, the skin on the stretched side of the body stores some of the energy that is then released as the skin recoils to help unbend the body and sweep the tail in the opposite direction.

The fish dermis often gives rise to dermal bone, and dermal bone gives rise to dermal scales of fishes. In addition, the surface of fish scales is sometimes coated with a hard, acellular *enamel,* of epidermal origin, and a deeper layer of *dentin,* of dermal and neural crest origin.

Ostracoderms, Placoderms, and Lampreys

In ostracoderms and placoderms, the integument produced prominent bony plates of dermal armor that encased their bodies in an exoskeleton. The skin of living hagfishes and lampreys (figure 4.4) departs considerably from that of primitive fossil fishes. Dermal bone is absent, and the surface of the skin is smooth and without scales.

🦎 **Lamprey Skin** If a piece of lamprey skin is available, it will feel smooth and pliable. Although hagfishes are not eels (teleosts), many "eel-skin" wallets and purses are made from the skin of hagfishes, which is similar in

texture to the skin of lampreys. Examine a cross section of lamprey skin under the microscope. Notice that the epidermis is composed of stacked layers of epidermal cells with **unicellular glands** interspersed among them. Unicellular glands include club cells and granular cells (figure 4.4). The dermis is highly organized into regular layers of fibrous connective tissue with pigment cells throughout the dermis. The hypodermis includes adipose tissue.

Chondrichthyes

In the cartilaginous fishes, dermal bone is absent, but surface denticles, termed *placoid scales,* persist. These scales are what give the rough feel to the surface of the skin (figure 4.5a). Recent evidence suggests that these tiny placoid scales reduce friction drag as water flows across the skin of a swimming fish. In the epidermis, numerous *secretory cells* are present, in addition to the stratified epidermal cells (figure 4.5c). The placoid scale begins in the dermis but projects through the epidermis to reach the surface. A cap of enamel forms the tip, dentin lies beneath, and a pulp cavity resides within (figure 4.5b). Chromatophores occur in the lower part of the epidermis and upper regions of the dermis.

🦎 Examine the surface of a shark skin with a hand lens. Try to identify individual **placoid scales** and their **spines** (figure 4.5a,b). Rub your finger back and forth across the skin. Note that the skin feels rougher in certain directions because the spines are curved.

Bony Fishes

The dermis of bony fishes is composed of loose and dense connective tissue with chromatophores sprinkled throughout. The most important structural product of the dermis

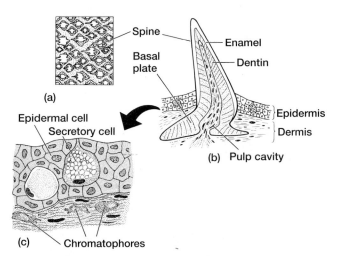

FIGURE 4.5 Shark skin. (a) Surface view of the skin showing regular arrangement of projecting placoid scales.
(b) Section through a placoid scale of a shark. The projecting scale consists of enamel and dentin around a pulp cavity. (c) Close-up of the epidermis, which contains numerous secretory cells.

(a) Based on H. M. Smith, 1960, Evolution of Chordate Structure, Holt, Rinehart and Winston, Inc.; (b) Redrawn from J. Bereiter-Han, et al., Biology of the Integument, Vol. 2, 1986 Springer-Verlag, Berlin.

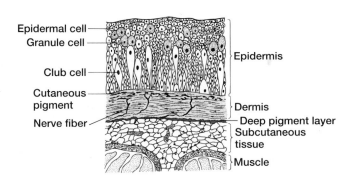

FIGURE 4.4 Lamprey skin. Among the numerous epidermal cells are separate unicellular glands that include granule cells and club cells. Note the absence of keratinization. The dermis consists of regularly arranged collagen and chromatophores.

is the scale. In bony fishes, *dermal scales* do not actually pierce the epidermis, but they are so close to the surface that they give the impression that the skin is hard (figure 4.6a). The covering epidermis includes a basal layer of cells above, which are *stratified epidermal cells* (figure 4.6b). As they move toward the surface, epidermal cells undergo a cytoplasmic transformation, but they do not become keratinized. Within these layered epidermal cells occur single *unicellular glands,* which along with epidermal cells, are the source of the surface "slime."

On the basis of their appearance, several types of scales are recognized among bony fishes (figure 4.3). One type is the *cosmoid scale,* which is present in some living sarcopterygians. This scale type is founded upon a double layer of bone, one layer of which is lamellar and the other, vascular (figure 4.7a). On the outer surface of this bone is a layer that is now generally recognized as dentin, and spread superficially on the dentin is a layer now recognized as enamel. The unusual appearance of these enamel and dentin coats inspired the respective names ganoin and cosmine, on the mistaken belief that ganoin was fundamentally a different mineral from enamel and cosmine from dentin. Although the chemical nature of these layers is now clear, the earlier names have stuck to give us the terms for distinctive scale types. In the cosmoid scale, there is a thick, well-developed layer of dentin (cosmine) beneath a thin layer of enamel.

The *ganoid scale,* found in gars and "the bichir" *Polypterus,* is characterized by the prevalence of a thick surface coat of enamel (ganoin), without an underlying layer of dentin (figure 4.7b). Dermal bone forms the foundation of the ganoid scale, appearing as a double layer of vascular and lamellar bone (in paleoniscoid fishes) or a single layer of lamellar bone (in other primitive actinopterygians) (figure 4.7b). Ganoid scales are shiny (because of the enamel), overlapping, and interlocking. Those from a gar are rhomboidal in shape. However, in most other lines of bony fishes, the scales are reduced through the loss of the vascular layer of bone and loss of the enamel surface. This produces, in teleosts, a rather distinctive scale.

The teleost scale lacks enamel, dentin, and a vascular layer of bone. Only lamellar bone remains, and it is acellular and mostly noncalcified (figure 4.7c). Two kinds of teleost scales are recognized. One is the *cycloid scale,* composed of concentric rings, or *circuli.* The other is the *ctenoid scale,* with a fringe of projections along its posterior margin (figure 4.7d). New circuli are laid down, like rings in a tree, as the fish grows. Annual cycles are evident in the groupings of these circuli and allow general aging of individual fishes from this pattern in their scales.

🦖 Compare the structure of **ganoid, cycloid,** and **ctenoid** scales using a hand lens or low power magnification on a microscope. Compare their overall shape, thickness, and ring pattern. If pieces of dried skin are available, notice how the scales are interconnected.

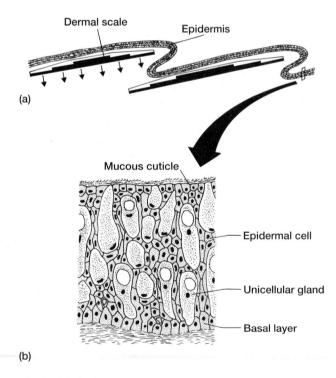

FIGURE 4.6 Bony fish skin. (a) Arrangement of dermal scales within the skin of a teleost fish (arrows indicate the direction of scale growth). (b) Enlargement of epidermis. Note epidermal cells and unicellular glands.

(a) After Spearman.

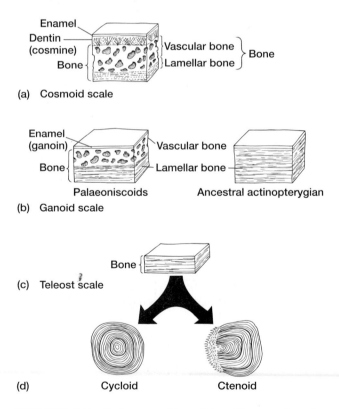

FIGURE 4.7 Scale types of bony fishes. Cross section of a (a) cosmoid scale, (b) ganoid scale, and (c) teleost scale. (d) Surface views of the two types of the teleost scale, cycloid and ctenoid scales.

Tetrapods

Although keratinization is unusual in fishes, among terrestrial vertebrates it becomes an extensive and a significant feature of the integument. Lipids are often added during the process of keratinization or spread across the surface from specialized glands. These lipids, plus the cornified layer, increase the resistance of the tetrapod skin to desiccation.

Multicellular glands are more common in the skin of tetrapods than of fishes. In fishes, the skin is coated by the products of unicellular glands at or near the surface. But in the skin of tetrapods, multicellular glands are usually sunk into the dermis. The products of these glands must pass through ducts that pierce the cornified layer to release their products on the surface.

Amphibians

Amphibians are of special interest because during their lives, they usually metamorphose from an aquatic to a terrestrial form. Phylogenetically, amphibians are also transitional between aquatic and terrestrial vertebrates. In most modern amphibians, the skin is specialized as a respiratory surface for *cutaneous respiration,* a form of gas exchange between the air and capillary beds in the lower epidermis and dermis. In fact, some salamanders lack lungs and depend entirely upon respiration through the skin to meet their metabolic needs.

Examine a cross section of frog or salamander skin at low magnification on a microscope. Distinguish between the upper **epidermis** and the lower **dermis.** Identify two types of glands that reside in the dermis but whose ducts extend upward through the epidermis (figure 4.8). The **mucous glands** tend to be smaller; each gland is made up of a little cluster of cells that release their product into a common duct. The **poison glands** (= granular glands) tend to be larger and often contain stored secretion within the lumen of each gland. You may need to use a magnification of about 400× to distinguish between these glands. The secretions of poison glands tend to be distasteful or even toxic to predators, but are usually of little concern to humans unless they are eaten or injected

into the blood. **Chromatophores** may occasionally be found in the amphibian epidermis, but most reside in the dermis. Capillary beds, restricted to the dermis in most vertebrates, reach into the lower part of the epidermis in amphibians, a feature supporting cutaneous respiration. If you are examining the skin of a larval amphibian, you may also find **Leydig cells,** thought to secrete antimicrobial products onto the skin surface.

Reptiles

The skin of reptiles reflects their greater commitment to a terrestrial existence. Keratinization is much more extensive, and skin glands are fewer than in amphibians. Scales are present, but these are fundamentally different from the dermal scales of fishes, which are built around bone of dermal origin. The reptilian scale usually lacks the bony undersupport or any significant contribution from the dermis. Instead, it is a fold in the surface epidermis; hence, it is an epidermal scale. The junction between adjacent epidermal scales is the flexible *hinge* (figure 4.9).

Reptiles shed pieces of the cornified layer of skin, often in large or complete sections such as in snakes. Because they shed only an upper layer of the epidermis, it is clearly improper to say that reptiles shed their skin. In turtles and crocodiles, sloughing of the epidermis is modest, comparable to birds and mammals, in which small flakes fall off at irregular intervals.

The shell of turtles is a composite structure. The *carapace* (see figure 5.12) that forms the dorsal half of the shell incorporates dermal bone fused with expanded ribs and vertebrae. Ventrally, the *plastron* represents fused dermal bones along the belly. On the surface of both carapace and plastron, keratinized plates of epidermis cover the underlying bone. Although not usually associated with scales, dermal bone is present in many reptiles. The *gastralia,* a collection of bones in the abdominal area, are examples.

Examine a shed "snake skin," noting the thin superficial nature of the cornified layer. Compare it to a piece of skin from a lizard or snake. On the reptile skin, note the shape and arrangement of the individual scales and find the flexible **hinges** between them. Identify the

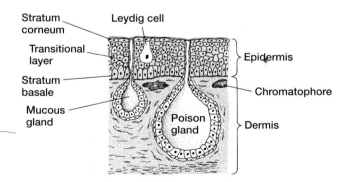

FIGURE 4.8 Amphibian skin. Diagrammatic view of amphibian skin, showing mucous and poison glands that empty their secretions through short ducts to the surface of the epidermis.

FIGURE 4.9 Reptile skin. Epidermal scales. The extent of projection and overlap of epidermal scales varies among reptiles and even along the body of the same individual. Snake body scales (top) and tubercular scales of many lizards (bottom) are illustrated. Between scales is a thinned area of epidermis, a "hinge" allowing skin flexibility.

After Maderson.

carapace, plastron, and **keratinized epidermal plates** of a turtle shell.

On a mounted alligator skeleton, identify the rib-like **gastralia** on the ventral side just caudal to the true ribs. Note that unlike typical ribs, the gastralia do not articulate with the vertebrae.

Birds

Basic Structure   The feathers of birds have been called nothing more than elaborate reptilian scales. This oversimplifies the homology, but probably not by much. Certainly the presence in birds of *epidermal scales* along the legs and feet testifies to the close phylogenetic relationship between birds and reptiles (figure 4.10a).

Examine a piece of skin from the leg of a bird to compare these **epidermal scales** to those of reptiles.

In the epidermis, two regions are usually recognized: the *stratum basale (stratum germinativum)* and the *stratum corneum.* Between them is the transitional layer of cells transforming into the keratinized surface of the corneum (figure 4.11). Chromatophores and pigment granules are interspersed throughout the epidermis and dermis (figure 4.11).

Feathers   Feathers distinguish birds from all other vertebrates. Feathers can be structurally elaborate and come in a variety of forms. Yet, feathers are essentially products of the skin, principally of the epidermis and the keratinizing system. They are nonvascular and nonnervous and are laid

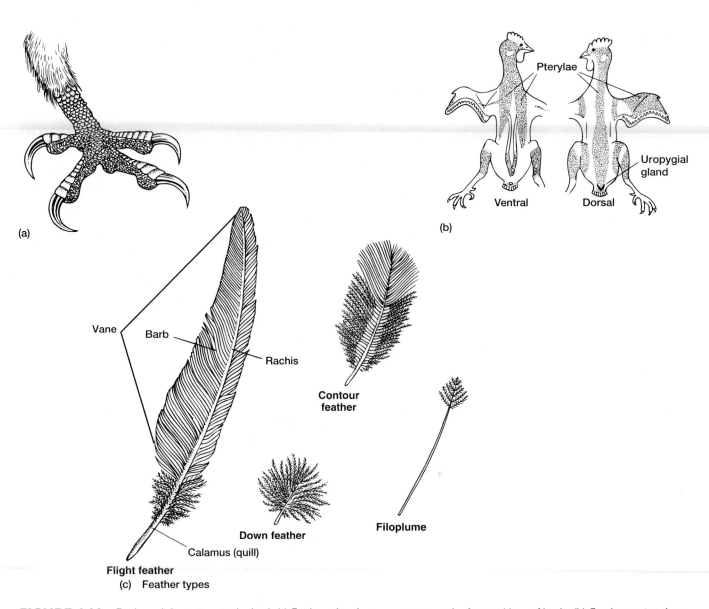

FIGURE 4.10   Epidermal derivatives in the bird. (a) Epidermal scales are present on the feet and legs of birds. (b) Feathers arise along specific pterylae (or feather) tracts. (c) Feather types. Flight feathers constitute the major locomotor surfaces. Down feathers lie close to the skin as thermal insulation. Contour feathers aerodynamically shape the surface of a bird. Filoplumes are often specialized for display.
After H. M. Smith.

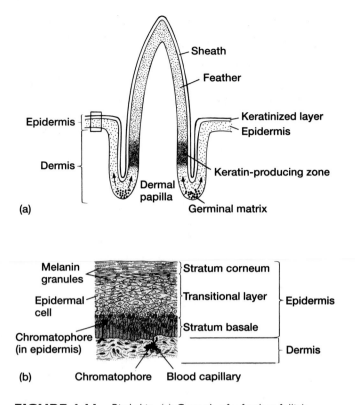

Sheath

Feather

Keratinized layer

Epidermis

Keratin-producing zone

Epidermis

Dermis

Dermal papilla

Germinal matrix

(a)

Melanin granules

Epidermal cell

Chromatophore (in epidermis)

Stratum corneum

Transitional layer

Stratum basale

Epidermis

Dermis

(b)   Chromatophore   Blood capillary

FIGURE 4.11   Bird skin. (a) Growth of a feather follicle. The feather itself forms within a sheath that, like the feather, is a keratinized derivative of the epidermis. (b) Section of skin showing the stratum basale and the keratinized surface layer, the stratum corneum. Cells moving out of the basal layer spend time first in the transitional layer before reaching the surface. This middle transitional layer is equivalent to the spinosum and granulosum layers of mammals.

(a) After Spearman.

out along distinctive tracts, termed *pterylae,* on the surface of the body (figure 4.10b). They are replaced during one or more molts each year. Feathers develop embryologically from feather *follicles,* invaginations of the epidermis that dip into the underlying dermis. Here the root, in association with a dermal pulp cavity, begins to form the feather. The feather itself grows outward in a sheathed case (figure 4.12a). Feather color results in part from pigments released by chromatophores within the epidermis. But light refraction on the feather also creates some of the iridescent colors that feathers display.

Examine a feather and identify the following components. Within the sheath, the central axis is divided into a distal **rachis** that bears **barbs** with interlocking connections, termed **barbules,** and a proximal **calamus** that attaches to the body (figures 4.10c and 4.12b). Flight feathers of the wings are characterized by a long rachis and prominent **vane,** the broad surface of the feather also important in display and insulation. The rachis and two attached vanes constitute the **spathe.**

Form & Function

Located just above the base of the tail of most birds (figure 4.10b), the uropygial gland secretes a lipid and protein product. Birds collect this secretion on the sides of their beaks and then smear it on their feathers. Preening coats the feathers with this oily mixture and makes them water repellent. It may also help moisturize the keratin of the feathers to keep them flexible, much as oil from glands in our scalp conditions human hair.

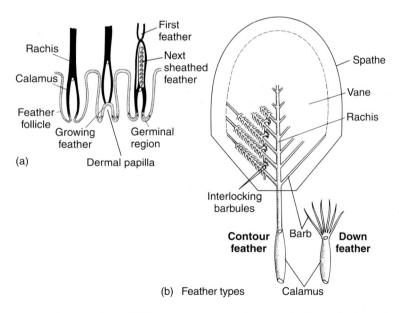

Rachis

Calamus

Feather follicle

Growing feather

Dermal papilla

Germinal region

First feather

Next sheathed feather

(a)

Spathe

Vane

Rachis

Interlocking barbules

Contour feather   Barb   **Down feather**

Calamus

(b)   Feather types

FIGURE 4.12   Feather growth and morphology. (a) Successive stages in the growth of new feathers. During molting of old feathers, new ones arise in the feather follicles to replace those shed. (b) General morphology of contour and down feathers.

(a) After Spearman.

Mammals

As in other vertebrates, the two main layers of the mammalian skin are epidermis and dermis. Beneath lies the hypodermis, or superficial fascia, composed of connective tissue such as fat.

Epidermis The epidermis may be locally specialized into hair, nails, or glands. Epithelial cells of the epidermis are *keratinocytes* and belong to the keratinizing system that forms the dead, superficial cornified layer of the skin. The surface keratinized cells are continually lost and replaced by cells arising primarily from the deepest layer of the epidermis, the *stratum basale (stratum germinativum)*. Cells within the stratum basale divide mitotically, producing cells that restore the stem cell population and epidermal cells that are pushed outward. As the epidermal cells move outward, they pass through keratinizing stages exhibited as distinct, successive layers toward the surface. Progressing toward the surface, several additional layers—*stratum spinosum, stratum granulosum,* often a *stratum lucidum,* and a *stratum corneum*—are encountered (figure 4.13). The process of keratinization is most distinct in regions of the body where the skin is thickest, as on the soles of feet. Elsewhere, these layers, especially the stratum lucidum, may be less distinct.

Melanophores are another prominent cell in the epidermis. They secrete granules of the pigment *melanin,* which are passed directly to epithelial cells and eventually carried into the stratum corneum or into the shafts of hair. Skin color results from a combination of the yellow stratum corneum—red contributed by underlying blood vessels, and the dark pigment granules secreted by melanophores.

Examine a section of mammalian skin under a microscope beginning with low power and moving to a higher magnification near 400×. Identify the five regions, extending from bottom to top: **stratum basale (stratum germinativum), stratum spinosum, stratum granulosum, stratum lucidum,** and **stratum corneum** of the epidermis. Also identify **melanophores** containing dark-brown granules. Keep the section of skin on

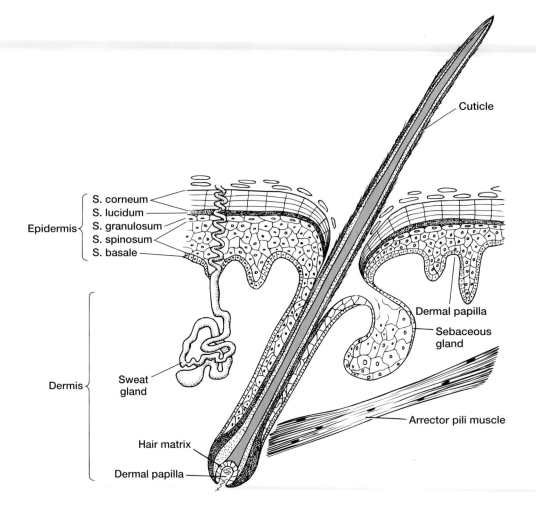

FIGURE 4.13 Mammalian skin. The epidermis is differentiated into distinct layers. As in other vertebrates, the deepest is the stratum basale, which through mitotic division produces cells that as they age become successively part of the stratum spinosum, stratum granulosum, often the stratum lucidum, and finally the surface stratum corneum. The dermis pokes up dermal papillae that give the overlying epidermis an undulating appearance. Sweat glands, hair follicles, and sensory receptors lie within the dermis. Notice that the sweat ducts pass through the overlying epidermis to release their watery secretions on the surface of the skin. Abbreviations: Stratum, S.

the microscope to examine the components of the dermis described next.

Dermis The mammalian dermis is double layered. The outer *papillary layer* pushes fingerlike projections, termed *dermal papillae,* up into the overlying epidermis (figure 4.13). These papillae increase the surface area between the epidermis and dermis and strengthen the connection between these two layers. The deeper *reticular layer* includes irregularly arranged fibrous connective tissue and anchors the dermis to the underlying fascia. *Blood vessels, nerves,* and *smooth muscle* occupy the dermis but do not reach into the epidermis. However, hair originates in the dermis and extends through the epidermis. The base of a hair is the *root;* its remaining length constitutes the *shaft.* The outer surface of the shaft often forms a scaly *cuticle* (figure 4.13). *Arrector pili muscles* in the dermis pull on the base of a hair to cause the hair shaft to thrust upward. The skin immediately surrounding the hair is also pulled up, forming a goose bump or goose pimple. This erect hair forms a thicker boundary layer of trapped warm air that better insulates the body when the skin is chilled. In addition, erect hair can also be a defensive maneuver to make a mammal appear larger.

The mammalian dermis still produces *dermal bones,* but these contribute to the skull and pectoral girdle and only rarely form *dermal scales* in the skin. One exception is the armadillo, in which a secondary development of dermal bone arises under the epidermis.

Oil and sweat glands are also typically found in the mammalian dermis. *Sebaceous glands* release an oily material onto the hair shaft to keep the hair flexible. Oil from sebaceous glands is also an important waterproofing mechanism, especially in aquatic mammals. Highly coiled *sweat glands* in the dermis release their product through ducts that extend through the epidermis and out sweat pores at the surface.

🦖 Return to the section of mammalian skin. In the dermis, identify the **papillary layer, dermal papilla,** deeper **reticular layer** with irregularly arranged fibers, **blood vessel, hair root, hair shaft, hair cuticle, arrector pili muscle, sebaceous gland,** and a **sweat gland.** If available, examine the unique integument of the armadillo, noting the firm nature of its dermal shell.

Specializations of the Integument

Nails, Claws, and Hooves

Nails are plates of tightly compacted, cornified epithelial cells on the surface of fingers and toes; thus, they are products of the keratinizing system of the skin. The *nail matrix* forms new nail at the nail base by pushing the existing nail forward to replace that worn or broken at the free edge (figure 4.14a). Nails protect the tips of digits from inadvertent mechanical injury. They also help stabilize the skin at the tips of the fingers and toes, so that on the opposite side, the skin can establish a secure friction grip on objects grasped.

Only primates have nails. In other vertebrates, the keratinizing system at the terminus of digits produces claws or hooves. *Claws* (figure 4.14b), or *talons,* are curved, laterally compressed, keratinized projections from the tips of digits. They are found in some amphibians and in most birds, reptiles, and mammals. *Hooves* are enlarged keratinized plates on the tips of the digits of ungulates (figure 4.14c).

🦖 Compare the relatively simple structure of talons, claws, and hooves that are available for study. Compare these structures to your own nails. Note how the shape of each contributes to specific functions that the animal performs.

Horns and Antlers

🦖 Examine the laboratory demonstrations of horns and antlers. Note the features of each as you read the descriptions that follow.

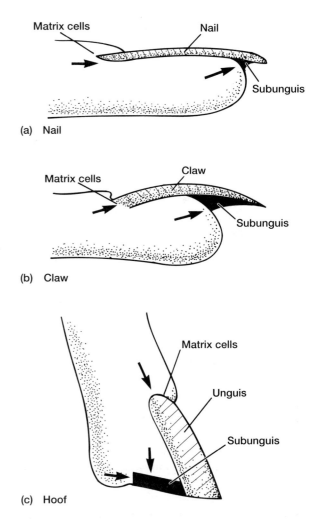

(a) Nail

(b) Claw

(c) Hoof

FIGURE 4.14 Epidermal derivatives. The unguis (nail, claw, or hoof) is a plate of cornified epithelium growing outward from a bed of proliferating matrix cells at its base and from a softer subunguis nearer its tip.

The skin, together with the underlying bone, contributes to both true horns and antlers. As these structures take shape, the underlying bone rises up, carrying the overlying integument with it. In *horns*, the associated integument produces a tough, cornified sheath that fits over the bony core (figure 4.15a). In *antlers*, the overlying living skin (called "velvet") apparently shapes and provides vascular supply to the growing bone. Eventually the velvet falls away to unsheath the bare bone, the actual material of the finished antlers (figure 4.15b).

True antlers occur only in members of the family Cervidae (e.g., deer, elk, moose). Typically, only males have antlers. These are branched and shed annually. There are notable exceptions. Among caribou, both sexes have

seasonal antlers. In deer, the antler usually consists of a main *beam,* from which branch shorter *tines,* or *points.* In yearling bucks, antlers are usually no more than prongs or spikes that may be forked. The number of tines tends to increase with age, although not exactly. In old age, antlers may even be deformed. In caribou and especially in moose, the main antler beam is compressed and *palmate,* or shovel-like, with a number of points projecting from the rim.

Among mammals, *true horns* are found among members of the family Bovidae (e.g., cattle, antelope, sheep, goats, bison, wildebeest). Commonly, horns occur in both males and females, are retained year-round, and continue to grow throughout the life of the individual. The horn is unbranched and formed of a bony core and a keratinized sheath (figure 4.16). Those of the males are designed to withstand the forces encountered during headbutting combat. In large species, females usually have horns as well, although they are not as large and curved as those in males. In small species, females are often hornless.

Unlike the true horns of bovids, horns of the pronghorn, family Antilocapridae, are forked in adult males (figure 4.17a). The old, outer cornified sheath, but not the bony core, is shed annually in early winter. The new sheath beneath, already in place, becomes fully grown and forked by summer. Female pronghorns also have horns in which their keratinized sheath is replaced annually, but these are usually much smaller and only slightly forked. The horns of giraffes are different still (figure 4.17b). They develop from separate, cartilaginous processes that ossify, fuse to the top of the skull, and remain covered with living, noncornified skin. The rhinoceros horn does not include a bony core, so it is exclusively a product of the integument. It forms from compacted keratinous fibers (figure 4.17c).

Baleen

The integument within the mouths of mysticete whales forms plates of *baleen* that act as strainers to extract krill from water gulped in the distended mouth. Although it is

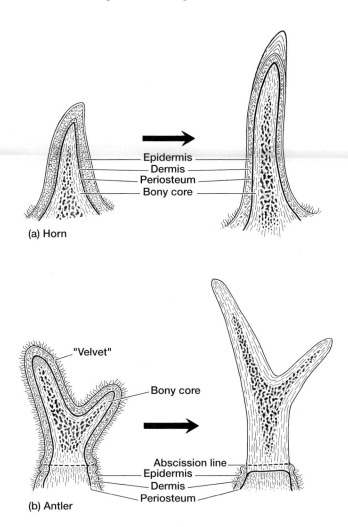

(a) Horn

"Velvet"

Bony core

Abscission line
Epidermis
Dermis
Periosteum

(b) Antler

FIGURE 4.15 Horns and antlers. (a) Horns appear as outgrowths of the skull beneath the integument, which forms a keratinized sheath. Horns occur in bovids of both sexes and are usually retained year-round. (b) Antlers also appear as outgrowths of the skull beneath the overlying integument, which is referred to as "velvet" because of its appearance. Eventually this overlying velvet dries and falls away, leaving the bony antlers. Antlers are restricted to members of the deer family and, except for caribou (reindeer), they are present only in males. Antlers are shed and replaced annually.

After Modell.

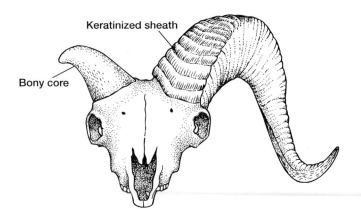

Keratinized sheath

Bony core

FIGURE 4.16 True horns of the mountain sheep (Bovidae). The cornified covering of the horn of the mountain sheep is removed on the right side of the skull to reveal the bony core.

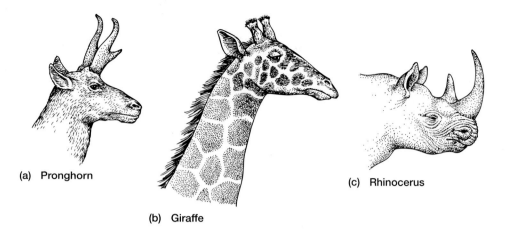

(a) Pronghorn

(b) Giraffe

(c) Rhinocerus

FIGURE 4.17 Other types of horns. (a) In pronghorns, the bony core of the horns is unbranched, but the cornified sheath is branched. (b) Giraffe horns are small ossified knobs covered by the integument. (c) Rhinoceroses have several horns that rest on a low knob on the skull, but these horns have no inner core of bone. As outgrowths of the epidermis alone, they are mainly composed of compacted keratinized fibers.

After Modell.

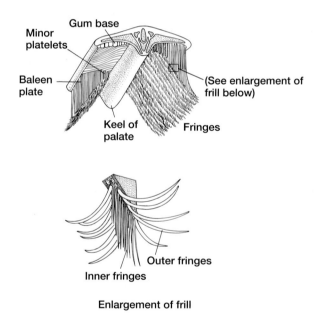

Gum base

Minor platelets

Baleen plate

Keel of palate

Fringes

(See enlargement of frill below)

Outer fringes

Inner fringes

Enlargement of frill

FIGURE 4.18 Baleen from a whale. The lining to the mouth includes an epithelium with the ability to form keratinized structures. Groups of outgrowing epithelium become keratinized and frilly to form the baleen.

Adapted from A. Pivorunas, 1979, "The feeding mechanisms of baleen whales," American Scientist, July/August.

sometimes referred to as "whalebone," baleen contains no bone. It is a series of keratinized plates that arise from the integument. During its formation, groups of dermal papillae extend and lengthen outward, carrying the overlying epidermis. The epidermis forms a cornified layer over the surface of these projecting papillae. Collectively, these papillae and their covering of epidermis constitute the plates of frilled baleen (figure 4.18).

🦖 Examine a piece of baleen noting the long central **keel** and lateral **fringes.**

5 Skeletal Systems

Tissues of the Skeletal System

Connective tissues include bone, cartilage, fibrous connective tissue, adipose tissue, and blood (figure 5.1). The extracellular *matrix* of connective tissues determines the physical properties of the tissue, and hence its functional role. This matrix is composed of protein fibers and a surrounding ground substance. Cartilage and bone are *specialized connective tissues* (figure 5.1) that constitute the skeletal system proper. *Generalized connective tissues* (figure 5.1) are dispersed widely throughout the body where they hold together other tissues, store nutrients, and perform other functions. *Tendons,* which join bone and muscle, and *ligaments,* which join bones, are examples of general connective tissues associated with cartilage and bone.

The fundamental cell of bone is the *osteocyte,* and of cartilage the *chondrocyte.* In bone, two other cell types can be identified under the microscope: *osteoblasts,* immature osteocytes involved in overall new bone deposition, and *osteoclasts,* involved in overall bone removal through the breakdown of bone matrix. Cartilage and bone also vary in the composition and abundance of the substances deposited within the matrix. These matrix materials give bone and the various types of cartilage unique mechanical properties.

All bone and most types of cartilage are surrounded by a similar appearing coat of fibrous connective tissue. This fibrous sheath is termed the *periosteum* (bone) or *perichondrium* (cartilage) depending upon which tissue it encloses.

Cartilage

Cartilage is made up of chondrocytes and a matrix containing two basic components: a *ground substance* and *protein fibers.* The ground substance is chemically composed of *polysaccharides,* most notably chondroitin sulfates and hyaluronate. Inorganic salts, such as calcium, may also be present. The protein fibers are usually *collagen fibers* or *elastin fibers* in various abundances, depending upon the type of cartilage.

Cartilage tissues do not receive a blood supply directly. Instead, the closest blood vessels are usually in the perichondrium. Thus, nutrients and gases must undergo distant diffusion from blood, through the matrix, to the chondrocytes. Similarly, no nerves penetrate cartilage directly. Cartilage may be heavily invested with calcium salts. Such calcified cartilage is found, for instance, in the shark skeleton and selectively in some stages of bone development.

🦖 Three types of cartilage are recognized primarily upon the basis of the type and abundance of protein fibers in the matrix. Examine representative cross sections of each of these tissues under the microscope and identify the structures indicated in figure 5.2.

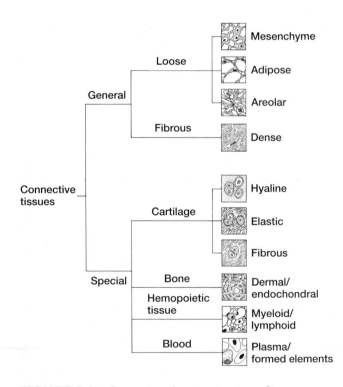

FIGURE 5.1 Categories of connective tissue. Bone, cartilage, fibrous tissue, adipose tissue, and blood are some of the body's connective tissues. Each type of connective tissue includes a distinctive cell type surrounded by an extracellular matrix.

Form & Function

Reach up with your hand and bend your ear. Let go, and your ear returns to its previous shape. Did this hurt? Will there be excessive bruising? *Imagine* doing this to a bone in your arm. Why would the bone react differently? Your ear contains elastic cartilage, with a matrix packed with elastic fibers that act like a rubber band, returning the ear to its original shape. In addition, your ear's cartilage lacks a direct blood and nerve supply, resulting in little to no pain. Bones, with a rigid calcium matrix and well supplied with blood and nerves, don't bend as easily and it will be painful if they are bent too far.

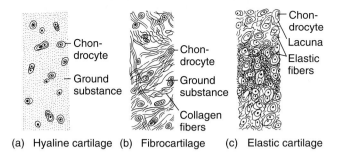

| (a) Hyaline cartilage | (b) Fibrocartilage | (c) Elastic cartilage |

FIGURE 5.2 Types of cartilage. The cartilage cell, or chondrocyte, is surrounded by a matrix composed of a ground substance and protein fibers. (a) Fibers are not apparent in the matrix of hyaline cartilage when it is viewed with a light microscope. (b) Collagen fibers are abundant in fibrocartilage, giving it mechanical resistance to tensile forces. (c) Elastin, the predominant protein fiber in elastic cartilage, makes it springy and flexible.

Hyaline Cartilage

🦖 Collagen fibers are present in the **matrix** surrounding **chondrocytes,** but never in sufficient abundance to be easily visualized with standard preparation in the microscope. Note the absence of blood vessels. Hyaline cartilage is the most common type of cartilage, found widely throughout the body. Hyaline cartilage is found at the tips of ribs, within developing bones, and at the ends of bones where it forms a smooth surface and thus assists the gliding motion of joints.

Fibrocartilage

🦖 Fibrocartilage is similar to hyaline cartilage except that the **collagen fibers** are present in much greater abundance so that their presence can usually be demonstrated routinely under microscopic examination. **Chondrocytes** are often lined up into short rows, formed because of the collagen fibers around them. The numerous strands of collagen give fibrocartilage its special physical property of being more resistant to various forces than other types of cartilage. Thus, fibrocartilage is found at sites where tensile and shear forces are thought to be significant, such as in the pubic symphysis and intervertebral disks.

Elastic Cartilage

🦖 As the name implies, elastic cartilage is flexible, a physical property attributable to the abundance of elastic fibers within the matrix. **Chondrocytes** are also found within the matrix. Elastic cartilage is found at sites where flexible support is required, such as the end of the nose, epiglottis, and pinna of the ear.

Bone

Bone is a calcified connective tissue, but one in which the calcium and other inorganic salts have been deposited in a regularly organized matrix around a central core. In compact bone

Vertebrates that reduce their usage of a limb experience a rapid decline in bone density (mass) in that appendage. The reduction of bone mass is usually measurable after as little as a week. For example, humans who injure an ankle joint favor the leg, use it less, and lose bone mass and muscle tone in the injured leg. Astronauts living in weightless space have similar consequences. The loss of gravity reduces the workload of the skeletal and muscular systems. Significant declines in bone mass routinely occur in their bodies after living for weeks or months in space. After returning to Earth, astronauts may not recover former bone mass if the loss in space was significant. This, of course, poses a serious medical problem that must be overcome if humans are to remain in gravity-free space vehicles for years at a time.

this repeated unit of bone is the *osteon.* Unlike cartilage, bone is vascular and supplied directly with nerves. It is responsive to mechanical stresses. Thus, if its use increases, it undergoes *hypertrophy* (increase in bone density and deposition); if use decreases, it undergoes *atrophy* (decrease in bone density).

Classification of bone is done by several criteria. If position is used, we recognize *cortical bone* in the outer boundary or cortex of a bone and *medullary bone* that lies within the core. If gross visual appearance is used, then two types of bone are recognized: *cancellous* (= spongy) *bone,* which is porous, and *compact bone,* which appears dense to the naked eye (figure 5.3). A sheet of fibrous connective tissue surrounds the outside of a bone forming the *periosteum.*

🦖 Examine a cross section of developing endochondral bone. Identify **cancellous bone,** composed of interconnected islands deep to **compact bone,** composed of repeated **osteon** units (figure 5.3). Identify the thin **periosteum** surrounding the bone.

If the embryonic origin is used to classify bone, then three types of bone, each with a distinctive embryonic history, are recognized. *Endochondral bone* development is perhaps the most elaborate. Basically, it involves the formation of a cartilage model of the future bone from mesenchyme and the subsequent replacement of this cartilage model by bone tissue. *Dermal bone* forms directly from mesenchyme without a cartilage precursor. *Sesamoid bones* form within tendons and are not preceded by a cartilage model. The patella bone of the knee and the pisiform bone of the wrist are examples.

Regardless of the method of embryonic development— endochondral, dermal, or sesamoid—the resulting bone may be cancellous or compact in appearance. Thus, the gross visual appearance of a bone will not tell you its method of embryonic origin. The skeleton is a composite of endochondral, dermal bones, and sesamoid bones. You should keep in mind which kind of development characterizes any given bone in the body.

Rather than memorize each bone and its developmental history in isolation, consider the general part of the skeletal system to which it belongs: vertebral column, skull, or

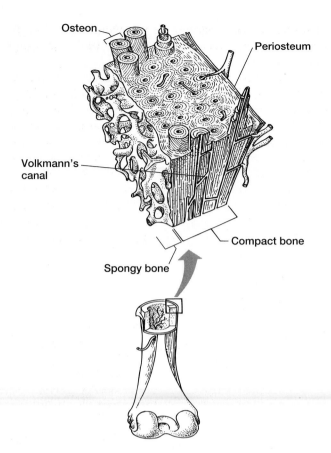

Osteon

Periosteum

Volkmann's canal

Compact bone

Spongy bone

FIGURE 5.3 Bone architecture. Osteons make up compact bone. Each osteon is a series of concentric rings of osteocytes and their matrix. Nerves and blood vessels pass through a central canal within each osteon. Diagonal connections, known as Volkmann's canals, allow blood vessels to interconnect between osteons. As new osteons form, they usually override existing, older osteons as part of the ongoing dynamic process of bone remodeling.

After Krstić.

TABLE 5.1 Patterns of the Embryonic Origin Within the Vertebrate Skeleton	
Endochondral	**Dermal**
Vertebrae	
Bones in limbs*	
Pelvic girdle	
Pectoral girdle	Pectoral girdle
(scapula, coracoid)	(all other bones)
Skull	
chondrocranium, splanchnocranium	Dermatocranium

*Except sesamoid bones (patella, pisiform)

TABLE 5.2 Common Anatomical Terminology and Directions

Cavities

  canal—an extensive, tubelike opening

  fenestra—an opening within a bony braincase

  fissure—a long, narrow cleft

  foramen—a perforation or hole through a tissue wall

  fossa—a large, hollowed cavity or depression

  sinus—a large cavity, usually within a bone

Projections

  apophysis—an outgrowth or bony protuberance

  condyle—a smooth articular end of a long bone

  head—general name for the proximal end of a long bone

  transverse process—a general term for a lateral process from a vertebra

  tuberosity—a large, roughened projection

Directions

  lateral—away from the midline of the body

  medial—toward the midline of the body

  proximal—toward the point of attachment of an appendage

  distal—away from the point of attachment of an appendage

  dorsal—toward the back or upper surface of the body

  ventral—toward the belly or lower surface of the body

  cranial/anterior—toward the head or front end of the body

  caudal/posterior—toward the tail or back end of the body

appendicular skeleton (table 5.1). All bones of the vertebral column are endochondral, as are most bones of the limbs, and all elements of the pelvic girdle. The pectoral girdle is a composite structure: scapula and coracoids are endochondral, but all other elements are dermal. The skull is also a composite structure: chondrocranium and splanchnocranium are endochondral; dermatocranium is dermal.

Divisions of the Skeletal System

You will find several ways to collectively group skeletal elements. Some persons group endochondral bones and cartilages associated with the mouth and pharynx into the *visceral skeleton.* All other endochondral bones and cartilages belong to the *somatic skeleton.* The somatic skeleton in turn is divided into the axial skeleton (skull and backbone) and appendicular skeleton (limbs and girdles). For convenience only, we divide laboratory material into *postcranial* and *cranial skeleton.* Table 5.2 provides a quick reference for basic terminology used to describe skeletal and general anatomical structure. Quickly learn to use this terminology, as it will provide essential clues to the structures you seek to identify. You may also wish to review the planes of the body in chapter 1, figure 1.5.

Postcranial Skeleton

Vertebrae and Ribs

288 291

Two general phylogenetic trends occur in the *vertebral column,* or backbone. First, the *notochord* becomes reduced as the vertebrae (segmental cartilaginous or bony elements) increase in prominence to surround and mostly replace the notochord. Second, the chains of vertebrae in the column differentiate (change in form and function), especially in later terrestrial vertebrates.

Typically each *vertebra* (pl., *vertebrae*) consists of a middorsal *neural arch* enclosing the *nerve cord,* a midventral *hemal arch* enclosing blood vessels, and a *centrum,* a spool-shaped body enclosing the notochord (figure 5.4a,b). Projecting beyond each arch is a *neural spine* and *hemal spine,* respectively.

Two types of segmental ribs, dorsal and ventral (figure 5.4a), occur primitively. Both form in *myosepta* separating muscles (figure 5.4c). *Ventral,* or *subperitoneal, ribs* lie beneath the axial muscles, just outside the peritoneum of the coelom. They are serially homologous with the hemal arch. *Dorsal ribs* form at the intersection of myosepta with a longitudinal sheet of fibrous connective tissue, the *horizontal skeletogenous septum.* Opinion varies as to the homology of fish and tetrapod ribs (figure 5.4d), although most now hold that tetrapod ribs derive from dorsal ribs of fish. Partly because of this uncertainty, the anatomical

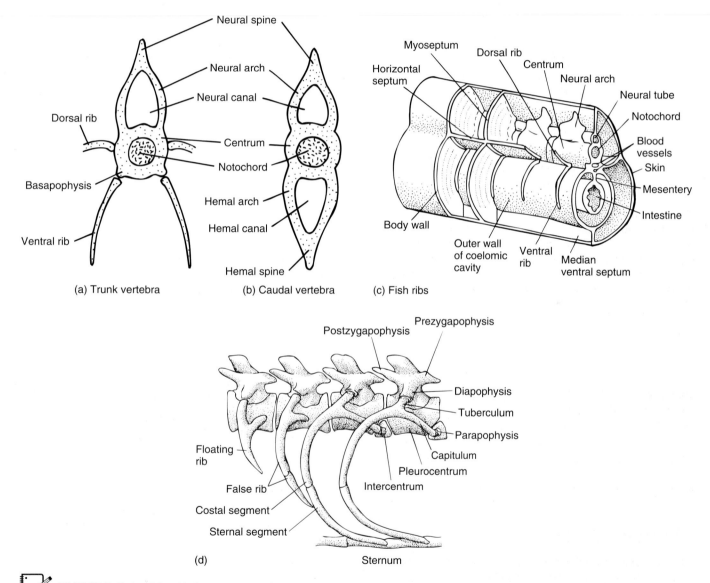

(a) Trunk vertebra   (b) Caudal vertebra   (c) Fish ribs

(d)

FIGURE 5.4 Ribs. (a) Cross section of trunk vertebra of a fish. (b) Cross section of caudal vertebra of a fish. (c) In fishes, dorsal ribs develop where myosepta intersect with the horizontal septum, and ventral ribs develop where myosepta meet the wall of the coelomic cavity. (d) Amniote ribs. Ribs are named on the basis of their articulation with the sternum (true ribs), with each other (false ribs), or with nothing ventrally (floating ribs). Primitively, ribs are bicipital, having two heads, a capitulum and a tuberculum, that articulate respectively with the parapophysis on the intercentrum or the diapophysis on the neural arch. The body of the rib may differentiate into a dorsal part, the vertebral rib or costal segment, and a ventral part, the sternal rib or segment that articulates with the sternum.

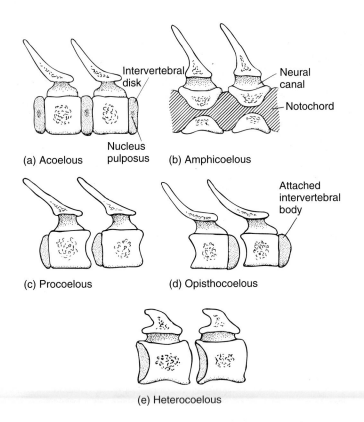

(a) Acoelous — Intervertebral disk / Nucleus pulposus

(b) Amphicoelous — Neural canal / Notochord

(c) Procoelous

(d) Opisthocoelous — Attached intervertebral body

(e) Heterocoelous

FIGURE 5.5 General centra shapes. Anterior is to the right. The shapes of articulating centra ends as viewed in sagittal section define specific anatomical types: (a) acoelous, both ends are flat; (b) amphicoelous, both ends are concave; (c) procoelous, anterior end is concave; (d) opisthocoelous, posterior end is concave; (e) heterocoelous, saddlelike articulating ends.

After George C. Kent and Larry Miller, Comparative Anatomy of the Vertebrates, *8th edition, 1997, McGraw-Hill Company, Inc., Dubuque, Iowa.*

descriptions usually do not attempt to specify whether a rib is dorsal or ventral, but term it simply "rib."

Because the function of the vertebral column in large part depends upon the shapes and articulations of vertebrae, much descriptive detail accompanies the centrum. Five principal centrum types are recognized on the basis of the shape of their ends (figure 5.5; table 5.3).

Generally, two regions and hence two vertebral types can be identified in the fish vertebral column: **trunk** and **tail** (or **caudal**) **vertebrae.**

Agnatha

Ostracoderms Ostracoderms are known only from fossils. Only trace impressions of vertebral elements have been found in fossils of some ostracoderm groups. These elements are probably small, unossified pieces of vertebrae surrounding a prominent notochord. Thus, among ostracoderms, a strong notochord provided the central mechanical axis for the body.

Cyclostomes—Lampreys Lampreys possess vertebral elements, but these are small, cartilaginous structures resting dorsally upon a very prominent notochord that provides axial support for the body (see figure 3.3).

TABLE 5.3   The Structure of Centra

Shape of the ends of centra

amphicoelous—both ends concave (amphi = "double"; coelo = "hollow")—sharks, teleosts, extant amphibians

acoelous (amphiplatyan)—both ends flat (platy = "flat")—some mammals

heterocoelous—both ends saddle shaped (hetero = "different")—necks of turtles and birds

opisthocoelous—anterior convex, posterior concave (opistho = "behind")—some mammals

procoelous—anterior concave, posterior convex (pro = "in front")—alligators

Common types of apophyses (processes) of centra

basapophyses—(sing., *basapophysis*) paired ventrolateral processes on centrum, the likely remnants of the hemal arch bases now the articulation with a ventral rib (or remaining hemal arch)

diapophysis and **parapophysis**—lateral projections from the centrum that articulate respectively with dorsal (tuberculum) and ventral (capitulum) heads of a dorsal rib

hypapophysis—unpaired midventral processes from centra

zygapophysis—articulations between successive vertebrae

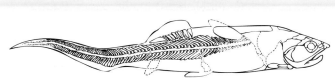

Coccosteus

FIGURE 5.6 Axial skeleton of the placoderm *Coccosteus*, with prominent notochord supporting dorsal and ventral vertebral elements.

After J. A. Moy-Thomas, 1971, Palaeozoic Fishes.

Placodermi

Some placoderms preserve evidence of a prominent notochord supporting ossified neural and hemal arches (figure 5.6).

Elasmobranchii—Shark

Examine segments of a shark vertebral column embedded in clear plastic blocks. Identify boundaries of each **centrum** (figure 5.7). Directly above each is the triangular **neural arch,** which supports a low ridge, the **neural spine.** Between successive neural arches are inverted triangular elements, the **interneural arches.** Together neural and interneural arches enclose the **neural canal** through which passes the spinal cord. If you are looking at a vertebral section from the tail, then **hemal arch** and **hemal spine** are present. If the vertebral section comes from the trunk, then lateral, projecting, ridgelike **basapophyses** are present and hemal arches and hemal spines are absent.

Within and through successive centra runs the **notochordal canal,** a canal constricted centrally, but

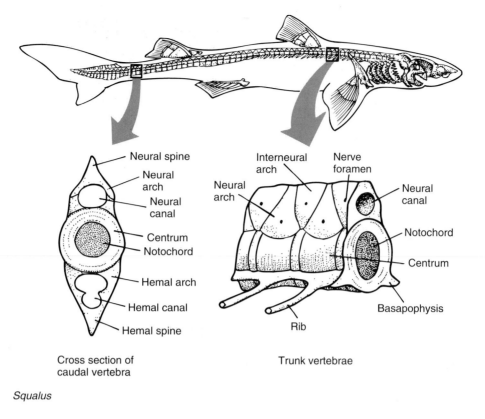

Neural spine
Neural arch
Neural canal
Centrum
Notochord
Hemal arch
Hemal canal
Hemal spine

Cross section of
caudal vertebra

Interneural arch
Nerve foramen
Neural arch
Neural canal
Notochord
Centrum
Basapophysis
Rib

Trunk vertebrae

Squalus

FIGURE 5.7   The axial skeleton in the shark *Squalus*. The vertebral elements tend to enlarge in elasmobranchs, surpassing the notochord as the major mechanical support for the body in modern sharks.

expanded on the ends of each centrum so the notochord, filling this canal, likewise is constricted and expanded.

Examine a preserved shark skeleton in tall jars to identify the same specific features just located. In addition, examine cross sections of a shark tail to appreciate the vertebral column set in its natural environment of muscles and septa. Get your bearings by first locating the neural arch, notochord, and so on. From either side of the centrum, a fibrous connective tissue sheet runs directly outward to the skin. This is the **horizontal skeletogenous septum,** which divides the body musculature into two regions. Muscles above this septum are **epaxial muscles;** those below are **hypaxial muscles.** A **dorsal skeletogenous septum** from the neural spine upward to the skin of the back and a **ventral skeletogenous septum** from the hemal spine downward to the skin further divide muscles into left and right sides. The concentric whorls of muscle are **myomeres,** each separated by myosepta.

Osteichthyes

Chondrostei—Sturgeon   Examine vertebrae from a sturgeon. Note the prominent notochord with cartilaginous elements riding upon it in dorsal and ventral positions (figure 5.8a). Homologies are presently in doubt, although the neural spine, neural arch, and rib seem clear enough.

Holostei—Amia (Bowfin)   Examine a mounted skeleton. The vertebrae are ossified. The notochord is not in evidence because it is much reduced in size and now is completely enclosed by the vertebrae. Follow ventral ribs caudally noting how they are serially homologous with hemal arches (figure 5.8b). Take a close look at vertebrae in the tail. Here two centra are present per segment: the **pleurocentrum** carries no arches, but the adjacent **intercentrum** articulates with neural and hemal arches. Try to find the forward point where the pattern of two centra ends and a single centrum per segment begins.

Teleostei—Perch   Again consult the mounted specimen looking for what should begin to be familiar structures: **centra, hemal arches** and **spines, neural arches** and **spines,** and so on (figure 5.8c). In the trunk, dorsal ribs articulate with small ventrolateral projections, **basapophyses.** Note their serial homologies.

Now examine an isolated, single, dried teleost vertebrae—**centrum, spines, arches,** and so on (figure 5.8d). If you have a tail vertebra, how do you determine dorsal and ventral (i.e., hemal and neural arches)? Compare with a mounted perch skeleton if necessary. Compare relative sizes of neural and hemal canals. The centrum is biconcave, and thus is what type of vertebra? Note the pin-sized hole in the center of the centrum. Through this the constricted **notochord** passes, then upon exit, expands to partially fill the concavities between successive vertebrae in a fashion similar to sharks.

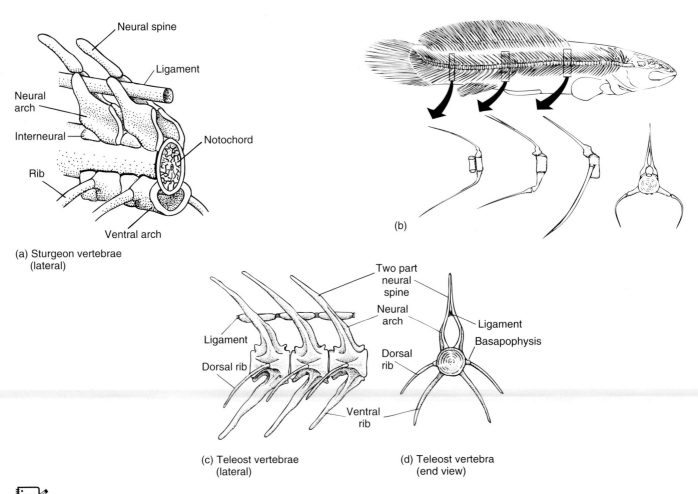

(a) Sturgeon vertebrae
(lateral)

(b)

(c) Teleost vertebrae
(lateral)

(d) Teleost vertebra
(end view)

FIGURE 5.8 Actinopterygian vertebrae. (a) Sturgeon vertebrae, lateral view. (b) Axial skeleton of the bowfin *Amia calva.* (c) Teleost vertebrae, lateral view. (d) Teleost vertebra, end view.

(a, c) After Jollie; (b) after Jarvik.

Amphibia

Urodela—*Necturus* Start with a mounted skeleton. Four sequential regions are present (figure 5.9a): the **cervical** (or neck) **region** consists of a single vertebra that joins the vertebral column to the skull; the long **trunk region** supports tiny, wishbonelike ribs (figure 5.9c); the **sacral region** consists of one vertebra whose ribs join with girdles of the hindlimbs; and the **caudal region** (figure 5.9b), whose vertebrae commonly possess **hemal arches,** lack ribs, and have short, lateral projections called **transverse processes.**

Seen to best advantage in isolated, disarticulated vertebrae, ribs of the trunk are **bicipital** (possessing two **heads**). Where they articulate with the vertebrae, ribs fork into a dorsal head, the **tuberculum,** and a ventral head, the **capitulum.** These heads meet respective articulation points on the **transverse processes:** a **diapophysis** (dorsal) and a **parapophysis** (ventral).

Near the base of the **neural spines,** each vertebra sends a pair of shelflike projections forward: **prezygapophyses** (articular face upward) to engage with paired, caudally projecting **postzygapophyses** (articular

Patterns & Connections

The pattern of two centra per body segment found in the tail of the bowfin, *Amia,* is not found in the modern tetrapods (amphibians, reptiles, birds, mammals). Each of these modern groups possesses but a single centrum per body segment. But why should we see this difference? Functionally, a vertebral column with fewer elements per body segment is more rigid and provides greater support. As tetrapods became increasingly adapted to life on land, the advantages of a firm vertebral column to carry the weight of the body became more important.

face downward) of the adjacent anterior vertebra. These prevent twisting or wringing of the vertebral column.

Anura—Frog On mounted frog skeletons, identify regions of the vertebral column (figure 5.10). The **cervical region** consists of a single vertebra. The **trunk region** is longest, usually seven to ten vertebrae depending upon species; **transverse processes,** low **neural spines, prezygapophyses** and **postzygapophyses** are present. The single vertebra constituting the **sacral**

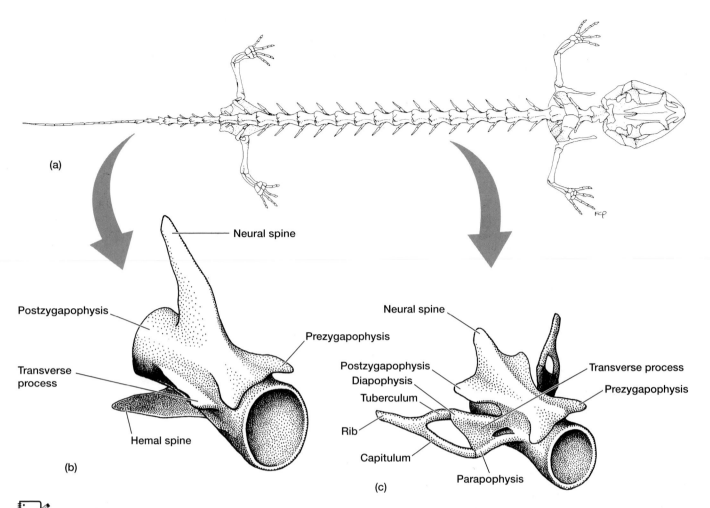

FIGURE 5.9 Skeleton of the mudpuppy, *Necturus*. (a) Skeleton, dorsal view. (b) Caudal vertebra. (c) Trunk vertebra with articulated ribs. Cranial is to the right in both (b) and (c).

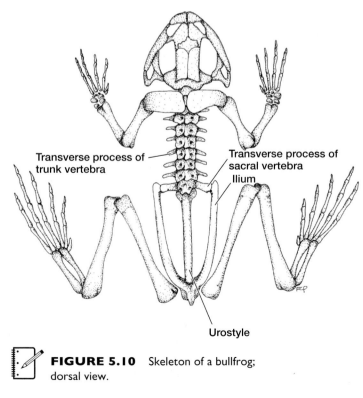

FIGURE 5.10 Skeleton of a bullfrog; dorsal view.

region attaches to the long **ilium** (pl., *ilia*) of the pelvic girdle by **transverse processes.** The vertebral column continues beyond the sacrum as a single, long, splint-like element, the **urostyle,** the stout bone between the two ilia of the pelvic girdle. Distinct **caudal vertebrae** are absent. Instead, caudal vertebra may have fused in the evolution of the urostyle.

Reptilia

Alligator (Mounted Skeleton) The vertebral column divides into five regions: **cervical, thoracic** (long ribs), **lumbar** (without ribs), **sacral,** and **caudal** (figure 5.11a). Because of their unique structure, complex embryology, and specialized function, the first two cervical vertebrae receive their own names, **atlas** and **axis,** numbers 1 and 2 cervical vertebrae, respectively (figure 5.11d). The atlas is a ring composed of four components: **intercentrum,** two bases of the **neural arch,** and part of the **proatlas** (an extra vertebra, now lost, save for its contribution here to the reptile atlas). The axis has a neural arch and broad neural spine. Attached to its centrum and projecting forward like a spike is the **odontoid process** (phylogenetically an outgrowth of the atlas

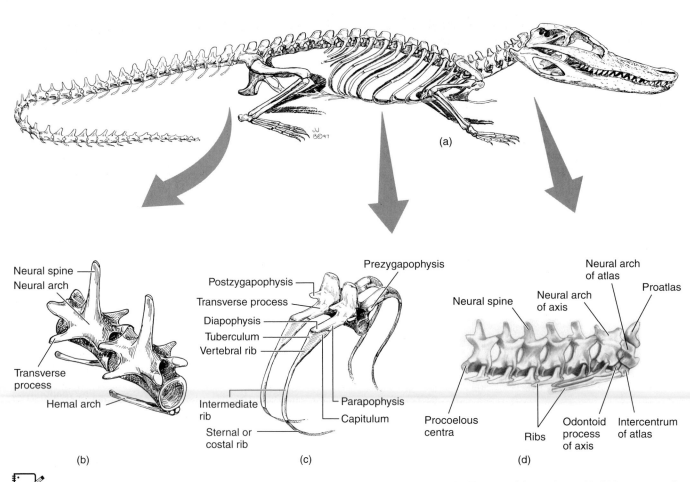

Neural spine
Neural arch

Transverse
process

Hemal arch

(b)

Postzygapophysis
Transverse process
Diapophysis
Tuberculum
Vertebral rib

Prezygapophysis

Intermediate
rib
Sternal or
costal rib

Parapophysis
Capitulum

(c)

Neural spine

Procoelous
centra

Neural arch
of atlas

Neural arch
of axis

Proatlas

Ribs

Odontoid
process
of axis

Intercentrum
of atlas

(d)

FIGURE 5.11 Skeleton of an alligator. (a) Lateral view of complete skeleton. (b) Two caudal vertebrae. (c) Oblique view of two thoracic vertebrae. (d) Lateral view of cervical vertebrae. Cranial is to the right in (b–d).

centrum and part of the axis itself), which will probably be obscured by adjacent vertebrae. A short pair of spine-like ribs belonging to the axis actually articulate with the odontoid process. The remaining cervical vertebrae are more conventional: neural arch and neural spine, prezygapophyses, postzygapophyses, and a single procoelous centra. Short, V-shaped ribs attach to **transverse processes.** Cervical and a first few thoracic vertebrae bear midventral bumps, **hypapophyses.**

Progressing caudally, the first vertebra to bear long ribs reaching the midventral region is the first **thoracic vertebra.** Approximately nine others follow, all bearing ribs on the ends of transverse processes (figure 5.11c).

The following five **lumbar vertebrae** carry broad transverse processes, but no ribs (figure 5.11a).

Two vertebrae, each with strong transverse processes attached at their tips to the pelvic girdle, compose the **sacral region** (figure 5.11a).

The **caudal vertebrae** follow: neural arch, neural spine, transverse processes, hemal arches (missing on

first caudal vertebra) that articulate almost between centra (figure 5.11b). Ribs are absent altogether.

Overall, note the prominence of neural spines and their basic similarity in shape and orientation throughout thoracic, lumbar, and most of the caudal region (figure 5.11a). Pre- and postzygapophyses occur through most regions. Cervical and thoracic ribs (phylogenetically, dorsal ribs) are **bicipital (tuberculum** and **capitulum),** with respective articulations to a **diapophysis** and a **parapophysis** riding on transverse processes. The two heads define a **transverse foramen.** As successive arches might define a passageway, the successive transverse foramina define the **vertebrarterial canal** parallel to the column. Blood vessels to the head pass through this canal. Each rib has three parts: **vertebral rib** including the two heads, **intermediate** part, and terminally the **sternal** or **costal cartilage,** which articulates midventrally with thin bone or dried cartilage (figure 5.11a,c). Depending upon the size of the individual, the first few sternal cartilages are likely to articulate with the sternum and the remainder of the costal cartilages with a midventral thin, dried membrane. Lying midventrally between the last sternal cartilage and the pelvic girdle are the **gastralia** (figure 5.11a). The gastralia are Y-shaped dermal bones of the belly, directly below the lumbar region, and are often slightly obscured by dried cartilage.

Turtle (Mounted Skeleton) Identification of homologous parts of the vertebral column is complicated in the turtle because elements fuse with dermal bone to form the protective dorsal **carapace. Cervical, trunk, sacral,** and **caudal regions** compose the vertebral column (figure 5.12).

An **atlas** and **axis** begin the cervical series of vertebrae. The remaining cervical vertebrae lack ribs.

Neural spines and neural arches of the trunk, sacral, and first caudal vertebrae are fused to the carapace. **Ribs** of trunk vertebrae, possessing but a single head **(capitulum)**, distally broaden and fuse with bony plates of the carapace.

Two vertebrae compose the **sacrum.**

The first caudal vertebra fuses with the sacrum. Remaining caudal members possess neural arches, hemal arches, and transverse processes of decreasing prominence.

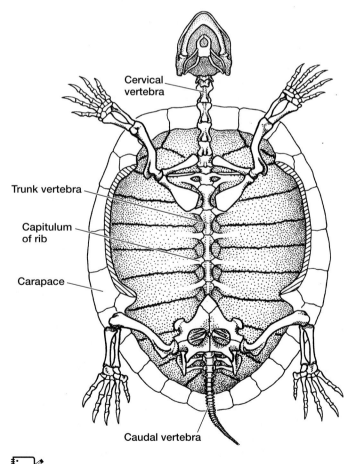

FIGURE 5.12 Skeleton of a painted turtle, ventral view. The plastron has been cut away to reveal the internal skeletal detail.

Aves—Pigeon or Chicken (Mounted Skeleton)

Fusion is the key feature of the bird skeleton. Slight differences in skeletal structures occur among species, but generally all regions of the vertebral column fuse except the first, the cervical region (figure 5.13a).

The cervical region contains numerous vertebrae (16 in the chicken, 13–14 in the pigeon). They are **heterocoelous,** saddle-shaped centra, that give great flexibility to this region, a fact you should confirm on disarticulated neck vertebrae (figure 5.13b). The **atlas** is small and ring shaped. The **axis** bears an **odontoid process.** The other cervical vertebrae possess zygapophyses, low neural arches and spines, hypapophyses, and reduced ribs. The **bicipital ribs** are lateral swellings that bear sharp, caudally directed spines. The base of each forked rib defines a **transverse foramen** and so contributes to the **vertebrarterial canal. Capitulum** and **tuberculum** fuse with **parapophysis** and **diapophysis** on the transverse process.

From their transverse processes, **thoracic vertebrae** support long ribs, which usually have two segments (figure 5.13a). In the middle of the ribs, the **vertebral section** usually carries a flat, caudal projection, the **uncinate process** (also present in some reptiles) resting over the lateral face of the next successive rib. These processes act primarily as lever arms for inhalatory muscles that flare the rib cage. The overlap between successive ribs may add some overall firmness that helps them act as a unit during lung ventilation and locomotion. The **sternal section** is the distal part of the rib that articulates with the sternum. Besides possessing an incomplete rib, the first **thoracic vertebra,** in most species, fuses with the next few successive vertebrae. Note fusion of centra, spines, transverse processes, and zygapophyses.

The fusion of the last **thoracic,** all **lumbar,** all **sacral,** and first few **caudal** vertebrae forms a composite piece, the **synsacrum** (figure 5.13c). In turn, the synsacrum fuses with the large ilium, which itself is fused with other pelvic elements. The number of truly

Form & Function

Birds are an interesting example of the close match between form and function within the vertebral column. Cervical vertebrae are flexibly articulated to give the head great freedom of movement and reach when a bird preens its feathers or probes for food. However, most of the vertebrae in the middle and caudal part of the vertebral column are fused to each other (synsacrum) and to the three fused bones (innominate) of the pelvic girdle. This adds rigidity and establishes a firm and stable axis for control while a bird is in flight. Indirectly, this fusion of vertebral elements decreases the weight of the body because less muscle is required to control flexible, individual vertebrae.

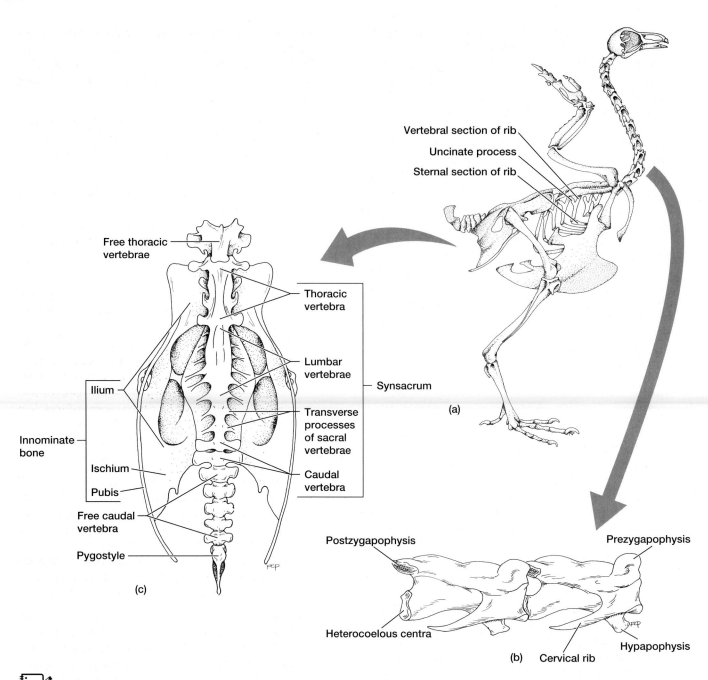

Vertebral section of rib
Uncinate process
Sternal section of rib

(a)

Free thoracic vertebrae

Thoracic vertebra

Lumbar vertebrae

Synsacrum

Ilium

Transverse processes of sacral vertebrae

Innominate bone

Ischium

Caudal vertebra

Pubis

Free caudal vertebra

Pygostyle

(c)

Postzygapophysis

Prezygapophysis

Heterocoelous centra

(b)   Cervical rib

Hypapophysis

FIGURE 5.13   Skeleton of a pigeon. (a) Articulated skeleton with the left member of paired structures not included for clarity. (b) Cervical vertebrae, lateral view. (c) Synsacrum, ventral view; cranial is up.

separate vertebrae involved can be counted by examining the ventral synsacrum (figure 5.13c) to count still distinct transverse processes. Only two vertebrae contributing to the synsacrum are truly sacral (except in ostriches, where three contribute).

A few free caudal vertebrae follow. The last several fuse into a small, solid terminal piece, the **pygostyle.** Examine disarticulated vertebrae from throughout the vertebral column to better appreciate the structure of the bones just listed.

Mammalia—Cat (Mounted and Disarticulated Skeletons)

🦖 Five distinct regions characterize the mammalian vertebral column: **cervical, thoracic, lumbar, sacral,** and **caudal** (figure 5.14a). Between vertebrae in living animals lie circular pads of fibrous cartilage: the **intervertebral cartilages,** or **intervertebral disks.** Centrally they contain a gelatinous material, the *nucleus pulposus,* the only adult derivative of the embryonic notochord in mammals. The nucleus pulposus is surrounded by the *annulus fibrosis,* a tough sheath of fibrous cartilage. The cartilage of intervertebral disks shrinks and deforms when it dries. For this reason, these disks are usually absent on mounted skeletons prepared by routine methods. On a mounted skeleton, note that adjacent vertebrae form an opening between them near the base of each neural arch (figure 5.14a). This opening is the **intervertebral foramen,** through which spinal nerves exit. Each vertebra defines half this foramen, and so bears an indentation, the **intervertebral notch.**

The special interest humans have in mammalian anatomy has given rise to additional terms for previously learned morphological names. These special descriptive terms are given in parentheses.

Now begin your examination of the individual, disarticulated vertebrae, observing regional changes in prominence of neural arch and spine, neural canal (vertebral foramen), zygapophyses (articular processes), transverse processes, and ribs. Often two regions of the neural arch are identified: **pedicle** (each foot of a neural arch where it stands on the centrum) and **lamina** (the section between the prezygapophysis and the base of the neural spine).

Cervical Vertebrae   The first two cervical vertebrae are the atlas and axis (figure 5.14b). Five more follow in almost all mammals. The **atlas** is ring shaped and carries lateral, paired, and winglike **transverse processes.** The base of each transverse process is pierced by a **transverse foramen,** a distinctive aperture in all cervical vertebrae. The neural arch lacks a spine. The base of the neural arch is perforated by holes for passage of nerves. Anteriorly, large, scooped, paired depressions, the **articular facets,** belonging to the prezygapophyses (= anterior articular processes) receive the occipital condyles of the skull.

Above its neural arch, the **axis** carries a broad neural spine (spinous process) that projects over the atlas. Caudally this spine forms the **postzygapophyses.** The **odontoid process** (dens) projects forward from the centrum into the interior base of the atlas. Lateral to the odontoid process are two swellings, the **prezygapophyses** (= anterior articular processes), which fit into the postzygapophyses (= posterior articular processes) of the atlas. Transverse processes pierced by **transverse foramina** lie along the sides of the centrum. Each remaining cervical vertebra possesses zygapophyses, a neural canal, a neural arch, neural spine (spinous process), a well-defined centrum, transverse processes, and transverse foramina (figure 5.14c).

Thoracic Vertebrae   Note the tall, caudally directed neural spines (spinous processes), neural arches, zygapophyses (articular processes), short transverse processes, and centra (figure 5.14d). Locate the two facets that articulate with each bicipital rib, the **full facet** on

the transverse processes (correctly diapophysis), and a **demifacet** on all centra where a capitulum articulates between centra. Between demifacet and prezygapophysis, the neural arch indents, forming the **intervertebral notch.**

Ribs   Most ribs consist of a bony **vertebral rib,** and a bar of hyaline cartilage, the **costal cartilage** (or sternal rib) (figure 5.14a,e). Ribs are classified on the basis of their distal attachment:

true rib—the costal cartilage articulates separately with the sternum
false rib—the costal cartilage unites with adjacent costal cartilages before finally articulating with the sternum
free (or **floating**) **rib**—false rib whose distal end has no attachment to the sternum

On a mounted specimen, note the sites of rib contact with the vertebrae. The rib is **bicipital** (figure 5.14e). The **capitulum** (head) articulates with the centrum, while the more distal **tuberculum** (tubercle) articulates with the transverse process. The small depression on the centrum where the capitulum articulates is the **facet.** If a rib inserts between two centra, each vertebra contributes part of the articular surface, a **demifacet.** The narrow part between the two heads is the **neck,** the rest of the rib the **body,** or **shaft;** the point of sharpest bend just distal to the tuberculum is the rib **angle.**

Lumbar Vertebrae   Note the change in orientation of neural spines, and forward-directed transverse processes (pleurapophysis by some because it is thought to be formed by the fusion of rib and diapophysis) (figure 5.14f). Above the articular face of the prezygapophysis lies a small knob, the **mammillary process** (metapophysis). Below and lateral to the postzygapophysis is a spinelike **accessory process** (anapophysis).

Sacrum   Individual sacral vertebrae fuse to form the **sacrum** so that individual transverse processes and ribs are indistinguishable (figure 5.14g). However, former boundaries can be determined by noting persisting individual neural spines and **intervertebral foramina** (dorsal and ventral sacral foramina). Three sacral vertebrae contribute in the cat and dog, four in the rabbit and pig, five in the horse and humans. Prezygapophyses and postzygapophyses occur on respective ends of the sacrum.

Caudal Vertebrae   Note the loss in prominence of arches, transverse processes, and zygapophyses caudally until just centra remain (figure 5.14h). Tiny, V-shaped hemal arches accompany some vertebrae, but are commonly lost during preparation of the skeleton. Numbers of caudal vertebrae vary with tail length. In humans, those present fuse into a single piece, the **urostyle,** or so as not to confuse with frogs, more commonly, the **coccyx.**

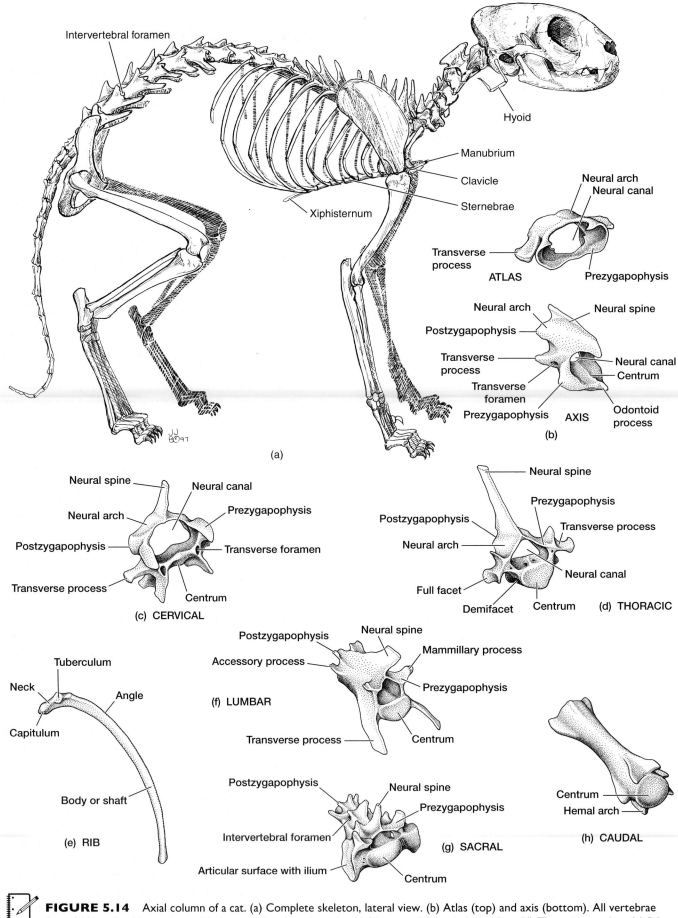

FIGURE 5.14 Axial column of a cat. (a) Complete skeleton, lateral view. (b) Atlas (top) and axis (bottom). All vertebrae (b–d, f–h) are in oblique view, with the cranial end to the right. (c) More caudal cervical vertebra. (d) Thoracic vertebra. (e) Rib. (f) Lumbar vertebra. (g) Sacral vertebra. (h) Caudal vertebra.

Girdles and Limbs

Introduction

319

The girdles are parts of the skeleton that support the fins or limbs (figure 5.15), and serve as sites of muscle attachment. The *pectoral girdle* supports anterior appendages and the *pelvic girdle* supports posterior appendages. Tetrapod limbs evolved from fish fins. Many of the anatomical changes you will study in appendages and girdles reflect changing functional demands of locomotion and support. Primitive fins were supported by a series of skeletal rods, *pterygiophores,* some proximal members of which enlarged and fused in later forms to give rise to the major elements of the girdles.

Pelvic Girdle and Posterior Appendages

Elasmobranchii—Shark Consult an articulated skeleton for pectoral and pelvic girdle position on the vertebral column. The distal portion of the flat fin projects

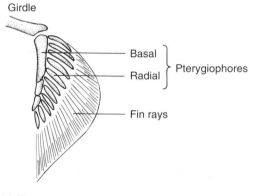

(a) Fin

(b) Limb

FIGURE 5.15 Basic components of the fin and limb. (a) The fin is composed of pterygiophores, basal and radials, and dermal fin rays. Fin rays are called lepidotrichia in bony fishes and ceratotrichia in elasmobranchs. (b) The limb, either fore- or hindlimb, includes three regions: stylopodium (upper arm/thigh), zeugopodium (forearm/shank), and autopodium (manus/pes).

Morphological term	Forelimb	Hindlimb
Stylopodium	Upper arm	Thigh
Zeugopodium	Forearm	Shank (crus)
Autopodium	Manus (wrist-palm-fingers)	Pes (ankle-sole-toes)

The pelvic girdles of tetrapods follow a basic pattern. In general, each girdle is formed by three pairs of bones: the pubic, iliac, and ischiac, seen laterally in a triangular configuration. At the top of the triangle is the iliac, usually identified as the dorsal portion of the pelvic girdle that makes contact with the sacral vertebrae. At one corner of the triangle are the pubic bones, usually directed ventrally and cranially. The ischiac bones are in the remaining corner, directed ventrally and caudally. Derivatives of this tripartite pattern exist, so stay alert to deviations from this pattern.

outward and is supported internally by slender fin rays, the **ceratotrichia,** giving it a streaked appearance (figure 5.16b). At the base of the fin lie larger cartilaginous elements, **pterygiophores.** The fins in turn articulate with a straight, single cartilaginous pelvic girdle, the **puboischiac bar.** Notice, on each end of this bar, a lateral projection, the **iliac process.** In the shark, two groups of the pterygiophores can be recognized. The first group is the **radials,** an outer series of elements arranged like the diverging blades of a hand-held fan. The second group is composed of two **basals,** enlarged medial cartilages probably derived phylogenetically by the fusion of several radials. The largest of the two basals is the **metapterygium,** a long, stout medial cartilage supporting the series of radials. The smaller basal is the **propterygium,** positioned just cranial to the series of radials, and probably enlarged from an anterior radial. There are sexual differences in the pelvic girdle (figure 5.16b). Only the male possesses prominent claspers, inserted into the female to facilitate sperm transfer, that are also likely derived from fused radial cartilages.

Amphibia—*Necturus* The pelvic girdle in urodeles is largely unossified. In *Necturus,* the single sacral vertebra bears, on its transverse process, a rib that articulates with the **ilium,** a bony rod joined ventrally to the **puboischiac plate** (figure 5.17a,b). This plate is shaped like an arrowhead directed anteriorly, and lacks any sutures to delineate regions clearly. Basically, the cranial pointed half is termed the **pubic cartilage,** and the caudal half the **ischiac cartilage.** Within the ischiac cartilage a pair of semicircular centers of ossification form, termed **ischia** (figure 5.17a,b). The **acetabulum** is an open depression that receives the rounded proximal end of the **femur.** It is formed where the pubic cartilage, ischium, and ilium join together. Medial to the acetabulum in the pubic cartilage is the small **obturator foramen** through which the obturator nerve passes.

The limb, from proximal to distal, is divided into thigh, shank, ankle, and foot. In order, the **femur** articulates with two parallel bones. The **tibia** lies cranial and the **fibula** caudal when the limb is held

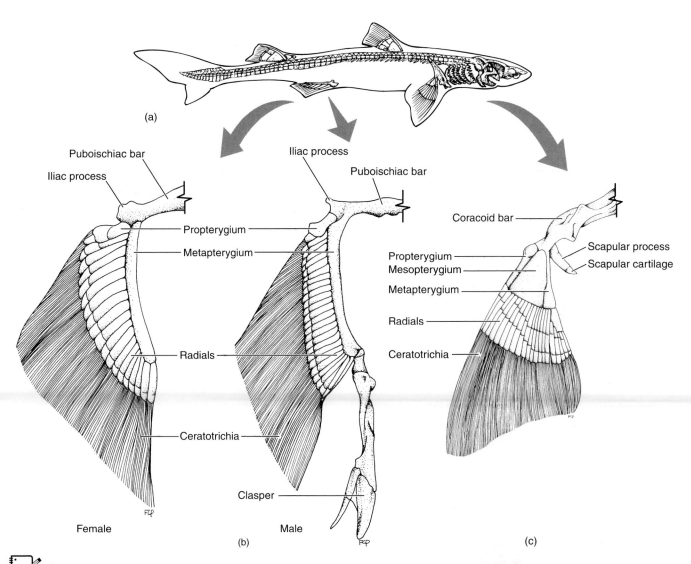

FIGURE 5.16   Right pelvic and right pectoral girdles of a shark. (a) Skeleton showing the general position of each girdle. (b) Female (left) and male (right) pelvic girdles, ventral view. Note the elongate clasper of the male used for sperm transfer. (c) Pectoral girdle (similar in males and females), ventral view.

extended at right angles to the body. These two bones join distally with six smaller bones, the **tarsals,** followed distally by **rays.** The first elongate bone in each ray is a **metatarsal,** the shorter two or three that end the ray are **phalanges** (sing., **phalanx**). Thus, tarsals form the ankle, metatarsals form the sole, and phalanges form the toes or digits of the foot. *Necturus* normally has but four toes on each foot, although most salamanders have four on the front feet and five on the back feet.

Reptilia—Turtle   The pelvic girdle of the turtle is tripartite—ilium, ischium, pubis (figure 5.18b). The expanded dorsal end of the **ilium** joins with three or more sacral ribs depending upon species. The anterior **pubis** bears a lateral projection in its middle, the **prepubic** (or **pectineal**) **process.** At the ventral midline, the pubis of

each side touches at the midline to form a **pubic symphysis,** which bears a forward-directed **epipubic cartilage** from this symphysis. The **ischium** caudal to the pubis also bears in its middle a caudally directed process (unnamed) and enters into an **ischiac symphysis** with the ischium from the other side. Ischium and pubis form the edges of an opening in the ventral girdle, the large **obturator foramen.** In a disarticulated skeleton, note the bones that contribute to the acetabulum.

Present in the limb are the **femur** with prominent head and a pair of ventral flanges, **tibia** (larger) and **fibula** (smaller), **tarsals, metatarsals, phalanges,** and **claws.** Fibulare (= calcaneum) and tibiale (= astragalus) (fused tibiale, intermedium, and centrale) can be identified by position to the fibula and tibia, respectively. Depending upon species, and even within the

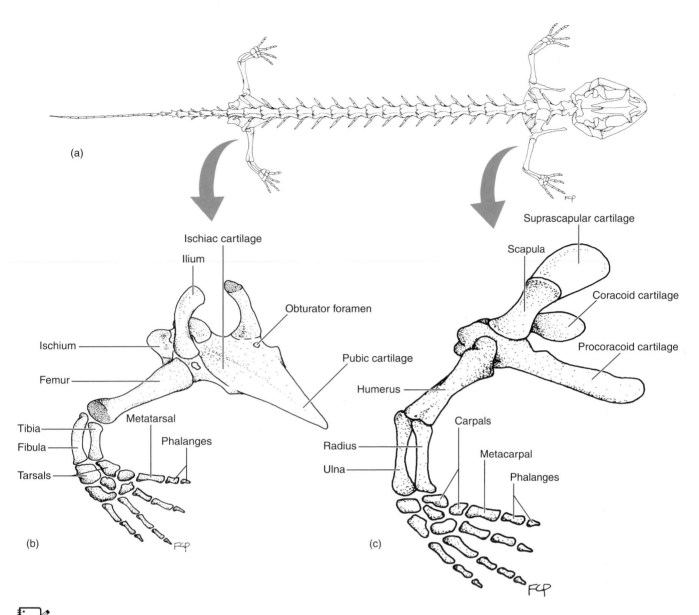

FIGURE 5.17 Right pelvic and right pectoral girdles of *Necturus*. (a) Skeleton showing the general position of each girdle. (b) Pelvic girdle, oblique view. (c) Pectoral girdle, oblique view.

same species, these ankle bones may fuse with each other producing large, single units in the ankle. The remaining tarsals can be simply termed **tarsalia,** which vary in size and number among species. **Metatarsals** and **phalanges** are also present.

Reptilia—Alligator Examine a mounted skeleton of an alligator (figure 5.19a). The pelvic girdle is ossified in the adult, although added cartilaginous components may associate. Like the turtle, the basic pelvic girdle is tripartite, consisting of three stout bones—ilium, ischium, pubis (figure 5.19b). The **ilium** joins dorsally with two sacral vertebrae. Ventrally it contributes with the ischium and pubis to form the **acetabulum,** which receives the proximal head of the femur. The **ischium** is posterior and passes to the ventral midline where its

expanded end joins the ischium of the opposite side in an **ischiac symphysis.** The **pubis** is anterior and also flattens and widens as it passes ventrally; here, where it touches its partner from the opposite side, a **pubic symphysis** is formed. Forward from this, it continues with gastralia, not formally part of the pelvic girdle. Between the pubis and ilium is the **obturator foramen.**

Proximally the **femur** is expanded into a prominent knob, the **head.** Distally it articulates with a robust **tibia** (anterior) and slender **fibula** (posterior). Beyond this follow the **tarsals,** the two most prominent being the **tibiale** and **fibulare,** which articulate with the tibia and fibula, respectively. There are four rays in the alligator hindlimb, each beginning with a long **metatarsal** and ending with one or more **phalanges** tipped with **claws.**

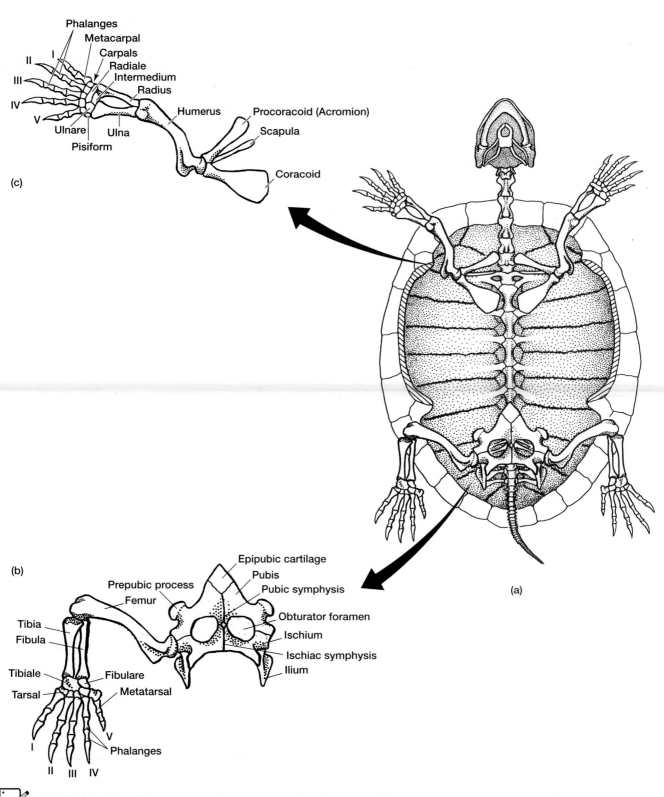

FIGURE 5.18 Right pelvic and right pectoral girdles of a turtle. (a) Skeleton with plastron removed, showing the general position of each girdle. (b) Pelvic girdle, ventral view. (c) Pectoral girdle, ventral view.

🐦 **Aves—Chicken** Fusion of bones is a hallmark of the bird skeleton (figure 5.20a). As seen when examining the vertebral column, components of the last thoracic, lumbar, sacral, and first few caudal vertebrae unite into a single piece, the **synsacrum** (see figure 5.13c). Straddling each side and sutured to it is the paired, broad hip bone, the **innominate,** part of the pelvic girdle (figure 5.13c). **Ilium, ischium,** and **pubis** fuse

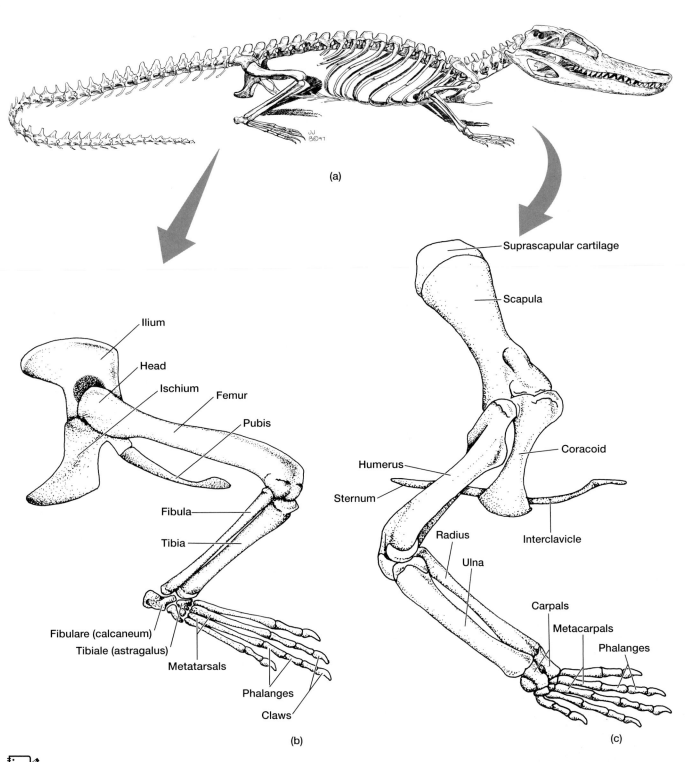

FIGURE 5.19 Right pelvic and right pectoral girdles of an alligator. (a) Skeleton showing the general position of each girdle. (b) Pelvic girdle, lateral view. (c) Pectoral girdle, lateral view.

to form it, thus obliterating fine boundaries between them in the adult bird. The perforated **acetabulum** marks the approximate point of their common union. The slender, riblike **pubis** runs backward from the acetabulum. The bladelike **ilium** extends forward and dorsally from the acetabulum and dorsally to articulate with the synsacrum. The cranial part of the ilium is concave, while its caudal half is convex. Between the

caudal ilium and pubis lies the **ischium.** A low ridge marks the point of fusion of the ischium with the caudal convex half of the ilium. Note that neither a pubic nor an ischiac symphysis forms. Nestled between anterior points of union of pubis and ischium is the small, round **obturator foramen,** and above this foramen and caudal to the acetabulum is the large, oval **ilio-ischiac foramen.** Consult embryonic stages in

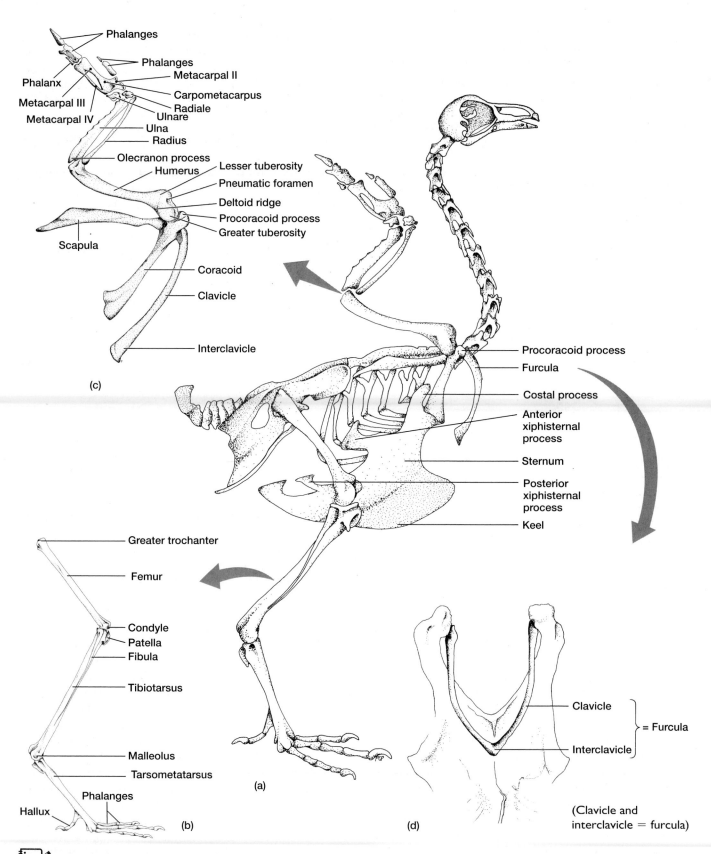

Phalanges

Phalanges
Metacarpal II
Phalanx
Carpometacarpus
Metacarpal III
Radiale
Metacarpal IV
Ulnare
Ulna
Radius
Olecranon process
Humerus — Lesser tuberosity
Pneumatic foramen
Deltoid ridge
Procoracoid process
Greater tuberosity
Scapula

Coracoid

Clavicle

Interclavicle

(c)

Procoracoid process
Furcula
Costal process
Anterior
xiphisternal
process
Sternum
Posterior
xiphisternal
process
Keel

Greater trochanter

Femur

Condyle
Patella
Fibula

Tibiotarsus

Malleolus
Tarsometatarsus
Phalanges
Hallux
(b)

(a)

Clavicle
= Furcula
Interclavicle

(d)

(Clavicle and
interclavicle = furcula)

FIGURE 5.20   Right pelvic and right pectoral girdles of a pigeon. (a) Skeleton showing the general position of each girdle. The left element of paired structures is not shown for clarity. (b) Hindlimb, lateral view. (c) Partial pectoral girdle and forelimb, lateral view. (d) Furcula, ventral view.

The two groups of dinosaurs, the Saurischia and Ornithischia, differ in their pelvic structure. In saurischians, the ilium, ischium, and pubis radiate outward from the center of the pelvis. The pubis is directed cranially and ventrally. However, in ornithischians, part of the pubis is directed caudally, paralleling the ischium. Compare these hip arrangements to a modern alligator and bird. The name Saurischia comes from the word roots *sauro* for lizard and *ischia* for hip. Ornithischia comes from the word root *ornitho* for bird, inspired by its similar appearance to bird hips. Note, however, that birds evolved within the Saurischian radiation, and hence the superficial similarities are an example of convergent evolution, producing an example of homoplasy.

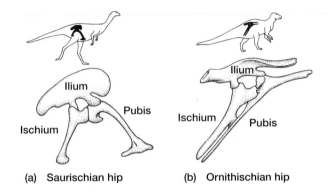

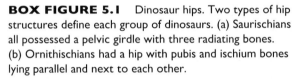

(a) Saurischian hip   (b) Ornithischian hip

BOX FIGURE 5.1 Dinosaur hips. Two types of hip structures define each group of dinosaurs. (a) Saurischians all possessed a pelvic girdle with three radiating bones. (b) Ornithischians had a hip with pubis and ischium bones lying parallel and next to each other.

development of the bird pelvis (figure 5.21a–d) to clarify your understanding of the bony contributions to the innominate.

Fusion of skeletal elements continues to varying degrees in the hindlimb (figure 5.20b). A lateral projection, the **greater trochanter,** is located on the lateral head of the **femur.** At its distal end, the femur carries raised, curved ridges, the **condyles,** and a depression between in which the **patella,** or kneecap, slides. In the shank, the large bone is the **tibiotarsus,** formed by fusion of the distal tibia with proximal tarsal bones. The small, splintlike bone accompanying the tibiotarsus is the reduced **fibula.** Proximally, the tibiotarsus has two **condyles** positioned opposite the **condyles** of the femur. Below these condyles on the tibiotarsus run two low ridges, or **crests,** to which muscles attach. Distally, the end of the bone swells into two ridges, or **malleoli** (sing., *malleolus*), that form a rocking articulation with the succeeding composite bone, the **tarsometatarsus.** As the name suggests, several tarsals and metatarsals fuse. Hence, tarsal bones are not separate, but are incorporated into this or the preceding tibiotarsus. The ankle joint

formed is thus within the tarsals, an intratarsal joint. The fifth ray is lost. **Metatarsals** one through four are incorporated into the tarsometatarsus and support respectively numbered toes with separate **phalangeal** series. The short first digit (sometimes called **hallux**) is on the medial side. Digits two through four are opposite it.

Embryonic Development of Bird Pelvic Girdle   The pelvic girdle appears first as a mesenchymal condensation that forms a single triradiate cartilage, at least in the chick (figure 5.21). Within this single cartilage, three areas of ossification eventually form that are destined to be the three bones of the pelvis: namely, *ilium, ischium,* and *pubis.* In some species of birds, as in reptiles generally, the cartilaginous elements formed from mesenchymal precursors are initially separate. That is, each has its own center of *chondrogenesis.* Later, as ossification begins, each cartilage will possess its own *center of ossification.* In the chick, the caudal ends of the ilium and ischium grow together, leaving only the *ilio-ischiac foramen* open between them. There is also a tendency for the pubic bones to fuse with the ventral edge of the ischium, except anteriorly where the *obturator foramen* persists. The three bones of the pelvic girdle so united with one another are collectively the *innominate bone.*

🦖 **Mammalia—Cat**   The three bones of the girdle fuse indistinguishably into the **innominate bone** (figure 5.22a–c). The large opening in the bone is the **obturator foramen.** The bladelike **ilium,** contributing to the anterior part of the innominate, articulates with the sacrum and has two major regions. The roughened region, the **articular scar,** is the site where the ilium joins the sacrum. The **body** of the ilium lies caudal to the articular scar. The **wing** of the ilium lies cranial to the scar and bears the **crest** of the ilium along its anterior rounded end. A small swelling, the **posterior inferior spine,** projects from the dorsal edge just above the scar. On this same edge, but posteriorly above the obturator foramen, protrudes a slightly more prominent process, the **ischial spine.** The **iliopectineal eminence** is a small rise along the ventral edge of the body of the ilium. The **ischium** contributes to the posterior girdle beyond the **acetabulum.** The end of the ischium, the **ischial tuberosity,** is blunt and roughened. The **pubis** lies ventral to the **acetabulum.** Two general regions of the ischium and pubis are recognized: the **bodies,** which contribute to the acetabulum and also include most of the rest of the bones, and the **rami** (singular, ramus), those remaining parts of each bone that define the ventral border of the obturator foramen. The rami meet along the ventral midline with rami of the respective ischium and pubis of the opposite side to form the **ischiac** and **pubic symphyses.**

The **femur** has a hemispheric **head** pedestaled upon a short, columnar **neck** that joins the rest of the long bone at about a 45° angle to its long axis (figure 5.22d,e). The **fovea capitis** is a shallow pit offset from the middle of the head. A **greater trochanter** projects lateral to the head. Between the base of

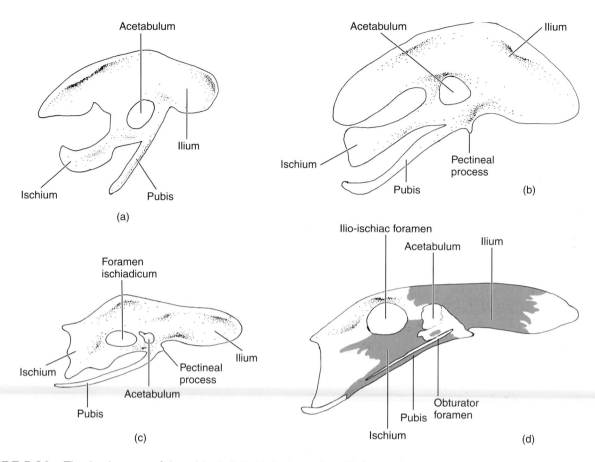

FIGURE 5.21 The development of the pelvic girdle in birds during their 21-day incubation period. Bone begins to replace cartilage after day 15. The acetabulum marks the junction of the three main components of the pelvic girdle: the ilium, pubis, and ischium. The origin of the name ilio-ischiac foramen is revealed by the two bones that develop around it. Note the migration of the pubis toward the ilium and the formation of the obturator foramen. (a) 7 days. (b) 8 days. (c) 13 days. (d) 18 days.

Redrawn with modification after Lebidinsky, 1913.

the greater trochanter and the neck of the femur is a cavity, the **trochanteric fossa**. The **lesser trochanter** is a smaller knob distal to and caudal to the head. The low, posterior ridge between trochanters is the **intertrochanteric line**. The posterior **linea aspera** is a low ridge coursing in a slow spiral from the proximal end of the femur distally along the shaft. The two large, rounded articulating swellings at the distal end of the femur are **lateral** and **medial condyles**. Between them resides a pit, the **intercondyloid fossa**. The roughened sides of the condyles are **lateral** and **medial epicondyles**. A **patella** (figure 5.22f) is present at the knee and slides in a low depression on the anterior face of the distal femur, the **patellar surface**.

In the shank is a slender **fibula** and robust **tibia** (figure 5.22g,h). The articular surfaces of the proximal end of the tibia are condyles (**lateral** and **medial**). Caudally they form a projecting shelf, whose indented underside is the **popliteal notch**. Along the cranial side of the tibia runs the raised **tibial crest** that begins in front of the condyles in a roughened **tibial tuberosity**. At the distal end of the tibia are two projections. The longer medial one is the **medial malleolus**. The shorter lateral process is the **dorsal projection**. The **lateral**

malleolus forms the distal articulating end of the fibula. Note how the proximal end, the **head** of the fibula, is expanded and articulates with the tibia.

Morphological terms for ankle bones help correlate these elements with homologues in other vertebrates already studied. But these bones in the cat are the same as those in humans, although of different shape, and so have more commonly been given the descriptive terms of their counterparts in human anatomy. These descriptive names will be used, with morphological equivalents in parentheses. The heel is a backward projection of the largest of the six ankle bones, the **fibulare** (calcaneum)[1] (figure 5.22i). The **tibiale** (= astragalus) articulates below with the fibulare (calcaneum), above with the malleoli of the tibia and fibula, and in front with the compressed, curved one or two **centralia** (navicular) reaching to the medial side of the foot. The **fourth** and **fifth tarsalia** (cuboid) lie in front of the calcaneus and articulate with the **fourth** and **fifth metatarsals**. Next, proceeding medially, is the **third tarsale** (lateral cuneiform) articulating with the **third metatarsal**. Next, the **second tarsale** (second or middle cuneiform) articulates with the

---

[1]In the human literature, "calcaneum" is often spelled "calcaneus."

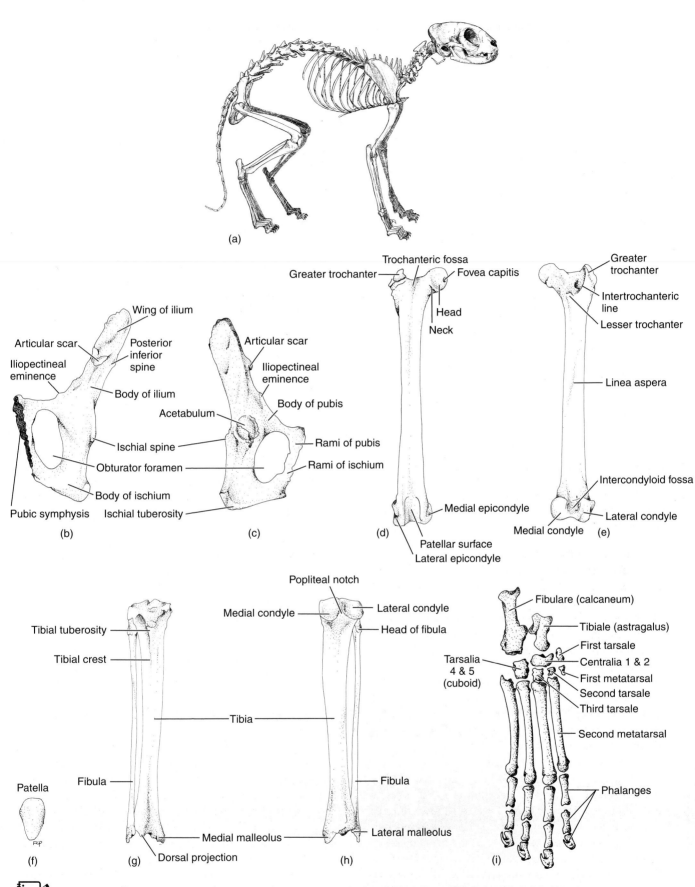

(a)

Trochanteric fossa
Greater trochanter
Fovea capitis
Head
Neck
Medial epicondyle
Patellar surface
Lateral epicondyle
(d)

Greater trochanter
Intertrochanteric line
Lesser trochanter
Linea aspera
Intercondyloid fossa
Medial condyle
Lateral condyle
(e)

Wing of ilium
Articular scar
Posterior inferior spine
Iliopectineal eminence
Body of ilium
Acetabulum
Ischial spine
Obturator foramen
Body of ischium
Pubic symphysis
Ischial tuberosity
(b)

Articular scar
Iliopectineal eminence
Body of pubis
Rami of pubis
Rami of ischium
(c)

Tibial tuberosity
Tibial crest
Tibia
Fibula
Patella
Medial malleolus
(f)
(g)
Dorsal projection

Popliteal notch
Medial condyle
Lateral condyle
Head of fibula
Fibula
Lateral malleolus
(h)

Fibulare (calcaneum)
Tibiale (astragalus)
First tarsale
Centralia 1 & 2
First metatarsal
Second tarsale
Third tarsale
Tarsalia 4 & 5 (cuboid)
Second metatarsal
Phalanges
(i)

FIGURE 5.22   Right pelvic girdle and right hindlimb of a cat. (a) Lateral view of cat skeleton. (b) Innominate bone, dorsal view. (c) Innominate bone, ventral view. (d) Femur, anterior view. (e) Femur, posterior view. (f) Patella. (g) Tibia and fibula, anterior view. (h) Tibia and fibula, posterior view. (i) Hindfoot, ventral view.

second metatarsal, and the small **first tarsale** (first or middle cuneiform) articulates with a rudimentary **first metatarsal. Phalanges** are present. The first toe **(hallux)** is reduced, and each toe ends in a **claw.**

Pectoral Girdle, Sternum, and Appendages

Like the pelvic girdle, the pectoral girdle arose by fusion and enlargement of simple cartilaginous elements and underwent later modification reflecting changes in locomotion and support. However, unlike the pelvic girdle, the basic endochondral bones were joined by dermal bones to give rise to a composite pectoral girdle made up of both. Thus, in addition to noting changes in bone shape, size, and presence from group to group, also be aware of the type of bone involved.

🐾 **Elasmobranchii—Shark**   The pectoral girdle is a semicircular, curved cartilage divided into regions (see figure 5.16c). The ventral portion between the bases of the fins is the **coracoid bar.** The continuations above the fin articulations are the **scapular processes,** tipped with separate cartilage points, the **scapular cartilages.**

Like the pelvic girdle, fins are supported by many dermal rays, **ceratotrichia,** and proximally by **pterygiophores.** Three large, basal pterygiophores support the fin at its articulation with the girdle. The central one is the large, triangular **mesopterygium.** Caudal to it lies the slender **metapterygium,** and in front a small cartilage wedge, the **propterygium.** Distal to these lie rows of aligned **radials.** No sternum is present.

🐾 **Primitive Osteichthyes—Amia and derived Teleost— Perch**   During early bony fish evolution, dermal bones became applied to the endochondral girdle and compose the most prominent part. In fact, the cartilage bones are difficult to see in most dried fish skeletons. Although impractical to see on most specimens, a **scapulocoracoid** is usually present. The approximate position of the dried scapulocoracoid can be found at the point where the base of the fin attaches to the girdle (figure 5.23). Occasionally, specific ossified pterygiophores can be seen as well, although these are also impractical to see. Further, *Amia,* but not the perch, has a tiny, sliverlike clavicle near the anterior, ventral end of the cleithrum (figure 5.23); however, this clavicle is also impractical to locate in commercially purchased specimens.

Examine a mounted skeleton to find the position of the pectoral fin and site of the pectoral girdle that supports it. Return now to an *Amia* skull. The pectoral girdle is covered by the large, disklike opercular bones at the back of the skull. The girdle attaches dorsally at the top of the skull and runs in an arched series of increasingly larger size to beneath the gill arches (figure 5.23). The gill arches support hairlike gill filaments that fill most of the posterior skull opening. The dorsal most **post-temporal** is the first bone in the series. Forked in most bony fish, here it occupies the

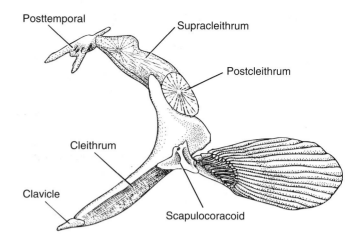

FIGURE 5.23   Left pectoral girdle of *Amia,* a primitive actinopterygian. Lateral view with cranial to the left.

upper, posterior corner of the skull and sends a strut from its undersurface down to meet a projection from the skull. Next follows the **supracleithrum,** which joins the extensive **cleithrum** in an overlapping articulation. Positioned on the caudal margin of this lap joint is a flakelike **postcleithrum.** No sternum is present.

🐾 **Amphibia—*Necturus***   Left and right sides overlap ventrally when in place, but do not fuse as occurs in the pelvic girdle (see figure 5.17c). Only the **scapula** is ossified. It is a short bone that bears a **suprascapular cartilage** above. Below the scapula is a concave depression, the **glenoid fossa,** that receives the head of the **humerus.** The ventral part of the girdle is the **coracoid cartilage,** a broad plate that enlarges like a disk toward the ventral midline where it may overlap with its partner. The slender anterior projection at a right angle to the coracoid cartilage is the **procoracoid cartilage.** The single **humerus** is the upper limb bone. Next, the forearm consists of the **radius** and **ulna,** which lie, respectively, anterior and posterior when the limb is outstretched. Six small **carpals** follow and support four projecting rays. Formally, the proximal element of each ray is termed a **metacarpal** and the remaining two, **phalanges.**

Care should be taken in generalizing the *Necturus* pectoral girdle to other amphibians, especially fossilized forms, because it is in many ways specialized. In early amphibians, the girdle was more completely ossified and usually possessed an interclavicle, clavicle, and cleithrum.

Reptilia

In primitive reptiles, the pectoral girdle consists of the dermal (interclavicle, clavicle, and remnants of the cleithrum) and cartilaginous (scapula, coracoid, and procoracoid) (see note 2)

---

[2]The procoracoid cartilage you first met in *Necturus* is likely homologous to the amniote shoulder bone of the same name, the procoracoid. However, the coracoid cartilage of *Necturus* is not homologous to the coracoid bone of amniotes but just a coincidence of similar names.

components. These have been lost or modified in the two specific reptiles examined here.

Turtle   A sternum is absent, although a **plastron** is present in its place. The outer surface of the plastron is covered by cornified material derived from the epidermis. On the inner surface of the plastron, dermal bones are fused along zigzag sutures. Anteriorly are two small medial bones called **epiplastra,** likely derived from the clavicles. Caudal to them lies the single central **entoplastron** from the interclavicle. The larger plates that follow may be **gastralia.**

Three bones form the pectoral girdle, which lies, unlike in other vertebrates, within the rib cage (figure 5.18c). The flat lower bone is the **coracoid.** The slender bone reaching dorsally into the carapace is the **scapula.** A third bone, in the same plane as the coracoid, is the **procoracoid,** also called by some the acromion or acromial process. The scapula and procoracoid fuse in the adult to form an L-shaped bone, which together with the coracoid contributes to the **glenoid fossa** that receives articulation with the humerus (see figure 5.18c). Forelimb bones are as you would expect. A single proximal **humerus** articulates proximally with the girdle and distally with the **ulna** and **radius.** When the limb is extended laterally, the ulna is more caudal than the radius. The bones are bowed to allow limb retraction and to accommodate restricted locomotion. (Be prepared for improperly mounted forelimbs that are actually reversed left and right.) The distal radius contacts the **radiale** within the wrist. In some species, the radiale fuses with a centrale. The distal end of the ulna makes contact with two small bones: the **ulnare** (outer) and **intermedium** (inner). The **centralia** reside in front of (distal to) these two bones and the five **carpalia,** which each support a **metacarpal** and two or three **phalanges.**

Alligator   Clavicles are absent. A bladelike **scapula** tipped dorsally with a **suprascapular cartilage** meets a dumb-bell-shaped **coracoid** laterally to form a posterior **glenoid fossa** that receives the proximal end of the **humerus** (see figure 5.19c). The left and right coracoids articulate medially with the sides of a thin, oval sternum. A slender **interclavicle** is usually difficult to see because it is fused with the ventral side of the **sternum,** although the spear-shaped head of the interclavicle can be seen projecting forward from the anterior edge of the sternum. The procoracoid, sometimes present in the embryo, does not differentiate into a distinct adult bone but becomes incorporated without sutures into the ventral part of the scapula.

The radius forms the medial margins and the larger ulna the lateral margins of the forearm. **Carpals, metacarpals,** and **phalanges** tipped with claws continue distally.

Aves—Chicken or Turkey   Three bones compose the adult pectoral girdle: a long, slender bone over the top of the rib cage, the **scapula;** a robust bone running downward to the sternum, the **coracoid;** and anterior to this the **furcula,** or wishbone (see figure 5.20a,c,d). The furcula is a composite bone formed by the fusion of two **clavicles** (forks) joined ventrally to the single **interclavicle** (compressed plate). The avian procoracoid is reduced to a process fused to the distal end of the coracoid, the **procoracoid process.**

The large **sternum,** between ventral tips of the ribs, bears a large ventral projection, the **keel,** or **carina,** for attachment of flight muscles (see figure 5.20a). At its anterior end, the sternum bears two large depressions or facets in which rest the coracoids. Medially, between these two facets arises a single projection, the **rostrum.**[3] Like a pair of curved horns, paired **costal processes** (see note 3) project anteriorly and dorsally from the sternum; they depart at a point between the coracoid and a series of small depressions for articulation with ribs, the **rib facets.** Caudal to these rib facets, a branched process sends an **anterior xiphisternal process** upward, and a slender **posterior xiphisternal process** (see note 3) arching backward above the sternum.

The **glenoid fossa,** formed by scapula and coracoid, receives the head of the **humerus** (see figure 5.20c). The **greater tuberosity,** on the ventral side of the **head** of the humerus, bears a large hole at its base, the **pneumatic foramen,** the portal of entrance into air sacs that fill much of the long bone's core. The **lesser tuberosity,** on the dorsal side of the head, is continuous with a peaked, lateral projection, the **deltoid ridge.** The curved **ulna,** larger than the straight, slender **radius,** possesses an **elbow** or **olecranon process,** a short projection at its articulation with the humerus. Carpals fuse and some are lost in the wrist, but there is no consensus as to which contribute to the two small bones, **ulnare** and **radiale,** that articulate with the respective forearm bones of the adult chicken. Other wrist bones fuse with metacarpals to form the **carpometacarpus,** the large composite bone that follows next distally. At the base of the carpometacarpus, metacarpal II supports the small first digit. Metacarpal III is the most prominent, and its two well-developed phalanges (singular, phalanx) continue beyond the distal end of the carpometacarpus. Metacarpal IV also supports a small phalanx.

Mammalia—Cat   The pectoral girdle of mammals typically includes the smallest number of elements, usually just a clavicle and scapula. The **clavicle** is small and resides in shoulder muscle. It will be located when muscles are dissected, but should be located now on a mounted skeleton where it is often wired near the joint between the forelimb and girdle. The **scapula** is a flat, large, and not quite triangular bone bearing a ventral, cup-shaped depression, the **glenoid fossa** (figure 5.24a–c). A prominent **spine** divides the lateral surface of the scapula

---

[3]In the pigeon, these processes are less prominent. The posterior xiphisternal process fuses with the sternum, but its hooklike tip can be seen as a small forward projection at the posterior end of the sternum.

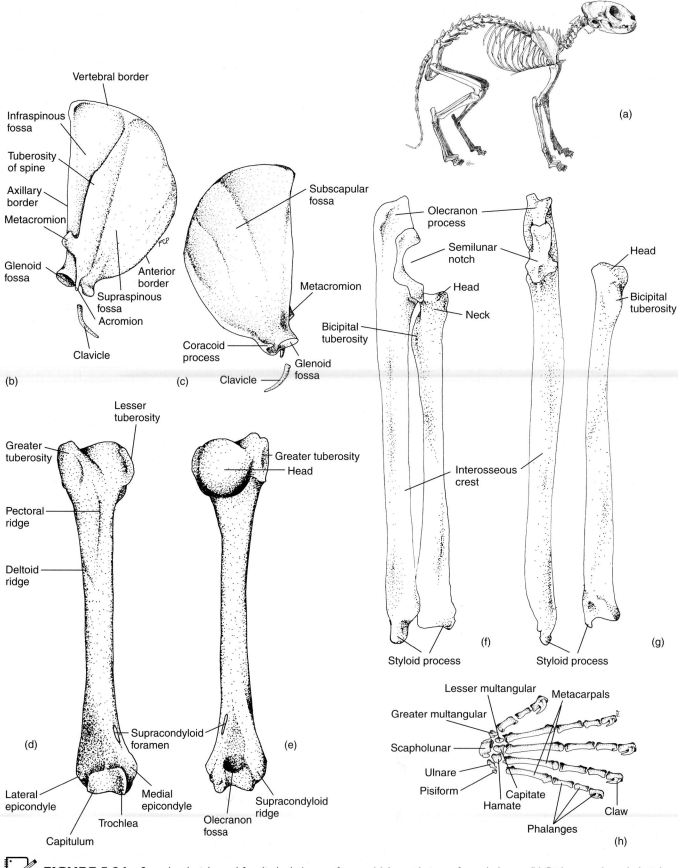

FIGURE 5.24 Scapula, clavicle, and forelimb skeleton of a cat. (a) Lateral view of cat skeleton. (b) Right scapula and clavicle, lateral view. (c) Right scapula and clavicle, medial view. (d) Right humerus, anterior view. (e) Right humerus, posterior view. (f) Articulated left radius and ulna, lateral view. (g) Left radius and ulna, anterior view. (h) Right forefoot, dorsal view.

Examine again the *Amia* pectoral girdle. The scapulocoracoid and proximal fin elements are endochondral but the other bones of the girdle are dermal. The pectoral girdles of later vertebrates retain this basic composition, but the dermal elements tend to be reduced as the endochondral elements become more prominent. In primitive amniotes, a new endochondral bone appears ventrally, the posterior coracoid, to join with the phylogenetically older scapulocoracoid bone. The three persist into primitive mammals. In marsupials and placental mammals, only the scapula and posterior coracoid (called just *coracoid*) persist. In modern reptiles and birds, the procoracoid remains distinct. In turtles, it is largely reduced or remains only a rudiment incorporated into or merged with a coracoid.

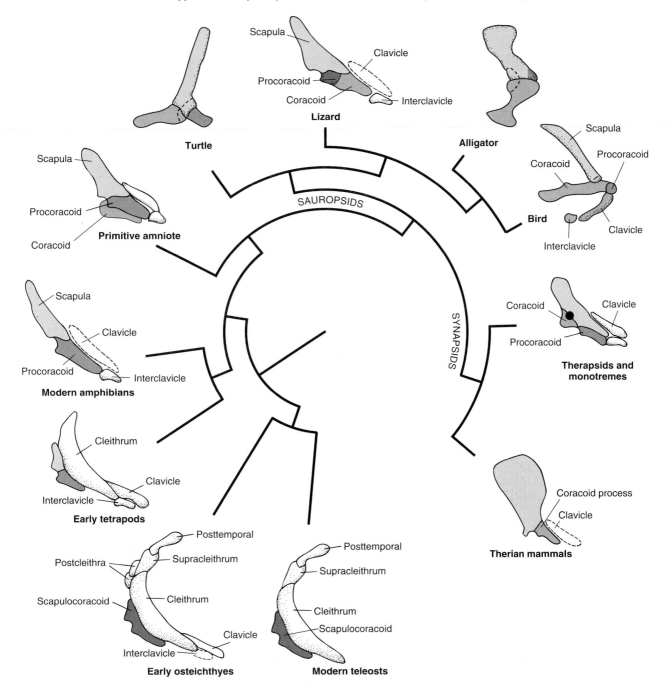

BOX FIGURE 5.2 Summary of pectoral girdle evolution. Notice that the dermal elements (no shading) of the girdle tend to be lost, and endochondral elements (shaded elements) tend to assume a greater prominence. A single endochondral element, the scapulocoracoid, is present in fishes, but in early tetrapods two distinct bones, the scapula and the procoracoid, are present. In primitive amniotes, a third endochondral bone appears, the coracoid, to join with the phylogenetically older scapula and procoracoid. The three persist into primitive mammals although only two remain in therian mammals (scapula and coracoid as a process). In modern reptiles and birds, the scapula and coracoid persist; the procoracoid is reduced or lost from the adult shoulder girdle.

into two regions: the anterior **supraspinous fossa** and posterior **infraspinous fossa.** The entire, undivided medial surface of the scapula facing the rib cage is called the **subscapular fossa.** The short, anterior edge of the scapula is the **anterior border,** the straight, posterior edge is the **axillary border,** and the curved, upper edge is the **vertebral border.** The middle, roughened border of the spine is the **tuberosity.** Ventral to the tuberosity, the spine forms a sharp, posterior, flat projection, the **meta-cromion.** The ventral end of the spine terminates in a downward-pointed **acromion,** receiving one end of the clavicle. The small, hooklike **coracoid process,** a remnant of the coracoid bone, forms embryologically as a separate center of ossification. It can be found fused with the anterior rim of the glenoid fossa.

The chain of eight **sternebrae** constitute the sternum and run in tandem between the ventral tips of the ribs (see figure 5.14a). The first sternebra in the series is the **manubrium,** and the last is the **xiphisternum** (often with a terminal sternebra **xiphoid** or **ensiform cartilage**). The six sternebrae between these constitute the **body** of the sternum.

The forelimb bones are those as expected, although a few distal elements are modified for the cat's locomotor requirements, and the forearm bones cross to direct the toes forward. The **head** of the **humerus** projects medially and is bounded on its sides by the **greater** (lateral) and **lesser** (medial) **tuberosities** (figure 5.24d,e). Named for the general muscle groups that insert below the greater tuberosity, the low **deltoid** (posterior) and **pectoral** (lateral) **ridges** appear like seams on the bone. The distal humerus is smooth and rounded where it articulates with forearm bones: The swollen medial part of the articular surface is the **trochlea,** and the lateral articular swelling is the **capitulum.** On the side and above each of these are the **epicondyles, medial** and **lateral,** respectively. The slit above the medial epicondyle is the **supracondyloid foramen.** Running out of the lateral epicondyle along the side of the humerus is the **supracondyloid ridge.** When the forelimb is extended, the elbow rests in the **olecranon fossa,** lying between humeral epicondyles.

In the forearm, the **ulna** is deeply indented, forming the **semilunar notch,** located where the ulna articulates with the trochlear surface of the humerus (figure 5.24f,g). Caudal to the semilunar notch is the **elbow,** or **olecranon process.** The distal end of the ulna terminates in a pointed **styloid** process. The **interosseous crest** runs along the shaft of the ulna. The cup-shaped **head** of the radius articulates with the capitulum of the humerus. Supporting the head is a short **neck.** The knoblike swelling on the side and just adjacent to the neck is the **bicipital tuberosity.** The radius and ulna each end distally in their own small projection, a **styloid process.**

In the wrist, the **scapholunar** (formed from the fusion of the radiale and intermedium) rides on the end of the radius, and the **ulnare** (triquetrum) rides on the ulna (figure 5.24h). Lateral to the ulnare is the small

pisiform bone, a sesamoid bone. Lying in a transverse row distal to these elements are, from medial to lateral, the **greater multangular** (first carpale), **lesser multangular** (second carpale), **capitate** (third carpale), and **hamate** (fourth and fifth carpale). Five **metacarpals,** their **phalanges,** and terminal **claws,** are present distally.

Skull

Introduction

The vertebrate skull receives three phylogenetic contributions: the *chondrocranium,* which, if ossified, is often known as the *neurocranium,* the *dermatocranium,* and the *splanchnocranium.*

Chondrocranium

Details of its embryonic formation differ among species, but, generally, head mesenchyme condenses into cartilage elements (figure 5.25), such as the *parachordals* (next to the notochord), *trabeculae* (anterior to the parachordals), the *occipitals* (posterior to the parachordals), and *cartilaginous capsules,* which surround chief sense organs (e.g., nasal, optic, otic). These elements fuse to form the embryonic chondrocranium. In elasmobranchs, this cartilaginous skull persists, develops a roof, and becomes the adult braincase. However, in most vertebrates, these cartilaginous plates are transitory. Centers of ossification soon appear (not necessarily in the same sites as the original cartilage elements), replace the cartilage, and give rise to specific bones of the mature skull. These bones include the *occipitals* (basi-, ex-, supra-), *sphenoids, mesethmoid, otic capsule* (prootic, opisthotic, epiotic), *optic capsule* (birds, reptiles), and *nasal capsule.*

Splanchnocranium

The splanchnocranium generally supports the gills and offers attachment for the respiratory muscles. Elements of the splanchnocranium contribute to the jaws, and *hyoid apparatus* of gnathostomes, and inner ear bones of mammals (figure 5.26).

Dermatocranium

The dermatocranium consists of dermal bones that contribute to the skull. Phylogenetically these bones first appear as enlarged bony armor in ostracoderms, sink inward, and become applied to the chondrocranium and splanchnocranium. They also contribute to the pectoral girdle. The dermal skull may contain a considerable series of bones joined firmly at sutures in order to box in the brain and other skull elements. As a convenience, we can group these series and recognize the most common bones in each (figure 5.27).

The Composite Skull

Within the embryonic chondrocranium, centers of ossification appear destined to form specific endochondral bones of the base of the skull. The specific bones so formed vary

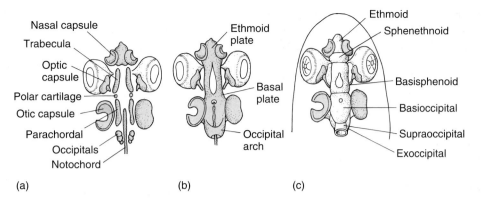

FIGURE 5.25 Embryonic development of the chondrocranium. Cartilage (shaded) appears first but in most vertebrates is replaced by bone (not shaded) later in development. The chondrocranium includes these cartilaginous elements that form the base and back of the skull together with the supportive capsules around sensory organs. Early condensation of mesenchymal cells differentiates into cartilage (a) that grows and fuses together to produce the basic ethmoid, basal, and occipital regions (b) that later ossify (c), forming basic bones and sensory capsules.

After de Beer.

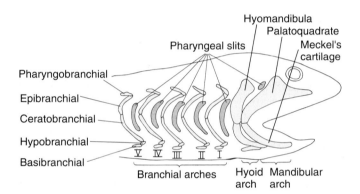

FIGURE 5.26 Primitive splanchnocranium. Seven arches are shown. Up to five elements compose an arch on each side, beginning with the pharyngobranchial dorsally and in sequence to the basibranchials most ventrally. The first two complete arches are named: mandibular arch for the first and hyoid arch for the second that supports it. The characteristic five-arch elements are reduced to just two in the mandibular arch, the palatoquadrate and Meckel's cartilage. The large hyomandibula, derived from an epibranchial element, is the most prominent component of the next arch, the hyoid arch. Caudal to the hyoid arch are variable numbers of branchial arches I, II, and so on.

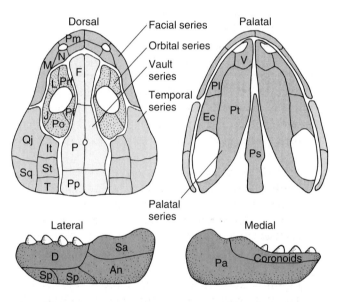

FIGURE 5.27 Major bones of the dermatocranium. Sets of dermal bones form the facial series surrounding the nostril. The orbital series encircles the eye, and the temporal series composes the lateral wall caudal to the eye. The vault series, the roofing bones, run across the top of the skull above the brain. Covering the top of the mouth is the palatal series of bones. Meckel's cartilage (not shown) is encased in the mandibular series of the lower jaw. Abbreviations: angular (An), dentary (D), ectopterygoid (Ec), frontal (F), intertemporal (It), jugal (J), lacrimal (L), maxilla (M), nasal (N), parietal (P), prearticular (Pa), palatine (Pl), premaxilla (Pm), postorbital (Po), postparietal (Pp), prefrontal (Prf), parasphenoid (Ps), pterygoid (Pt), quadratojugal (Qj), surangular (Sa), splenial (Sp), squamosal (Sq), supratemporal (St), tabular (T), and vomer (V).

somewhat among groups, and they often in turn fuse to form composite bones (e.g., temporal bone, occipital bone). The more common bones to form within the chondrocranium are shown in figure 5.28a. The nasal capsule often remains unossified, except in mammals where it forms the coiled *turbinate bones* (= turbinales, = nasal conchae).

Bones derived from the splanchnocranium are also endochondral (figure 5.28b,c). The *palatoquadrate* element of the first *mandibular arch* gives rise to the *quadrate* and *epipterygoid*. *Meckel's cartilage* gives rise to the *articular*. The *hyomandibula* and its homologues *(stapes)* arise from

the second hyoid arch. The ventral elements of the splanchnocranium contribute to the *hyoid bone*.

Generally, dermal bones come to encase the endochondral elements of the skull (figure 5.28d). The six regions of

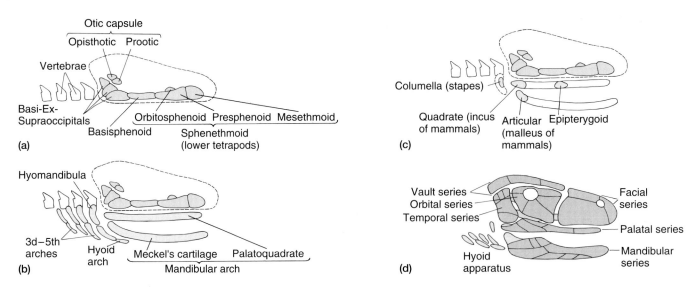

FIGURE 5.28 Contributions to the skull. The chondrocranium (a) establishes a supportive platform that is joined by contributions from the splanchnocranium (b), in particular the epipterygoid (c). Other parts of the splanchnocranium give rise to the articular, quadrate, and hyomandibular, as well as to the hyoid apparatus (d). The dermatocranium encases most of the chondrocranium together with contributions from the splanchnocranium.

TABLE 5.4 Six Series of Bones Forming the Dermatocranium

Facial series

nasal, premaxilla, maxilla

Vault series

frontal, parietal, postparietal

Orbital series

prefrontal, postfrontal, postorbital, jugal, lacrimal

Temporal series

intertemporal, supratemporal, tabular, squamosal, quadratojugal

Palatal series

vomer, palatine, ectopterygoid, pterygoid, parasphenoid

Mandibular series

dentary, splenials, coronoids, angular, prearticular, surangular

dermal bones identified earlier (see figure 5.27) form natural groupings that make learning their names easier. These groups, the *facial, vault, orbital, temporal, palatal,* and *mandibular* series, are listed in table 5.4 with their components.

Cranial Skeleton

Elasmobranchii—Shark

The shark skull is entirely cartilaginous, being derived from the chondrocranium with gill arch contributions from the splanchnocranium. Consequently, it does not air-dry well for inspection. Thus, plastic or wax models are usually the handheld "skulls" actually examined. This is sufficient for most structures, but be sure to inspect fluid-preserved, bottled shark skulls for some structures that are broken or missing from the models.

Begin with a model of the shark chondrocranium. The skull forms by a dramatic enlargement of the embryonic chondrocranium, although this is hard to appreciate until the more modest chondrocranium of other vertebrates is studied. (After having examined the *Necturus* skull, this enlargement in sharks will be more apparent.)

Dorsal Surface

The prominent, prowlike scoop projecting forward is the **rostrum.** Its rounded floor is the **precerebral cavity** (figure 5.29a). The base of the rostrum lies around the sides and base of an opening, the **precerebral fenestra,** through which one can peer into the **cranial cavity,** occupied in life by the brain. Lateral to the base of the rostrum are the paired **nasal capsules.** They imprint poorly into models. Thus, what in life are marble-sized swellings, each pierced anteriorly by two slitlike **nares,** usually turn out to be missing in wax models. Often only a remnant of the base of the walled capsules remains. Caudal to the nasal capsules, the lateral edge of the dorsal skull first forms an outward process, the **antorbital process.** This edge then dips sharply inward, forming the **supraorbital crest,** and finally swings laterally again as a prominent projection, the **postorbital process.** The supraorbital crest is thus the edge of the dorsal skull between bases of the antorbital and postorbital processes, and in life lies over the median half of the eyeball.

Many holes pierce the dorsal surface of the skull. Two openings lie along the midline. Anteriorly, the single median hole caudal to the rostrum and opening to the cranial cavity directly below is the **epiphyseal foramen,** filled in life by the **epiphysis** or **pineal gland.** Posteriorly, a large, medial depression with bilaterally paired openings at its base can be seen. The border of the opening into

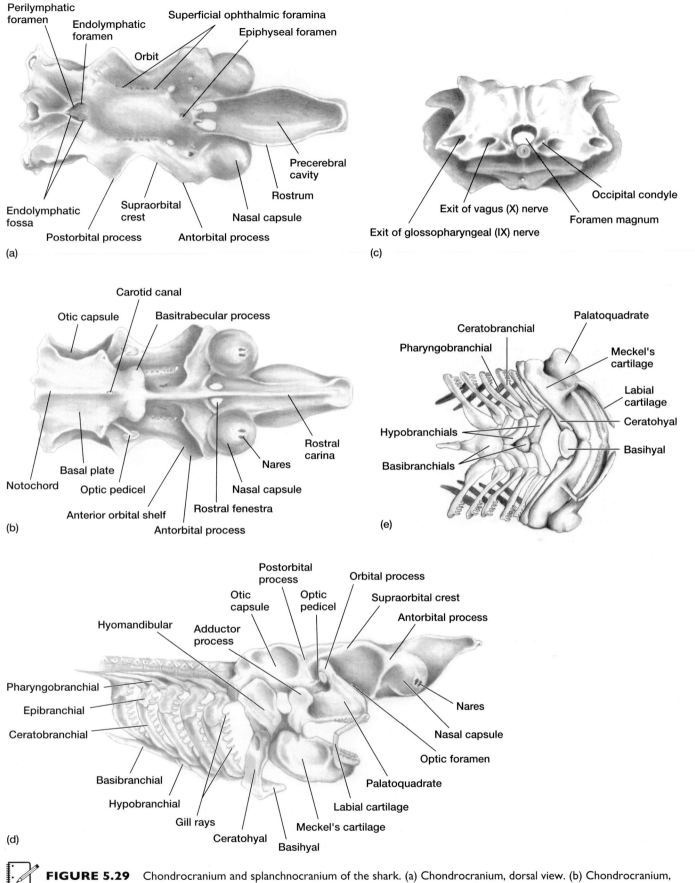

FIGURE 5.29 Chondrocranium and splanchnocranium of the shark. (a) Chondrocranium, dorsal view. (b) Chondrocranium, ventral view. (c) Chondrocranium, posterior view. (d) Chondrocranium with attached splanchnocranium, lateral view. (e) Splanchnocranium, ventral view.

the depression is called the **endolymphatic fossa.** On the floor of the depression the small (anterior) paired pores are the **endolymphatic foramina,** and the paired, large (posterior) pores are the **perilymphatic foramina,** which open to a deep, lateral tunnel leading into the **auditory,** or **otic capsule.** Along the base of the supraorbital crest run, bilaterally, a series of small pores, the **superficial ophthalmic foramina.** In front of them is a large pore opening into the orbit, and in front of it is a final opening at the base of the nasal capsule. Collectively these are generally termed **foramina,** through which branches of cranial nerves (trigeminal, V, and facial, VII) exit.

Ventral Surface

🦕 The **rostral carina** is a keel running the midventral length of the rostrum (figure 5.29b). On either side of its base are two prominent openings: the **rostral fenestrae** and, lateral to these, the **nasal capsules** (or their remnants). A lateral projection of the chondrocranium forms the anterior wall of the orbit, the **anterior orbital shelf.** Dorsally, it carries the **antorbital process.** Ventrally, the anterior orbital shelf ends in an unnamed, pointed process. The chondrocranium narrows between the orbits, then posteriorly swells into the paired **basitrabecular processes.** Immediately caudal to these processes is the single medial foramen for entrance of the internal carotid arteries, the **carotid canal.** The posterior chondrocranium slopes away from the basitrabecular processes to widen into a flat area, the **basal plate,** bounded laterally by the **otic capsules.** The median swelling in the basal plate is the **notochord.**

Return to a preserved specimen and identify the **nasal capsules.** Also locate, near the rear of the orbit, a thin stalk with a cup-shaped end. This is the **optic pedicel,** resembling a small golf tee in some respects.

Posterior End

🦕 The single, medial large opening is the **foramen magnum** (figure 5.29c). Next, lateral to it is the foramen for exit of the vagus nerve (X) and, further laterally, the foramen through which the glossopharyngeal nerve (IX) passes. Below the foramen magnum is the cupped surface receiving articulation of the first vertebra and, lateral to it, paired sharp processes, the **occipital condyles.**

Lateral Surface

🦕 The major cavity is the **orbit,** which in life holds the eyeball. Note the projecting **postorbital process, antorbital process,** and shelflike **supraorbital crest** between them (figure 5.29d). The back wall of the orbit is pierced by many pores for nerves and blood vessels. The two largest are the openings for the optic nerve, the **optic foramen** (anterior) and trigeminal nerve (posterior, directly below postorbital process). An **optic pedicel,** identified previously in the ventral view, projects forward from this same wall in front of the trigeminal foramen, but is usually broken off in skull models. Try

to remember to look for the optic pedicel later in the course when you dissect the muscles of the eye.

Splanchnocranium

🦕 Study a preserved specimen with gill arches in place (figure 5.29d). Seven gill arches are present in series, although the first two are highly modified. The last five (3–7) in sequence, known as **gill** or **branchial arches,** are similar. Each of these arches consists of up to five elements. Dorsal to ventral, these elements are: **pharyngobranchial,** a posteriorly directed, splintlike element that lies next to the vertebral column; **epibranchial,** a stout piece that together with the next, longer **ceratobranchial,** bear fingerlike **gill rays** to support gills; and a **hypobranchial,** represented by three short pieces on the anterior three branchial arches. The **basibranchials,** seen to best advantage in ventral view (figure 5.29e), are unpaired. Two exist between the tips of the gill arches, 3–7.

The second arch, or **hyoid arch,** is modified. A single, midventral piece, the **basihyal,** articulates on each end with a slender bar, the **ceratohyal.** Above each of these is the robust **hyomandibular.** The hyomandibular articulates with the otic region of the skull. The hyomandibular acts as a major site of attachment of the visceral skeleton to the chondrocranium, and is thus said to serve as **suspensor** of the lower jaw.

The first gill arch, or **mandibular arch,** is the most modified member of the gill-arch series, and forms the tooth-bearing upper and lower shark jaws. The paired ventral elements forming the ventral half of the mandibular arch are the large **Meckel's cartilages.** Rows of teeth run along their inner margins. Often preserved and still attached to Meckel's cartilage is a slender, antennaelike process, the **labial cartilage.** The dorsal half of the first arch consists of a pair of substantial pieces, the **palatoquadrates,** which form the upper jaw of a shark. The anterior end of each palatoquadrate supports the **orbital process,** projecting upward along the back wall of the orbit. A shorter **adductor process** (= **quadrate process**) for muscle attachment projects dorsally from the rear of the palatoquadrate.

Osteichthyes—*Amia* (Bowfin)

Dermatocranium

🐦 Gill arches and gill rays are a dried mass occupying the center of the skull, seen best in posterior view. The chondrocranium is present but obscured by the extensive covering of dermal bones.

Internally, on the roof of the mouth, three pairs of bones are recognized by the stand of small teeth they bear: **ectopterygoid, palatine,** and **vomer.** However, it is impractical to distinguish these three bones separately in dried specimens. Further, the snout is often coated with dried tissue, obscuring some bones beneath. Let's start with what can be seen. The most external row of upper teeth is carried on two pairs of bones: the medial **premaxillae** and the **maxillae,** which define the lateral margins of the mouth. Riding atop each maxilla is the wedge-shaped **supramaxilla.** Passing over the premaxillae are the **anterior nostrils** and between them is the unpaired **rostral** (figure 5.30a,b). Just posterior to the rostral are the paired **nasals,** and lateral to them, the **antorbitals** (figure 5.30b). Parts of both bones may be obscured by dried tissue, as may be the posterior nostril. Running laterally from the antorbital is the larger **lacrimal,** and the **jugal,** actually two bones that define the lower rim of the orbit. Caudal to the orbit are two large bones: the **postorbitals** (infraorbito-suborbitals). In the recess beneath their posterior rims is the **hyomandibula.** Within this recess, the ventral border of the hyomandibula articulates with the **quadrate.** The posterior edges of the hyomandibula and quadrate meet the **preopercular,** a curved, vertical bone that comes out of the recess to support a series of platelike bones on its posterior border. The largest and most dorsal of these is the **opercular,** joined ventrally to the **subopercular,** and below these, the **interopercular.** Like overlapping blades of an oriental fan, the **branchiostegal** bones unfurl in a series across the ventral throat region. In front of them is the single, shield-like **gular,** which is bordered laterally by the lower jaws.

The lower jaw is made up of three dermal bones viewed laterally. The **dentary** forms the middle and anterior half of the lower jaw; it bears the external row of lower teeth. Just posterior to the dentary, at the back of the jaw, is the triangular **angular.** Dorsal to the angular is the small sliver of bone, the **surangular** (= supraangular). Again find the quadrate, which is deeply recessed below the hyomandibula. Note that it extends laterally to articulate with the medial, posterior corner of the angular.

On the dorsal surface of the skull, two pairs of roofing bones follow the nasals, defined by irregular suture lines (figure 5.30b). Caudal to the nasals lie the **parietals,** followed caudally by the **postparietals.** The **extrascapulars,** dermal additions at the end of the roofing series, occupy the posterior rim of the skull. Adjacent to the roofing bones in the temporal region are the paired **intertemporal,** lateral to the parietal, and **supratemporal,** lateral to the postparietal.

Other bones attached to the back of the skull belong to the girdle, which was addressed earlier.

Amphibia—*Necturus*

Dermatocranium

🐦 A dissection microscope may assist you in the study of these small skulls. In the dried *Necturus* skull, notice the position of the **premaxilla, frontal,** and **parietal,** across the dorsal surface (figure 5.31a). Ventrally, the **premaxilla, vomer, pterygoid** (= palatopterygoid), and **parasphenoid** can be seen (figure 5.31b). The bone covering the outer side of the lower jaw is the **dentary** (figure 5.31c).

Chondrocranium

🐦 An isolated chondrocranium embedded in plastic offers an opportunity to identify this part of the skull without dermal bones obscuring the study. Once examined, however, try to visualize how it fits within the casing of the dermatocranium (figure 5.31a).

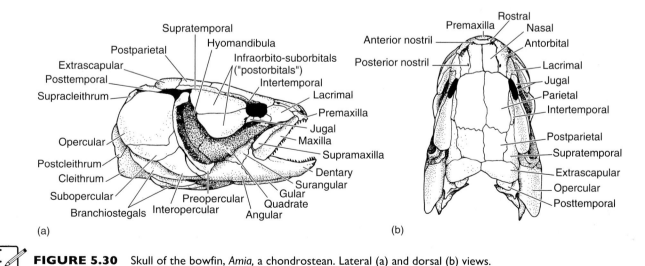

(a)

(b)

FIGURE 5.30 Skull of the bowfin, *Amia,* a chondrostean. Lateral (a) and dorsal (b) views.

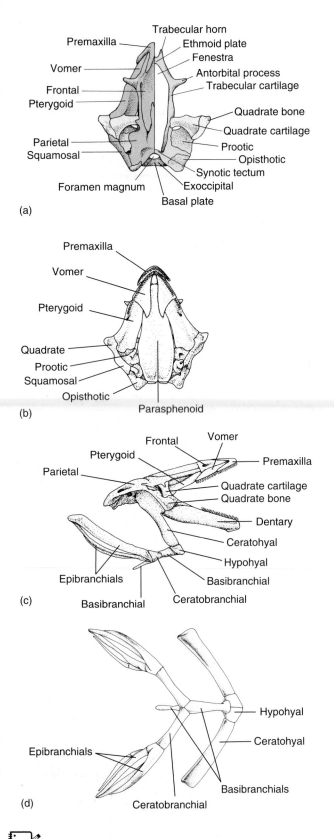

(a)

(b)

(c)

(d)

FIGURE 5.31 Skull of *Necturus*, a modern amphibian. (a) Superficial skull bones are indicated on the left. These bones have been removed to reveal the chondrocranium and derivatives of the splanchnocranium on the right. (b) Ventral view of skull with jaw removed. (c) Lateral view of skull with splanchnocranium attached. (d) Splanchnocranium, ventral view.

Compare the chondrocranium to figure 5.31a. Note that **trabeculae** send forth lateral projections, the **antorbital processes,** and fuse anteriorly via an **ethmoid plate** from which two short **trabecular horns** project. At the opposite end, the **foramen magnum** is surrounded ventrally by the tiny **basal plate,** laterally by the ossified **exoccipitals** and bulging **otic capsules,** and above by the **synotic tectum.** Within the otic capsule two sites of ossification occur: **prootic** (anteriorly) and **opisthotic** (posteriorly). Again, be sure to note how the chondrocranium fits with the dermatocranium.

Splanchnocranium

Compare figure 5.31c to a preserved or embedded skull with gill arches intact. The lateral **quadrate cartilage** ossifies, forming the **quadrate bone.** Identify the corresponding elements in the various gill arches as indicated here and in figure 5.31d:

a. hyoid arch representatives: hypohyal, ceratohyal
b. 3rd, 4th, and 5th gill arches: basibranchials, ceratobranchials, and epibranchials

Reptilia

Alligator

Note that the face is drawn out into a flattened snout. Second, the temporal region of the skull is pierced by two large paired openings: the **infratemporal** and **supratemporal fenestrae.**

Dorsal Surface The paired **supratemporal fenestra** are best viewed dorsally on the elevated top of the skull caudal to the orbits (figure 5.32a). The edge of the shelf to the side of each fenestra is the **supratemporal arcade.** Lateral and the next level below is the **infratemporal arcade.** The opening in this shelf is the **infratemporal fenestra,** just caudal to the orbit. The paired **external nares** open at the front of the snout. The **premaxillae** at the tip of the snout, and the **maxillae** that follow laterally, are the only two pair of upper jawbones that bear teeth. Down the dorsal midline, beginning caudal to the external nares, run the paired **nasals,** followed in succession by the unpaired **frontal,** and finally the unpaired **parietal** between the supratemporal fenestrae. The lateral wall of the supratemporal fenestra is formed by the **postorbital** (anterior) and **squamosal** (posterior). These two particular bones lying between the infra- and supratemporal fenestrae are what make this a **diapsid** skull. Processes of the postorbital and **jugal** below join to form a **postorbital bar** defining the rear of the orbit. The long, stout jugal anteriorly meets the maxilla and medially the **lacrimal,** penetrated by a **lacrimal canal** best seen where it opens to the orbit (figure 5.32c). Between lacrimal and frontal lies the **prefrontal,** completing the circle of bones that define the orbit. One additional bone, the **palpebral** (= superciliary), lies within the anterior orbit

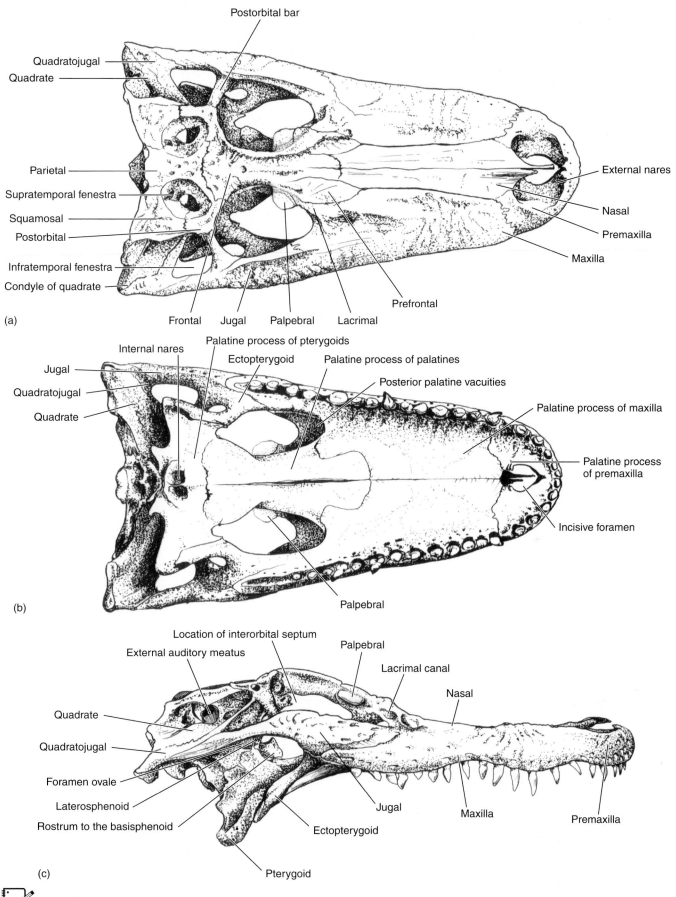

Postorbital bar

Quadratojugal

Quadrate

Parietal

Supratemporal fenestra

Squamosal

Postorbital

Infratemporal fenestra

Condyle of quadrate

External nares

Nasal

Premaxilla

Maxilla

Prefrontal

(a)

Frontal   Jugal   Palpebral   Lacrimal

Palatine process of pterygoids

Internal nares

Jugal

Quadratojugal

Quadrate

Ectopterygoid

Palatine process of palatines

Posterior palatine vacuities

Palatine process of maxilla

Palatine process of premaxilla

Incisive foramen

(b)

Palpebral

Location of interorbital septum

External auditory meatus

Palpebral

Lacrimal canal

Nasal

Quadrate

Quadratojugal

Foramen ovale

Laterosphenoid

Rostrum to the basisphenoid

Jugal

Maxilla

Premaxilla

Ectopterygoid

(c)

Pterygoid

FIGURE 5.32   Skull of alligator. (a) Dorsal view. (b) Ventral view. (c) Lateral view. See page 74 for figures 5.32d, e, and f. Palpebral not shown.

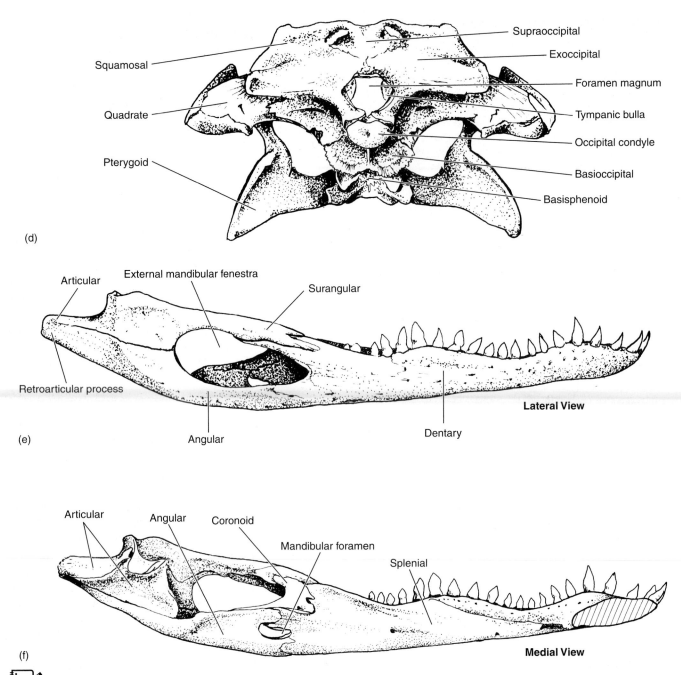

(d)

(e)

Lateral View

(f)

Medial View

✏️ **FIGURE 5.32** *Continued.* (d) Posterior view. (e) Lateral view of jaw. (f) Medial view of jaw.

attached to the prefrontal. However, the palpebral may have been lost in preparation of the skull (figure 5.32a–c). The **quadrates** project laterally at the back of the skull. Their ventral, rounded surfaces, **condyles,** articulate with the lower jaw. Compressed between quadrates and jugal is the **quadratojugal.**

🦖 **Ventral Surface**   The roof of the mouth forms by the inward growth of processes that collectively join along the midline. These processes form the **secondary palate,** separating air and food passageways (figure 5.32b). That part of the premaxilla, maxilla, palatine, and pterygoid[4] contributing to the palate is called its **palatine process.** Thus, the

palatine processes of the premaxillae contribute to the roof of the mouth and also form the margins of the small, unpaired **incisive foramen** (= anterior palatine vacuity) covered in life by an epithelium. The posterior medial openings are the **internal nares,** which open within the **palatine processes of the pterygoids.** The two large, oval openings beneath the orbits are **posterior palatine vacuities.** Between them lie the **palatine processes of the palatines.** Follow the remaining regions of bones contributing to the secondary palate to identify their associations.

────────

[4]Collectively, these four, paired bones plus the ectopterygoids constitute the **maxillary arch** in the alligator.

The palpebral bone rides upon the anteriomedial surface of the orbit. Look at an alligator skull from a lateral view and note how the nostrils and eyes are raised on the head. Consider the living animal, swimming just below the surface, with only its external nares and eyes above the water. In this position, the palpebral functions like a "cowcatcher" on a train, colliding with debris floating on the surface before it strikes a damaging cut to the eye.

Look again at the alligator skull that you have been studying. Notice the long, extended snout. A longer face can evolve in two ways: (1) by adding new bones at the end of the snout, or (2) by stretching the components that are already present. Long snouts in alligators have evolved by reshaping the ancestral bones and not by the addition of new elements. Evolutionary remodeling of existing structure, and not new construction, is a common theme in morphology. Look for other examples as you compare and study the different skulls.

Between posterior maxilla and lateral margins of the pterygoid lies the **ectopterygoid.** Note position and associations of jugal, quadratojugal, and quadrate.

Lateral Surface   The space between the orbits, in life, is closed by the cartilaginous interorbital septum (figure 5.32c). Although the septum is usually gone in prepared specimens, the paddlelike process projecting in posteriorly can be seen. This process is the **rostrum** to the **basisphenoid,** most of which is otherwise concealed beneath the pterygoid. Above and bordering the rostrum is the irregularly shaped and thin **laterosphenoid** (= pleurosphenoid), which reaches up to the undersides of the frontal and postorbital. The laterosphenoid forms the anterior rim and the quadrate the posterior rim of an opening, the **foramen ovale,** through which the large trigeminal nerve (V) exits from the cranial cavity. The two laterosphenoids do not meet, thus leaving an irregular medial cleft between them through which pass the optic nerves and olfactory tracts. Tucked under the supratemporal arcade is a large, deep pit, the **external auditory meatus,** down the core of which runs a slender, needlelike bone, the **stapes** (frequently missing in specimens).

Posterior Surface   Peering through the **foramen magnum,** the two lateral bulges, **tympanic bullae,** can be identified (figure 5.32d). The occipital ring includes the paired, winglike **exoccipitals** lateral to the **foramen magnum,** the unpaired **supraoccipital** (dorsal), and **basioccipital** (ventral), which bears a single **occipital condyle** (with processes of exoccipitals lending some support). Between basioccipital and pterygoids is squeezed the **basisphenoid.** We shall not name them here, but do at least notice the several paired foramina on the posterior face or surface of the skull. These include a slitlike pair of foramina laterally between exoccipital and quadrate, and four pairs (two large, two small) on the exoccipital closer to the midline. Through the lower set of larger foramina adjacent to the occipital condyle, the internal carotid arteries pass to enter the skull and distribute to the brain. The other foramina are openings through which cranial nerves exit from the cranial cavity. Note the single, unpaired foramen between basioccipital and crowded basisphenoid.

Lower Jaw   Each half of the lower jaw is a **ramus** (pl., *rami*) composed of six bones (figure 5.32e,f). Only the **dentary** bears teeth. The two dentaries unite anteriorly to form a cartilaginous **symphysis.** In lateral view (figure 5.32e), the large oval opening is the **external mandibular fenestra,** bounded above by the **surangular** (= supraangular) and below by the **angular.** The **articular,** a cartilage bone derived from the posterior of Meckel's cartilage, occupies the rear of the ramus. The articular bone ossifies in the posterior end of Meckel's cartilage and is thus the only cartilage bone of the lower jaw. The articular contributes to the **retroarticular process** (retro = behind) caudal to (behind) the articular at the back of the jaw. The large **splenial** lies along the inner anterior half of the lower jaw. Posteriorly, it joins the angular along the ventral edge of the jaw. Directly between them opens the small internal **mandibular foramen** (= **Meckelian fenestra**). Above this foramen sits the crescent-shaped **coronoid** (= **prearticular**). The coronoid forms the anterior border and the articular the posterior border of the **mandibular adductor fossa.** In life, Meckel's cartilage lies along the bottom of this fossa. This cartilage continues forward to the mandibular symphysis within the appropriately named Meckelian (= mandibular) canal, formed laterally by the dentary and medially by the splenial bones.

Hyoid Apparatus   The hyoid apparatus (see mounted skeleton) derives from parts of the hyoid arch and remaining gill arches. The broad cartilaginous plate constitutes the **body of the hyoid,** and the pair of processes constitutes the **horns** (= cornua).

Turtle

The following description is of an alligator snapping turtle. Although the pattern of bones is similar in related species, the path of suture lines will vary from one species to another.

Two major landmarks on the skull are the **external nares** (fused into a single aperture) and the two **orbits** (figure 5.33a–c). The small, paired **premaxillae** lie at the

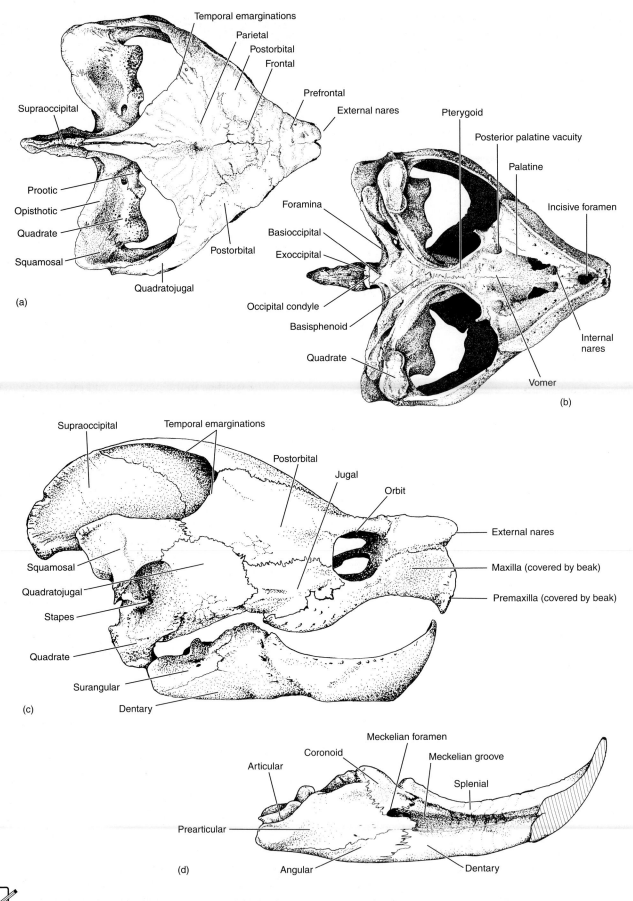

FIGURE 5.33 Skull of turtle. (a) Dorsal view. (b) Ventral view. (c) Lateral view of skull and jaw. (d) Medial view of jaw.

very tip of the snout below the external nares. The **maxilla** follows laterally, contributing to the side of the external naris and anterior border of the orbit. Often a large, cornified **beak** covers the entire upper jaw, obscuring the sutures between the premaxilla and maxilla. Caudal to the maxilla follows the long, broad **jugal** (= zygomatic), whose anterior end contributes to the lower, posterior rim of the orbit. The rear of the jugal joins the **quadratojugal.** A ventral process of the quadratojugal meets the **quadrate** below a deep depression, the **otic notch.** In life, this notch holds the **external auditory meatus,** which is covered externally by the tympanum of the ear. Note that in this specimen, as in most modern turtles, the nasal and lacrimal bones are absent. In their expected positions instead reside the bones interpreted to be the prefrontals, identified next.

The expected nasals are lost. Instead, caudal to the external naris and between the orbits lie the **prefrontals** (figure 5.33a), followed posteriorly, in turn, by **frontals** (sometimes fused) and then **parietals.** At the posterior midline, the parietals narrow into a sharp, backward-directed process completed at its tip by the dorsal blade of the **supraoccipital.** Lateral to the frontal and parietal lies the **postorbital,** which defines the rear of the orbit (figure 5.33c). The **squamosal** follows, forming the outer posterior projecting corner of the skull above the otic notch. Some specimens retain the needlelike **stapes** deep within the auditory meatus.

In the alligator snapping turtle shown in figure 5.33, the posterior edges of the parietal and squamosal bones form the rims of two enormous notches pushed forward from the back of the skull. Because these notches are open, they are not properly temporal fenestrae but rather **temporal emarginations.** In life, adductor muscles passing to the lower jaw bulge from these great forward notches. These adductor muscles rest on a broad laterally projecting shelf, the **otic capsule.** Otic bones, described next, roof over this capsule and fuse broadly with the lateral squamosal and quadrate.

🐾 **Posterior Skull** Above the **foramen magnum** resides the supraoccipital, already identified by its contribution to the middorsal posterior projection of the skull (figure 5.33a–c). Note that the **occipital condyle** is formed by three separate, contributing bones (figure 5.33b). The lower, central part arises on

Identify the internal nares in the roof of the mouth of an alligator skull and a cat skull. Here, the air taken in through the external nares reaches the pharynx. Note how the palatine processes of the premaxillae, maxillae, and palatines create a bony shelf, the secondary palate, which separates the oral and nasal cavities. Why has the separation of these cavities been adaptive for these animals? Many reptiles and mammals chew or hold food in their mouths for extended periods, making breathing through the mouth difficult or impossible. The secondary palate permits the passage of air around an occupied oral cavity. Snakes, which swallow large prey whole, solve this problem a different way. Snakes push the trachea under and in front of the prey during swallowing, thereby snorkeling in air to sustain breathing during the time prey otherwise makes its slow passage down the esophagus.

the **basioccipital;** the lateral two contributions are made by the paired **exoccipitals.** The body of each exoccipital is pierced by small **foramina.** Dorsally, the exoccipital swings upward to form the lower, outer margins of the foramen magnum. Lateral to the exoccipital and running across the medial base of the otic capsule lie two bones: the **prootic** (lateral to the base of the parietal), and the **opisthotic** (posterior and adjacent to the supraoccipital) (figure 5.33a).

🐾 **Ventral Surface** In front of the basioccipital lies the unpaired, triangular **basisphenoid** (figure 5.33b). Lateral and forward to it lie the broad pair of **pterygoids** bearing crescent-shaped emarginations on their lateral borders. Even more anterior lie the **palatines.** The posterior ends of the palatines contribute to the medial rim of small, lateral, paired openings adjacent to the pterygoids, the small, slanted, **posterior palatine vacuities.** You may need to tilt the snout downward to notice these vacuities. Anteriorly, between palatines, lies the unpaired **vomer.** The center of the vomer rises to form interior openings, the **internal nares,** leading to a short canal that opens to the external nares (not to the orbit). Just in front of the internal nares, the premaxillae cooperate in forming a small, medial opening, the **incisive foramen** (= **anterior palatine vacuity**), perhaps covered by the cornified beak.

🐾 **Lower Jaw** The lower jaw is composed of two bones, the two **mandibles,** left and right (figure 5.33c,d). They meet at the midline in a **symphysis,** a type of joint where bones appear fused to each other. The following description pertains to either mandible.

Articulating with the rounded, ventral end of the quadrate is the cup-shaped **articular.** Lateral to and above the articular is the slender **surangular** (= supra-angular), forming the outer rear corner of the jaw. The large **dentary** on the anterior edge of the surangular forms almost all of the outer side of the jaw. The very slender **angular** runs along the rear, ventral edge of the jaw. Often, in some turtle species, articular, surangular, and angular fuse indistinguishably in the adult.

Look around the room at the heads of your classmates. Notice how much of their heads is dedicated to housing their brains. Next return to the *Necturus,* alligator, and turtle skulls. If you haven't done so already, figure out precisely where the brain is located. It soon becomes clear that the skulls of these animals are primarily dedicated to many roles other than housing the brain. Look at these three skulls more closely to identify the various regions of the skulls not associated with the brain and examine their associated functions.

On the medial surface, the triangular **prearticular** forms the posterior half of the inner jaw surface. The edge of its posterior corner contributes to the medial articular surface, which meets the quadrate. The anterior corner of the prearticular runs under a small foramen, the **Meckelian foramen,** and its anterior dorsal corner supports a small cap of bone, the **coronoid.** Anteriorly, the inner surface bears a longitudinal furrow, the **Meckelian groove,** filled in life by Meckel's cartilage. Often a dried remnant persists. Above this groove is the **splenial,** but its sutures are difficult to define.

In the turtle skull, follow the line of action of the jaw adductor muscle from the back of the skull downward to its attachment to the lower jaw primarily in the region of the coronoid.

Aves

The following description is of a chicken skull (figure 5.34a–e). Although the pattern of bones is similar in related birds, the path of suture lines will vary among species.

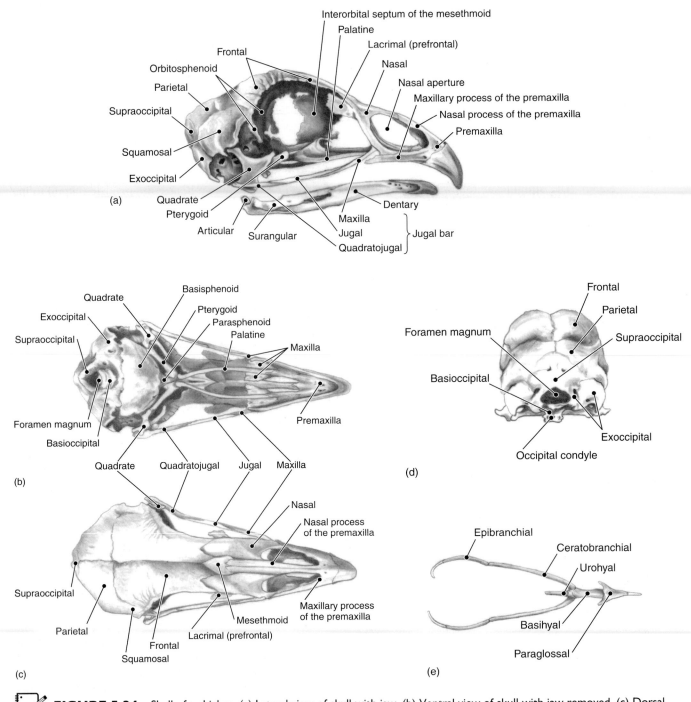

FIGURE 5.34 Skull of a chicken. (a) Lateral view of skull with jaw. (b) Ventral view of skull with jaw removed. (c) Dorsal view of skull with jaw removed. (d) Posterior view of skull with jaw removed. (e) Hyoid apparatus.

Bones of the bird skull fuse before hatching and many sutures helping to delineate elements are lost. The bird skull is also quite specialized, and this has led to considerable alteration in bone structure.

Dorsal and Lateral Views

🦖 Forming the tip of the beak, the **premaxilla** lies where you would expect (figure 5.34a). The **nasal process of the premaxilla** defines the dorsal rim and the **maxillary process of the premaxilla** the ventral rim of the **nasal apertures,** the two large openings in front of the orbits. The rear of each nasal aperture is defined by the forked **nasal** bone. The **lacrimal** (prefrontal) projects ventrally from the lateral side of the nasal, and defines the anterior edge of the **orbit.** Return to the dorsal surface of the skull. Just medial to the rear of the nasals are the fused **mesethmoids** (figures 5.34c), which also contribute an **interorbital septum** between the orbits (figure 5.34a). Just caudal to the mesethmoids and nasals, the **frontals** lie between the orbits and form the anterior roof of the skull. More caudally, the frontals fuse with the **parietals** to form most of the posterior vault of the cranium.

Laterally, the parietal meets the **squamosal** bone (figure 5.34a,c). Anterior and medial to the squamosal bones, the **orbitosphenoids** contribute to the cupped rear of the orbit below the frontals (figure 5.34a). The ventral edge of the squamosal articulates with the dorsal end of the forked **quadrate.** The ventral end of the quadrate supports the end of the slender **jugal bar** (= arch), passing forward beneath the orbit. Three bones contribute to the **jugal bar: quadratojugal** (posterior), **jugal** (middle), and **maxilla** (anterior where the arch joins the nasal bone).

The lower jaw contains six or seven fused bones (figure 5.34a). The **articular** articulates with the ventral end of the quadrate. The lateral face of the lower jaw consists of the **surangular** and more anteriorly the **dentary.** It is impractical to see, but splenial, angular, prearticular, and occasionally coronoid also contribute to the dermal casing.

Ventral and Posterior Views

🦖 At the posterior end of the skull, the occipital ring that forms the **foramen magnum** includes the fused **supraocccipital** (dorsally), **exoccipitals** (laterally), and the **basioccipital** (ventrally), which also bears the **occipital condyle** (figure 5.34b,d). Seen ventrally just anterior to the basioccipital, the **basisphenoid** forms much of the floor of the cranium (figure 5.34b). Just anterior to the basisphenoid is a small and unpaired **parasphenoid.** The elongate **pterygoids** are lateral to the basisphenoid and articulate caudally with the quadrates. Anteriorly, the pterygoids angle medially where they join the long and thin, paired **palatines.**

Sclerotic Ring and Hyoid

🦖 Within the orbit but not included in figure 5.34 is a circular series of bones forming the **sclerotic ring.** The hyoid bones, supporting the tongue, are naturally positioned in the floor of the mouth, but are often mounted out of place. See figure 5.34e to identify the bones contributing to the hyoid.

Mammalia—Cat

🦖 Generally the skull divides into two regions: the **facial** region bearing eyes and nose, and the more posterior **cranial** region enclosing the brain and the middle and inner ears (figure 5.35a,b). The cheekbone, or **zygomatic arch,** bows laterally between these two regions. The **external nares** in life are separated by a cartilaginous septum. On each side of the head, a **postorbital process** projects down from the roof of the skull and up from the zygomatic arch (figure 5.35a). These processes incompletely separate the **orbit** (anterior) from the more caudal **temporal fenestra.** At the posterior base of the zygomatic arch is an opening, the **external auditory meatus** (figure 5.35b). This opening continues into a hollow cavity, the **middle ear,** contained in the prominent smooth swelling, the **tympanic bulla.** Immediately caudal to the external auditory meatus is a pin-sized opening, the **stylomastoid foramen,** which is partially covered by a small, ventrally directed knob, the **mastoid process.** About the same size, at the back of the tympanic bulla, projects the ventrally directed **jugular process.** Connecting the bullas is a ridge arching up across the back of the skull, the **nuchal line,** to which neck muscles attach.

Dorsal Surface

🦖 The **premaxillae** bear teeth and lie at the front of the face. They send their **frontal processes** upward along the sides of the nares (figure 5.35a). Directly above the external nares lie the **nasals.** Just posteriorly lie the **frontals,** each bearing a lateral **postorbital process.** Caudal to the frontals are the **parietals,** which form most of the roof over the cranial cavity. Look closely to find a small, medial, triangular bone, the **interparietal,** at the posterior ends of the parietals before the back of the skull starts up into the nuchal line. The **occipital bone** forms the back of the skull, including the **foramen magnum** and paired **occipital condyles.**

Lateral Surface

🦖 In the upper jaw, only the **premaxillae** and **maxillae** bear teeth (figure 5.35b). The part of the **maxilla** bearing teeth is called the **alveolar process.** The upward, bladelike extension of the maxilla in front of the orbit is the **orbital** (= frontal) **process,** and the medial shelf contributing to the roof of the mouth is the **palatine process.** Finally, the lateral projection of the maxilla that begins the zygomatic arch is the **zygomatic process.** Penetrating the anterior base of the zygomatic process is the **infraorbital canal** for passage of blood vessels and nerves from the orbit. The zygomatic

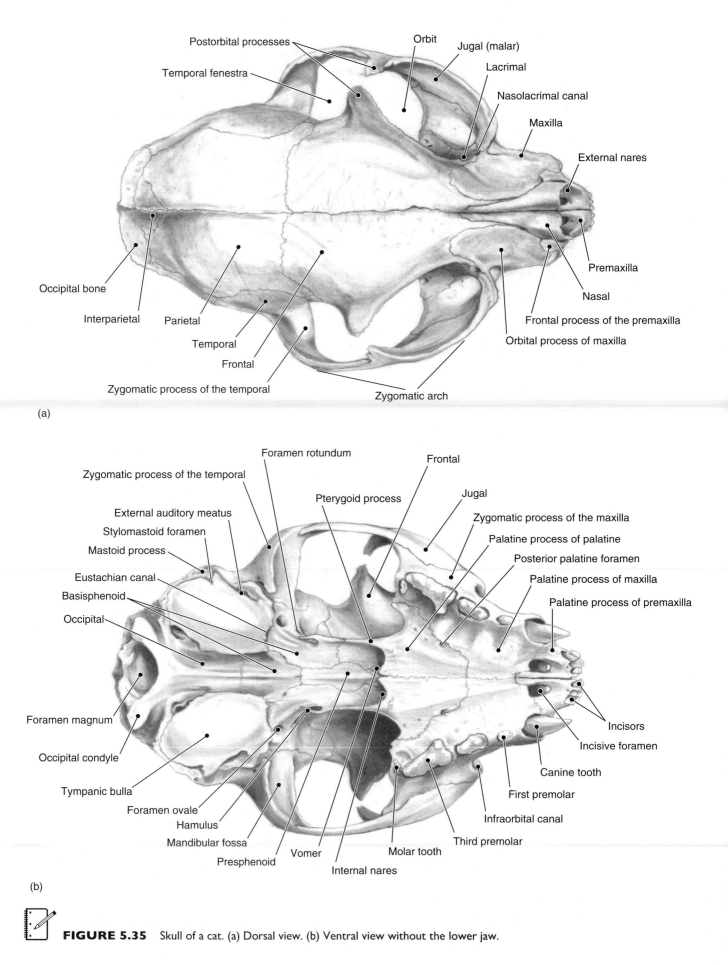

Postorbital processes
Temporal fenestra
Orbit
Jugal (malar)
Lacrimal
Nasolacrimal canal
Maxilla
External nares
Premaxilla
Nasal
Frontal process of the premaxilla
Orbital process of maxilla
Occipital bone
Interparietal
Parietal
Temporal
Frontal
Zygomatic process of the temporal
Zygomatic arch

(a)

Foramen rotundum
Frontal
Zygomatic process of the temporal
Pterygoid process
Jugal
Zygomatic process of the maxilla
External auditory meatus
Stylomastoid foramen
Palatine process of palatine
Mastoid process
Posterior palatine foramen
Eustachian canal
Palatine process of maxilla
Basisphenoid
Palatine process of premaxilla
Occipital
Incisors
Foramen magnum
Incisive foramen
Occipital condyle
Canine tooth
Tympanic bulla
First premolar
Foramen ovale
Infraorbital canal
Hamulus
Third premolar
Mandibular fossa
Molar tooth
Presphenoid
Vomer
Internal nares

(b)

FIGURE 5.35 Skull of a cat. (a) Dorsal view. (b) Ventral view without the lower jaw.

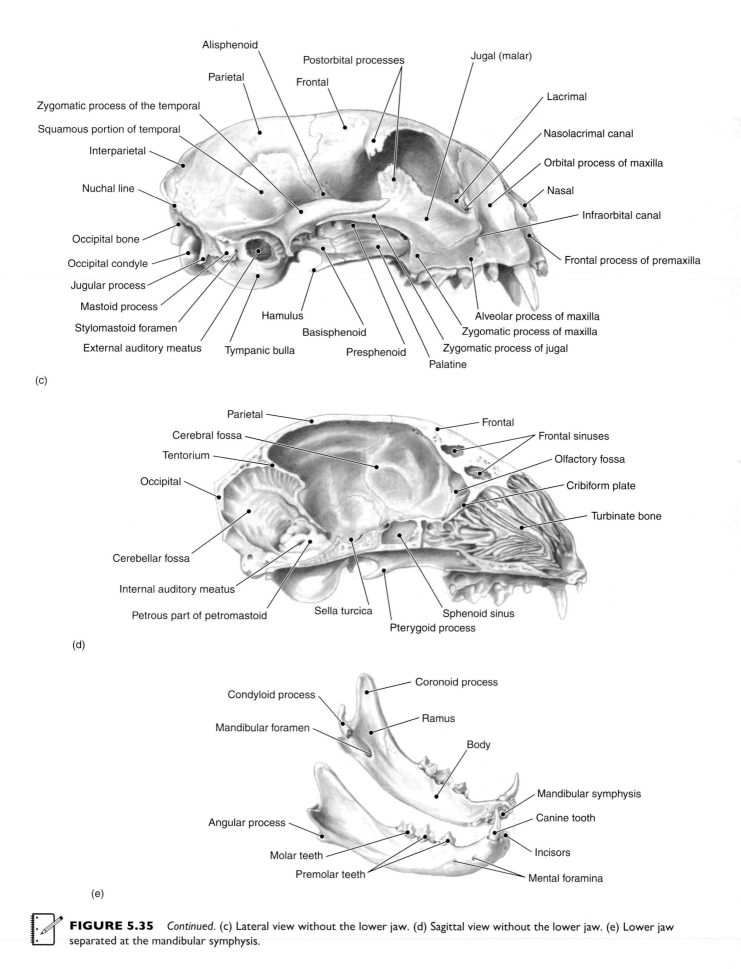

(c)

(d)

(e)

FIGURE 5.35 *Continued.* (c) Lateral view without the lower jaw. (d) Sagittal view without the lower jaw. (e) Lower jaw separated at the mandibular symphysis.

process supports the **jugal** (= malar), which bears the upward-directed postorbital process. The slender, posterior end of the jugal, the **zygomatic process of the jugal,** forms a scarf joint (slanted, overlapping) with the **zygomatic process of the temporal bone,** which completes the zygomatic arch. The smooth depression on the ventral base of the zygomatic process, the **mandibular fossa,** receives the lower jaw. The zygomatic process projects from a plate of bone lying above the tympanic bulla and below the parietal, the **squamous portion of the temporal.** The **temporal bone** is a composite bone in which the squamosal, otic elements, and tympanic bulla are fused.

Most of the back wall of the orbit forms by a downward growth of the frontal. The ventral edge of the frontal meets a series of bones at the base of the orbit. The most anterior of this series is the **lacrimal,** positioned caudal to the orbital process of the maxilla, and with a pore, the **nasolacrimal canal,** leading to the nasal cavity. Next, posteriorly lies the **palatine,** bearing a **palatine process** that contributes to the roof of the mouth. The bone is pierced by two openings: the **sphenopalatine foramen** (large, upper) and **posterior palatine canal** (pin sized, lower).

The sharp, posteriorly directed process that projects ventrally below the basisphenoid is the paired **hamulus.** Between the frontal bone and the tail of the palatine can be found the unpaired **presphenoid,** and caudal to it the single and larger **basisphenoid.** A dorsal, winglike projection of the basisphenoid, the **alisphenoid,** rises laterally between frontal and temporal bones. A row of four foramina run across the presphenoid and basisphenoid. From anterior to posterior, they are the **optic foramen** (in the presphenoid), the **orbital fissure** (between presphenoid and basisphenoid), the **foramen rotundum** (basisphenoid), and the **foramen ovale** (basisphenoid). Branches of various cranial nerves exit through these foramina. Just medial and posterior to the foramen ovale is the **Eustachian canal,** which runs over the anterior end of the tympanic bulla, slanting outward to connect with the middle ear.

Patterns & Connections

Place a cat and alligator skull side by side in front of you. In both animals, compare the series of bones that form the top of the skull. Begin with the nasal bone immediately caudal to the external nares and continue to the foramen magnum. Notice that the names and sequences are generally the same. These similarities reflect part of the morphological pattern common to many vertebrate skulls. Now compare the ventral series of bones from foramen magnum to incisive foramen. By comparing similar regions of vertebrate skulls, you can start to see consistent patterns and make connections that make learning these names easier.

Ventral View and Disarticulated Skull

Use a whole skull and a disarticulated skull, preferably simultaneously, to work through the tightly interlocking bones along the base of the skull (figure 5.35c). The **palatine processes of the premaxillae, maxillae,** and **palatines** cooperate in the formation of the **hard palate** continued caudally in life by a fleshy soft palate. The arched, internal nares open at the back of the hard palate. The paired, oval **incisive foramina** (= anterior palatine foramina) lie in the premaxillae. The **posterior palatine foramina** are a pair of pin holes in the suture between palatine processes of the maxillary and palatine bones. Down the midline runs the unpaired **vomer** dorsal to and partially hidden by the palatine processes of the palatines, followed posteriorly by the slender presphenoid, the broader basisphenoid, and finally occipital bones between the tympanic bullas. The **pterygoid** in some mammals (e.g., dogs) is a separate bone. In the cat, it fuses with the basisphenoid as the **pterygoid process,** with some delineating sutures difficult to see. Generally, it is a triangular process whose long base runs along the posterior half of the presphenoid and anterior body of the basisphenoid. It flares laterally, bearing the slender, ventrally directed hamulus at its apex. Its anterior edge meets the posterior end of the palatine along an almost obliterated outward suture even with about the middle of the presphenoid. The presphenoid sends a winglike process into the orbit that carries the optic foramen (first in series of four foramina). This process is thought to be homologous to the orbitosphenoid of lower vertebrates, although here in the cat it is properly a part of the composite sphenoid. Compare the other disarticulated bones with those in place.

Sagittal View

Bones already identified should be further followed in this revealing preparation. The **frontal sinus** is the spacious cavity within the frontal bone (figure 5.35d). Delicate,

Patterns & Connections

You may have learned previously that the smallest bones in your body are the incus, malleus, and stapes located in your middle ear. It's true. But where evolutionarily did these bones come from? These are not found as a trio in the fish, amphibians, reptiles, or birds that you have studied. You will recall that the jaws of *Necturus,* an alligator, a turtle, and a bird articulate with the quadrate bone at the rear of the skull (a pattern that makes learning this bone a bit easier). Further, you found the articular bone in the lower jaws of these animals but not in mammalian jaws. And finally, the stapes in these same animals was already part of the middle ear. The quadrate (incus) and articular (malleus), have joined the stapes in the middle ear of mammals, where they have assumed a specialized role in sound conduction. Mammals hear with "old" bones in relatively new positions!

shell-thin **turbinate bones** (= nasal conchae) fill the **nasal cavity.** There are three compartments to the cranial cavity. The largest is the extensive **cerebral fossa,** formed mostly by frontal, parietal, and temporal bones. The more posterior **cerebellar fossa** is partially separated from the cerebral fossa by a downward septum called the **tentorium.** Anteriorly, and continuous with the cerebral fossa, is the small, projecting **olfactory fossa,** just ventral to the frontal sinuses. The **cribriform plate** is the part of the ethmoid through which tiny, numerous olfactory nerves pass toward the brain. The medial depression pressed into the top of the basisphenoid is the **sella turcica,** which houses the pituitary. The presphenoid contains a dorsal cavity, the **sphenoid sinus.**

In the ventral rear of the skull, the **petromastoid** (= periotic) forms from fused otic bones and itself contributes to the temporal bone. The petromastoid bears an outer, lateral mastoid process projecting over the external surface on the side of the tympanic bulla. Internally, the **petrous** part of the petromastoid rests on top of the bulla and bears an opening, the **internal auditory meatus,** through which the auditory nerve (VIII) passes to reach the small **inner ear cavity** contained within.

Lower Jaw

🦖 The mandible consists of only a single pair of bones, the **dentaries,** fused anteriorly into a **mandibular symphysis** (figure 5.35e). The longitudinal part of the dentary bearing teeth is the **body,** and the caudal part is the **ramus.** The **coronoid process** extends upward from the rear of the dentary. The **condyloid process** expands transversely into the rounded, articulating surface of the mandible, which fits into the mandibular fossa at the base of the zygomatic arch. Below the condyloid process, the rear corner of the dentary projects as the **angular process.** Caudally, on the inner surface of the dentary, the **mandibular foramen** permits entrance of a nerve that passes forward but within the bone and finally exits on the outer, anterior end through two **mental foramina.**

Hyoid Apparatus

🦖 On a mounted skeleton, locate the hyoid apparatus (see figure 5.14a). This is formed by remnants of the hyoid and other gill arches. The single transverse bar is the **body,** and paired projecting extensions are the **horns: anterior** (long) and **posterior** (short).

Teeth

Teeth are unique among vertebrate animals in both form and function. They are usually capped with *enamel,* a mineralized coat found only in vertebrates. Teeth help catch and hold prey and can offer strong opposing surfaces that can crush hard shells of prey.

In mammals and a few other vertebrates, *mechanical digestion* begins in the mouth. After each bite, the tongue and cheeks collect food and place it between the upper and lower tooth rows. The teeth break down the bolus mechanically, reducing it to smaller chunks to make swallowing easier. By breaking the large bolus into many smaller pieces, *mastication* (chewing) also increases the surface exposed to chemical digestion. Even in vertebrates that do not chew their food, sharp teeth puncture the surface of the prey, creating sites through which digestive enzymes penetrate when food reaches the gut. For vertebrates that feed on insects and other arthropods, punctures through the chitinous exoskeleton are especially important in giving proteolytic enzymes access to the digestible tissues within.

Tooth Anatomy

The part of the tooth projecting above the gum line, or *gingiva,* is the *crown;* the region below is the *base.* If the base fits into a hole, or *socket* (alveolus), within the jaw bone, the base is referred to as a *root.* Within the crown, the pulp cavity narrows when it enters the root, forming the *root canal,* and opens at the tip of the root as the *apical foramen.* Mucous connective tissue, or *pulp,* fills the pulp cavity and root canal to support blood vessels and nerves that enter the tooth via the apical foramen. The *occlusal surface* of the crown makes contact with opposing teeth. The *cusps* are tiny, raised peaks or ridges on the occlusal surface (figure 5.36).

In lower vertebrates, teeth are usually *homodont,* similar in general appearance throughout the mouth. Modern turtles and birds lack teeth altogether, but mammals have *heterodont* teeth that differ in general appearance throughout the mouth (figure 5.37). Most lower vertebrates have *polyphyodont* dentition; that is, their teeth are continuously replaced. A polyphyodont pattern of replacement ensures rejuvenation of teeth if wear or breakage diminishes their function. However, most mammals are *diphyodont,* with just two sets of teeth. The first set, the *deciduous dentition,* or "milk teeth," appears during early life (figure 5.37).

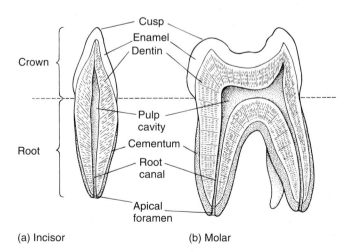

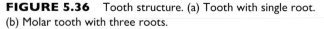
FIGURE 5.36 Tooth structure. (a) Tooth with single root. (b) Molar tooth with three roots.

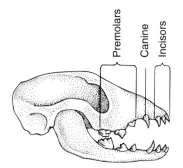

(a) Puppy

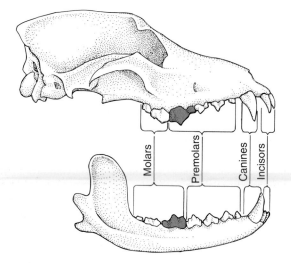

(b) Adult dog

FIGURE 5.37 Deciduous (a) and permanent teeth (b) in a dog. The carnassials (shaded teeth) are specialized carnivore teeth derived from last premolar (upper) and first molar (lower).

It consists of incisors, canines, and premolars, but no molars. As a mammal matures, these are shed and replaced by the *permanent dentition,* consisting of a second set of incisors, canines, and premolars and now molars (figure 5.37).

Sharks—Homodont and Polyphyodont Teeth

☞ Examine a set of dried shark jaws. The teeth in some predators, such as sharks, have sharp, knifelike cutting edges along their sides to help pierce skin (figure 5.38a). For slicing chunks from flesh, these edges are further serrated, like those on a bread knife, to cut the tough skin. Notice the many rows of teeth that roll dorsally and forward to replace those routinely lost along the anterior margin.

Amphibia/Salamander—Homodont and Polyphyodont Teeth

In larval salamanders, most teeth are pointed cones, but the teeth of metamorphosed adults often show specializations (figure 5.38b). The crowns in some species are bicuspid,

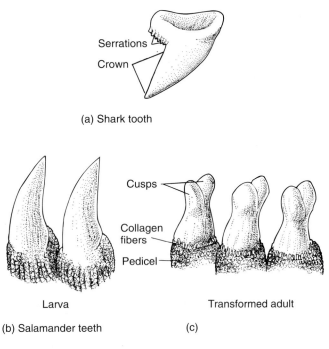

(a) Shark tooth

Larva   Transformed adult

(b) Salamander teeth   (c)

FIGURE 5.38 Specializations of teeth. (a) Shark tooth (blacknose shark, *Carcharhinus acronotus*). The pointed crown has a nearly smooth edge for piercing prey; the base is serrated for cutting flesh. This tooth can be more than ten times larger than the salamander teeth in (b) and (c). (b) Teeth before (larval) and after (adult) metamorphosis in the northwestern salamander (*Ambystoma gracile*). Larval teeth are pointed. Those of the transformed adult have divided cusps that articulate with a basal pedicel. The cusps are thought to inflect with the struggling prey, thus discouraging its escape from the mouth.

having two cusps, and the crown sits upon a basal pedicel to which it is attached by collagenous fibers.

☞ Examine a *Necturus* skull under a dissection microscope and examine the type of teeth. Are the teeth in *Necturus* pointed cones or bicuspid?

Reptiles/Alligator—Homodont and Polyphyodont Teeth

☞ Examine an alligator's teeth still embedded within the upper and lower jaws. These sharp, pointed teeth vary in size. They are well adapted to grasping but are relatively ineffective at chewing or cutting the integument of most prey. A tooth **socket** (alveolus) may be examined in places along the jaw where teeth have fallen out.

Mammal—Heterodont and Diphyodont Teeth

In mammals, the teeth not only capture or clip food but they are also specialized to chew it, producing a complex and distinctive dentition. In fact, the dentition in different groups is so distinctive that it is often the basis for identifying living animals and fossil species. Not surprisingly, an elaborate terminology has grown up to describe the particular features of mammalian teeth.

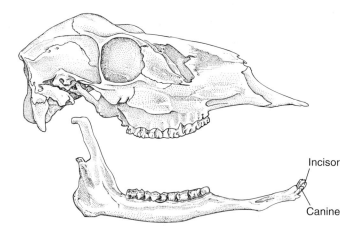

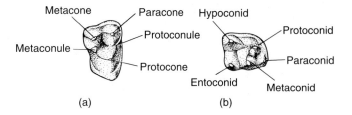

FIGURE 5.40 Molar anatomy in placental mammals. Dorsal view of crushing surface. (a) Upper right molar. (b) Lower left molar.

Modified from A. S. Romer and T. S. Parsons, 1985, The Vertebrate Body, Saunders College Publishing.

FIGURE 5.39 Skull of the mule deer (*Odocoileus hemionus*) with lower jaw detached. The dental formula for the upper row is 0-0-3-3, and the formula for the lower is 3-1-3-3. The absence of upper incisors and canines is normal, and is indicated in the dental formula by zeros. The lower incisors and canines are present, and the canines are adjacent to the incisors at the front of the mandible.

Form and Function

Examine the teeth of a ruminant cow, sheep, goat, or deer (figure 5.39). What is unusual about the front incisor teeth? Something is missing! The top incisors and canines of ruminants are replaced by a thick, dense dental pad (pulvinus). The absence of these upper teeth may permit greater range of motion of the tongue, used to select forage when grazing. Further, during remastication, ruminants will bring swallowed forage up from their stomachs to be rechewed, to further break down the cell walls of the plant material. The dental pad is involved during the subsequent turning and shaping of this food bolus in the mouth before swalllowing it again.

The **heterodont** dentition of mammals includes four types of teeth within the mouth: *incisors* at the front, *canines* next to them, *premolars* along the sides of the mouth, and *molars* at the back. The number of each type differ among groups of mammals. The *dental formula* is a short-hand expression of the number of each kind of tooth on one side of the head for a taxonomic group. For example, the dental formula of the coyote (*Canis latrans*) is:

I 3/3, C 1/1, PM 4/4, M 2/3.

This means that there are three upper and three lower incisors (I), one upper and one lower canine (C), four upper and four lower premolars (PM), and two upper and three lower molars (M), 21 per side or 42 total teeth. Often the dental formula is written as 3-1-4-2/3-1-4-3, the first four numbers indicating the upper teeth and the second four the lower teeth (in this example, for the coyote). The dental formula for the mule deer (*Odocoileus hemionus*) is 0-0-3-3/3-1-3-3 (figure 5.39). Notice that the missing upper incisors and canines are indicated by zeros.

The cusp patterns of mammalian cheek teeth are so distinctive that they are also used to identify species. Cusps are termed *cones*. Major cones (figure 5.40) are identified by adding the prefixes *proto-* (first), *para-* (beside), *meta-* (beyond), *hypo-* (under), or *ento-* (within); minor cusps are indicated by the suffix *-ul(e)*.

🐾 Examine the teeth in the upper and lower jaws of a cat. The dental formula for a cat is 3-1-3-1/3-1-2-1. Using this formula, calculate the number of each type of tooth and identify them in the cat. Notice the gap, or **diastema,** between the canines and premolars. Notice that the last premolar *(of the upper row)* and first molar *(of the lower row),* the **carnassials,** form the primary shearing surface along the lateral side of the jaw. These shearing forces can be best appreciated by articulating a jaw to the skull and noting the precise alignment, similar to scissor blades in form and function.

Examine other mammalian skulls before you. When available, carefully place the lower jaw in position and rotate it slowly to close the mouth. Notice how closely the teeth mesh in a very precise occlusion. Determine the dental formula for each of these skulls and verify them with your instructor. If an adult human skull is available, compare the structure of the premolar teeth (typically 2 cusps) to molar teeth (4 cusps) and determine the adult human dental formula.

Pattern of Evolutionary Change

Evolutionary change occurs in organisms by several mechanisms. The previous skeletal material and the examples in the boxed essays can illustrate a few of these ways. First, change can occur through the renovation of existing structures (figure 5.41). As indicated in the Patterns & Connections boxed essays, anterior gill bars evolved into jaws, and several jawbones evolved into middle ear bones. These anatomical changes may also be accompanied by a change in function. For example, gill bars that support the respiratory apparatus shift to elements of the jaw associated with feeding. In turn, some of these jaw elements have shifted form

The teeth of vertebrates are one of the best examples of the tight relationship between form and function. A close examination of the teeth of an unknown skull will provide insight into the diet of the animal. This is how paleontologists can make informed hypotheses about the diets of dinosaurs. This figure below indicates the relationship between tooth shape and diet in three

mammals. (a) A carnivore skull showing position of carnassial teeth (shaded). Carnassials function like scissors to slice through soft but pliable foods. (b) Artiodactyl skull showing the position of the grinding teeth (shaded). Corrugated occlusal surfaces of these teeth help grind fibrous foods. (c) Primate skull showing the position of the compression teeth (shaded) that pulverize hard foods.

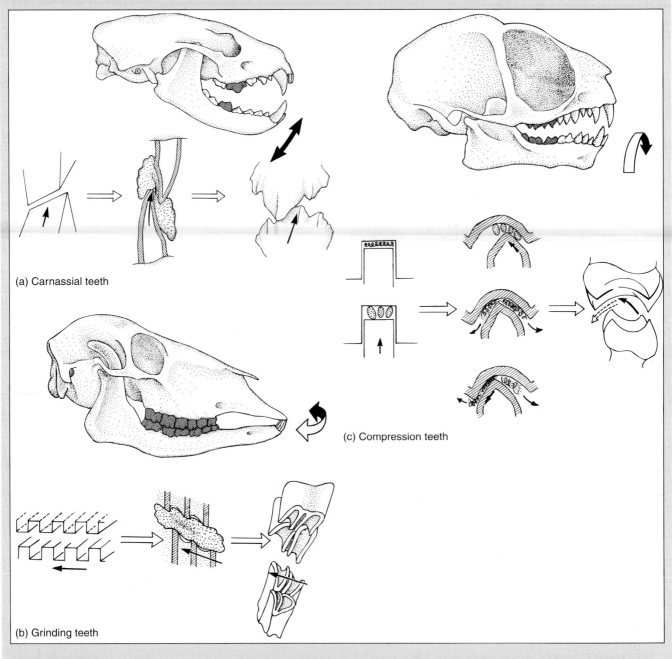

(a) Carnassial teeth

(c) Compression teeth

(b) Grinding teeth

BOX FIGURE 5.3 Mastication in mammals. (a) Carnivore skull showing position of carnassials (shaded). Carnassials function like scissors to slice through soft but resilient foods. (b) Artiodactyl skull showing position of grinding teeth (shaded). Corrugated occlusal surfaces of these teeth grind fibrous foods. (c) Primate skull showing position of compression teeth (shaded) that pulverize hard foods.

After M. Hildebrand et al., eds., 1985, Functional Vertebrate Morphology, Belknap Press of Harvard University Press.

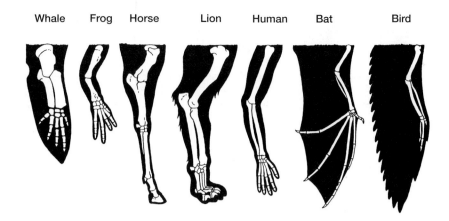

Whale   Frog   Horse   Lion   Human   Bat   Bird

FIGURE 5.41   The forelimbs of seven vertebrates show great diversity, but all are modifications of a common underlying pattern.

Form & Function

Mastication in mammals has been accompanied by very precise tooth occlusion to serve the mechanical breakdown of food. Precise, strong occlusion requires a firm skull, so mammals have lost the cranial kinesis of reptiles. In addition, the mammalian mandibular condyle often fits into a very restrictive articulation with the squamosal bone. When jaws close about this joint, upper and lower rows of teeth are placed in very precise alignment. This allows specialized teeth to function properly. To avoid disruption of the occlusial tooth row, teeth in most mammals are not continually replaced. Instead, most mammals exhibit diphyodonty, giving adults a "permanent" set of teeth and thereby a permanently functional occlusial tooth row.

and function from prey capture to the conduction of sound as part of the middle ear!

Second, change can proceed by the loss of some elements or by the enlargement of others. For example, the vertebral columns of amphibians and amniotes (reptiles, birds, and mammals) possess just a single centrum per body segment, resulting from the loss of either the intercentrum or pleurocentrum in the fish vertebral column, as illustrated in *Amia* (see figure 5.8). By contrast, alligator snouts are elongated by drawing out ancestral facial and palatal bones and not by the addition of new elements.

Evolutionary changes in structure do not run under their own momentum. Instead, structural adaptations result from changes in a species' environment and lifestyle. For example, the major transitions from water to land or from land to air have required major modifications in the basic vertebrate body. As early tetrapods and amniotes became more terrestrial, the number of vertebral elements was reduced (the loss of either the intercentrum or pleurocentrum noted earlier). In addition, the demands for greater support for flight favored the fusion of vertebral elements in the sacrum of a bird and other parts of the skeleton. Lesser shifts in lifestyle have resulted in more subtle changes in anatomy.

In the laboratory you concentrated on identifying structures. This is valuable because a basic understanding of anatomy is an important foundation for many disciplines, including physiology, medicine, and even natural history. But if you step back from the task of identification, you will see that the various morphological forms are much more than a conglomeration of anatomical parts. Instead, the forms are adaptive responses to changes in function and environmental setting. Form, function, and environment must all be considered to better understand evolutionary change. Thus, in addition to finding the anatomy, take time to consider the comparative function of the parts and the likely environmental setting to which they are adapted.

6

Muscular Systems and External Anatomy

Introduction

Skeletal, smooth, and cardiac muscles help regulate physiological activities and permit movements. Skeletal muscles are the active companion of the skeleton, often considered together as a functional unit, the musculoskeletal system. Together, muscles, bones, and the connective tissues that join them permit body movements and locomotion vital to a vertebrate's existence. Smooth muscles affect the activity of the viscera. For example, muscles wrapping the tubular digestive tract constrict in peristaltic waves to mix and move the food within. Smooth muscles also form sphincters that control the passage of materials out of blood vessels and tubular portions of the digestive tract. Sheets of smooth muscle within the walls of the respiratory tract affect the flow of air to and from the lungs by changing the size of air passageways. Cardiac muscle generates rhythmic contractions to propel blood through an organism's vessels. Further, some unusual muscles have very specialized functions, including heat generation and the production of electrical fields useful in navigation, defense, and predation. Chapter 6 addresses only part of the muscular system, the skeletal muscles.

Terminology

A muscle is technically identified by where its ends attach—its origin and insertion (table 6.1). Its name may reflect a combination of the origin and insertion, in that order, into a compound word, such as **pubofemoralis** (**pubis** = "origin"; **femur** = "insertion"). Many muscle names in vertebrate anatomy are carried over from human anatomy. These names are often based not on attachments, homologies, or actions, but upon shape (**clavotrapezius: -trapez** = "table"), size (**gluteus maximus: = -maxim** = "largest"), or imaginary resemblances (**gastrocnemius: gastro-** = "stomach/belly").

Actions

Muscles can affect themselves and the skeletal system in many ways. For example, muscles can act in support of each other (synergists) or in opposition to each other (antagonists). The effect of muscles on the skeletal system has required a set of terminology (included in table 6.1) to specifically identify the consequences of muscle actions (figure 6.1).

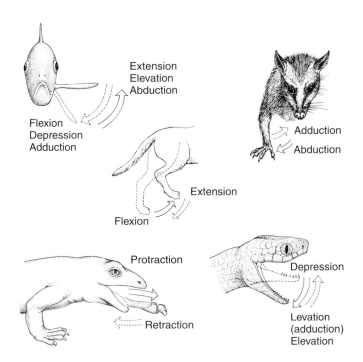

FIGURE 6.1 Muscle actions. Muscle adduction draws an appendage toward the ventral midline, and muscle abduction moves it away. Although these terms apply to tetrapod limbs and fish fins, the terms **depression** and **flexion** are sometimes used synonymously with adduction in fishes; **extension** and **elevation** are synonymous with abduction. In tetrapods, flexion means bending a part, extension means straightening it. Protraction sends a part outward from its base, retraction pulls it back. Opening the jaws is depression or abduction, and closing them is elevation, levation, or adduction.

TABLE 6.1 Mini-Glossary of Muscle Terminology

abduct—to move a limb away from the midline of the body

adduct—to move a limb toward the midline of the body

aponeurosis—flat, sheetlike tendon

depress—to move a part inferiorly, such as opening the jaws

extend—to straighten one part relative to another

flex—to bend one part relative to another

head—the part of the muscle that takes part in its origin

insertion—attachment site that moves relative to the origin

levate—to move a part superiorly, such as closing the jaws

ligament—connective tissue band or sheet joining bone to bone

origin—relative fixed site of muscle attachment

protract—to extend or protrude a part

retract—to draw back or pull in a part

tendon—connective tissue band joining muscle to bone

General Muscle Groups

Parietal or Somatic Musculature

Derived from myotomes, these muscles are innervated by spinal nerves, and form voluntary muscles covering most of the body and appendages.

Hypobranchial Musculature

These muscles derive from myotomes that grow downward caudal to the head and then forward into the area beneath the gills. They are also innervated by spinal nerves.

Branchial or Branchiomeric Musculature

These derive from somitomeres and are innervated by cranial nerves.

Dissection

Dissection (dis-sékt-shên) n. The careful exposure of anatomical parts, allowing students to discover and master the extraordinary morphological organization of an animal to understand the processes these parts perform and the remarkable evolutionary history from which they came.

The Process

Dissection is not a competitive sport that consists of rapidly cutting the animal into pieces. The primary objective of a muscle dissection is to expose and study a muscle's origin, insertion, shape, and function by carefully separating it from the surrounding tissues. A good dissection does this while minimizing damage to the muscles and surrounding tissues, including blood vessels and nerves. Dissection requires patience, delicacy, and skill. Skeletal muscles are typically connected to each other and the skin by fascia, light colored and interwoven sheets of fibrous connective tissue that may look like spider webs. Fascia and fat typically obstruct the muscle boundaries and may render muscles difficult to see.

To dissect well, one needs little more than a good, blunt probe (often a gloved finger is best) and a pair of grasping forceps. Scissors are necessary only to cut skin and reflect muscles by cutting them perpendicularly along their midlength (reflecting a muscle in this way produces two long ends that can be found and brought together more easily than if the muscle is cut near its insertion or origin). Scalpels can cut quickly and easily along unnatural boundaries, permitting rapid and disastrous consequences in untrained hands. Their use is therefore discouraged.

Progress through your muscle dissection by (1) reading the detailed text description of the muscle's location, (2) searching for the reference points on the animal, (3) looking for the boundaries of the muscle (slight shifts in muscle fiber direction, identifying a distinct insertion or origin, etc.), and (4) verifying the boundaries of the muscle through gentle probing. Once the boundaries are identified, use your probe and forceps to remove fascia and fat around the muscle from insertion to origin. Be careful not to break nerves and blood vessels that traverse these regions.

While dissecting, be careful that important structures do not dry out. Keep them moist with a spray or dowsing of preservative. When finished for the day, package the animal in a closed plastic bag or immerse it in suitable brine (e.g., sharks). If skin has been reflected during the dissection, refold it back into place before packing up the animal at the end of the day. Sometimes moist paper towels wetted with preservative can be wrapped around exposed parts of the specimen to keep it fresh. Respect these special animals and the valuable education they provide, so they last the full term of the course.

General Safety Precautions

Your instructor will inform you of risks and safety precautions associated with the chemicals used to fix and preserve animals that you dissect. Good progress has been made by most biological suppliers in reducing your exposure to noxious and potentially toxic chemicals. However, prudent behavior is still required.

When dissecting preserved tissues, care should be taken to avoid contacting or breathing the vapors of chemicals used in the preservation process. Your exposure can be reduced by wearing thin, flexible gloves and working in a well-ventilated area. In addition, wash with cold water the external and internal surfaces of your animal exposed through your dissection. Frequently wash away the fluids that seep from your animal into its dissection tray. Soft, gas-permeable contact lenses can act as absorbent surfaces that hold the vapors against your eye. Avoid wearing such lenses when dissecting. Furthermore, protective eyewear should be worn by anyone observing (not just conducting) a dissection as small amounts of fluid are frequently splashed into the air (especially when turning an animal in its tray). Closed-toe shoes are strongly advised for all laboratory work. Additional precautions and rules governing your dissection will be provided by your instructor.

External Anatomy

Take time to examine the external anatomy of your specimens before beginning your dissection of the muscular systems. Descriptions of the external anatomy precede the muscular dissection of the shark, *Necturus*, and cat.

Shark Dissection

External Anatomy

Major Anatomical Components

🗡 Find the **pectoral** and **pelvic** fins examined previously in the skeletal system. Note the sex of your specimen by determining if the pelvic fin bears claspers (see figures 5.16b and 6.7). The body is generally divided into three regions: the **head** extends from the tip of the snout to the cranial border of the pectoral fin; the **trunk** of the body is between the pectoral and pelvic fins; and the **tail** is caudal to the pelvic fin. Note that the dorsal side of the trunk is darker than the ventral side, a common pattern called **countershading.** Light from above casts a shadow on the ventral regions, producing a uniformly shaded shark more likely to blend into its environment.

Sharks have two unpaired **dorsal fins,** each with a prominent spine on the leading edge. (These spines, primarily used for defense, are often trimmed off before you receive your animal to prevent damage to you and the storage bags.) The dorsal lobe of the tail fin is larger than the ventral lobe, forming a **heterocercal** tail configuration.

Sensory Systems

Chemoreception   Just cranial and ventral to the eyes are paired **nares,** which open into blind-ended **olfactory sacs** for chemoreception (smelling) (figure 6.2).

Lateral Line System   Sharks and many other aquatic vertebrates (but not aquatic birds or mammals) sense vibrations and movement in the water by a **lateral line system** composed of bundles of sensory cells and canals located within the skin (figure 6.3). These canals are concentrated along the head and extend along the sides of the body onto the tail. The largest canal, the **lateral line,** is a light, thin stripe extending just above the most lateral side of the trunk.

Electroreceptors   Gently squeeze the shark's snout and note the thick fluid exuding from pores on the dorsal and lateral sides of the head. These pores are openings into the **ampullae of Lorenzini,** a modified portion of the lateral line system sensitive to electrical fields.

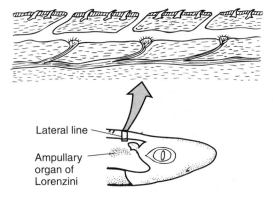

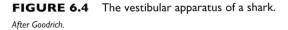

FIGURE 6.3   The lateral line system. Section through the skin of a shark showing the sunken lateral line canal opening to the surface via small pores.
After Goodrich.

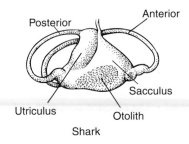

FIGURE 6.4   The vestibular apparatus of a shark.
After Goodrich.

Form & Function

Position yourself in a location that allows you to safely close your eyes. Now slowly tilt your head back and forth from side to side. How can you tell that your head is tilted? Like sharks and all other vertebrates, we have a vestibular apparatus that provides similar information about the position and movement of our heads. In addition, stretch receptors in your neck muscles let you know that the muscles on one side of your head are stretched more than other neck muscles, providing additional positional information.

Vestibular Apparatus   The **vestibular apparatus** (figure 6.4) is a paired balancing organ that arises phylogenetically from the lateral line system. It is suspended in the **otic capsule,** identified previously when studying the skeletal system (see figure 5.29). A pair of **endolymphatic ducts** connect the vestibular apparatus to the outside of the body through **endolymphatic pores.** Although difficult to see without a dissection microscope, these pores are located along the dorsal midline just caudal to the eyes. The three **semicircular canals,** part of the vestibular apparatus, are positioned in different planes to better sense specific movements. As the body moves, fluid in these canals shifts, triggering sensory signals conveying information about the nature

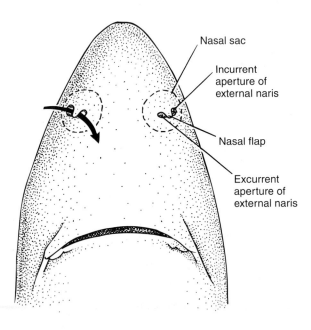

FIGURE 6.2   Nasal sacs of a shark. Ventral view of a shark head showing the direction water flows (solid arrow) through the nasal sac and across the olfactory epithelium.
After M. H. Wake (ed.), 1979, Hyman's Comparative Vertebrate Anatomy, University of Chicago Press

of the movement. The semicircular canals are embedded in calcified cartilage of the otic capsules. They will be identified when the nervous system is dissected.

Respiratory System

Sharks respire by ventilating their gills with water pulled into the mouth as the floor of the mouth is depressed and the pharynx is expanded. As the floor of the mouth is elevated, the water is forced through the pharynx into separate gill chambers containing gills richly supplied with blood. Some pelagic sharks use their forward motion to ventilate the gills in the process of ram ventilation. The water next moves away from the gills and out of the body through separate gill slits. Identify the five **gill slits** located on the posterior and lateral sides of the head. An additional opening, the **spiracle** (seen in figure 6.6), is technically the first gill slit in dogfish sharks (thus six gill slits in total). It is a rounded opening, lacking gills, and located just caudal to the eye on the dorsal side of the head. In bottom-dwelling skates and rays, the ventral mouth may be partially buried. The dorsally placed spiracle is thus in an unobstructed position to allow water to be drawn in for respiration. However, for most sharks, the function of the spiracle is unknown.

Cloaca

Examine the ventral side of the body between the cranial edges of the pelvic fins to find the medial depression, the **cloaca** (cloaca = "sewer"). The urinary and digestive tracts of both sexes empty into the cloaca, giving significance to its literal meaning of "sewer." The urinary system enters the middle of the cloaca through the tip of the nipplelike **urinary papilla** (female) or **urogenital papilla** (male). The uterus of the female empties separately into the cloaca, and the reproductive tract of a male sends sperm along the **clasper.**

Skinning the Shark

Position your shark in the dissection tray so that the ventral side of the trunk is directed toward you. To begin skinning the shark, you may need to use scissors to make a small incision midway between pelvic and pectoral fins. (A slit in the ventral musculature may have already been made to permit the injection of latex into the blood vessels of the shark.) At the cut edge, use a probe and forceps to separate and pull a flap of skin away from the shark. Then when a good grip can be had, pull the skin away from the underlying muscle. Pull while using a probe or finger to hold the musculature down at the junction where the skin is still attached. Use caution to ensure that muscle is not torn out as the skin is removed. The skinning process can be continued by gently pulling on the skin and pushing down on the muscle with finger or probe.

Try to leave large flaps of skin still attached near the dorsal midline. These flaps can be repositioned to protect the underlying musculature during storage. Continue

to reflect skin to expose the muscles on the ventral and right sides of the head and trunk up to the dorsal midline. Completely remove skin from the dorsal side of the head forward to the tip of the snout. The most fragile musculature will be encountered on the ventral and lateral sides of the head caudal to the jaws and around the gill slits. Take extra care to minimize damage to these delicate muscles. Work slowly and carefully. The entire skinning of the shark may take two to four hours. Thin but quite tough **fascia** (connective tissue) covering the muscles should also be cleared, but just sufficiently to reveal underlying muscles. A thorough job of removing fascia can wait until the underlying musculature is identified in the dissection.

Musculature

Insertion, Origin, and Function

Knowing the origin and insertion of a muscle makes identification easier, especially if the muscle's fibers closely parallel one or more of its neighbors. Use table 6.2 to assist your identification of the following muscles and to help you relate their position and shape to their functions.

Axial Musculature

The axial musculature arises from myotomes that differentiate from somites (figure 6.5). These myotomes grow and expand along the sides of the body, forming the musculature associated with the vertebral column, ribs, and lateral body wall. On the shark, the slanted blocks of muscle are **myomeres** separated from each other by **myosepta** (figure 6.6). The long, horizontal line is the outer margin of the **horizontal septum** running

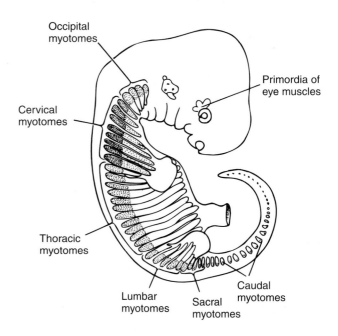

FIGURE 6.5 Muscles derived from embryonic myotomes in a reptile (lizard). During embryonic development, the myotomes expand into respective areas of the body.

TABLE 6.2   The Origin, Insertion, and Function of Shark Musculature

Muscle	Origin	Insertion	Function
Fin Musculature			
pectoral extensors	scapular process and adjacent fascia	radials and ceratotrichia	elevates pectoral fin
pectoral flexors	coracoid bar	radials and ceratotrichia	depresses pectoral fin
pelvic extensors			
superficial part	iliac process of puboischiac bar	radials and ceratotrichia	elevates pelvic fin
deep part	metapterygium	radials and ceratotrichia	elevates pelvic fin
pelvic flexors			
proximal part	linea alba and puboischiac bar	metapterygium	depresses pelvic fin
distal part	metapterygium	radials and ceratotrichia	depresses pelvic fin
Superficial Branchial Constrictor Muscles			
adductor mandibulae	posterior part of the palatoquadrate cartilage	Meckel's cartilage	closes the mouth
dorsal and ventral constrictors 3–6	vertical raphe	vertical raphe	compresses gill pouches
interhyoideus	midventral raphe	ceratohyal	compresses gill pouches
intermandibularis	Meckel's cartilage and fascia of the adductor mandibulae	midventral raphe	raises the floor of the mouth
levator hyomandibulae	otic capsule	hyomandibular	compresses gill pouches
preorbitalis	midventral surface of the chondrocranium	adductor mandibulae	closes the mouth
spiracularis	otic capsule	palatoquadrate cartilage	elevates the palatoquadrate
Levator Series			
cucullaris	dorsal longitudinal bundle fascia	epibranchial cartilage and scapular process	elevates the gill arches and pectoral girdle
interarcuals	pharyngobranchials	epibranchials	expands the pharynx
levator palatoquadrati	otic capsule	palatoquadrate cartilage	raises the upper jaw
Hypobranchial Musculature			
common coracoarcuals	coracoid bar	coracohyoid	opens the mouth
coracomandibularis	fascia of common coracoarcuals	Meckel's cartilage	opens the mouth
coracohyoid	common coracoarcuals	basihyal	opens the mouth
Extrinsic Eye Muscles			
inferior oblique	anterior end of orbit	ventral side of eyeball	rotates eye ventrally
inferior rectus	posterior end of orbit	ventral side of eyeball	rotates eye ventrally
lateral rectus	posterior end of orbit	lateral side of eyeball	rotates eye posteriorly
medial rectus	posterior end of orbit	medial side of eyeball	rotates eye cranially
superior oblique	anterior end of orbit	dorsal side of eyeball	rotates eye dorsally
superior rectus	posterior end of orbit	dorsal side of eyeball	rotates eye dorsally

from tail to head. This divides the body musculature into **epaxial** (above) and **hypaxial** (below) **musculature** within which specifically named muscle groups are recognized. The epaxial muscles form the **dorsal longitudinal bundles** and appear to be a single mass externally, but in cross section are two or three distinct bundles. Follow these bundles forward to their insertion at the back of the chondrocranium (see figure 6.9b).

The hypaxial musculature divides into the more dorsal **lateral longitudinal bundle** (just below the horizontal skeletogenous septum) and **ventral longitudinal bundle** forming most of the ventral surface of the trunk (figure 6.6). Follow the lateral longitudinal bundle forward to its attachment on the scapular process (figure 6.6). A thin line, the **linea alba,** occurs where the paired ventral longitudinal bundles meet along the ventral midline (see figure 6.9a).

Fin Musculature

🖝 If you have not already done so, clear skin from both sides of the pelvic and pectoral fins. The dorsal surface of the fins is covered by **extensors** of the fins, the

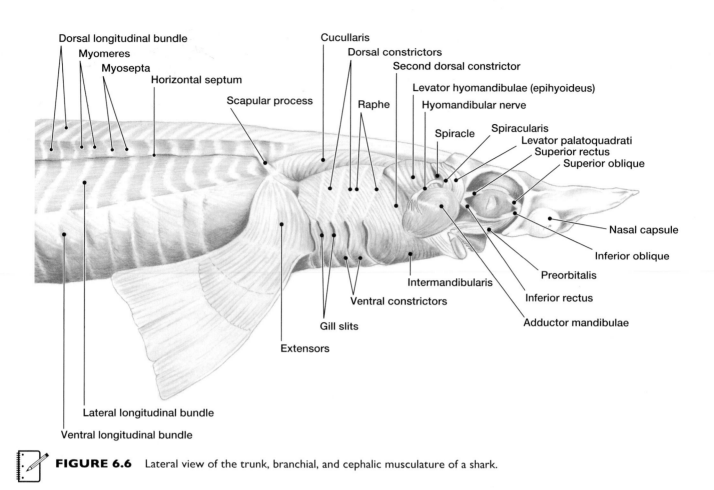

Dorsal longitudinal bundle
Myomeres
Myosepta
Horizontal septum
Scapular process
Cucullaris
Dorsal constrictors
Second dorsal constrictor
Levator hyomandibulae (epihyoideus)
Raphe
Hyomandibular nerve
Spiracle
Spiracularis
Levator palatoquadrati
Superior rectus
Superior oblique
Nasal capsule
Inferior oblique
Preorbitalis
Inferior rectus
Adductor mandibulae
Intermandibularis
Ventral constrictors
Gill slits
Extensors
Lateral longitudinal bundle
Ventral longitudinal bundle

FIGURE 6.6 Lateral view of the trunk, branchial, and cephalic musculature of a shark.

ventral surface by **flexors** of the fins (figures 6.6, 6.7, and 6.9). In males, the **siphon** covers the ventral side of the pelvic fin (figure 6.7). Peel it away on one pelvic fin and note that the flexor is divided into proximal and distal halves near the edge of the metapterygium.

Branchial Musculature

🦈 **Constrictor Series**   The **superficial constrictor musculature** covers the gill region and consists of six dorsal and six ventral constrictors, above and below the external gill slits, respectively (figure 6.8). The first constrictor muscle in this series is highly modified, as is the first gill arch it moves.

The first dorsal constrictor consists of three muscles (see figure 6.6). The **adductor mandibulae** (quadratomandibularis) is the massive hemisphere of muscle caudal to the eye. The **spiracularis** runs tightly along the anterior margin of the spiracle and inserts on the palatoquadrate. Clear fascia to find it and another muscle that closely parallels it anteriorly, the **levator palatoquadrati,** actually part of the levator series traced later. The third component of the first dorsal constrictor is the **preorbitalis,** adjacent to the deep ventral surface of the eyeball.

The first ventral constrictor consists of the paired **intermandibularis** (figure 6.9a) on the ventral surface

of the throat running between Meckel's cartilages and divided by a **median raphe.**

The second dorsal constrictor consists of a broad sheet of muscle between the spiracle and raphe of the first complete external gill slit (see figure 6.6). The anterior half of this muscle is the **levator hyomandibulae** (epihyoideus), which extends caudally to the posterior margin of the adductor mandibulae. Fibers of the posterior half of the second dorsal constrictor sweep downward past the rear of the adductor mandibulae to below the gills. This posterior half constitutes the **second dorsal gill constrictor,** proper.

The second ventral constrictor, the **interhyoideus,** is a sheet of muscle just dorsal to the intermandibularis. To find it, cut a window in the belly of the thin intermandibularis to reveal the parallel fibers of the interhyoideus running at a slightly different angle (figure 6.9a).

Dorsal and ventral constrictors 3–6 follow, separated from each other by a **raphe** (figure 6.6).

🦈 **Levator Series**   The first levator, **levator palatoquadrati,** is closely applied to the anterior face of the spiracularis (figure 6.6). The second levator, levator hyoideus, is impractical to distinguish because it fuses with the medial face of the levator hyomandibulae.

Levators 3–8 fuse into a single long, thin wedge of muscle, the **cucullaris,** lying in front of the scapular process between the top of the gill pouches and the dorsal longitudinal bundles (figures 6.6 and 6.9b).

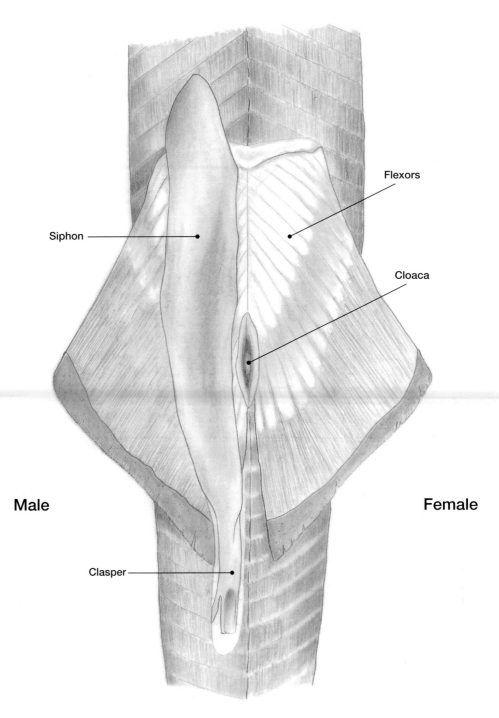

Siphon

Flexors

Cloaca

Male

Female

Clasper

FIGURE 6.7 Ventral view of the pelvic fins of a male (left) and female (right) shark.

➤ **Interarcual Series** Separate the upper margin of the gill pouches from the lower edge of the cucullaris (figure 6.9b). Spread the two regions until the deep pharyngobranchial and epibranchial gill-arch elements come into view. Between these cartilages lie the series of small **interarcuals,** which move the gill-arch cartilages.

Hypobranchial Musculature These muscles run on the ventral side of the shark from the pectoral girdle to the mandible (figure 6.9a). They lie deep to the ventral constrictors identified previously. In the ventral midline, just anterior to the coracoid bar, lies a paired muscle

triangular in shape, the **common coracoarcuals.** Cut along the lateral edge of the left (apparent right) common coracoarcual just deep enough to clear the intermandibular and interhyoideus (figure 6.9a). Carefully fold back the flaps to expose the muscles but do not cut deeper than directed as this endangers delicate blood vessels traced later.

The thin muscle (actually a paired muscle tightly fused together) directly above (dorsal to) the median raphe is the **coracomandibularis,** on its way to the mandible. Above (dorsal) to it are the paired **coracohyoids,** continuous caudally

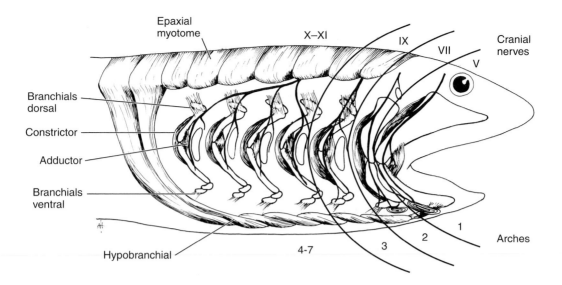

Arch	Cranial Nerves	Constrictors		Levators
1st	V	**Dorsal** Spiracularis Adductor mandibulae Preorbitalis **Ventral** Intermandibularis		Levator palatoquadrati
2nd	VII	**Dorsal** Levator hyomandibulae Second dorsal constrictor **Ventral** Interhyoideus (deep to intermandibularis)		Levator hyoideus (medial to the levator hyomandibulae)
3rd	IX	**Dorsal** Superficial constrictors **Ventral** Superficial constrictors		Cucullaris*
4–7th	X	**Dorsal** Superficial constrictors **Ventral** Superficial constrictors		Cucullaris* Interarcuals (deep to cucullaris)

*The cucullaris receives separate contributions from muscles in several arches.

FIGURE 6.8 Branchiomeric musculature. Jaw muscles and their cranial nerve supply tend to stay with their respective branchial arch during the course of subsequent evolution. Each arch has levator and constrictor muscles that, respectively, elevate and close the articulated elements. Cranial nerves V, VII, IX, and X–XI supply muscles of arches 1, 2, 3, and 4–7, respectively. Fidelity of muscles, nerves, and arches generally was maintained as the branchial arches evolved and subsequently became modified into components of the jaws. The specific derivatives of the levator and constrictor series of muscles in each arch of a shark are shown together with their cranial nerve supply.

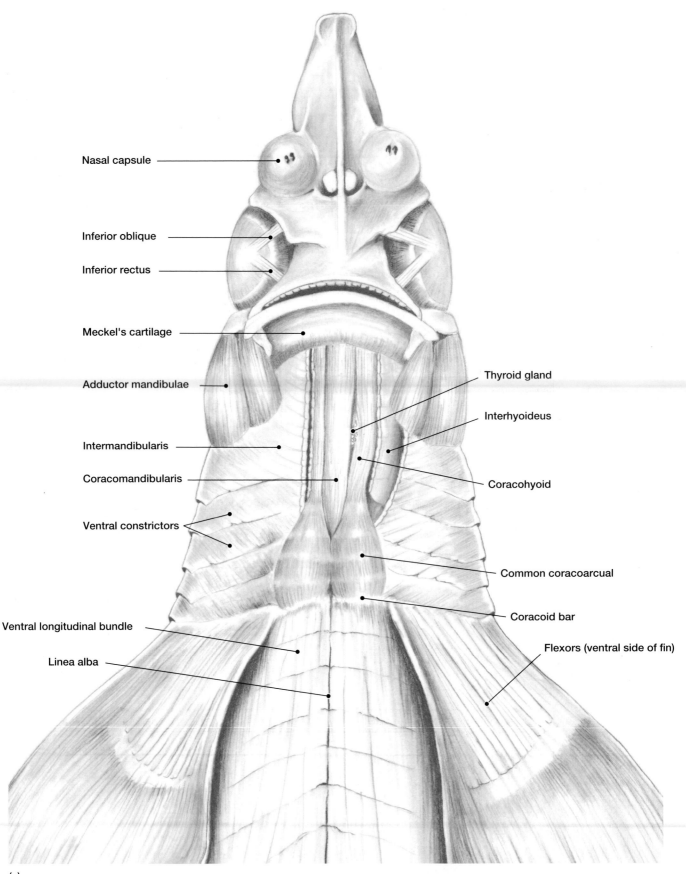

Nasal capsule

Inferior oblique

Inferior rectus

Meckel's cartilage

Adductor mandibulae

Intermandibularis

Coracomandibularis

Ventral constrictors

Ventral longitudinal bundle

Linea alba

Thyroid gland

Interhyoideus

Coracohyoid

Common coracoarcual

Coracoid bar

Flexors (ventral side of fin)

(a)

FIGURE 6.9 Trunk, branchial, and cephalic musculature of a shark. (a) Ventral view.

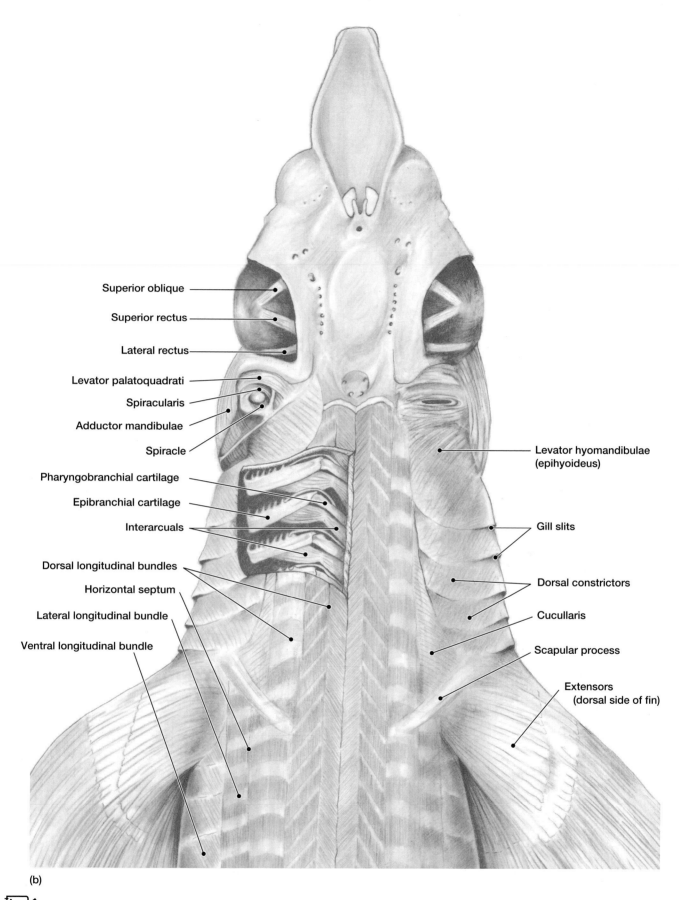

Superior oblique

Superior rectus

Lateral rectus

Levator palatoquadrati

Spiracularis

Adductor mandibulae

Spiracle

Pharyngobranchial cartilage

Epibranchial cartilage

Interarcuals

Dorsal longitudinal bundles

Horizontal septum

Lateral longitudinal bundle

Ventral longitudinal bundle

Levator hyomandibulae
(epihyoideus)

Gill slits

Dorsal constrictors

Cucullaris

Scapular process

Extensors
(dorsal side of fin)

(b)

FIGURE 6.9 *Continued.* (b) dorsal view.

with the common coracoarcuals. The **thyroid gland** appears as a squashed, dark mass slightly to the left of center, between the coracohyoids and coracomandibularis.

Extrinsic Eye Muscles

🦖 Expose the left eyeball by removing the supraorbital crest over the orbit. Leave the right eye undisturbed until the nervous system is dissected. Clean out debris, taking care not to break any delicate nerves (white threadlike structures) or muscles. Viewed from above, four thick ribbons of muscle

leave from a common site on the posterior medial wall of the orbit (figures 6.6, 6.9, and 6.10). These four rectus muscles of eye diverge and attach to the top, bottom, medial, and lateral sides of the eyeball. These are the **superior** (dorsal), **inferior** (ventral), **medial,** and **lateral rectus muscles.** In addition, two oblique muscles extend from the anterior medial wall of the orbit (thus, six eye muscles in total). The **inferior oblique** inserts below the eyeball and the **superior oblique** above the eye near the insertion of the superior rectus. The larger white cords are major nerves: the **superior ophthalmic** coursing along the back

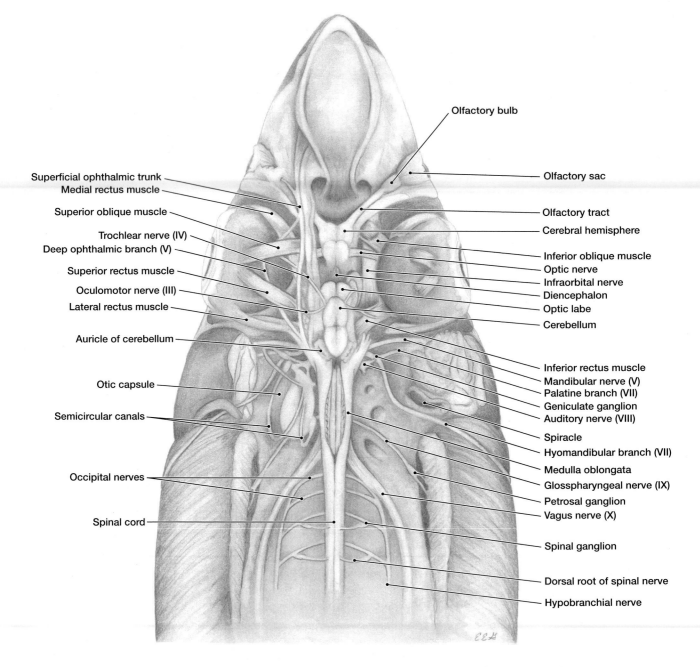

Olfactory bulb

Superficial ophthalmic trunk
Medial rectus muscle
Superior oblique muscle
Trochlear nerve (IV)
Deep ophthalmic branch (V)
Superior rectus muscle
Oculomotor nerve (III)
Lateral rectus muscle
Auricle of cerebellum
Otic capsule
Semicircular canals
Occipital nerves
Spinal cord

Olfactory sac
Olfactory tract
Cerebral hemisphere
Inferior oblique muscle
Optic nerve
Infraorbital nerve
Diencephalon
Optic labe
Cerebellum
Inferior rectus muscle
Mandibular nerve (V)
Palatine branch (VII)
Geniculate ganglion
Auditory nerve (VIII)
Spiracle
Hyomandibular branch (VII)
Medulla oblongata
Glosspharyngeal nerve (IX)
Petrosal ganglion
Vagus nerve (X)
Spinal ganglion
Dorsal root of spinal nerve
Hypobranchial nerve

FIGURE 6.10 Extrinsic eye musculature of the shark (dorsal view). The extrinsic eye muscles are derived from somitomeres and rotate the eyeball within the orbit to direct the gaze. The roof of the chondrocranium has been removed and the auditory region dissected on the left, to reveal the semicircular canals. On the right, the superior oblique, superior rectus, medial rectus, and lateral rectus muscles have been cut to reveal the inferior oblique and inferior rectus muscles.

of the orbit, the **optic nerve** projecting medially from the back of the eye between medial and inferior oblique muscles, and the **infraorbital nerve** running across the preorbitalis muscle (figures 6.6 and 6.10). (Preserve the integrity of the other nerves in this region for easier identification in the later section on the nervous system.)

Between the diverging rectus muscles lies the cartilaginous **optic pedicel,** shaped somewhat like a golf tee. Finally, expose the **nasal capsule** just anterior to the eye for a firsthand view (figure 6.6 and 6.9a).

Necturus

External Anatomy

Mudpuppies (genus *Necturus*) are common inhabitants of clear streams in the midwestern and eastern United States. Unlike most salamanders, *Necturus* species retain many larval features into the adult stage, a condition called *paedomorphosis*. For example, most salamander larvae are aquatic, hatching from eggs laid in water. Usually after several months the larvae *transform* (undergo *metamorphosis*) into terrestrial juveniles, which grow and develop on land into adults. However, *Necturus* does not transform. Instead, it remains aquatic throughout its life.

The large, bushy, and dark **external gills** located at the rear of the head are a good example of a larval characteristic retained into adulthood. Small, dark **eyes** reside on the front dorsal side of the head. At the tip of the snout are the paired **external nares** through which air can be drawn into the mouth. Air moving through the external nares passes a chemosensory organ before entering the **buccal cavity** (mouth) through the internal nares. The sharp **teeth,** identified previously when studying the skull, are bound laterally by thick **lips.**

The expansive **trunk** occupies most of the body between the pectoral and pelvic girdles. A distinct **gular fold** (figure 6.11) occurs on the ventral side of the neck

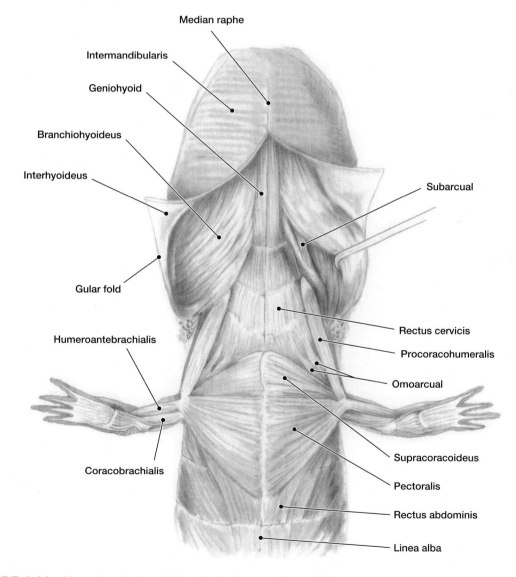

FIGURE 6.11 Ventral trunk, branchial, and cephalic musculature of *Necturus*.

where the trunk merges with the head. The **cloaca** should next be found along the ventral midline at the caudal end of the trunk. As in the shark, this chamber is the location where the digestive, urinary, and reproductive tracts deposit their products. Surrounding the cloacal chamber is a large hump, the **cloacal gland,** which releases products used during reproduction (see figure 6.14).

Dissection

🦖 Begin separating the skin from underlying muscles at the edge of a cut along the ventral trunk. (As in the shark, a slit may already have been made in this region to inject blood vessels.) Completely remove the skin from the ventral surface and right side of the body to the dorsal midline. Cut a circle around the external gills and leave them in place. While pulling the skin off the external gills, take care not to damage the delicate subarcual muscles. Pull the skin over each limb, leaving it inside out, like removing a sweater.

To minimize damage to the muscles, continue to practice the precautions used for the shark. Muscles running from body to limbs and around the gills are especially delicate, usually fan-shaped, and often adhere to the integument. Thus, in removing the skin, be especially careful not to tear these muscles away.

Musculature

Insertion, Origin, and Function

🦖 As noted previously with the shark, knowing the origin, insertion, and function of a muscle makes identification easier. Use table 6.3 to assist your identification of the

TABLE 6.3 The Origin, Insertion, and Function of *Necturus* Musculature

Muscle	Origin	Insertion	Function
Ventral Pectoral Muscles			
omoarcual	rectus cervicis	humerus	draws humerus craniad
pectoralis	linea alba and coracoid cartilage	humerus	adducts forelimb
procoracohumeralis	procoracoid cartilage	humerus	draws humerus craniad
supracoracoideus	coracoid cartilage	humerus	adducts forelimb
Ventral Throat Muscles			
branchiohyoideus	first branchial arch	ceratohyal cartilage	moves branchial arches and hyoid
geniohyoid	medial mandible	second basibranchial cartilage	depresses lower jaw and draws hyoid cranially
interhyoideus	fascia covering the branchiohyoideus	median raphe	elevates floor of throat
intermandibularis	rami of mandible	median raphe	elevates floor of throat
rectus cervicis	rectus abdominis	branchial cartilages	retracts tongue and lowers head
subarcual	gill cartilage	gill cartilage	depresses gills
transversis ventralis	rectus cervicis	last gill cartilage	depresses gills
Arm and Forearm Muscles			
coracobrachialis	coracoid cartilage	humerus	adducts forelimb
humeroantebrachialis	humerus	radius	flexes forearm
triceps brachii	glenoid fossa, humerus, and coracoid cartilage	ulna	extends forearm
Dorsal and Lateral Muscles of the Head and Gills			
depressor mandibulae	third branchial arch	posterior end of mandible	depresses lower jaw
dilatator laryngis	fascia near dorsal midline	lateral cartilages of larynx	dilates larynx
levator mandibulae anterior	frontal and parietal bones	lateral dentary	elevates lower jaw
levator mandibulae externus	frontal and parietal bones	lateral dentary	elevates lower jaw
levatores arcuum	fascia of epibranchial muscles	epibranchial cartilages	elevates external gills
Lateral Pectoral Muscles			
cucullaris	fascia of epibranchial muscles	scapula	elevates the pectoral girdle
dorsalis scapulae	suprascapular cartilage	humerus	elevates the forelimb
latissimus dorsi	fascia of epibranchial muscles	shoulder and proximal end of humerus	draws humerus caudad
pectoriscapularis	epibranchial cartilage	scapula and procoracoid cartilage	draws shoulder and forelimb craniad

TABLE 6.3 Continued. The Origin, Insertion, and Function of Necturus Musculature

Muscle	Origin	Insertion	Function
Dorsal-Epaxial Trunk Muscles			
dorsalis trunci	myosepta and transverse processes of vertebrae	myosepta and transverse processes of vertebrae	flexes the vertebral column laterally
Ventral–Hypaxial Trunk Muscles			
external oblique	transverse processes of trunk vertebrae	connective tissue of rectus abdominis	compresses the coelom and flexes and rotates the trunk
internal oblique	transverse processes of trunk vertebrae	connective tissue of rectus abdominis	compresses the coelom and flexes and rotates the trunk
rectus abdominis	anterior margin of the pubic cartilage	transverse inscription	flexes the trunk ventrally
transversus	transverse processes of trunk vertebrae	connective tissue of rectus abdominis	compresses the coelom and flexes and rotates the trunk
Pelvic Girdle and Hindlimb Muscles			
caudocruralis	posterior border of the puboischiotibialis	haemal arches of caudal vertebrae 3–6	flexes tail
caudofemoralis	haemal arches of caudal vertebrae 3–6	femur	flexes tail and draws hindlimb caudad
ilioextensorius	ilium	tibia and fascia covering the knee	extends the shank
iliofibularis	ilium	fibula	extends the shank and draws the hindlimb caudad
iliotibialis	ilium	tibia	abducts the hindlimb
ischiocaudalis	ischium	haemal arches of caudal vertebrae 3–6	flexes tail
ischioflexorius	ischium	fascia of foot	flexes shank and toes
pubo-ischio-femoralis externus	puboischiac plate	femur	adducts the hindlimb
pubo-ischio-femoralis internus	internal surface of the puboischiac plate	femur	draws the femur craniad
pubo-ischio-tibialis	posterior puboischiac plate	tibia	adducts the hindlimb
pubotibialis	anterior puboischiac plate	tibia	adducts the hindlimb

Necturus muscles and to help you relate the position and shape of the muscles to their functions.

Ventral Pectoral Girdle, Forelimb, and Throat

Begin by finding the fan-shaped **pectoralis** muscles ventrally, between the forelimbs (figure 6.11). Just caudal is the **rectus abdominis,** which originates on the pubic cartilage, and runs forward interrupted at regular intervals by myosepta, cross inscriptions marking myotomal boundaries. The **linea alba** is the connective tissue line that runs down the middle of the rectus abdominis at the central midline. Anteriorly, it inserts on a strong, transverse inscription just dorsal to the coracoid cartilages. Anterior to the pectoralis lies the **supracoracoideus,** riding on the coracoid cartilage. To help distinguish it from the pectoralis, grasp the most superficial coracoid cartilage with tweezers and slide it forward and back. The line between muscles can then be better established. The elongate procoracoid cartilage runs forward on the lateral side of

the neck and is covered by the **procoracohumeralis** muscle passing backward to insert on the humerus. Probe this muscle to uncover the procoracoid cartilage. Between these paired cartilages lies the ventral neck muscles, the **rectus cervicis,** segmented by inscriptions, but lacking a linea alba. The rectus cervicis, essentially a continuation of the rectus abdominis, originates on the strong, transverse inscription on the body above the coracoid cartilages and runs into the throat. A small, short muscle, the **omoarcual,** inserts on the procoracoid and lies between the supracoracoideus and procoracohumeralis. It is apparently derived from the rectus cervicis.

Across the ventral surface of the upper arm lie two small muscles, the **humeroantebrachialis** (anterior) and **coracobrachialis** (posterior). The ventral surface of the forearm contains several flexors but these will not be followed individually.

Ventrally, the throat is covered by transverse fibers between the mandibles and the **median raphe.** The anterior part of this sheet is the **intermandibularis,** which indistinguishably merges into the **interhyoideus,** forming the posterior half. Make a slit along the median raphe and open the flaps. Beneath is the paired, slender **geniohyoid,** passing from the chin posteriorly to a basibranchial cartilage (figure 6.11). Lateral to the geniohyoids are the large, paired **branchiohyoideus** muscles arching forward to the chin. Retract the branchiohyoideus muscles laterally to expose the cartilaginous gill arches. Deep between the gill arches are the **subarcuals,** short wedges of longitudinal muscles running between successive gill arches. Note the attachment of the rectus cervicis medially to the gill arches. Dorsal to the rectus cervicis and between it and the last gill cartilage runs the **transversis ventralis** (not shown in a figure).

Lateral Pectoral Girdle and Forelimb

🦎 Turn the specimen so its exposed right side faces you and pull forward the external gills to reveal the several muscles above the limb (figure 6.12a). Above the external gills are two muscles that may at first appear as one. The anterior, the **levatores arcuum,** inserts on the arches. The posterior, the **dilatator laryngis** (partially covering the origin of cucullaris), inserts deeper (figure 6.12a,b). Beginning along the ventral edge of the dilatator laryngis and above the procoracohumeralis are four muscles, in the following order (figure 6.12a): Most anterior is the thin **pectoriscapularis** followed by the **cucullaris,** the broader **dorsalis scapulae,** and the **latissimus dorsi.** Each of these four muscles converges on the shoulder near the junction of the scapula and humerus.

Across the dorsal surface of the upper arm lies the large **triceps brachii** (= anconeus). The dorsal surface of the forearm contains several **extensors** that will not be identified individually.

Lateral Head

🦎 Again identify the branchiohyoideus, which forms a large mass immediately in front of the gills (figure 6.12a). Along the dorsal and anterior margin of the branchiohyoideus is the lateral edge of the **depressor mandibulae.** Both muscles originate on gill bars, but the depressor mandibulae passes forward to the rear of the mandible, while the branchiohyoideus passes ventrally and forward to the chin. To separate them, trace the anterior edge of the branchiohyoideus dorsally. Note how its fibers arch posteriorly, and part the two muscles along this line.

Dorsally, between the eye and dorsalis trunci, resides a large muscle mass, the **adductor mandibulae** (figure 6.12b). It is divisible into a medial **levator mandibulae anterior** and a larger and more lateral **levator mandibulae externus.**

Trunk and Tail Musculature

🦎 Epaxial and hypaxial musculature along the trunk is separated by a horizontal line, the outer margin of the **horizontal septum** (figure 6.13). The superficial part of the epaxial muscles forms the **dorsalis trunci,** whose fibers run horizontally between myosepta and deep to transverse processes of the vertebrae (figures 6.12a and 6.13).

The hypaxial muscles form three major lateral layers (figure 6.13). Superficially lies the **external oblique,** with fibers slanting caudad and ventrad. Cut a window in this thin sheet to reveal the intermediate **internal oblique,** slanting in the opposite direction. Deeper still is the **transversus,** a very thin layer reached just before penetrating into the body cavity and identified by its slightly different fiber slant, somewhat transversely around the body. The ventral most hypaxial muscle is the paired **rectus abdominis,** extending from the anterior edge of the pubic cartilage to the base of the rectus cervicis near the pectoral girdle (figure 6.14). Along the ventral midline between the rectus abdominis is a thin connective strip, the **linea alba** (figure 6.14).

Pelvic Girdle and Hindlimb

🦎 Rubbery cloacal glands surround the cloaca. Carefully probe between these and the more dorsal musculature to remove and discard one side of the gland (figure 6.14). Laterally, on either side of the cloaca is a downward-projecting shelf of hypaxial myotomes. Between these hypaxial muscles and the cloacal glands are three pairs of muscles: two ventral and one dorsal. The most lateral of the ventral pair is the **caudocruralis,** which runs between the tail and the **pubo-ischio-tibialis** muscle, which itself covers the posterior half of the puboischiac plate (figure 6.14). The medial member of the pair is the **ischiocaudalis,** which runs between the tail and the ischium. Deflect (pull aside) the bellies of the ischiocaudalis and caudocruralis medially. This should slightly expose dorsally yet a third caudal muscle, the **caudofemoralis,** also seen by looking laterally (figure 6.13). The insertion of this third muscle should be confirmed by following it forward to the proximal end of the femur.

The pubo-ischio-tibialis, just identified, narrows laterally, and as the name suggests, inserts on the tibia (figure 6.14). Lying anterior to and occupying the front part of the pubic cartilage is the **pubo-ischio-femoralis externus,** whose fibers converge at their insertion on the femur. Note how these last two muscles partially overlap at their origins. Next identify the muscles of the thigh. On the ventral surface, from posterior to anterior, is the flat **ischioflexorius,** pubo-ischio-tibialis, already described, and the round **pubotibialis.** Separate these three muscles from each other near their insertions and confirm origins and insertions implied by their names.

Continuing forward from the anterior edge of the pubotibialis and across the dorsal side of the thigh are

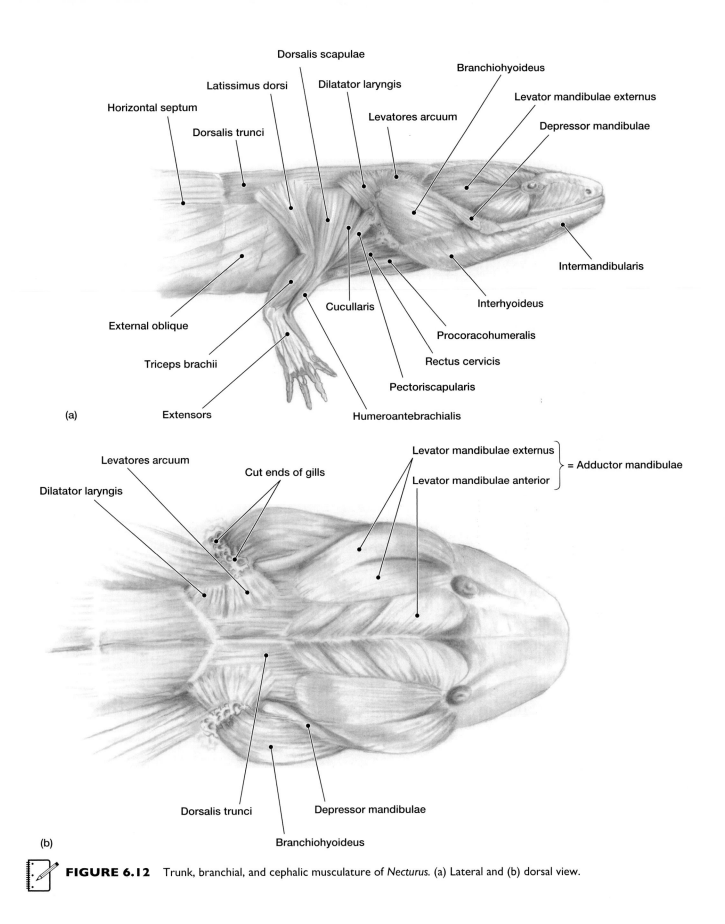

FIGURE 6.12 Trunk, branchial, and cephalic musculature of *Necturus*. (a) Lateral and (b) dorsal view.

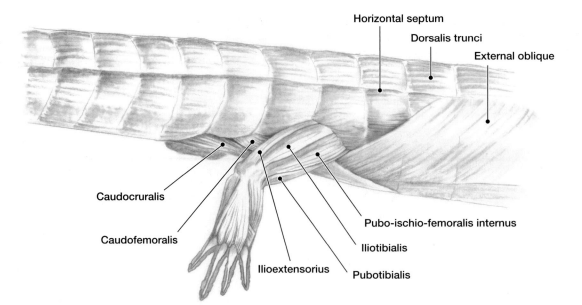

FIGURE 6.13   Lateral trunk and hindlimb musculature of *Necturus*.

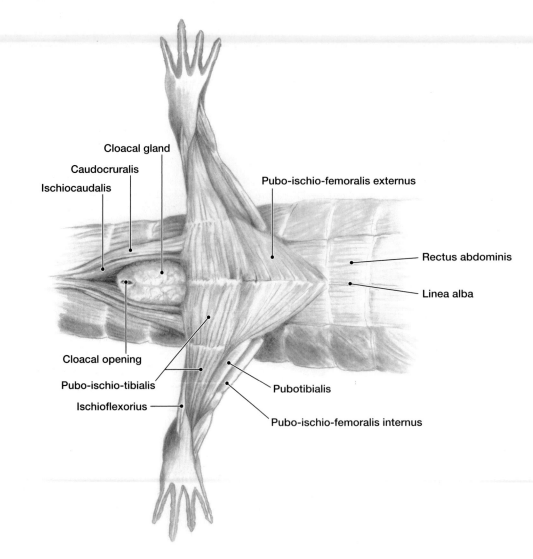

FIGURE 6.14   Ventral trunk and hindlimb musculature of *Necturus*.

four muscles (figure 6.13). Lying along the anterior edge of the pubotibialis is the **pubo-ischio-femoralis internus,** then the more dorsal **iliotibialis,** the more posterior **ilioextensorius,** and finally caudal to it the **iliofibularis** (not pictured as it occurs on the most posterior edge hidden in dorsal and lateral views). Following around the posterior (postaxial) side of the thigh returns you to the ischioflexorius identified earlier. Spend time now delineating with blunt probe or forceps each muscle and confirm attachments suggested by the name of each muscle.

Cat

External Anatomy

🐾 Most students have a general familiarity with cat anatomy. However, some features, their names, and their specific functions are not well known. The **pinnae** of the ears are the external, flexible portions that initially collect sound. On the snout, long hairs, or **vibrissae,** provide tactile information. Numerous **tori** form thick pads that cushion the feet.

Sexing Your Cat

🐾 At the caudal end of the body, identify the **anus,** directly below the tail. Look along the midventral line just anterior to the anus. Male cats have a single, fur-covered **scrotum** forming a noticeable lumpy projection (see figure 6.23). Just cranial to the scrotum is the **penis,** likely withdrawn and difficult to detect. Female cats will instead have a second opening, the **urogenital sinus,** in the position of a male's scrotum. (This is no accident, as the genitals of male and female mammals develop from common embryonic structures.) Further details of the reproductive systems will be addressed later in chapter 9, which is devoted to the reproductive system.

Approach

The cat has the same basic muscle regions as the shark and *Necturus,* except that there is a greater number of named muscles, probably reflecting the greater complexity of motion and refinement of action. The right side of the cat will be saved for later study of the circulatory system. Thus, confine your deep muscle dissection to the cat's *left side.* Even here, however, be sure to avoid unnecessarily destroying nerves or injected blood vessels.

Reflecting Muscles

Remember that when you need to expose a deeper structure covered by a muscle, *reflect* the muscle by cutting it through its belly (middle) so you can later reposition severed ends of the muscle for study and review.

Skinning the Cat

🐾 The skinning should be done so that, at the end of each laboratory, the skin can be refolded to its original position. This helps retain moisture and prevents drying of the tissues between dissections.

Most instructors have specific preferences for the removal of the skin. (After all, "there is more than one way to skin a cat.") In general, use a pair of scissors to make a shallow incision along the ventral midline from the chin to the groin. (As noted in the shark and *Necturus,* an incision may have already been made, often laterally, to inject a portion of the circulatory system. If so, begin with this incision instead of making another.) On the left side, extend the midventral incision laterally along the fore- and hindlimbs to the base of the paws. At the terminus of these incisions into the limbs, extend the cut circularly all the way around the ends of the paw. Now the skin can and should be freed from underlying tissue all the way to the dorsal midline—taking care not to pull attached muscle with it. Those with female cats will encounter the thick, glandular **mammary glands** along the belly. These should accompany the skin as it is worked free from the abdominal region.

Between the skin and muscle lies **fascia,** which includes the cobweblike **superficial fascia** and the **deep fascia,** often impregnated with yellow fat deposits that cling to the surface of the muscles. This should all be removed to reveal muscular details. As you remove skin, you may notice a very thin layer of muscle sometimes attached to the skin, or clinging to the underlying musculature. This is the *dermal muscle.* In the head and neck it is called the **platysma,** but elsewhere the **cutaneous maximus.** It acts to move the skin and in the head contributes to changes in facial expression.

General Regions

🐾 The white streak down the ventral midline along the abdomen is the **linea alba.** Along the back, middorsally over the lumbar region, is an extensive area covered with white sheets of connective tissue. The sheets are collectively the **lumbodorsal fascia.** The **axilla** (or armpit) is the angle between the base of the forelimb and body. The angle between the medial base of the hindlimb and abdominal wall is the **inguinal region.**

Musculature

Use table 6.4 to assist your identification of the following muscles and to help you relate their positions and shapes to their functions.

Abdomen, Ribs

🦕 The first set of muscles to identify covers the ribs and abdominal regions. Turn the cat on its back and locate the following six muscles:

(1) Rectus abdominis

The rectus abdominis runs along the abdomen adjacent to and parallel with the linea alba (figure 6.15). It originates on the pubic symphysis and inserts on the sternum and costal cartilages. To better reveal the rectus

TABLE 6.4	The Origin, Insertion, and Function of Cat Musculature		
Muscle	**Origin**	**Insertion**	**Function**
Abdomen and Ribs			
external intercostal	rib	rib	draws ribs craniad
external oblique	lumbodorsal fascia and ribs	aponeurosis on the linea alba	compresses abdominal viscera
internal intercostal	rib	rib	draws ribs caudad
internal oblique	pelvis and lumbodorsal fascia	aponeurosis on the linea alba	compresses abdominal viscera
rectus abdominis	pubic symphysis	sternum and costal cartilages	flexes trunk and compresses abdominal viscera
transversus	ilium, lumbar vertebrae, and posterior ribs	linea alba	compresses abdominal viscera
Chest Musculature			
clavobrachialis	clavicle	proximal end of ulna	draws humerus craniad
pectoantebrachialis	manubrium	fascia of forelimb near the elbow	adducts and rotates the forelimb
pectoralis major	cranial sternebrae	proximal half of humerus	adducts and rotates the forelimb
pectoralis minor	caudal sternebrae	proximal half of humerus	adducts and rotates the forelimb
xiphihumeralis	xiphoid process	proximal end of humerus	adducts and rotates the forelimb
Neck and Throat			
cleidomastoid	clavicle	mastoid process	rotates head
cricothyroid	cricoid cartilage	thyroid cartilage	moves the larynx
digastric	mastoid and jugular processes of occipital	mandible	depresses the lower jaw
geniohyoid	mandible	hyoid	draws hyoid craniad
hyoglossus	hyoid	tongue	retracts the tongue
mylohyoid	mandible	median raphe	elevates floor of mouth
sternohyoid	manubrium of sternum	hyoid	draws hyoid caudad
sternomastoid	manubrium of sternum	lambdoidal ridge and mastoid process	rotates head
sternothyroid	manubrium of sternum	thyroid cartilage	draws larynx caudad
thyrohyoid	thyroid cartilage	hyoid	draws larynx craniad
Lateral Jaw			
masseter	zygomatic arch	mandible	elevates lower jaw
temporalis	lambdoidal crest and zygomatic process of frontal bone	coronoid process of mandible	elevates lower jaw
Dorsal Shoulder			
acromiodeltoid	acromion process of scapula	humerus	flexes and rotates humerus
acromiotrapezius	neural spines of cervical and first thoracic vertebrae	metacromion process and spine of scapula	draws scapula dorsally and medially
clavotrapezius	nuchal line of skull and dorsal midline of neck	clavicle	draws humerus craniad
infraspinatus	infraspinous fossa of scapula	greater tuberosity of humerus	rotates the humerus
latissimus dorsi	lumbodorsal fascia and neural spines of the last thoracic and lumbar vertebrae	medial surface of humerus	elevates the arm and draws it caudad
levator scapulae ventralis	occipital bone and transverse process of atlas	metacromion process	draws scapula craniad
rhomboideus	neural spines of thoracic vertebrae	ventral border of scapula	draws scapula dorsally
rhomboideus capitis	superior nuchal line	angle of scapula	draws scapula craniad

Muscle	Origin	Insertion	Function
spinodeltoid	spine of the scapula	deltoid ridge of humerus	elevates and rotates the humerus
spinotrapezius	neural spines of the thoracic vertebrae	fascia around scapula	draws scapula dorsally and caudad
splenius	mid-dorsal fascia of the neck	nuchal line	turns and elevates head
supraspinatus	supraspinous fossa	greater tuberosity of humerus	extends the humerus
teres minor	axillary border of scapula	greater tuberosity of humerus	rotates humerus
Ventral Shoulder			
coracobrachialis	coracoid process	proximal end of humerus	adducts humerus
scalenus posterior	rib 3	cervical vertebrae	flexes neck, draws ribs craniad
scalenus medius	ribs 6–9	cervical vertebrae	flexes neck, draws ribs craniad
scalenus anterior	ribs 2 and 3	cervical vertebrae	flexes neck, draws ribs craniad
serratus ventralis	first ten ribs	vertebral border of scapula	draws scapula ventrally and craniad
subscapularis	subscapular fossa	lesser tuberosity of humerus	draws humerus medially
teres major	axillary border of scapula	medial surface of humerus	draws humerus medially
transversus costarum	sternum	first rib	draws sternum craniad
Upper Arm—Brachium			
biceps brachii	border of glenoid fossa	tuberosity of radius	flexes forearm
brachialis	lateral surface of humerus	below semilunar notch of ulna	flexes forearm
epitrochlearis	lateral surface of latissimus dorsi	olecranon process of ulna	rotates ulna
triceps brachii	shaft of humerus	olecranon process of ulna	extends forearm
lateral head	deltoid ridge of humerus	olecranon process	extends forearm
medial head	axillary border of scapula	olecranon process	extends forearm
long head	humerus	olecranon process	extends forearm
Forearm—Antebrachium Dorsal Side			
abductor pollicis longus	lateral surface of radius and ulna	first metacarpal	abducts the forelimb
brachioradialis	middle of humerus	distal end of radius	supinator of the manus
extensor carpi radialis brevis	distal portion of the humerus	third metacarpal	extends the manus
extensor carpi radialis longus	distal end of humerus	second metacarpal	extends the manus
extensor carpi ulnaris	lateral epicondyle of the humerus	fifth metacarpal	extends the manus
extensor digiti secundi	proximal end of ulna	middle phalanx of second digit	extends second digit
extensor digitorum communis	distal portion of the humerus	digits II–V	extends the manus
extensor digitorum lateralis	distal portion of the humerus	digits II–V	extends the digits of the manus
extensor pollicis	proximal end of ulna	middle phalanx of first digit	extends first digit
supinator	lateral epicondyle of humerus	radius	supinator of the manus
Forearm—Antebrachium Ventral Side			
flexor carpi radialis	medial epicondyle of humerus	digits II and III	flexes the manus
flexor carpi ulnaris	medial epicondyle of humerus and ulna	carpal pisiform bone	flexes the manus
flexor digitorum profundus	five heads: 1 head ulna, 3 heads humerus, and 1 head radius	digits I–V	flexes the manus
flexor digitorum superficialis	medial epicondyle of humerus and surface of flexor digitorum profundus	digits II–V	flexes the manus
pronator teres	medial epicondyle of humerus	medial border of radius	pronates the manus
Muscles of the Lateral Thigh			
biceps femoris	ischial tuberosity	proximal third of tibia	flexes the shank, abducts thigh
caudofemoralis	proximal caudal vertebrae	patella	abducts thigh
gluteus maximus	last sacral and first caudal vertebra	fascia lata	abducts thigh

Continued

Muscle	Origin	Insertion	Function
gluteus medius	iliac crest, sacral, and first caudal vertebrae	greater trochanter of femur	abducts thigh
tensor fasciae latae	ilium	fascia lata	extends the fascia lata
tenuissimus	second caudal vertebra	proximal tibia	flexes the shank
Muscles of the Medial Thigh			
adductor femoris	pubic symphysis	ventral surface of femur	adducts thigh
adductor longus	pubic symphysis	femur	adducts thigh
gracilis	pubic symphysis	fascia of lower thigh	adducts thigh
iliopsoas	lumbar vertebrae and ilium	lesser trochanter of femur	flexes and rotates thigh
pectineus	pubis	shaft of femur	adducts thigh
rectus femoris	ilium	patella	extends lower leg
sartorius	iliac crest	patella and proximal tibia	adducts and rotates thigh
semimembranosus	ischium	medial epicondyle of femur and proximal tibia	extends thigh
semitendinosus	ischium	tibia	flexes lower leg
vastus intermedius	femur	patella	extends lower leg
vastus lateralis	femur	patella	extends lower leg
vastus medialis	femur	patella	extends lower leg
Muscles of the Shank			
extensor digitorum longus	lateral epicondyle of femur	digits II–V	extends digits
flexor digitorum longus	middle tibia and head of fibula	digits I–V	flexes digits
flexor hallucis longus	tibia and fibula	tendon of flexor digitorum longus	flexes digits
gastrocnemius	femur and tendon of plantaris	calcaneum	extends foot
peroneus group	fibula	metatarsals	some flex foot / some extend foot
plantaris	patella and femur	calcaneum	flexes digits
soleus	proximal fibula	calcaneum	extends foot
tibialis anterior	proximal tibia and fibula	first metatarsal	dorsiflexes foot
tibialis posterior	tibia and fibula	tarsals	extends foot

abdominis, carefully remove the tough connective tissue formed by the aponeuroses of the oblique muscles identified next.

The next three muscles are most easily accessed on the ventral, lateral wall of the abdomen, although they form extensive, thin sheets spreading through the general body wall.

(2) **External oblique**

The superficial layer of muscle lateral to the rectus abdominis that forms the side of the abdomen is the external oblique. Its fibers slant downward in a dorsocranial to a ventrocaudal direction.

(3) **Internal oblique**

Separate a few fibers of the external oblique with a dissecting needle until fibers passing in another direction are encountered. This is the internal oblique, which

also connects to the ventral midline by a broad but more lateral sheet of fascia (figure 6.15).

(4) **Transversus**

Again pick away a few fibers from the internal oblique until fibers running in yet a third direction are found. This is the deepest of the three muscle layers of the abdomen, the transversus.

(5–6) **External** and **internal intercostals**

Cut an opening in the thin external oblique covering the posterior and ventral part of the rib cage (figure 6.15). This exposes the ribs and the short fibers that run between adjacent ribs. The superficial muscular layer between ribs is the **external intercostal.** The deeper layer, reached through the external intercostals and running at a different slant, is the **internal intercostal** muscle.

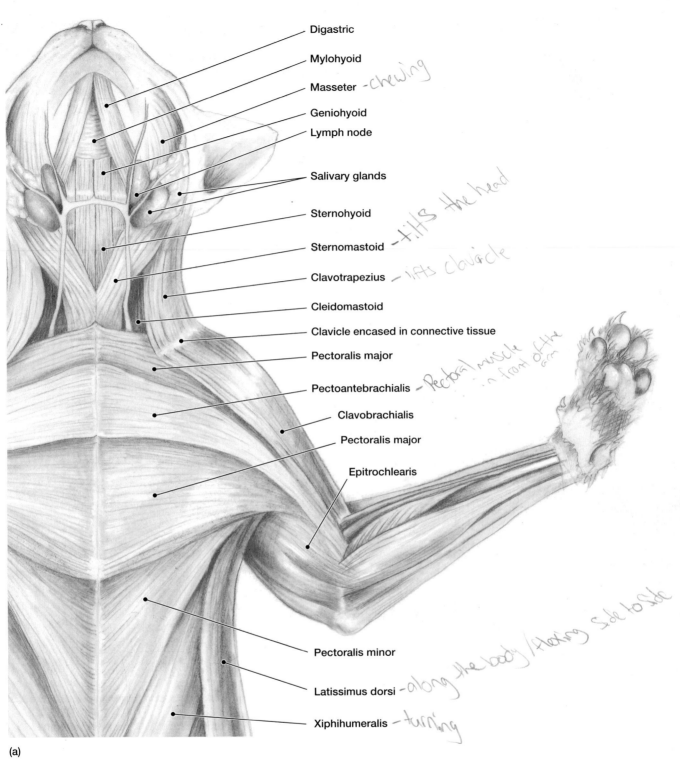

Digastric

Mylohyoid

Masseter — *chewing*

Geniohyoid

Lymph node

Salivary glands

Sternohyoid

Sternomastoid — *tilts the head*

Clavotrapezius — *lifts clavicle*

Cleidomastoid

Clavicle encased in connective tissue

Pectoralis major

Pectoantebrachialis — *Pectoral muscle in front of the arm*

Clavobrachialis

Pectoralis major

Epitrochlearis

Pectoralis minor

Latissimus dorsi — *along the body/flexing side to side*

Xiphihumeralis — *turning*

(a)

FIGURE 6.15   (a) Superficial ventral trunk, brachial, and cephalic musculature of a cat.

To confirm your identification, notice that the layer of internal intercostals persists all the way to the ventral tips of the ribs, whereas the external intercostals stop short of the rib tips.

Superficial Chest Musculature

🦖 Spread the forelimbs, tying them back if necessary, to bring the exposed muscles of the chest into better view.

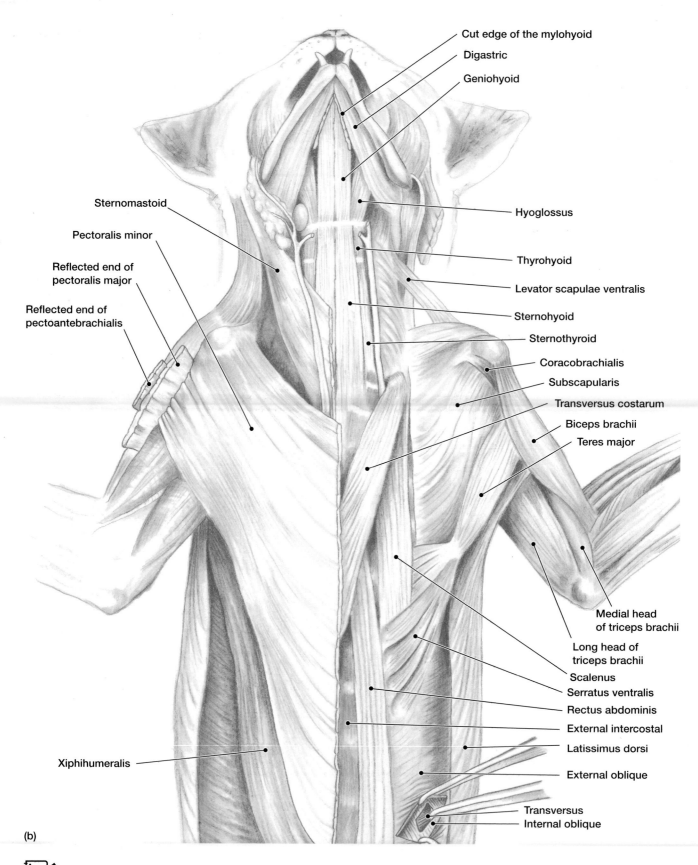

Cut edge of the mylohyoid
Digastric
Geniohyoid

Hyoglossus

Thyrohyoid

Levator scapulae ventralis

Sternohyoid

Sternothyroid

Coracobrachialis

Subscapularis

Transversus costarum

Biceps brachii

Teres major

Medial head
of triceps brachii

Long head of
triceps brachii

Scalenus

Serratus ventralis

Rectus abdominis

External intercostal

Latissimus dorsi

External oblique

Transversus
Internal oblique

Sternomastoid

Pectoralis minor

Reflected end of
pectoralis major

Reflected end of
pectoantebrachialis

Xiphihumeralis

(b)

FIGURE 6.15 *Continued.* (b) Deep ventral trunk, brachial, and cephalic musculature of a cat.

(1–2) Clavotrapezius and clavobrachialis

Clearly expose and follow the long slab of muscle that runs from the forearm, over the shoulder, into the neck, and to the head (figure 6.15). Run a blunt probe along this muscle as it passes over the shoulder and into the neck. You will discover here a hard bone within the tissue of the muscle, the free-moving **clavicle.** The portion of this muscle from clavicle to the head is the **clavotrapezius** and from clavicle to forearm, the **clavobrachialis.**

The muscle bands running from the chest out to the arm constitute the **pectoralis group** of muscles (figure 6.15a). In the cat, the pectoralis group is differentiated into four named muscles, discussed next. This differentiation is not at first easy to see (all originate upon different parts of the sternum). It is best to rely upon insertions to help in the separation and upon slightly different orientation of fibers within each muscle.

(3) Pectoantebrachialis (= pectoantibrachialis)

This muscle inserts via a tendon on the upper part of the forearm. It is also the most superficial of the four pectoralis muscles, applied to the middle surface of the pectoralis major.

(4) Pectoralis major

Slip a blunt probe under the pectoantebrachialis to separate it along its length from the underlying **pectoralis major.** Part of the broad pectoralis major also lies covered by the clavobrachialis. Follow the pectoralis major to its long insertion along the anterolateral side of the humerus.

(5) Pectoralis minor

This muscle follows next posteriorly and is in turn partially covered by the more anterior and superficial pectoralis major. Trace the pectoralis minor cranially to its insertion on the humerus near the insertion of pectoralis major.

(6) Xiphihumeralis

This is the last of the pectoralis group and one of the most delicate and difficult to separate. It originates along the xiphoid process of the sternum. It slants into the limb passing dorsally above the pectoralis minor to insert on the humerus. The posterior edge of the xiphihumeralis is often ill defined because connective tissue joins it with a quite extensive and unrelated muscle, the **latissimus dorsi.** The latissimus dorsi originates in the area of the lumbodorsal fascia and inserts on the humerus. This muscle will be addressed again when discussing the forelimb.

Deep Chest and Shoulder Musculature

🐾 Cut through the bellies of the pectoralis muscles so the ventral end of the scapula can be pulled away from the body wall—be careful not to damage the abundant and prominent blood vessels and nerves (figure 6.15).

(1) Serratus ventralis

The fan-shaped muscle on the wall of the rib cage facing the scapula is the serratus ventralis (figure 6.15b). It converges dorsally to insert on the vertebral border of the scapula.

The smaller serratus dorsalis lies in the body wall, runs only over the ribs, and does not pass to the limb. It will not be followed.

(2) Scalenus

Running across the origin of the fan-shaped serratus ventralis is the scalenus (figures 6.15b and 6.19). It is actually composed of three separate muscles that originate on separate ribs and converge as they pass forward to insert on the cervical vertebrae. From dorsal to ventral, they are the **scalenus posterior, scalenus medius,** and **scalenus anterior.** The largest is the scalenus medius, which extends more caudally to ribs 6 through 9.

(3) Transversus costarum

This single strap of muscle lies deep to the pectoralis minor (figures 6.15b and 6.19). It originates from the sternum via an aponeurosis and inserts on the first rib adjacent to the point of entrance and exit of major blood vessels and nerves from the thoracic cavity. Notice that the aponeurosis of the rectus abdominis lies beneath (deep to) the origin of transversus costarum on the sternum.

(4) Subscapularis

The subscapular fossa on the medial side of the scapula is occupied by the subscapularis muscle (figure 6.15b). It is a pinnate muscle whose fibers converge ventrally on a tendon to the lesser tuberosity of the humerus.

(5) Coracobrachialis

At about the point where the converging fibers of the subscapularis pass across the edge of the glenoid fossa on their way to the humerus, the tendon is crossed by a tiny muscle, the coracobrachialis (figure 6.15b).

(6) Teres major

The teres major occupies the ventrolateral border of the scapula. Separate it clearly from the more medial and larger subscapularis. Confirm its origin from the axillary border and insertion (with the latissimus dorsi) on the medial surface of the humerus (figure 6.15b).

Neck and Throat Musculature

🐾 To expose these muscles you may need to further free the overlying skin (figures 6.15 and 6.16a). Be sure to preserve blood vessels, nerves, and glands as best you can for later dissection.

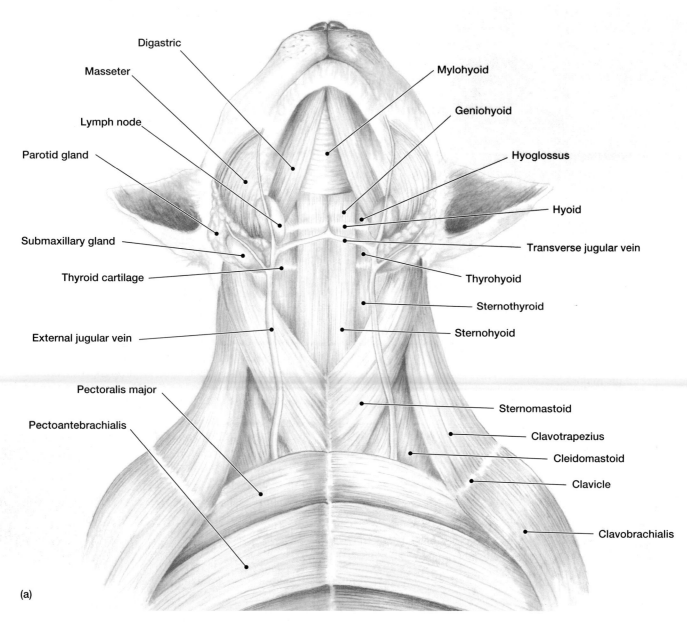

Digastric

Masseter

Lymph node

Parotid gland

Submaxillary gland

Thyroid cartilage

External jugular vein

Pectoralis major

Pectoantebrachialis

Mylohyoid

Geniohyoid

Hyoglossus

Hyoid

Transverse jugular vein

Thyrohyoid

Sternothyroid

Sternohyoid

Sternomastoid

Clavotrapezius

Cleidomastoid

Clavicle

Clavobrachialis

(a)

FIGURE 6.16 Head and neck musculature of a cat. (a) Ventral view.

(1) **Sternomastoid**

This muscle originates on the manubrium of the sternum and on the median raphe (figures 6.15 and 6.16a). It sweeps forward along the neck to insert on the lambdoidal ridge and mastoid process of the skull. Before following it to its insertion, note: (a) it is crossed superficially by the blue latex-injected external jugular vein, and (b) two salivary glands are also in contact with its superficial surface. The largest gland beneath the pinna of the ear is the **parotid gland,** and the smaller one ventral to it is the **submaxillary gland** (figures 6.16a and 6.17).

(2) **Cleidomastoid**

Lateral to the sternomastoid and beneath the clavotrapezius lies the slender cleidomastoid (figure 6.16a). As

the name suggests, the cleidomastoid originates on the clavicle (cleido = "clavicle") and inserts on the mastoid process. The clavicle was previously identified between the clavotrapezius and clavobrachialis muscles.

(3) **Sternohyoid**

This paired muscle runs between the two sternomastoids directly forward from the sternum along the ventral midline to the body of the hyoid (figure 6.16a). At about this insertion it is crossed by the transverse jugular vein. Free the sternohyoid along its length to confirm its origin and insertion.

(4) **Geniohyoid**

Parallel fibers of the geniohyoid continue on from the body of the hyoid forward to the chin, where they take

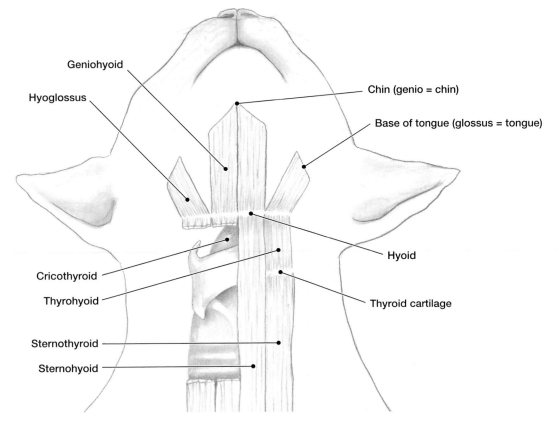

Geniohyoid

Hyoglossus

Chin (genio = chin)

Base of tongue (glossus = tongue)

Hyoid

Cricothyroid

Thyrohyoid

Thyroid cartilage

Sternothyroid

Sternohyoid

(b)

 FIGURE 6.16 *Continued.* (b) Diagram indicating the origination of the names of some neck muscles based upon their origins and insertions.

origin (figure 6.16a). A transversely crossing sheet of fibers often hides this origin. This obscuring sheet is the **mylohyoid.**

(5) **Mylohyoid**

This transverse sheet of muscle joins the two tips of the mandibular rami (figure 6.16a). A faint **median raphe** is often evident. Cut this muscle along its raphe to better expose the geniohyoid.

(6) **Digastric**

Running parallel with and along the ventral edge of the mandible is the digastric (figure 6.16a). Along one side, follow it deep to confirm its origin from the jugular and mastoid processes of the occipital bone of the skull.

(7–8) **Hyoglossus** and **thyrohyoid**

Lateral to and parallel with the geniohyoid is the broader hyoglossus (figure 6.16a). It will at first look like a continuous muscle sheet from the larynx up into the tongue, but investigation with a blunt probe will confirm the presence of the hyoid in the middle of this muscle mass. The stretch from the hyoid forward into the base of the tongue is the **hyoglossus.** The part from the hyoid posteriorly to the thyroid cartilage of the larynx is the **thyrohyoid** muscle. The thyroid cartilage is named for the thyroid gland that is directly applied to the ventral surface of this cartilage. Avoid damaging

the thyroid gland, explored again in chapter 7. Separate the hyoglossus from surrounding muscles, then grasp it with tweezers, and give a gentle pull. The tongue, into which it inserts, should move.

(9–10) **Sternothyroid** and **cricothyroid**

The expanded **larynx,** onto which the thyrohyoid inserts, begins the windpipe, or **trachea,** a ringed, flexible tube. Just posterior to the larynx, on the sides of the trachea, lies the paired **thyroid gland.**

The **sternothyroid** is a slender muscle running just lateral to the sternohyoid from the sternum to the base of the larynx. The **cricothyroid** is a small patch of muscle on the larynx itself near the insertion of the sternothyroid.

As you work through the neck muscles, remember that their names are often taken from the structures they attach to—origin, then insertion (table 6.4 and figure 6.16b).

Lateral Jaw Musculature

(1) **Masseter**

On one side, pull the skin away from the head. Cut a circle around the base of the **pinna** (= external ear) if necessary to free the skin. Be careful of blood vessels, nerves, and glands.

The large, domed masseter muscle lies below the pinna in the cheek region (figure 6.16a).

(2) **Temporalis**
The temporalis muscle primarily lies above the base of the pinna and is covered by a shiny tendon (figure 6.17).

Forelimb Musculature

Dorsal/Lateral Shoulder—Superficial Musculature
Lay your cat on its side with the exposed muscles of the left side facing up toward you. Refer to figures 6.17 and 6.18a as you identify these superficial muscles.

(1) **Latissimus dorsi**
The broad, exposed curtain of muscle covering the side of the chest and abdomen is the latissimus dorsi (figure 6.17). It originates from the **lumbodorsal fascia** in the lumbar region plus adjacent neural spines and passes forward converging upon the axilla. It inserts on the medial surface of the humerus.

(2) **Clavotrapezius**
Refamiliarize yourself with this muscle identified earlier, but here seen from a different position (figure 6.17). The clavotrapezius is the first of three

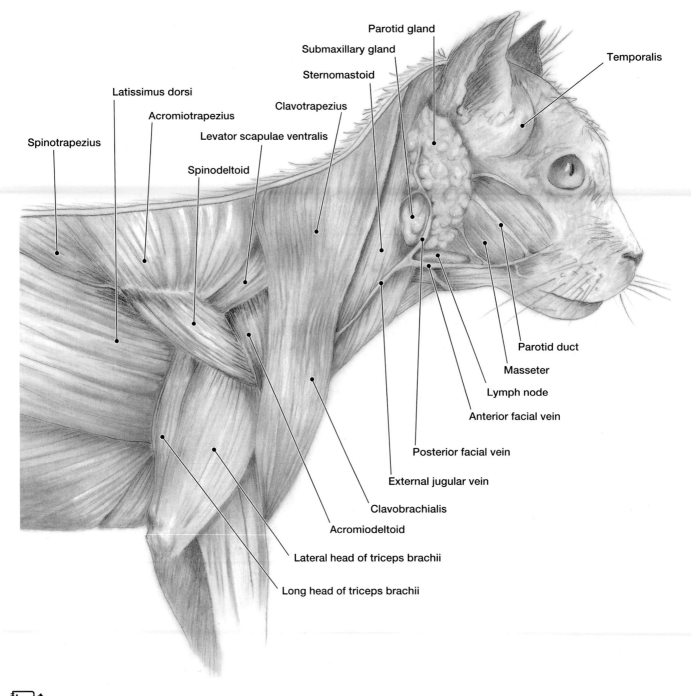

FIGURE 6.17   Superficial neck and shoulder musculature of a cat, lateral view.

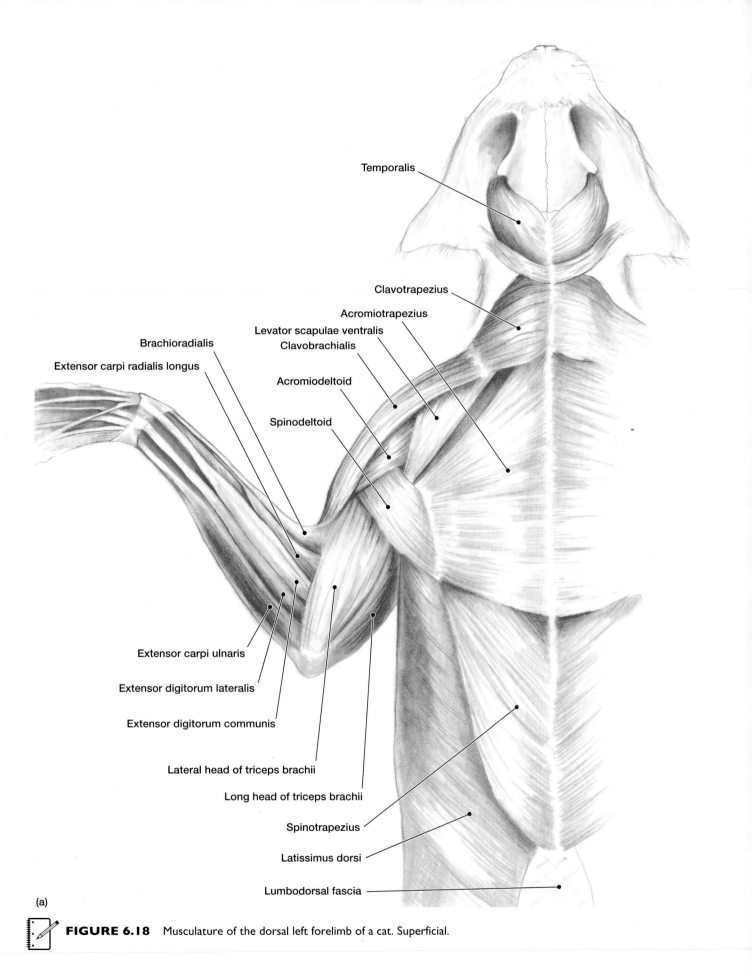

Temporalis

Clavotrapezius

Acromiotrapezius

Levator scapulae ventralis

Clavobrachialis

Brachioradialis

Extensor carpi radialis longus

Acromiodeltoid

Spinodeltoid

Extensor carpi ulnaris

Extensor digitorum lateralis

Extensor digitorum communis

Lateral head of triceps brachii

Long head of triceps brachii

Spinotrapezius

Latissimus dorsi

Lumbodorsal fascia

(a)

FIGURE 6.18 Musculature of the dorsal left forelimb of a cat. Superficial.

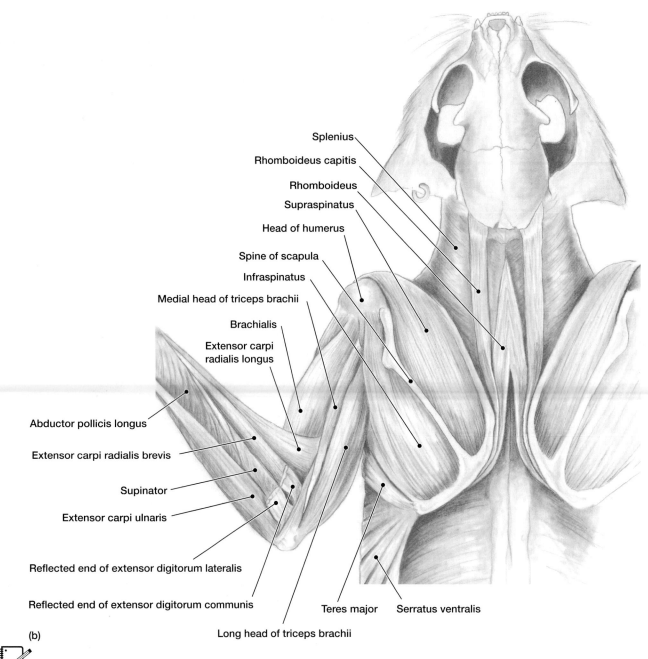

Splenius

Rhomboideus capitis

Rhomboideus

Supraspinatus

Head of humerus

Spine of scapula

Infraspinatus

Medial head of triceps brachii

Brachialis

Extensor carpi radialis longus

Abductor pollicis longus

Extensor carpi radialis brevis

Supinator

Extensor carpi ulnaris

Reflected end of extensor digitorum lateralis

Reflected end of extensor digitorum communis

Teres major   Serratus ventralis

Long head of triceps brachii

(b)

FIGURE 6.18   *Continued.* Musculature of the dorsal left forelimb of a cat. Deep.

muscles of the trapezius group. Note its broad origin from the superior nuchal line and median dorsal line of the neck.

(3) **Acromiotrapezius**

The origin of this muscle lies immediately caudal to the clavotrapezius along neural spines of the cervical and first thoracic vertebrae (figure 6.17). Its fibers converge to insert on the metacromion process and spine of the scapula.

(4) **Spinotrapezius**

The spinotrapezius is the smallest and most posterior of the three trapezius muscles (figure 6.17). It lies on the posterior edge of the acromiotrapezius. Arising from neural spines of the thoracic vertebrae, it passes

beneath the acromiotrapezius, and inserts on the spine of the scapula and adjacent fascia.

(5) **Levator scapulae ventralis**

This is a small wedge of muscle that lies between the clavotrapezius and acromiotrapezius (figures 6.15b, 6.17, and 6.18a). Confirm that the levator scapulae ventralis originates from the transverse process of the atlas and occipital bone. It inserts on the metacromion process and adjacent fascia.

In the cat, two[1] members of the **deltoid group** are present.

---

[1]Some consider the clavobrachialis to be a third deltoid present in the cat and so prefer the name *clavodeltoid* for this muscle.

The ventral musculature of the chest and throat of these three vertebrates reveals a common pattern. Notice, for example, that each of these animals has a set of muscles that extends along the midline from the throat to the chin (coracomandibularis and coracohyoideus in the shark and the geniohyoid muscles of *Necturus* and the cat). In addition, each animal has another set of muscles running perpendicularly across the throat (intermandibularis and interhyoideus in the shark and *Necturus* and the mylohyoid in a cat). Although these three animals are quite different, they share morphological patterns based upon a common ancestral design. Thus, these muscles are homologous. Table 6.5 includes these and other homologous muscles of the cranial musculature.

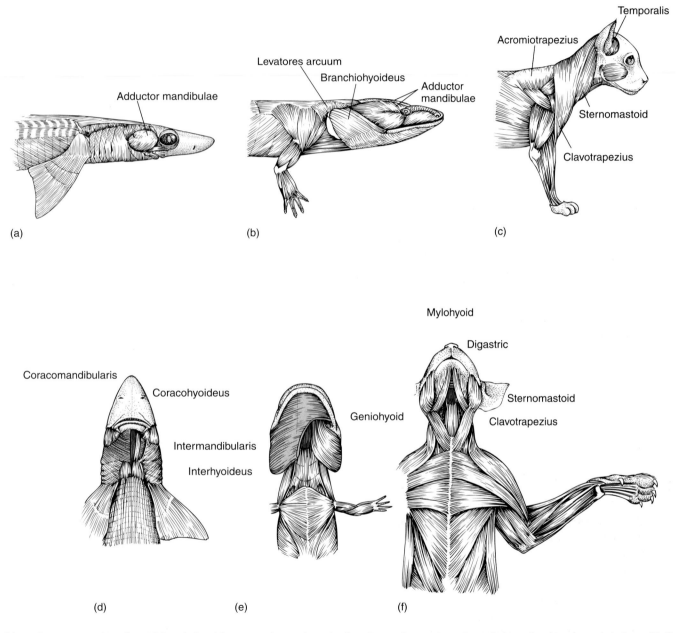

Homologous muscles of cranial and shoulder musculature in a shark, salamander, and cat. lateral views (a–c) and ventral views (d–f).

TABLE 6.5 Homologies of Cranial Musculature

Arch	Cranial Nerve Supply	Shark	*Necturus*	Cat or Mink
Branchiomeric Musculature				
I	V	levator palatoquadrati spiracularis adductor mandibulae preorbitalis	adductor mandibulae (levator mandibulae)	masseter temporalis pterygoids tensor veli palati tensor tympani
		intermandibularis	intermandibularis	mylohyoid anterior digastric
2	VII	levator hyomandibulae	depressor mandibulae branchiohyoideus	stapedius platysma and facial muscles (part)
		interhyoideus	interhyoideus	platysma and facial muscles (part) posterior digastric stylohyoid
3–7	IX, X XI* XI*	cucullaris interarcuals	cucullaris levatores arcuum —	trapezius complex sternocleidomastoid complex —
		superficial constrictors and interbranchials	dilatator laryngis subarcuals transversi ventrales depressors arcuum	some intrinsic muscles of the larynx and pharynx
Hypobranchial Musculature				
	Hypobranchial XII*	coracoarcuals coracohyoid	rectus cervicis	sternohyoid omohyoid thyrohyoid
		coracomandibularis	genioglossus geniohyoid	geniohyoid, others of the tongue and larynx
		coracobrachialis	—	—

*In tetrapods

(6) **Acromiodeltoid**

The short, acromiodeltoid lies just ventral to the insertion of the levator scapulae ventralis and dorsal to the clavotrapezius (figure 6.17). Confirm that it arises on the acromion and inserts on the humerus.

(7) **Spinodeltoid**

This muscle lies just posterior to the acromiodeltoid, which partially obscures its insertion on the deltoid ridge of the humerus (figure 6.17). It originates on the spine of the scapula next to the insertion of the acromiotrapezius.

🐾 **Dorsal/Lateral Shoulder—Deep Musculature** Turn your cat onto its belly so the exposed muscles of the upper part of the shoulder face you. Reflect the acromiotrapezius across its middle (figure 6.18b).

(1–2) **Infraspinatus** and **supraspinatus**

The **infraspinatus** and **supraspinatus** occupy their respective fossae on the lateral side of the scapula, namely the infraspinous fossa and supraspinous fossa

(figure 6.18b). You may need to spread the spinotrapezius and latissimus dorsi to expose the infraspinatus.

(3) **Teres minor**

Reflect the spinodeltoid to reveal the deep muscles below. Fibers of the infraspinatus converge and pass to their insertion on the humerus. Between the infraspinatus and two heads of the triceps lies the teres minor. This is a small muscle usually clinging to the edge of the infraspinatus. Separate it and locate its approximate origin and insertion.

(4) **Rhomboideus**

Slightly pull the scapula away from the body wall. The extensive, fan-shaped muscle emanating from an insertion on the vertebral border of the scapula and taking origin extensively along the neural spines of the neck is collectively the **rhomboideus** (figures 6.8b and 6.19). The first (anterior) strap of this muscle is somewhat separate from the rest and takes origin from the superior nuchal line. This single strap of muscle is termed the **rhomboideus capitis** (figures 6.18b and 6.19).

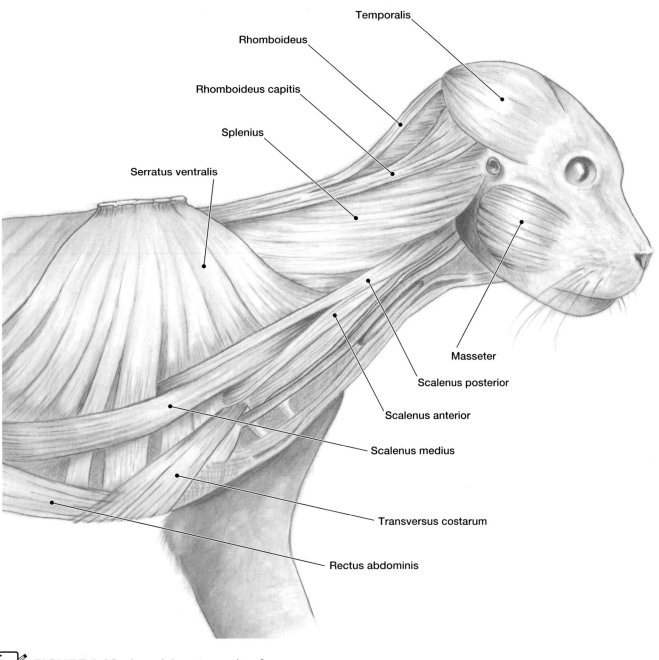

Temporalis

Rhomboideus

Rhomboideus capitis

Splenius

Serratus ventralis

Masseter

Scalenus posterior

Scalenus anterior

Scalenus medius

Transversus costarum

Rectus abdominis

FIGURE 6.19   Lateral thoracic muscles of a cat.

(5) **Splenius**
The **splenius** forms a broad sheet deep to
the clavotrapezius and rhomboideus muscles
(figures 6.18b and 6.19). It extends from the middorsal fascia of the neck to the nuchal line of the skull.

Upper Arm (Brachium) Musculature   Return the
cat again to its side with exposed muscles of the forelimb facing you.

(1) **Triceps brachii**
There are three heads that occupy the posterior side of
the upper arm and collectively insert on the olecranon
(figure 6.18). The **lateral** head of the triceps lies along

the lateral side of the upper arm. The equally prominent **long** head of the triceps occupies the posterior side
of the upper arm. Spread the long and lateral heads to
find the smaller **medial** head of the triceps deep to the
prominent radial nerve and lying against the humerus.

(2) **Brachialis**
The brachialis is parallel with, but anterior to, the
lateral head of the triceps along the lateral side of the
humerus (figure 6.18b). Its origin on the humerus lies
hidden beneath the acromiodeltoid. Follow the brachialis, under the lateral head of the triceps, to its insertion
on the forearm next to the insertion of the biceps brachii described in number (4).

(3) Epitrochlearis

Again lay your cat on its back to examine the medial surface of the arm. The medial face of the long head of the triceps is covered by a broad, flat muscle, the epitrochlearis (figure 6.15a and 6.21a). It originates from the latissimus dorsi and triceps. Insertion is by an aponeurosis to the olecranon. Cut through the belly of this muscle to discover the shiny medial surface of the long head of the triceps.

(4) Biceps brachii

Parallel with the long head of the triceps and the epitrochlearis, but along the anterior margin of the humerus, is the biceps brachii (figure 6.15b). Follow this muscle to its insertion on the forearm via a narrow, but strong tendon.

🦖 **Forearm (Antebrachium) Musculature—Dorsal Side**

Ten muscles form the dorsal surface of a cat's forearm

(figures 6.18 and 6.20). Remove skin on the dorsal and ventral surfaces of the foot (= manus) leaving the tendons intact. Near the wrist, a connective tissue band, a **retinaculum** (pl., **retinacula**) encircles tendons passing over the wrist. Forearm muscles will first be identified by their relative positions on the forearm, but it is essential that you clear and extend your dissection out onto the digits of the foot to allow you to verify the insertion of each muscle's tendons. This is the only way to confirm your initial identification and to recognize basic action.

Most of the muscles on the dorsal side extend elements of the foot or wrist; most ventral muscles flex these elements. After following a tendon to its insertion, grab the tendon firmly and give it a short tug to simulate the muscle's function (table 6.4).

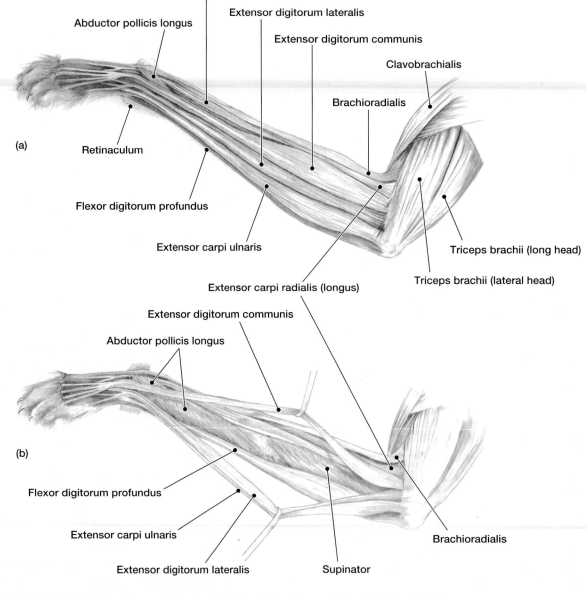

📝 **FIGURE 6.20** Musculature of the dorsal left forelimb of a cat. (a) Superficial and (b) deep.

(1) **Brachioradialis**

This slight muscle is often removed inadvertently or damaged during removal of the skin. It emerges along the anterior margin of the lateral head of the triceps brachii, then turns and travels down the medial surface of the upper arm to insert on the distal end of the radius (figures 6.18a and 6.20a). When dissected free of other muscles, it hangs loosely away from the forearm.

(2) **Extensor carpi radialis (longus)**

This muscle originates near the distal end of the humerus and also seems to emerge from beneath the lateral head of the triceps brachii. It is the most medial member of the tightly grouped forearm muscles (figures 6.18 and 6.20). ("Medial" means toward the side of the first digit.) Its single and thin tendon forms near the distal third of the radius and distinguishes it from adjacent muscles.

(3) **Extensor digitorum communis**

Next, moving laterally near the elbow joint, is the extensor digitorum communis (figures 6.18 and 6.20). Near the distal third of the radius it sends four long, tightly packed, and thin tendons under the retinaculum. The tendons then diverge to separately attach to digits II–V. Following these four tendons into the digits will confirm identification of this important muscle.

(4) **Extensor carpi radialis brevis**

Separate the "longus" and "communis" muscles just identified. Between them lies the deeper extensor carpi radialis brevis (figures 6.18b and 6.20a), a thicker muscle which, as you separate it from the "longus," will seem to tear away. This muscle can be confirmed by the fact that upon approach to the wrist it joins the thickened tendon of the "longus."

(5) **Extensor digitorum lateralis**

This muscle is about the same size as the "communis" and parallels it along its lateral side (figures 6.18a and 6.20a). The "lateralis" also produces three very thin tendons that run collectively to the same digits, but under the tendons of the "communis."

(6–7) **Abductor pollicis longus** and **supinator**

Deep between the "communis" and "lateralis" (the two muscles with three and four tendons each), are two more muscles (figures 6.18b and 6.20b) both tightly applied to the ulna and radius. The **abductor pollicis longus** originates as a broad muscle wrapped tightly around the distal half of the ulna and radius. It lies beneath forearm tendons, which should be parted to find it. From this origin, its fibers narrow to a tendon passing to the medial side of the foot. It acts on the thumb in cats, as it does in humans, but in the cat may also insert on the wrist. Just proximal to the "longus," in the same position between the "communis" and "lateralis," is the **supinator.** Between these two muscles is the partially exposed solid bone (ulna). Tap it with your blunt probe to confirm the boundary.

(8) **Extensor carpi ulnaris**

Just lateral to the proximal end of the "lateralis" is the extensor carpi ulnaris, approximately the same

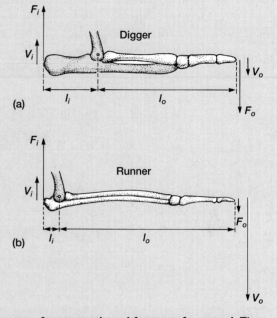
diameter as the "lateralis" but ending in a single, broad, tendon that inserts upon the base of the fifth metacarpal bone (figures 6.18 and 6.20a). Just lateral to the "ulnaris," forming the most lateral margin of the ventral forearm, is the ulnar head of the flexor digitorum profundus, described more extensively in the section on the dorsal flexor muscles of the forearm.

(9–10) **Extensor pollicis** and **extensor digiti secundi**

Directly below the "ulnaris" is a pair of muscles that are collectively smaller in diameter than any of the muscles previously examined. The more medial **extensor pollicis** (extensor digiti I) has a longer and larger tendon that passes to the first digit. The tendon of the very thin and more lateral **extensor digiti secundi** (extensor digiti II) merges with the tendon of the "longus." Neither muscle is shown in a figure.

🦖 **Forearm (Antebrachium) Musculature—Ventral Side**

A considerable sheet of superficial fascia overlies much of the ventral surface and must be reflected carefully to spare the musculature below.

High and Low Gear Muscles. Compare the structures and functions of the medial gluteus (gluteus medius) and semimembranosus muscles isolated in this figure of the hindlimb of a deer. Certainly during vigorous limb oscillations both muscles contribute to running by moving the hindlimb in the same direction. However, each muscle possesses a different mechanical advantage. Like the gears of an automobile shifting from low to high, the semimembranosus moves the limb with greater output force, most useful during acceleration. The mechanical advantage of the medial gluteus favors speed, most useful in sustaining the high velocity of the limb. Because the lever arm in (l_i) must vary to produce differences in gearing, a single muscle cannot have short and long input lever arms acting on the limb with equal force and speed advantage. Two or more muscles are required. This is why during your dissections you may find two muscles with similar actions on the same structure, each with a different mechanical advantage. (See also the Form & Function box on parallel and pinnate muscles on page 128.)

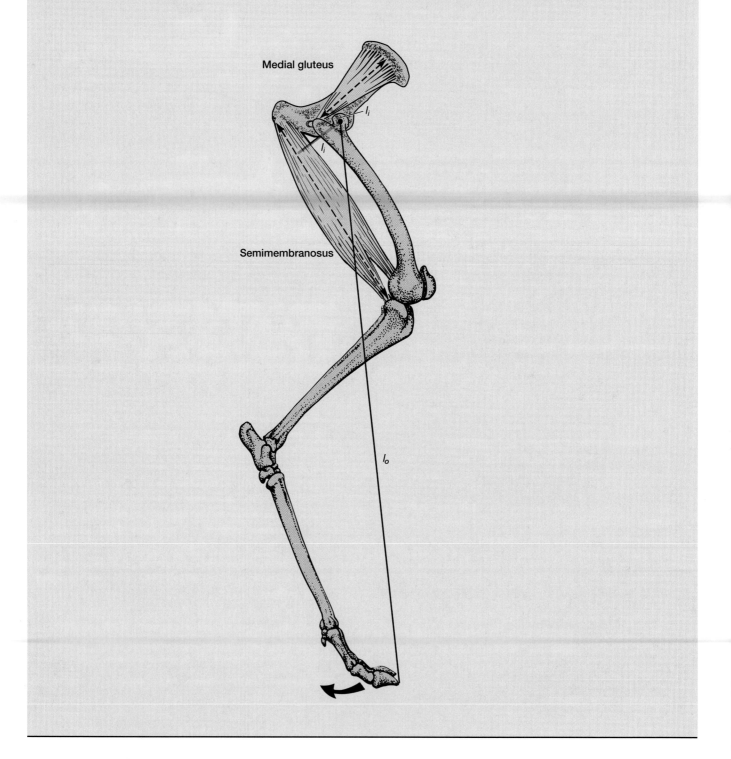

Medial gluteus

l_i

l_i

Semimembranosus

l_o

(1) **Flexor digitorum superficialis (long head)**
This is the most prominent superficial muscle, forming a broad, thin sheet along the medial portion of the forearm (figure 6.21a). Its distal tendon travels under the retinaculum and subdivides into four tendons that attach to digits II–V. A deeper, second muscle described below, the flexor digitorum profundus, also sends tendons directly into the digits.

(2) **Flexor carpi ulnaris**
This muscle, with two heads, is positioned lateral to the "superficialis" previously identified, which partially covers it (figure 6.21a). The smaller, more lateral head overlaps a deeper, larger head. Both heads run together soon forming a strong tendon inserting on the pisiform bone in the wrist.

(3) **Flexor carpi radialis**
This muscle is found on the opposite (medial) side of the "superficialis" (figure 6.21a). Sometimes a head of the flexor digitorum profundus (not yet described) extends between the flexor carpi radialis and the flexor

digitorum superficialis. The long slender tendon of the spindle-shaped flexor carpi radialis passes under the retinaculum and attaches to the wrist. To further confirm the flexor carpi radialis, identify the pronator teres, described next.

(4) **Pronator teres**
This wedge-shaped muscle primarily located along the proximal half of the forearm resides on the medial border of the flexor carpi radialis (figure 6.21a). The muscle fibers of the pronator teres initially run diagonally from their origin on the humerus.

(5) **Flexor digitorum profundus**
Pull the "superficialis" and "radialis" apart from the "ulnaris" (figure 6.21b). (Alternately, you may cut and reflect these muscles along their midlengths.) Most of what you see are the five heads of the flexor digitorum profundus: one ulnar head, one radial head, and the three humeral heads arising from the medial epicondyle of the humerus. The tendons of these five heads converge and collectively run below the retinaculum

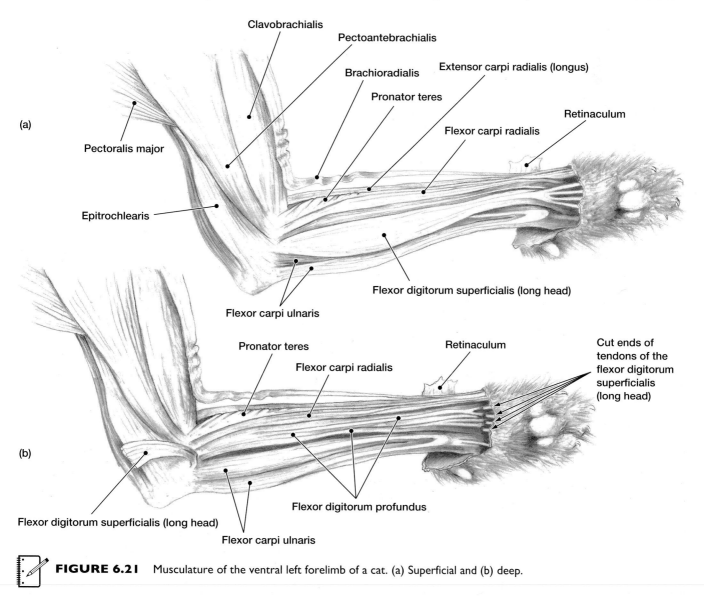

FIGURE 6.21   Musculature of the ventral left forelimb of a cat. (a) Superficial and (b) deep.

to insert on the five separate digits. The "profundus" should be recognized as a whole with numerous heads contributing to tendons into the digits. For our purposes, it is not necessary to distinguish the separate heads. However, you may find it helpful in delineating the full extent of the "profundus" to identify several heads. The most medial of the humeral heads was noted previously, as it may sometimes appear at the surface between the "superficialis" and the "radialis." The ulnar head is the largest head located along the lateral margin of the forearm. It was noted previously lateral to the extensor carpi ulnaris when the extensors were examined (figure 6.20a). The radial head of the flexor digitorum profundus is tightly applied to the distal third of the ulna.

(6) **Lumbricales**
These small muscles are found only on the hand and will not be identified separately.

Hindlimb Musculature

If you have not already done so, extend the midventral incision through the cat's skin posteriorly to the anus. Then, beginning at this incision, continue the cut outward along the inner (medial) side of one hindlimb. Note that one of the limbs was used as the site of latex injection into the blood vessels. You may wish to select the opposite for this dissection of the hindlimb musculature, or perhaps perform a superficial dissection on one side and a deep dissection on the other. Be careful now, as throughout the dissection, not to destroy any of the major blood vessels or nerves coursing through the limb. In addition, use extra caution along the ventral midline to preserve the spermatic cords in male cats.

Once the proper incisions have been made, use your fingers or a blunt probe to loosen the skin and gradually work it loose from the underlying muscles all the way to the tip of the foot. Free the skin completely from the limb, cutting it near the claws if necessary, but leave the flap connected at the dorsal midline on the back so that it might be replaced over the limb when you finish each day's dissection.

With the skinning of the limb completed, place the cat on its belly in the tray before you. The first part of the dissection will begin on the lateral, outer surface of the thigh. Consult figure 6.22 for general orientation, then return to this point in the description.

Muscles of the Lateral Thigh The thigh is covered by a superficial fascia, a connective tissue that covers the limb muscles. Use your fingers to carefully remove this fascia, being certain that any clinging muscle fibers are not removed along with it. Large and very tough connective tissue sheets, such as the large **lumbodorsal fascia** along the dorsal midline and the **fascia lata** on the lateral sides of the legs should not be removed.

(1) **Sartorius**
This thin strap of a muscle is found along the anterior margin of the thigh (figure 6.22). Most of the sartorius is actually found on the medial side, away from you as you now view it. However, it will make a useful landmark during the dissection. The next four muscles are to be found in this region.

(2) **Tensor fasciae latae**
This muscle is just posterior and deep to the sartorius. It is a triangular-shaped muscle encased within a tough, tendinous sheet, the **fascia lata,** which spreads ventrally across the limb toward the knee (figure 6.22). The tensor fasciae latae and fascia lata wrap the anterior margin of the thigh. They will be identified again when the medial side of the thigh is examined. Just dorsal to the tensor fasciae latae is a tough connective sheet that extends over the next muscle and inserts on the ilium. Make an incision in the middle of this dorsal fascia parallel to the axis of the body and reflect the fascia back.

(3) **Gluteus medius**
This muscle is to be found as the large, triangular muscle at the most anterior and dorsal part of the thigh just exposed by the reflection of the dorsal fascia (figure 6.22). In cats, the gluteus medius is larger than the gluteus maximus, which follows caudally.

(4) **Gluteus maximus**
This smaller muscle is found immediately caudal to the gluteus medius along the dorsal midline (figure 6.22).

(5) **Caudofemoralis**
This is a long, thin muscle, about the diameter of the gluteus maximus. It is found between the gluteus maximus and biceps femoris, found next (figure 6.22).

(6–7) **Biceps femoris** and **tenuissimus**
The biceps femoris is the largest of the muscles so far described and forms the posterior and lateral muscle mass of the thigh (figure 6.22). By pulling the biceps femoris laterally, you can find the sciatic nerve (and its branches) that pass deep to the biceps femoris. This is a good opportunity to locate the **tenuissimus.** It is a thin muscle, about the size of the nerve, but can be distinguished from the nerve by its browner color and that it shows no distal branching. Once you find

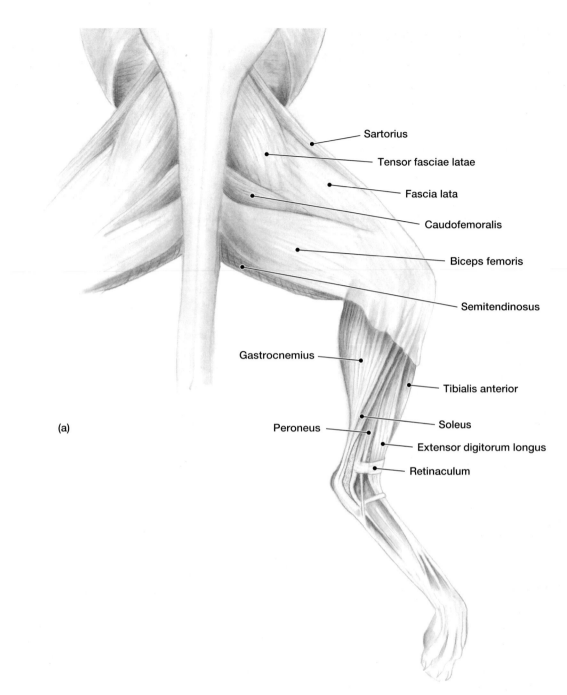

Sartorius

Tensor fasciae latae

Fascia lata

Caudofemoralis

Biceps femoris

Semitendinosus

Gastrocnemius

Tibialis anterior

Peroneus

Soleus

Extensor digitorum longus

Retinaculum

(a)

FIGURE 6.22 (a) Lateral musculature of cat hindlimb. Superficial.

these underlying structures, you may wish to reflect the biceps femoris to better appreciate their shape and position.

(8) **Vastus lateralis**

Lift up the thin tensor fasciae latae near its junction with the fascia lata to reveal the underlying vastus lateralis (figure 6.22).

Muscles of the Medial Thigh

(1–2) **Gracilis** and **sartorius**

Turn the cat over onto its back. The inner (medial) side of the thigh can now be seen and any fascia should be

cleared away. Take care not to destroy the major nerves and blood vessels that run along the middle of the thigh, as they are useful landmarks. Two thin straps of muscles lie on either side of these blood vessels and nerves: the **gracilis,** posterior, and the **sartorius,** anterior (figure 6.23). Reflect the sartorius at its middle to better expose the muscles beneath it. Note the posterior margins of the tensor fasciae latae and fascia lata that wrap the anterior margin of the thigh just anterior to the sartorius.

(3–4) **Rectus femoris** and **vastus medialis**

Two muscles can now be seen deep to the sartorius and anterior to the blood vessels and nerves. The

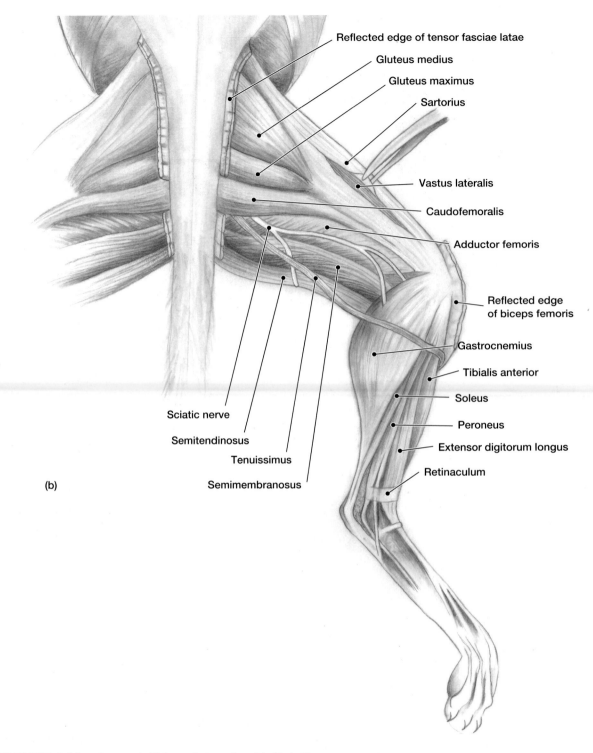

Reflected edge of tensor fasciae latae

Gluteus medius

Gluteus maximus

Sartorius

Vastus lateralis

Caudofemoralis

Adductor femoris

Reflected edge
of biceps femoris

Gastrocnemius

Tibialis anterior

Soleus

Peroneus

Extensor digitorum longus

Retinaculum

Sciatic nerve

Semitendinosus

Tenuissimus

Semimembranosus

(b)

FIGURE 6.22 *Continued.* (b) Lateral view of cat hindlimb. Deep.

rectus femoris lies near the ventral midline and is overlapped along its most lateral half by the more caudal **vastus medialis** (figure 6.23).

(5) **Vastus lateralis**
Just anterior to the rectus femoris, and under the medial portions of the tensor fasciae latae, is the medial margin of the vastus lateralis (identified previously on the lateral side of the thigh).

(6) **Vastus intermedius**
By widely spreading the rectus femoris from the vastus medialis, this deep muscle wrapped partly around the shaft of the femur can be seen.

The following six muscles are described in order as they extend from the medial to the most posterior margins of the thigh. To best expose them, reflect the middles of the gracilis and sartorius.

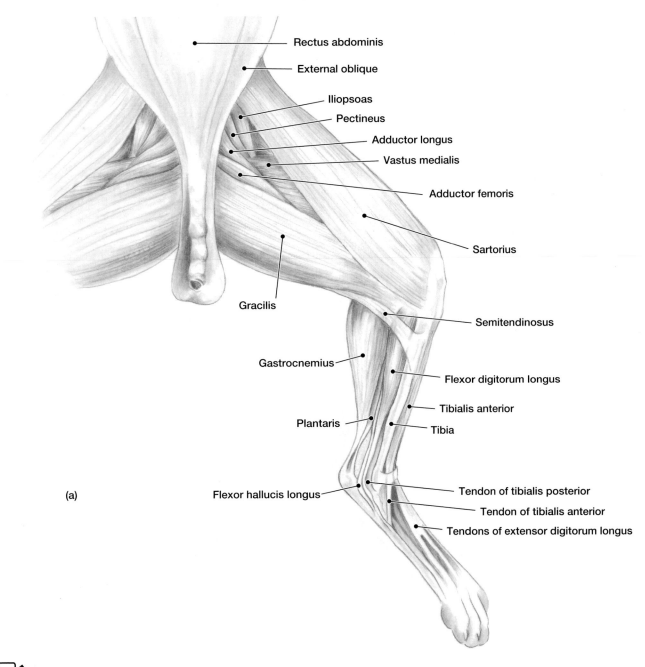

Rectus abdominis

External oblique

Iliopsoas

Pectineus

Adductor longus

Vastus medialis

Adductor femoris

Sartorius

Gracilis

Semitendinosus

Gastrocnemius

Flexor digitorum longus

Tibialis anterior

Plantaris

Tibia

(a)

Flexor hallucis longus

Tendon of tibialis posterior

Tendon of tibialis anterior

Tendons of extensor digitorum longus

FIGURE 6.23 (a) Medial musculature of cat hindlimb. Superficial.

Patterns & Connections

You have just identified the rectus femoris and three "vastus" muscles. Collectively, these form the quadriceps muscles, commonly referred to as "the quads" in athletes (unfortunately, all too often in regard to an injury). All four insert onto the patella (and adjacent ligaments) and act to extend the shank. Thus, they are important in powerful lifting (pumping) of the leg when running. When an athlete "pulls a quad," one or more of these muscles or their tendons has been ripped. The injury may take weeks for a full recovery. Properly warming up and stretching reduces the likelihood of such an injury.

(7) **Iliopsoas**
Running parallel to the axis of the body, the iliopsoas crosses the origins of the rectus femoris and vastus medialis.

(8) **Pectineus**
About the same size, the pectineus lies just medial and caudal to the iliopsoas.

(9) **Adductor longus**
Just caudal to the large femoral artery and vein lies the adductor longus, angled parallel to the axis of the femur. Just like the much larger adductor femoris that follows it, this slender muscle originates from the pubis, inserts on the femur, and adducts the thigh.

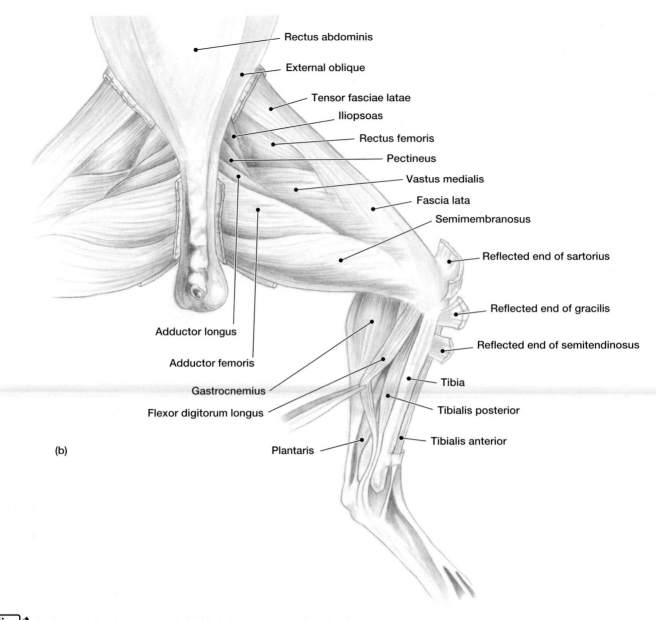

Labels in figure (clockwise from top):
Rectus abdominis
External oblique
Tensor fasciae latae
Iliopsoas
Rectus femoris
Pectineus
Vastus medialis
Fascia lata
Semimembranosus
Reflected end of sartorius
Reflected end of gracilis
Reflected end of semitendinosus
Tibia
Tibialis posterior
Tibialis anterior
Plantaris
Flexor digitorum longus
Gastrocnemius
Adductor femoris
Adductor longus

(b)

FIGURE 6.23 *Continued.* (b) Medial musculature of cat hindlimb. Deep.

Patterns & Connections

Athletes often refer to the hamstring muscles, or "hams," the collective term for posterior thigh muscles that include the biceps femoris, semimembranosus, and semitendinosus. They originate on the ischium and collectively insert on the knee (patella, femur, and tibia, respectively). The name "hamstring" comes from the terminology of butchers who used this strong tendon to hang hams during the smoking process.

(10) **Adductor femoris**

This much larger muscle lies along the caudal margin of the adductor longus.

(11) **Semimembranosus**

The semimembranosus is the largest of the medial thigh muscles. Its insertion is covered by the gracilis.

(12) **Semitendinosus**

This most posterior of the thigh muscles usually clings tightly to the semimembranosus. The semitendinosus inserts on the tibia, below the knee, where its tendon may be located and followed back to its belly.

🦖 **Muscles of the Shank** Some of the best aids in identification of shank muscles are the sites of insertion of tendons. Thus, clean away fascia from the shank so that the cordlike tendons to the heel, ankle, and digits can be seen. The insertions of the gracilis and biceps femoris should not be removed, but it will be necessary to pull them back as far as possible to reveal underlying muscles. Do this as needed as the dissection progresses.

(1–2) **Tibialis anterior** and **extensor digitorum longus**

Find the tibia. Its anterior *medial* side will be free of muscle, but on its anterior *lateral* side is a slab of muscle closely applied to and taking origin from it, the **tibialis anterior** (figure 6.23). It is very likely that you, at first, have also located the extensor digitorum longus. To be certain, check the sites of insertion of the long tendons emanating from each. The tibialis anterior inserts by a single tendon that passes to the medial side of the foot. The **extensor digitorum longus** forms a stout tendon that if followed into the foot divides into four tendons that pass along the top of the foot; each inserts on a digit (figure 6.22). (*Note:* Often a ligamentous band is preserved in the dissection that passes transversely across the tendons of both muscles near the end of the tibia. This is another retinaculum, and serves to prevent the tendons from bowing away from the ankle during foot flexion [figure 6.23].)

(3) **Peroneus muscles**

Return to the lateral side of the shank (figure 6.22). These muscles are partly covered by the extensor digitorum longus just identified. Although it is impractical to individually separate each of the peroneus muscles, their presence is suggested by the several slender tendons that pass over the lateral side of the ankle to insert on the bottom of the foot.

(4) **Gastrocnemius**

This is the large muscle forming the posterior muscle mass of the shank (figure 6.22). It constitutes most of the so-called calf. By a strong tendon (Achilles) it inserts on the calcaneum. It is more-or-less divisible into two large portions, lateral and medial.

(5) **Soleus**

If orientation is changed, return again to the lateral side of the shank (figure 6.22). This muscle runs internal to the lateral gastrocnemius and is in contact with the peroneus muscles ventrally. The soleus tapers to a tendon that joins the tendon of the gastrocnemius.

Turn the cat onto its back to next examine muscles of the medial side of the shank. Clean away surface fascia to reveal slender muscles and tendons between the gastrocnemius and tibia.

(6) **Plantaris**

On the medial side of the shank, locate the medial portion of the gastrocnemius (figure 6.23). Cut the middle of the semitendinosus, fold it back, and clean away fascia. Pull the medial portion of the gastrocnemius away from the tibia, peer inside the opened space, and note that the plantaris is enclosed. Its identification may be confirmed by lifting it from the gastrocnemius and noting its shiny aponeurosis. Confirm its partial fusion with the lateral portion of the gastrocnemius. It joins the tendon of the gastrocnemius, further confirmation of the plantaris.

(7–8) **Flexor digitorum longus** and **flexor hallucis longus**

Find the tibialis anterior (located previously); next to this is the tibia; next comes the **flexor digitorum longus,** and finally the fleshy **flexor hallucis longus,** tightly fused to the exposed medial surface of the flexor digitorum longus muscle (figure 6.23). Follow the flexor hallucis longus distally to the heel where it becomes a tendon passing to the foot. Pull on this stout tendon to confirm its action on the foot.

(9) **Tibialis posterior**

This muscle lies between the flexor digitorum longus and the tibia (figure 6.23). To find it, use a blunt probe to split the seemingly single tendon of the flexor digitorum longus. A long tendon is separated that can be found to pass upward (proximally) between the flexor digitorum longus and the tibia to the belly of tibialis posterior.

Form & Function

Compare the orientation of the muscle fibers in the gastrocnemius and sternohyoid muscles. Which muscle has a parallel and which has a pinnate fiber arrangement? Notice how the converging fibers of the gastrocnemius insert onto the calcaneum. This strong muscle is used to extend the foot and lift the body. The sternohyoid fibers are not arranged for strength. Instead, they contract to hold the hyoid in position or to move the hyoid caudally. From the study of these two muscles, we would predict that the forces needed to move the hyoid are much less than the forces needed to extend the foot.

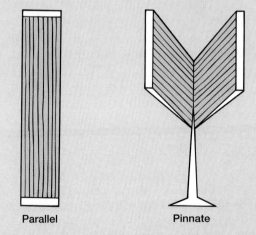

Parallel   Pinnate

Parallel and pinnate muscles. The pinnate orientation permits the packing of more fibers in the same volume than the parallel arrangement does. Thus, pinnate muscles are generally better suited for moving heavy loads short distances.

Most of the active muscle mass (the muscle fibers) that moves a giraffe's legs is located proximally, while tendons running out along the leg apply the force distally, at the hoof. By moving most of the muscle mass closer to the body, overall limb inertia is reduced. In cursorial animals, this allows the legs to move faster because the limbs have overall less mass (inertia) to accelerate/ decelerate back and forth during limb oscillations. The muscles of a cat's legs show this same feature. Examine the overall shape of the flexor digitorum longus. Notice that the greatest bulk of the muscle is positioned proximally, just below the knee. The long tendon transmits the contractile force distally to insert on the ventral side of the digits.

Limb tendons of a giraffe. Tendons distribute the forces of muscle contractions to sites distant from the muscle itself. The limb muscles of a giraffe are located close to the body, but tendons of these muscles extend outward along the leg bones and deliver their forces at the giraffe's hooves.

7 Digestive Systems

Introduction

The business of turning a meal into usable energy begins with the digestive system. Food is broken down mechanically by chewing, churning, or grinding, and chemically by the action of digestive enzymes released into the digestive tract as food passes. The form of the digestive system can be viewed in terms of these functions. If teeth and grinding jaws are present, mechanical breakdown of food, or *digesta,* begins in the buccal cavity. In birds and alligators, an especially muscularized part of the stomach, the *gizzard,* holds ingested grit (figure 7.1). The *gizzard* mechanically grinds food swallowed whole into small pieces, increasing the surface area for the action of chemicals. Digestive enzymes are added directly from microscopic glands within the walls of the digestive tract. Larger glands, outside the tract, also pour their chemical products into the digestive tract through ducts.

One way to increase the effectiveness of digestion is to prolong the time food is exposed to mechanical and chemical breakdown. For example, *spiral valves* in many types of fish (figure 7.2) turn digesta along a lengthened course, *ceca* temporarily sequester food, and longer intestines extend the time within the tract.

As food is broken down into fundamental components—carbohydrates, fatty acids, protein—it is absorbed through the walls of the *alimentary canal,* picked up by the circulatory system, and delivered to active parts of the organism or into sites of storage. The circulatory system is covered in chapter 8, but here we should notice the general position of part of the digestive system within the blood vascular system of supply (*arteries*) and drainage (*veins*). The amount of blood supply is generally an indication of the level of activity of the particular part of the digestive system and the route by which breakdown of products of digestion is carried away.

Shark

Pleuroperitoneal Cavity

Viscera

➤ Being careful that the deep blade of your scissors does not catch viscera beneath the belly wall, make an incision from cloaca forward through the pelvic girdle up to the pectoral girdle, keeping just to the left of the ventral midline. You may wish to make short, lateral cuts at each end of this incision so that the flaps of the body wall open like doors to reveal the viscera packed in the

Form & Function

The liver of sharks contains large amounts of oil called squalene. In some sharks, the oil alone can constitute 16–24% of the body weight. Oils are less dense than water, so they make the shark more buoyant, saving the shark energy that would otherwise be spent keeping it from sinking.

exposed **pleuroperitoneal cavity.** Posteriorly, this cavity communicates with the exterior through a pair of channels, the **abdominal pores,** that open at the bases of the pelvic fins on either side of the cloaca. Pass a blunt probe along the inside of the caudal end of the pleuroperitoneal cavity to confirm the position and course of an abdominal pore.

Down the middle of the pleuroperitoneal cavity runs the gut, with an S-shaped curve halfway along its length (figure 7.3). To either side lie the long **left and right lobes of the liver,** and anteriorly where they meet, a short **median lobe** of the liver, thus three lobes in all. Along the right lateral wall of the median lobe is a collapsed, dark green **gallbladder.** From the gallbladder runs a cord, the **common bile duct,** accompanied by blood vessels (figure 7.4). The bile duct runs to the gut, specifically the **duodenum** at one end of the S-shaped curve. Lying ventrally on the duodenum is the rounded **ventral lobe of the pancreas.** Slightly spread the bend of the duodenum to discover the long **dorsal lobe of the pancreas** passing posteriorly; the short connection between dorsal and ventral lobes is the **isthmus.** Formally, the duodenum begins just following a constriction in the gut, the **pyloric sphincter,** the last part of the stomach. The stomach region just before this sphincter is the **pylorus,** which leads to a great bend in the organ. This bend and the more anterior part of the stomach are the **body.** The right and left sides of the stomach are respectively termed the **lesser curvature** (right) and **greater curvature** (left). The wedge-shaped **spleen** rides posteriorly along the greater curvature of the stomach. No external demarcation marks the boundary between stomach and esophagus. However, by making a lengthwise incision in the body of the stomach forward to the still intact pectoral girdle, the distinctive internal structure permits certain identification of both as follows: The wall of the stomach forms longitudinal folds, **rugae;** the wall of the **esophagus** possesses stumpy, fingerlike projections, **papillae.**

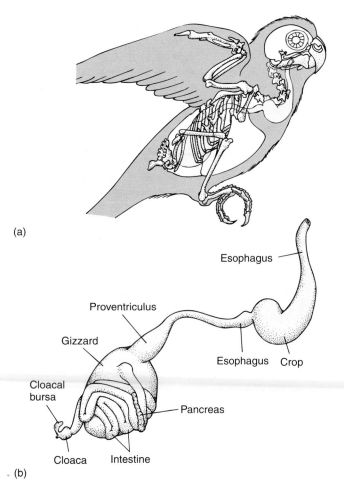

(a)

(b)

FIGURE 7.1 Alimentary canal of a parakeet. (a) Approximate position of the alimentary canal within the bird. (b) Alimentary canal enlarged.

After H. E. Evans, 1982, "Anatomy of the Budgerigar" in Diseases of Cage and Aviary Birds, ed. Margaret L. Petrak, 2d ed., Lea & Febiger, Philadelphia.

Return now to the duodenum (figure 7.4). Passing posteriorly, the duodenum enlarges, continuing into the **valvular intestine.** The outer walls of the intestine are circularly creased by an internal subdivision, the **spiral valve** (figure 7.2b). Cut open, lengthwise, the wall of the intestine to identify the spiral valve, the defining feature of the valvular intestine. The duodenum lacks a

spiral valve; thus its specific boundaries, between spiral valve and pyloric sphincter, can be settled.

Posterior to the valvular intestine, the gut narrows into the **colon,** which lacks a spiral valve and ends in the slitlike **cloaca** (figure 7.3). Dorsal to the colon, and joining it near the entrance to the cloaca, is the fusiform **rectal gland.**

Anteriorly, dorsal to the liver lobes are the paired gonads (**testes** or **ovaries**). Check the sex of your shark (note pelvic fins), then examine gonads of a shark of the opposite sex.

Mesenteries

The entire pleuroperitoneal cavity is lined by the shiny *parietal peritoneum* (figure 7.5d). Poke a blunt probe slowly into the liver, noting that a thin surface membrane must break before the probe enters the liver tissue. This lining membrane, covering all viscera, is the *visceral peritoneum* (serosa; figure 7.5d). Probe the parietal peritoneum similarly. The sheets of membranes suspending viscera and joining them at places with each other are the *mesenteries*. Vertebrate embryos commonly possess a broad *dorsal mesentery* joining the viscera to the dorsal body wall (figure 7.5d), and a similarly long *ventral mesentery* joining viscera to the ventral body wall (figure 7.5d). Only portions of the ventral mesentery persist into the adult stages of most vertebrates. However, broader regions of the dorsal mesentery occur frequently in adult vertebrates.

Dorsal Mesentery Pull the viscera to one side, noting that most mesenteries attach to the dorsal midline of the pleuroperitoneal cavity by a prominent, thin, and translucent sheet, the **dorsal mesentery.** Specific regions of the dorsal mesentery can be recognized. That part of the dorsal mesentery supporting the esophagus and stomach is the **mesogaster** (greater omentum). The part to the anterior half of the valvular intestine is the **mesentery**[1] proper. Pull the spleen from its nestled position on the stomach to reveal the **gastrosplenic ligament** joining the two organs. The short, isolated expanse of dorsal mesentery suspending the rectal gland is the **mesorectum.**

---

[1] An unfortunate term, because here it is used in a limited sense, while elsewhere **mesentery,** in a broad sense, means the whole class of such membranes supporting viscera.

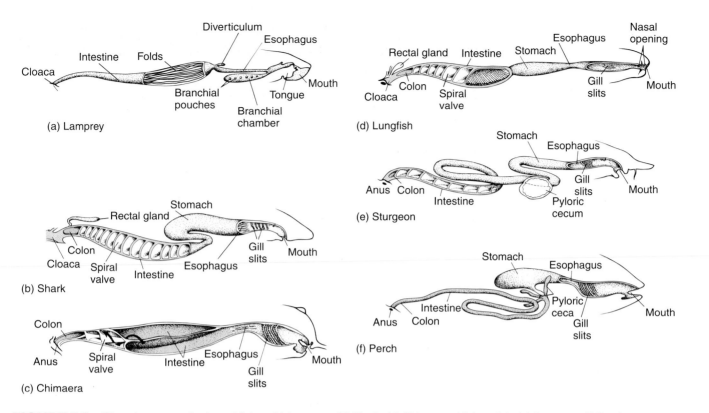

FIGURE 7.2 Digestive tracts of selected fishes. (a) Lamprey. (b) Shark. (c) Chimaera. (d) Lungfish. (e) Sturgeon. (f) Perch.
From Dean.

Form & Function

Why do vertebrates have mesenteries? Mesenteries, like those just examined, keep organs in position and define spaces in which active organs can operate more freely. In addition, mesenteries restrict the amount of tensile force experienced by the many blood and lymphatic vessels that extend between the body wall and internal organs. Without mesenteries, shifting organs could wrap these vessels or pull them apart.

🦖 **Ventral Mesentery**   The ventral mesentery is represented at a few specific sites (see figure 7.3). Return to the bile duct near its departure from the gallbladder. The **gastrohepatoduodenal ligament** (lesser omentum) runs along with blood vessels in a thin, ribbonlike mesentery extending to the liver. Carefully spread the organs to notice that this ligament actually branches upon its approach to the gut. One fork, the **hepatoduodenal ligament,** goes to the body of the duodenum. The other fork, the **gastrohepatic ligament,** goes to the body of the stomach along its lesser curvature.

Return to the anterior blunt end of the liver. A **transverse septum** (figure 7.3) separates the pleuroperitoneal cavity from the pericardial cavity. The anterior end of the liver attaches to the septum by the **coronary ligament,** which, strictly speaking, is a derivative of and thus part of the transverse septum. This very short mesentery may be seen by gently pulling the liver caudally away from the transverse septum. The **falciform (suspensory) ligament** is a short mesentery that attaches the cranial end of the liver to the ventral body wall (figure 7.5d). It appears as a ventral continuation of the coronary ligament. However, the falciform ligament may be partially torn by the initial midventral cut made when opening the pleuroperitoneal cavity. The common opening, or **ostium,** of the oviducts lies within the falciform ligament.

🦖 **Mesenteries of the Reproductive Tract**   The **mesorchium** suspends the testis, the **mesovarium** the ovary, and in mature females, the **mesotubarium** holds the oviduct.

Pericardial Cavity

🦖 The pericardial cavity lies immediately anterior to the coracoid bar, dorsally over the common coracoarcual muscles (figure 7.5a). Carefully remove these muscles to expose the **parietal pericardium,** a white membrane lining the **pericardial cavity.** Cut through the parietal pericardium to reveal the heart within. The heart is covered with a very thin **visceral pericardium.** The chambers of the heart will be identified with the circulatory system.

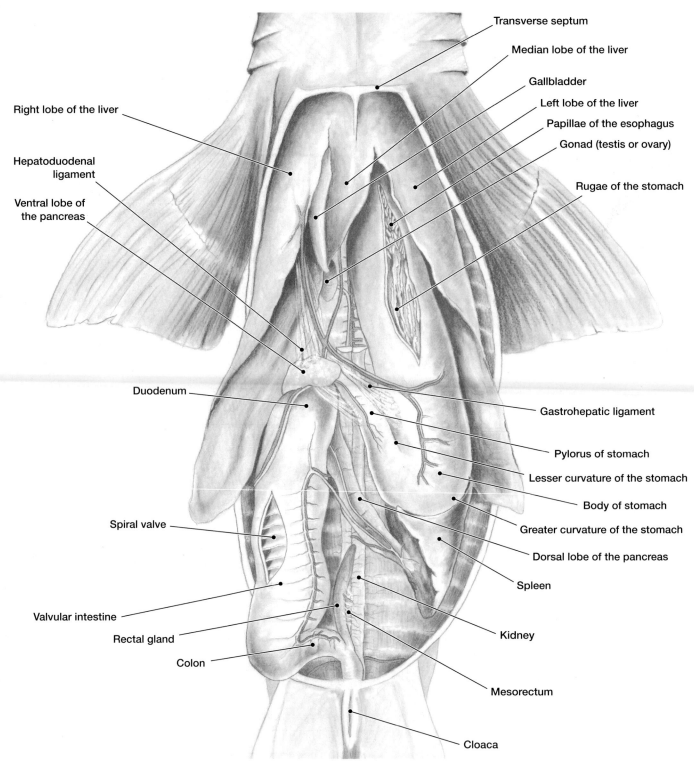

Transverse septum

Median lobe of the liver

Gallbladder

Left lobe of the liver

Papillae of the esophagus

Gonad (testis or ovary)

Rugae of the stomach

Right lobe of the liver

Hepatoduodenal
ligament

Ventral lobe of
the pancreas

Duodenum

Gastrohepatic ligament

Pylorus of stomach

Lesser curvature of the stomach

Body of stomach

Greater curvature of the stomach

Dorsal lobe of the pancreas

Spleen

Spiral valve

Valvular intestine

Rectal gland

Kidney

Colon

Mesorectum

Cloaca

FIGURE 7.3   Ventral view of the pleuroperitoneal cavity of a shark.

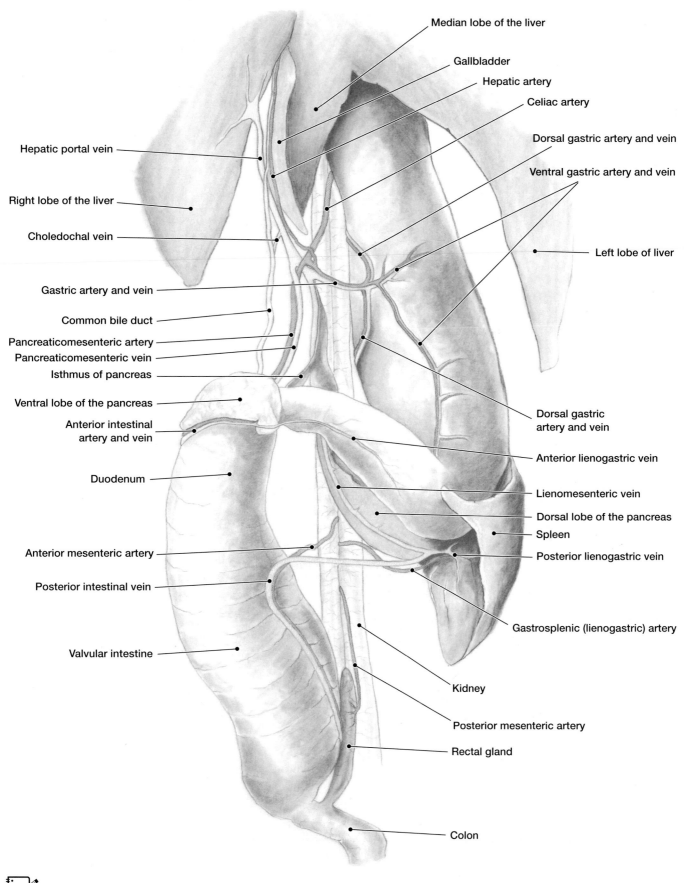

Median lobe of the liver

Gallbladder

Hepatic artery

Celiac artery

Dorsal gastric artery and vein

Ventral gastric artery and vein

Hepatic portal vein

Right lobe of the liver

Choledochal vein

Left lobe of liver

Gastric artery and vein

Common bile duct

Pancreaticomesenteric artery

Pancreaticomesenteric vein

Isthmus of pancreas

Ventral lobe of the pancreas

Anterior intestinal
artery and vein

Dorsal gastric
artery and vein

Anterior lienogastric vein

Duodenum

Lienomesenteric vein

Dorsal lobe of the pancreas

Spleen

Anterior mesenteric artery

Posterior lienogastric vein

Posterior intestinal vein

Gastrosplenic (lienogastric) artery

Valvular intestine

Kidney

Posterior mesenteric artery

Rectal gland

Colon

FIGURE 7.4 Blood supply to alimentary canal of a shark.

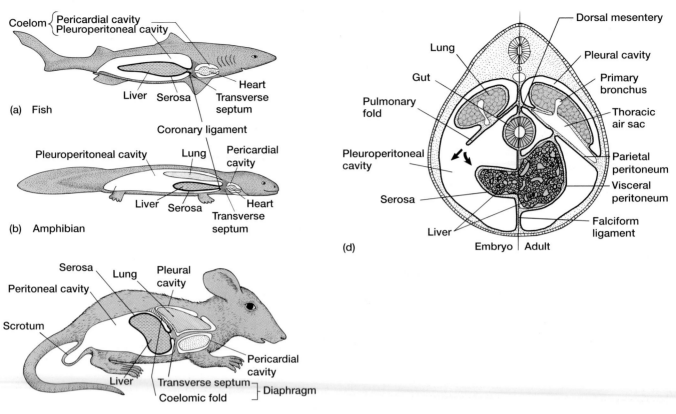

FIGURE 7.5 Body cavities. The coelom, arising in somitomeres, becomes divided by a fibrous transverse septum into pericardial and pleuroperitoneal cavities in fishes (a), amphibians (b), and mammals (c). In all three animals, the small coronary ligament attaches the cranial end of the liver to either the transverse septum or diaphragm. (d) Cross section of a bird illustrating the embryonic (left) and adult (right) cavities. In the embryo, the pulmonary fold grows obliquely to establish contact with the liver and the body wall. This confines the lung to its pleural cavity.

Necturus

Pleuroperitoneal Cavity

Viscera

To inject blood vessels with colored latex, the supply house preparing *Necturus* specimens must make a midventral incision to reach vessels in the **pleuroperitoneal cavity** (figure 7.5b). Usually the dark brown to black **liver** protrudes through this incision (figure 7.6). Extend this incision forward to the pectoral girdle and posteriorly through the pelvic girdle, **cloacal gland,** and into the cloaca. Take care not to damage viscera with the deep blade of your scissors. Keep slightly to the left of the ventral midline, noting now, and attempting to keep intact, the delicate **falciform ligament** from liver to ventral body wall. (Recall that this same ligament was likely damaged when examining the shark.)

Lift the posterior end of the liver to reveal the pea-sized and green **gallbladder** (figure 7.6). Gently push the liver to the right side of the pleuroperitoneal cavity to better reveal the more dorsal, tan **stomach,** and the dark red **spleen** on the left lateral side of the stomach.

The slender, saclike, and paired **lungs** are nestled along the dorsolateral sides of the pleuroperitoneal cavity.

Next trace the parts of the digestive tract. The cranial end of the stomach is nestled in the most anterior end of the pleuroperitoneal cavity (figure 7.6). It extends caudally and laterally until it abruptly narrows and ends in a constriction, the **pyloric sphincter.** Taking care not to damage mesenteries, cut open the stomach to expose the internal, lengthwise folds, or **rugae.** Beyond the pyloric sphincter, the digestive tract continues as the long, but highly folded **small intestine.** Cut into the lumen of the small intestine at a convenient point to confirm the absence of a spiral valve, but the presence of small, wavy folds, called **plicae.** The **duodenum** is the first short section of the small intestine. The light, pink **pancreas** is found cradled within the first bend of the duodenum. The short **cystic duct** from the gallbladder joins **hepatic ducts** draining the liver to form the **common bile duct,** which finally empties into the duodenum. However, tracing these small ducts in a *Necturus* usually requires patience and a dissection microscope. Near the end of the small intestine the gut straightens and widens to form the **large intestine.** A single, shriveled

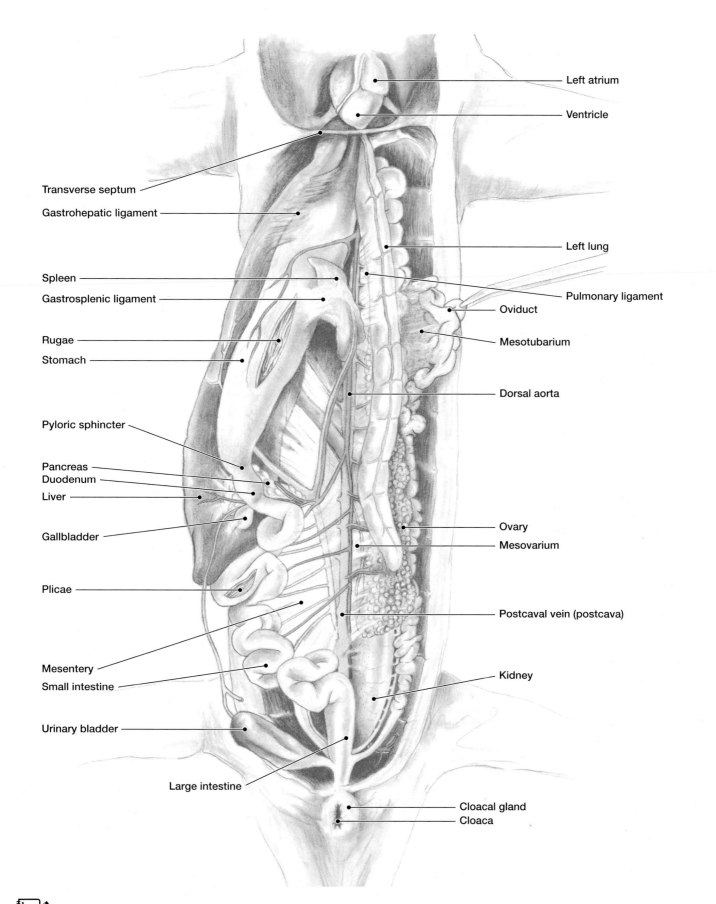

Left atrium

Ventricle

Transverse septum

Gastrohepatic ligament

Left lung

Spleen

Pulmonary ligament

Gastrosplenic ligament

Oviduct

Rugae

Mesotubarium

Stomach

Dorsal aorta

Pyloric sphincter

Pancreas

Duodenum

Liver

Gallbladder

Ovary

Mesovarium

Plicae

Postcaval vein (postcava)

Mesentery

Small intestine

Kidney

Urinary bladder

Large intestine

Cloacal gland

Cloaca

FIGURE 7.6   Ventral view of the pleuroperitoneal cavity of a female *Necturus*.

urinary bladder is found along the ventral midline ventral to the large intestine. The urinary bladder and large intestine empty separately into the cloaca, identified previously during the muscle dissection.

The kidney lies above the large intestine, tightly applied to the dorsal midline at the caudal end of the pleuroperitoneal cavity (figure 7.6). Solid, elongate testes or much larger, granular ovaries are found lateral to the kidneys of sexually mature specimens. The kidneys, gonads, and their ducts constitute the urogenital system and will be examined in greater detail in chapter 9 on reproductive systems.

Mesenteries

🦖 As in the shark, the pleuroperitoneal cavity is lined by the parietal peritoneum, and the organs are covered by the visceral peritoneum. Specific regions of the dorsal mesentery are recognized (figure 7.6). The mesogaster supports the anterior half of the stomach. The gastrosplenic ligament joins the spleen to the stomach. On the left side of the body is a very narrow but long mesentery, the pulmonary ligament, running along most of the length of the lung, and attaching the lung to the mesogaster. However, the pulmonary ligament of the right lung joins not the mesogaster, but a separate mesentery, the hepatocavopulmonary ligament. As the name suggests, this mesentery supports the liver (hence hepato), a large blood vessel called the postcaval vein (hence cavo), and receives attachment of the mesentery from the right lung (hence pulmonary). The small intestine is supported by the mesentery proper, which is continuous with the mesorectum supporting the large intestine.

As in the shark, only parts of the ventral mesentery are present (figure 7.6). The falciform ligament, identified earlier, runs from the liver to the ventral body wall. The gastrohepatic ligament joins the anterior end of the stomach to the dorsal surface of the liver. It is best seen from the left side by spreading the two organs. The hepatoduodenal ligament joins the duodenum and caudal part of the liver, but wraps the middle portion of the pancreas, thus making identification challenging. The urinary bladder attaches by the median ligament of the bladder to the ventral body wall. Finally, the mesorchium suspends the testis, a mesovarium suspends the ovary, and a mesotubarium suspends the oviduct.

As in the shark, a transverse septum forms the anterior end of the pleuroperitoneal cavity (figure 7.5b). The short coronary ligament, arising from the surface of the liver, attaches to this septum (figure 7.5b).

Pericardial Cavity

🦖 Pick away the triangular block of hypobranchial muscles between the procoracohumeralis muscles to expose the deep parietal pericardium, a membrane enclosing the pericardial cavity, which contains the heart. The very thin covering over the heart is the visceral pericardium.

The transverse septum divides the pericardial and pleuroperitoneal cavities. To see it better, extend forward the midventral incision used to open the pleuroperitoneal cavity. Spread the pectoral girdle at the caudal end of the pericardial cavity. Be careful not to split or damage the large blood vessels coming from the liver. Snip connective tissues to relieve tension and to better expose the septum. In Necturus, the transverse septum is a composite of peritoneum and muscle. Check again the attachment of the coronary ligament to the septum.

Cat

Salivary Glands

🦖 Five pairs of salivary glands are present in the cat (figure 7.7). The largest gland is the parotid gland, which lies just beneath the skin at the base of the pinna of the ear. (If the parotid is absent, check the reflected skin to see if it pulled away during skinning.) Free skin up to the ear to expose this gland. Its duct, the white, cordlike parotid duct, passes across the middle of the masseter muscle to empty into the mouth opposite the last premolar. Not to be confused with this duct are thin branches of the facial nerve that run superficially at the dorsal and ventral portions of this muscle. The round submaxillary gland (mandibular gland) touches the ventral corner of the parotid gland and is crossed by the posterior facial vein, the most dorsal of three tributaries to the external jugular vein running along the side of the neck. The long, middle tributary is the transverse jugular vein, passing under the throat and joining external jugulars of both sides. The submaxillary duct leaves the anterior margin of its gland, courses forward, and disappears beneath the lower edge of the masseter. A small glandular wedge, the sublingual gland, rides along the submaxillary duct. The sublingual and submaxillary ducts run together to the mouth. Spend some time confirming the boundaries of these three glands, their ducts, and distinguishing them from lymph nodes and branches of nerves.

The molar gland lies on the external surface of the mandible between the corner of the mouth and anterior facial vein. Its several short ducts, impractical to see, empty on the inside of the cheek.

The infraorbital gland lies against the anterior edge of the zygomatic arch, at the bottom of the orbit, but will not be examined at present.

Oral Cavity

🦖 The buccal cavity, or mouth, extends from the lips (labia) to the end of the hard palate (figure 7.8a). The bony roof of the buccal cavity is roughened by washboardlike ridges, palatal rugae. The pharynx is the space extending caudal to the hard palate to the level of the opening into the larynx. Its roof is the soft palate. The part of the buccal cavity between lips and teeth is the vestibule.

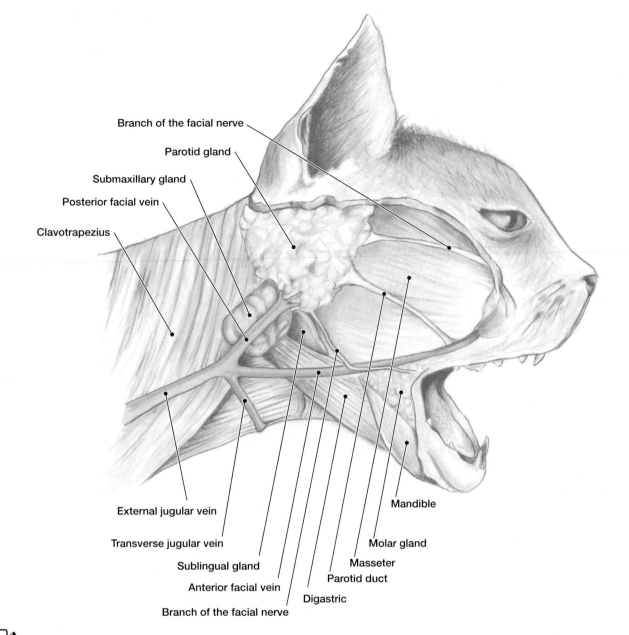

Branch of the facial nerve

Parotid gland

Submaxillary gland

Posterior facial vein

Clavotrapezius

External jugular vein

Transverse jugular vein

Sublingual gland

Anterior facial vein

Branch of the facial nerve

Digastric

Parotid duct

Masseter

Molar gland

Mandible

FIGURE 7.7   Lateral view of the neck and face of a cat, showing salivary glands.

Patterns & Connections

Many of these buccal structures can be found in your own mouth. Probe your tongue upward between your top front teeth and upper lip. There you should feel a thin membrane, the labial frenulum. A lingual frenulum attaches the anterior base of your tongue to the floor of your mouth; however, a mirror will likely be needed to see this connection. In humans, as in the cat, the submandibular glands empty near the lingual frenulum. But unlike the cat, the sublingual glands in humans empty separately through many openings in the floor of the buccal cavity.

The roof of our buccal cavity is similarly composed of a hard and soft palate. However, the junction of these two regions is difficult to find without deep probing with a clean finger (don't try this in lab).

Below the nose, pull the lips away from the teeth to find a vertical, medial flap, the **labial frenulum,** attaching lips to the **gingiva** (gums), and the median **lingual frenulum** attaching the anterior tongue to the floor of the buccal cavity. At the base of this frenulum, near its free edge, locate the small, pimplelike swelling, the **sublingual papilla,** that in the cat bears the openings for the ducts from both sublingual and submaxillary glands.

Preparing the Pharyngeal Region

To examine features near the larynx, you must prepare the right side of the oral cavity. This is done by rotating the right mandible, making a major cut along the right side, and laying open the right side of the face and neck.

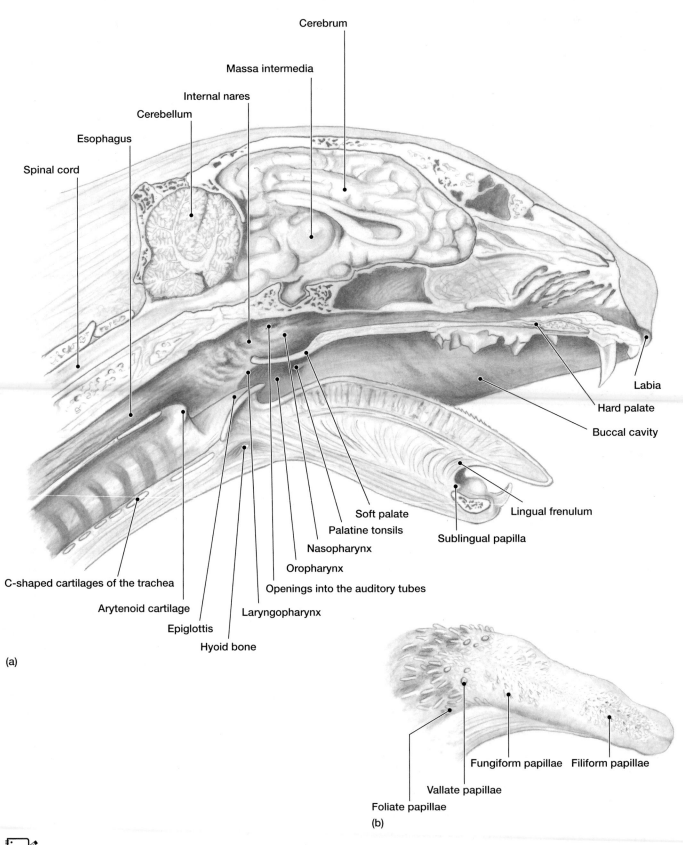

Cerebrum

Massa intermedia

Internal nares

Cerebellum

Esophagus

Spinal cord

Labia

Hard palate

Buccal cavity

Lingual frenulum

Soft palate

Palatine tonsils

Sublingual papilla

Nasopharynx

Oropharynx

Openings into the auditory tubes

C-shaped cartilages of the trachea

Arytenoid cartilage

Laryngopharynx

Epiglottis

Hyoid bone

(a)

Fungiform papillae   Filiform papillae

Vallate papillae

Foliate papillae

(b)

FIGURE 7.8   Oral and pharyngeal structures in bisected head and neck of a cat. (a) Midsagittal section. (b) Dorsal three-quarter view of the tongue.

Proceed as follows: With bone or heavy scissors, cut through the mandibular symphysis to divide the lower jaw. Next cut muscles, skin, and the soft tissues lining the floor of the buccal cavity, close to their attachments to the right mandible. This should free the mandible so it can then be rotated laterally, out of the way. Finally, with one blade of the scissors inside, cut posteriorly, following the oral cavity. Continue this cut about 1 inch or until the tip of the trachea can be easily seen (figure 7.8a). Spread the margins of the cut to identify the following structures.

Buccal and Pharyngeal Regions

🦖 The tongue is covered by four types of papillae (figure 7.8b). The **filiform papillae,** located across the anterior half of the tongue, usually bear tiny, spinelike projections. These spines are used in grooming and to remove soft tissue from bones. More posteriorly along the lateral margins of the tongue are less numerous **fungiform papillae,** which lack spines. Further posteriorly along the lateral margins of the tongue are larger, softer, and more rounded **foliate papillae.** Medial to the foliate papillae at the base of the tongue are several oblique rows of flattened **vallate papillae** recessed in pits.

At the anterior end of the hard palate, posterior to the incisor teeth, lies a pair of small openings for the **nasopalatine ducts** (incisive ducts) coming from the vomeronasal organ, small in cats.

Three regions of the pharynx are recognized (figure 7.8a). However, do not expect to see pronounced anatomical markers delineating the boundaries of each region of the pharynx. Instead, think of the pharynx as a passageway with three general regions. Roughly, the **oropharynx** lies ventral to the soft palate, the **nasopharynx** above the soft palate. These two regions join caudally to form the **laryngopharynx,** between the posterior edge of the soft palate and the end of the larynx. Just anterior to the free border of the soft palate, on the dorsolateral wall, is a pair of **palatine tonsils,** each slightly bulging from a pit, the **tonsillar fossa.** Paired folds in the lateral wall of the pharynx immediately bound the tonsillar fossa: the **palatoglossal arch** in front, the **palatopharyngeal arch** behind. These folds are important landmarks. The palatoglossal arch marks the official end of the **oral cavity.** The passage through this arch is the **fauces.**

Make a midsagittal cut through the soft palate to expose the nasopharynx, which opens anteriorly through two **internal nares** into the **nasal cavity** (figure 7.8a). On the dorsolateral walls, about halfway along the nasopharynx, are a pair of slitlike openings into the **auditory** (Eustachian) **tubes.** Carefully probe this region until your blunt probe can be gently inserted into one of these slits.

Just past the rear of the tongue is the cranial end of the **larynx** bearing a prow-shaped projection, the **epiglottis** (figures 7.8a and 7.9b). Continuing beyond the larynx is the **trachea,** supported by regular, C-shaped cartilages. Specifically, the laryngopharynx lies above the larynx and gives way caudally to the **esophagus.** The opening into the larynx is termed the **glottis.** Free the larynx from surrounding tissue to better expose the large cartilaginous elements composing it. The epiglottis rests on the large, incomplete **thyroid cartilage** (figure 7.9b), which reaches around the larynx to touch a pair of small, dorsal **arytenoid cartilages.** At the base of the larynx lies the more caudal **cricoid cartilage,** forming a complete but uneven ring. Cut the larynx along its dorsal midline to find the **vocal cords** inside, lateral folds running between arytenoid and thyroid cartilages. In some carnivores, such as the cat, **false vocal cords** are present as well, accessory folds between arytenoids and the base of the epiglottis.

The ventral glandular tissue just posterior to the cricoid cartilage and applied externally to the trachea is the **thyroid gland.** The paired lateral lobes are joined ventrally by a narrow band of tissue, the **isthmus.** Parathyroid glands are present, buried in the thyroid tissue, but are impractical to see.

Structures of the Thoracic Region

🦖 Palpate the sternum to discover its most posterior extent beneath the muscles of the chest. With tweezers, lift the end of the xiphisternum and, keeping to the cat's right of the midline, cut forward through the ribs, taking care not to catch the scissors on organs deep within the cavity. Next, at the caudal ends of the thoracic cavity, cut laterally on both sides, just anterior to the **diaphragm,** which forms a thin, muscular transverse wall. Muscle fibers contributing to the diaphragm take origin from ribs, sternum, and vertebrae and run inward toward the center of the muscle to insert on a round **central tendon,** about the size of a dime. The rib cage can thus be opened to expose three coelomic cavities without risking damage to the diaphragm. The **pericardial cavity** includes the heart and its enclosing membranes (figure 7.5c). The **left** and **right pleural cavities** contain the **lungs** on either side of the pericardial cavity (figure 7.5c). Note the thin partition, the **mediastinal septum,** between the heart and ventral medial wall. It is formed from the medial walls of left and right pleural cavities.

The thoracic cavity can now be opened further to examine the internal organs (figure 7.9). The inside walls of the pleural cavities are lined by smooth **parietal pleura.** The surface of the soft, spongy lungs is covered by the thin **visceral pleura.** The left lung consists of the **anterior, middle,** and **posterior lobes.** The right lung similarly consists of these lobes (anterior, middle, posterior) plus a fourth, the **accessory lobe.** To find it, lift the apex (blunt posterior end) of the heart to expose a small membranous compartment beneath. Cut through these membranes to discover the accessory lobe within. Finally, the **mediastinum** is a single and central cavity that encloses the heart, blood vessels, esophagus, and

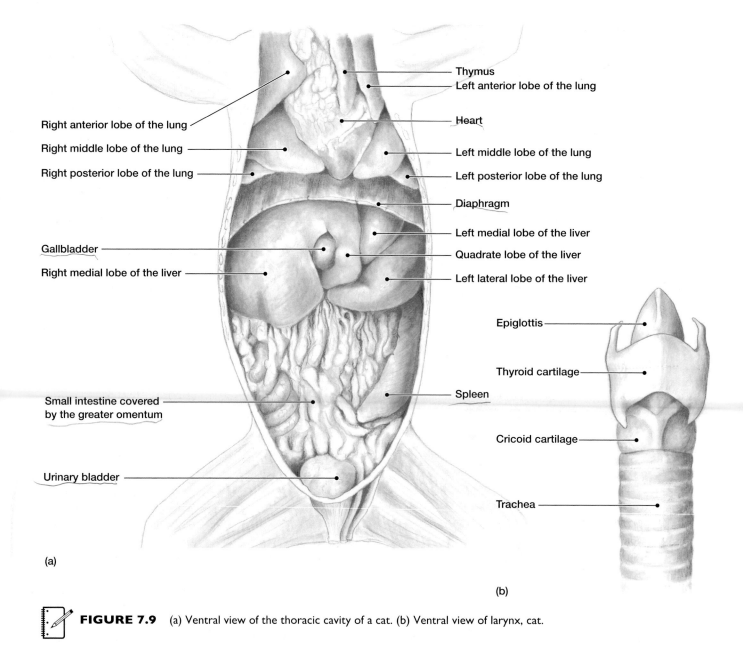

Right anterior lobe of the lung

Right middle lobe of the lung

Right posterior lobe of the lung

Gallbladder

Right medial lobe of the liver

Small intestine covered by the greater omentum

Urinary bladder

Thymus

Left anterior lobe of the lung

Heart

Left middle lobe of the lung

Left posterior lobe of the lung

Diaphragm

Left medial lobe of the liver

Quadrate lobe of the liver

Left lateral lobe of the liver

Spleen

Epiglottis

Thyroid cartilage

Cricoid cartilage

Trachea

(a)

(b)

FIGURE 7.9 (a) Ventral view of the thoracic cavity of a cat. (b) Ventral view of larynx, cat.

trachea. It is continuous ventrally with the mediastinal septum, which connects to the ventral body wall.

Externally the heart is wrapped in **parietal pericardium** (pericardial sac). Cut through this sac. The surface of the heart is covered by the **visceral pericardium.** The small space between parietal and visceral pericardia is, formally, the **pericardial cavity** (figure 7.5c).

Cranial to the heart, between anterior lobes of the lungs, lies a mass of glandular tissue, the **thymus** (figure 7.9a), enmeshed in blood vessels. Large in young individuals, it undergoes involution (reduction in size) as an animal grows older. Clearly expose and separate the thymus from the large blood vessels, taking special care not to damage these vessels, which will be identified later.

To expose structures in the dorsal pleural cavity, pull the right lung and heart to the left side of the thoracic cavity. The pulmonary ligament (possibly indistinct), part of the pleura, attaches the lung along most of its length to the dorsal body wall. Medially, a tangle of arteries, veins, and bronchi, the **root of the lung,** or **radix,** enter the middle of each lung. Spread the heart and right anterior lobe of the lung to reach the large blood vessel, injected red or pink, running directly forward beneath the thymus. Parallel, but dorsal to it is the tubular **trachea** ribbed with cartilage rings, which feel rough when stroked with the blunt probe (figure 7.9b). Dorsal to it is the collapsible and smooth-walled **esophagus.**

Turn to the left side of the cat and lift the left lobes of the lung to expose structures along the dorsal midline. A large blood vessel (dorsal aorta) passes along the dorsal wall. Next to the dorsal aorta you can again find the esophagus, which passes dorsal to the trachea and should be followed caudally as it passes through the diaphragm and into the abdominal cavity.

Abdomen

Extend the midventral incision caudally along the abdomen back to the inguinal (groin) region. Be sure to keep to the right of the midline. The cavity so opened is the **peritoneal cavity** (abdominal cavity; figures 7.5c and 7.10). Make a pair of lateral cuts along the caudal margin of the diaphragm, such that the body walls may each be folded back. An additional pair of lateral cuts may be made along the posterior margin of the abdomen.

As in the shark and *Necturus,* the **parietal peritoneum** lines the inner walls of the abdominal cavity, and the **visceral peritoneum** covers the surfaces of organs. The large **spleen** lies unpaired, along the left abdominal wall (figure 7.10). The **liver** fits into the dome-shaped diaphragm. Spread the liver and diaphragm to find the **falciform ligament** connected between them. In some specimens the ventral free edge of this ligament is thickened. This is the **round ligament,** an adult remnant of the embryonic umbilical veins that in the fetus carried blood from the placenta. The **coronary ligament** is connected to the dorsal margin of the falciform ligament. The coronary ligament connects the liver to the central tendon, identified previously as part of the diaphragm.

The falciform ligament joins the liver in a cleft that divides the liver into left and right halves. In turn, each half is divided into medial and lateral lobes (figure 7.10). Of the two, the **left lateral lobe** is larger than the **left medial lobe.** On the right the opposite is true; the **right medial lobe** is larger and covers the smaller **right lateral lobe.** Additionally, the right lateral lobe splits into two attached parts: an **anterior part** and a **posterior part,** which rests on the cranial end of the right **kidney.** The saclike **gallbladder** lies within a pocket of the right half of the liver between its medial lobe and the smaller **quadrate lobe,** which occupies a midventral position. Tilt the left and right halves of the liver cranially, spreading them from the stomach and coiled intestine. Finally, from the inner bend of the stomach will emerge the small, pointed **caudate lobe** of the liver, covered by a thin membrane, part of the lesser omentum.

This is also a good time to rediscover the esophagus emerging from the diaphragm. Follow it to where it ends in the expanded **stomach,** with several regions identifiable in gross structure (figure 7.10). The inner, right side of the stomach is its **lesser curvature,** the left side the **greater curvature.** The large, domed cranial end is the **fundus,** the large more caudal part is the **body,** or **corpus,** and the narrowed end the **pyloric region,** which terminates in a constriction, the

Patterns & Connections

The duodenum is positioned very closely to the gallbladder and pancreas in the shark, , and cat. In each of these animals, bile produced in the liver and secretions from the pancreas drain into the duodenum. Bile emulsifies fats. Pancreatic secretions contain proteolytic enzymes, amylases for carbohydrate digestion, and lipases for fat digestion. Thus, the close proximity of these organs reflects their interrelated functions.

---

pyloric sphincter (pylorus). The point in the stomach where contents from the esophagus first enter is a very short (few millimeters) **cardia,** but no gross features exist to approximate boundaries. Instead, the histological character of digestive glands in its walls serve to define the cardia. By cutting into the stomach, longitudinal folds, or **rugae,** can be identified.

Beyond the pyloric sphincter continues the **small intestine** folded upon itself (figure 7.10). It, like much of the rest of the gut, is draped in mesenteries streaked with fat, often considerable amounts in overfed cats. Sections of this mesentery will be identified shortly, but first delineate the portions of the gut and associated structures. With your fingers, gently lift this mesentery and examine the first segment of the intestine, the **duodenum,** extending from the pyloric sphincter to about the second caudal bend of the small intestine. In the first bend of the small intestine, along the side of the duodenum, lies the **head** of the **pancreas.** The **tail** of the pancreas extends away from the intestinal wall toward the spleen. The small intestine beyond the duodenum is the **jejunum** and then **ileum,** but no distinct gross features mark their boundaries.

The ileum ends at its juncture with an expanded region of the gut, the **colon,** or **large intestine** (figure 7.10). At this juncture, the end of the colon forms a blunt, small, fingerlike projection, the **cecum** (represented by the rudimentary **vermiform appendix** in humans). Three sections of colon are recognized: the **ascending colon,** the short, first part along the right side and arching cranially; the **transverse colon** crossing to the left; and the **descending colon,** the final and longest section continuing caudally. The descending colon ends in a short, straight portion, the **rectum,** bounded by the pelvic girdle. Parts of the urogenital system also converge upon or reside at this point in the abdomen. Thus, do not cut or attempt to expose any more of the gut at this point in the dissection. The **urinary bladder** appears as a muscular bag on the ventral side of the most caudal end of the descending colon.

Make small, short slits in the colon and small intestine, wash out debris, and then compare the textures of the internal walls. The wall of the small intestine is velvety due to the presence of a mat of tiny projections, called **villi.** These are absent from the colon and stomach.

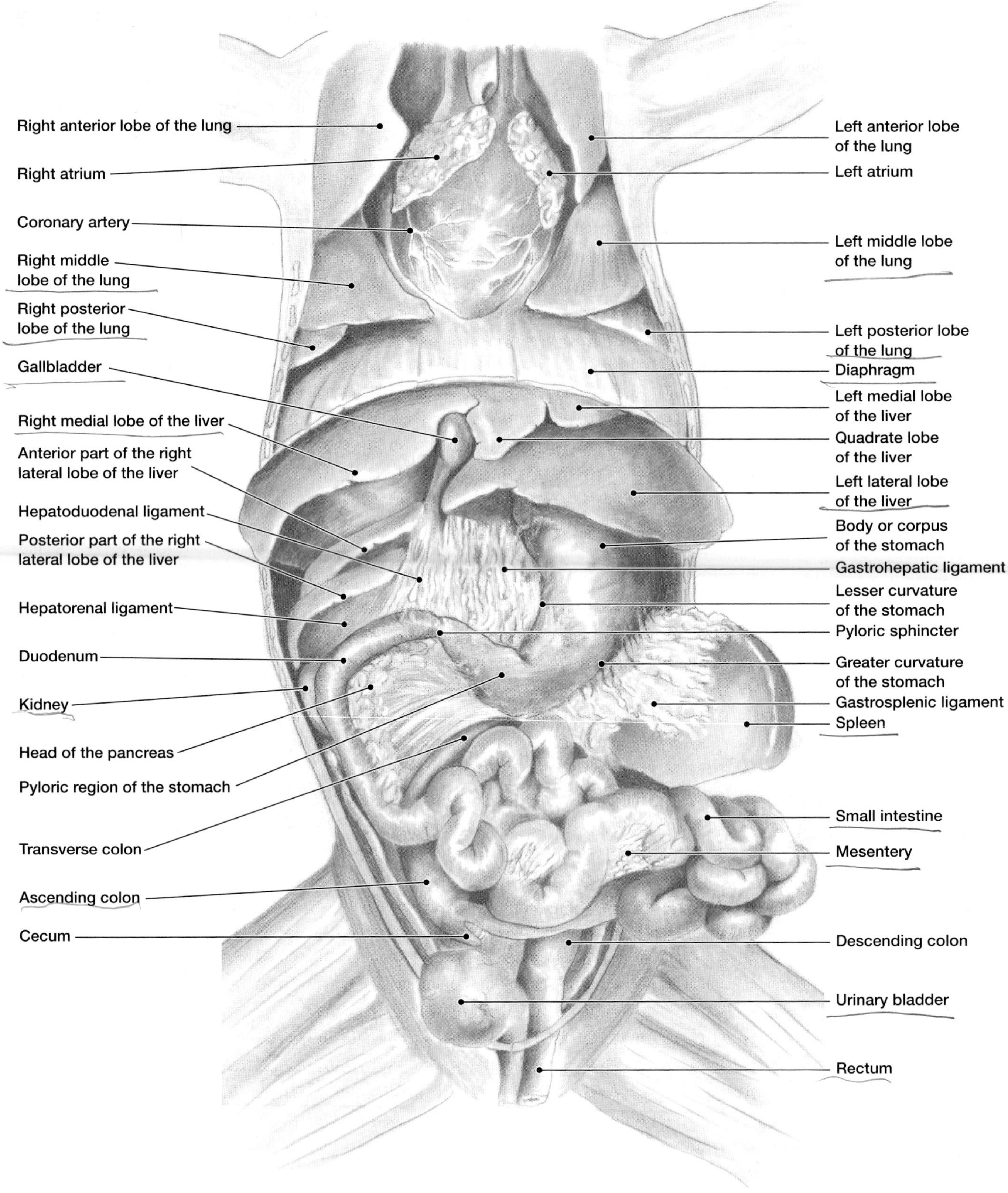

Right anterior lobe of the lung

Right atrium

Coronary artery

Right middle lobe of the lung

Right posterior lobe of the lung

Gallbladder

Right medial lobe of the liver

Anterior part of the right lateral lobe of the liver

Hepatoduodenal ligament

Posterior part of the right lateral lobe of the liver

Hepatorenal ligament

Duodenum

Kidney

Head of the pancreas

Pyloric region of the stomach

Transverse colon

Ascending colon

Cecum

Left anterior lobe of the lung

Left atrium

Left middle lobe of the lung

Left posterior lobe of the lung

Diaphragm

Left medial lobe of the liver

Quadrate lobe of the liver

Left lateral lobe of the liver

Body or corpus of the stomach

Gastrohepatic ligament

Lesser curvature of the stomach

Pyloric sphincter

Greater curvature of the stomach

Gastrosplenic ligament

Spleen

Small intestine

Mesentery

Descending colon

Urinary bladder

Rectum

FIGURE 7.10 Abdominal structures in a cat, viewed ventrally upon initial inspection.

Mesenteries

🦖 The mesentery from the liver to the duodenum and lesser curvature of the stomach is the **lesser omentum,** divisible into two parts: the **hepatoduodenal ligament** from liver to duodenum, and the **gastrohepatic ligament** from liver to lesser curvature of the stomach (figure 7.10). The mesentery attached to the opposite side of the duodenum and greater curvature of the stomach is the **greater omentum,** a derivative of part of the mesogaster (figure 7.9). It drapes over the small intestine. Because the greater omentum is modified in most mammals into a saclike pouch, it is often called an **omental bursa.** The stretch of greater omentum between the spleen and stomach is the **gastrosplenic ligament** (gastrolienic ligament).

Find again the posterior part of the right lateral lobe of the liver dorsal to the small intestine (figure 7.10). From this lobe the **hepatorenal ligament** extends to parietal peritoneum near the right kidney (figure 7.10). Note that this large liver lobe extends forward into a small pocket. Ventral to the lobe at this cranial point is the free edge of part of the hepatoduodenal ligament supporting pancreas and blood vessels. This free edge defines the **epiploic foramen,** the entrance into the omental bursa. Confirm this by returning to the omental bursa and penetrating it with a blunt probe, projecting its point dorsally to exit from the epiploic foramen.

The small **gastrocolic ligament** passes from the greater omentum dorsal to the spleen to the mesentery suspending the intestines. The dorsal mesentery supporting the gut is subdivided into three regions: the **mesoduodenum** supports the duodenum, the **mesocolon** the large intestine, and the **mesentery** proper the jejunum and ileum (figure 7.10). Note the numerous, smooth glandular **lymph nodes** on the mesentery. These are found elsewhere throughout the body as well.

The urinary bladder is supported by the single **median ligament** extending to the midventral abdominal wall, and by paired **lateral ligaments** at its base.

Return to the region of the duodenum to trace the ducts entering this region of the gut. Care and extra patience will now be helpful. The gallbladder narrows into a duct, the **cystic duct,** which joins ducts coming from various lobes of the liver, **hepatic ducts.** Cystic and hepatic ducts merge into a single **common bile duct,** which passes to the intestine in the hepatoduodenal ligament. When this duct reaches the duodenum, it slightly enlarges, forming the **hepatopancreatic ampulla** (ampulla of Vater). Joining the common bile duct in this ampulla is the **main pancreatic duct.** Pick away tissue to follow this duct a short distance into the pancreas. A second, **accessory pancreatic duct** enters the duodenum independently about 1–2 cm farther along (caudally), but is not easy to locate.

Patterns & Connections

Have you thanked a few trillion of your closest prokaryotic friends lately? In addition to tremendous population sizes, the digestive tracts of mammals typically host hundreds to thousands of symbiotic prokaryotic species with metabolic capacities particular to each host and segments of their host's digestive tract. Herbivorous mammals are in fact dependent upon gut microbes, which help extract most of the energy acquired from ingested plant materials. In addition to calories, resident gut microbial populations also help their hosts obtain nutrients and resist diseases of the digestive tract caused by pathogenic microorganisms. Additional health benefits of these symbiotic gut residents are being added to a long and growing list. In fact, the Human Microbiome Project is an ongoing international research effort to characterize the communities of microbes living in various parts of the human body. The coevolution of vertebrates and their resident microbial populations has undoubtedly been significant throughout vertebrate history, as new species with new diets evolved.

8 Circulatory and Respiratory Systems

Introduction

Respiratory System

The primary function of a respiratory system is to serve as the site of gaseous exchange between an animal and its environment. But because gases diffuse slowly through vertebrate body tissues, vertebrate circulatory and respiratory systems have coevolved to help distribute gases throughout the body. In addition to their primary roles in aiding the diffusion of gases, respiratory systems serve as sites of excretion, promote osmoregulation (gills), and help detect scents (chemoreception), among other functions. Here we focus on the functions related to gas exchange within the circulatory system. These two systems are thus addressed together in this chapter in respect of their interrelated functions.

Circulatory System

Blood vessels, especially veins, can show considerable variation in branching pattern within a species. The descriptions that follow are based upon the most common arrangements. The best criterion to identify a blood vessel is the specific area it supplies (arteries) or drains (veins). Thus, when working through this laboratory manual, be sure to note the region served as well as the general shape of each blood vessel. Wherever color is used in figures, red indicates arteries, blue indicates veins, and yellow indicates the hepatic portal system. However, color does not necessarily indicate relative oxygen content.

Peripheral arteries and veins tend to pass together in many parts of the body. Thus, you will find that veins often bear the same name as their companion arteries (e.g., ventral gastric artery/ventral gastric vein). Besides helping with naming, this tip should also help you locate veins that may be poorly injected.

Shark

Pericardial Cavity

🦅 When examining the peritoneum, the pericardial cavity was reached through hypobranchial muscles. Return to the pericardial cavity seen through this previous entry and identify the four chambers of the heart: in order, sinus venosus, atrium, ventricle, and conus arteriosus (figure 8.1). Using heavy scissors, cut the coracoid bar and transverse septum and clear ventral musculature as needed to reveal the region of the common cardinal and heart (figure 8.2). The triangular **ventricle** lies ventrally

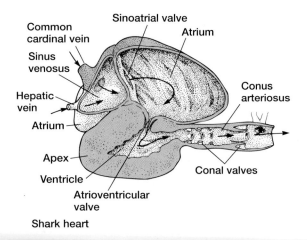

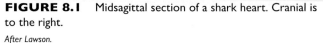

FIGURE 8.1 Midsagittal section of a shark heart. Cranial is to the right.

After Lawson.

and is continuous with a large tubular chamber, the **conus arteriosus,** which passes cranially through the wall of the pericardial cavity. Dorsal to the ventricle lies the **atrium.** Pull the apex (blunt posterior corner) of the ventricle forward to observe the **sinus venosus** where it empties into the atrium—the sinus venosus lies in the transverse septum.

Note the small artery on the lateral walls of the conus arteriosus and extending across the ventricle's surface. This is the **coronary artery** supplying the heart and should be disrupted no more than necessary to expose the following vessels.

Respiratory System and Branchial Arteries

Afferent Branchial Arteries

🦅 Where the conus arteriosus emerges from the pericardial cavity, it becomes the **ventral aorta** and immediately gives off lateral branches, the paired **afferent branchial arteries** (figure 8.3). Note that the afferent branchial arteries course between distinct bands of muscles, the **coracobranchials,** not identified previously. Part these muscles as you trace the afferent branchial arteries. The most caudal afferent branchial immediately divides into the **4th** and **5th afferent branchial arteries.** Make a single cut into the overlying musculature along the ventral midline to follow the ventral aorta cranially. Keep in mind that the ventral aorta is not injected and so is more fragile than similarly sized vessels so far dissected.

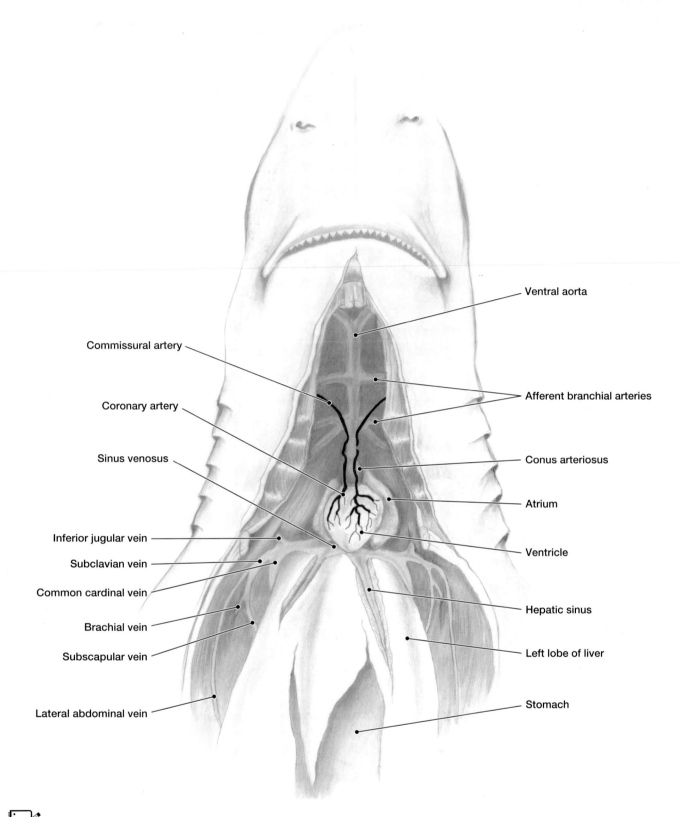

Commissural artery

Coronary artery

Sinus venosus

Inferior jugular vein

Subclavian vein

Common cardinal vein

Brachial vein

Subscapular vein

Lateral abdominal vein

Ventral aorta

Afferent branchial arteries

Conus arteriosus

Atrium

Ventricle

Hepatic sinus

Left lobe of liver

Stomach

FIGURE 8.2 Ventral view of the shark heart and associated blood vessels. The ventral aorta and afferent branchial arteries are likely uninjected in laboratory specimens.

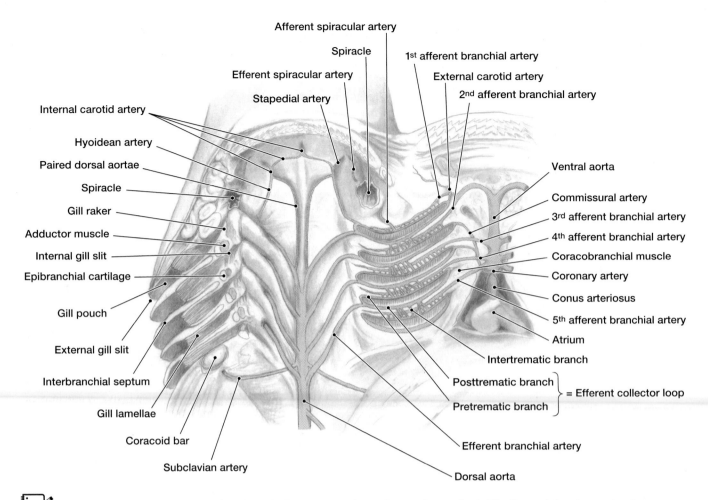

Afferent spiracular artery
Spiracle
Efferent spiracular artery
Stapedial artery
1st afferent branchial artery
External carotid artery
2nd afferent branchial artery

Internal carotid artery
Hyoidean artery
Paired dorsal aortae
Spiracle
Gill raker
Adductor muscle
Internal gill slit
Epibranchial cartilage
Gill pouch
External gill slit
Interbranchial septum
Gill lamellae
Coracoid bar
Subclavian artery

Ventral aorta
Commissural artery
3rd afferent branchial artery
4th afferent branchial artery
Coracobranchial muscle
Coronary artery
Conus arteriosus
5th afferent branchial artery
Atrium
Intertrematic branch
Posttrematic branch
Pretrematic branch
} = Efferent collector loop
Efferent branchial artery
Dorsal aorta

FIGURE 8.3 Ventral view of the branchial structure and vascular supply in a shark. The floor of the pharynx has been cut free caudal to the heart and along the right pharyngeal side. The ventral aorta and afferent branchial arteries are likely uninjected in laboratory specimens.

Shortly, the **3rd afferent branchial artery** exits, and finally the ventral aorta terminates in a bifurcation, which should be followed laterally on the left side. Eventually this fork itself divides, forming the **1st** and **2nd afferent branchial arteries** to supply the first two gills.

Gill Structure

To expose the respiratory system and arteries associated with the gills, you will need to make a major cut along the right side of the head and across the throat, to swing open the lower jaw like a door (in this analogy, the hinge of the door will be the gill region along the left side of the head). Begin at the right corner of the mouth and extend the cut longitudinally and caudally through the middle of the right gills. Then turn the scissors medially and ventrally to cut across the throat just cranial to the coracoid bar. Deviate slightly at the heart to pass through the conus arteriosus, then continue to the base of the opposite gills on the left side of the head. Force open the jaws and observe the transected gills on the right side.

Each gill is supported at its base by an **epibranchial cartilage** and an **adductor muscle** just medial to it (figure 8.3). Just lateral to the gill-arch cartilage are

Form & Function

The many fingerlike processes of gill lamellae increase the surface area of this respiratory surface. More surface area favors gaseous exchange between the capillaries in the gill lamellae and the water ventilating the gills.

several small arteries. On the anterior side of the gill-arch cartilage reside fingerlike processes, **gill rakers,** guarding the **internal gill slit,** the opening between adjacent gill bars. Spread two gills to reveal the **gill pouch,** a pocket that opens laterally through the **external gill slit.** Note the soft, fan-shaped **gill lamellae** extending laterally from the epibranchial cartilage. Trace the path of water as it is drawn into the mouth, forced through the internal gill slit, past the lamellae, and out the external gill slit.

Each gill is considered a complete gill, **holobranch,** if gill lamellae cover both sides. The first gill is the exception, for lamellae lie only on its caudal side. Thus, the first gill is a **hemibranch** or half gill. On the holobranch gills, note that gill lamellae are separated by, and supported on, an **interbranchial septum** that continues laterally to form the division between external gill slits.

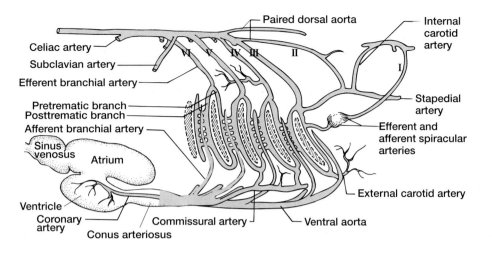

FIGURE 8.4 Aortic arches and related blood vessels in a shark. Roman numerals indicate aortic arches.

After George C. Kent and Larry Miller, Comparative Anatomy of the Vertebrates, 8th edition, 1997. McGraw-Hill Company, Inc., Dubuque, Iowa.

Patterns & Connections

Spelling counts! You may have heard this before, but in some of the circulatory and respiratory terms, a single letter can make a big difference. Compare the terms "branchial" and "brachial." The word "branchial" addresses structure associated with gills. But remove the "n," and the reference is instead to the chest or arms. Within the circulatory system, you have seen the terms "afferent" and "efferent." The "af" in afferent indicates an association with or toward something, while the substitution of an "ef" for the "af" suggests just the opposite, a direction away from something.

Efferent Branchial Arteries

🦖 Now expose the efferent arteries. With tweezers, pull away the mucous membrane on the roof of the mouth taking care not to also pick up underlying blood vessels. If you grasp only the loose mucous membrane, you can be fairly certain to avoid damaging the arteries. Four pairs of large blood vessels, the **efferent branchial arteries,** are so exposed (figure 8.3). These carry blood, oxygenated in the gills, posteriorly. Eventually they merge to form the single **dorsal aorta.** Note that the thin **paired dorsal aortae** arise from the anterior pair of efferent branchial arteries and pass rostrally on each side of the dorsal midline. Clear away connective tissue over all efferent branchial arteries, then follow them all to the cut right side.

Tracing the Blood Flow Past the Gills

🦖 The efferent branchial arteries disappear dorsal to gill cartilages. On the right side of the head, cut away these protruding medial ends of the gill cartilages, bit by bit, until the efferent arteries can be visually traced back to their respective gill lamellae. Each efferent branchial artery begins in the dorsomedial corner of the gill by fusion of two vessels: the thin **pretrematic branch** from gill lamellae on the rostral side of a gill pouch, and the

slightly thicker **posttrematic branch** from the lamellae on the caudal wall of the same gill pouch (figure 8.3). Spend time carefully exposing these branches.

Now turn your attention to the intact left side of the head. Make a neat and informative dissection of afferent and efferent vessels to these gills. Pick one of the middle efferent branchial arteries (on the left, intact side) and follow it to the gill, clearing away connective tissue and cartilage as you did for the right side. As you reach and clear the pretrematic and posttrematic branches, notice that, in fact, they form a complete arterial loop around the gill cleft, the **efferent collector loop.** Pretrematic and posttrematic branches riding on the same interbranchial septum occasionally interconnect directly by a few short **intertrematic branches.** As you reach the ventromedial corner of the gill, you should find the corresponding uninjected afferent branchial artery from the ventral aorta. Complete your dissection of the branchial region so that the relationship of one corresponding afferent and efferent branchial artery and collecting loop is clearly revealed.

Associated Branchial Arteries

🦖 **First Collector Loop** The first collector loop is the source of several arterial branches (figures 8.3 and 8.4). From its dorsomedial corner comes the **hyoidean artery,** which passes along the roof of the mouth medial to the spiracle. Each hyoidean artery is joined by the rostral end of a branch of the paired dorsal aorta, identified previously. Anterior to this juncture with the paired dorsal aorta, the hyoidean artery becomes the **internal carotid artery.** The internal carotid continues rostrally and turns medially where it gives off a small lateral branch, the **stapedial artery.** (Only the base of the stapedial is typically visible.) The paired internal carotids next join at the dorsal midline to form a single internal carotid vessel that enters the chondrocranium through the carotid foramen.

The **afferent spiracular artery** originates from the middle of the pretrematic branch of the first collector loop and passes to capillary beds in the spiracle (figures 8.3 and 8.4). The efferent spiracular artery continues rostrally from the spiracle (eventually uniting with the internal carotid within the cranium), but is impractical to see here as it is rarely sufficiently injected.

Finally, from the ventromedial corner of the first collector loop exits the short **external carotid artery** supplying the lower jaw region (figure 8.4).

Second Collector Loop The **commissural** (= hypobranchial) **artery** arises from the ventral corner of the second collector loop. Often it is joined by additional commissural arteries that arise similarly from the third and fourth collector loops. The commissural arteries enter the pericardial cavity to become the **coronary arteries,** identified earlier, to supply the heart (figure 8.4).

Subclavian Arteries Return to the efferent branchial arteries. Follow them caudally as they arch inward to join the unpaired dorsal aorta passing to the pleuroperitoneal cavity. Just before the last pair of efferent branchial arteries join the dorsal aorta, the paired **subclavian arteries** emerge laterally from the dorsal aorta (figures 8.3 and 8.4). You will have to probe deeply under the transverse septum to find them. The subclavian artery supplies the lateral body wall and pectoral fin, but these branches will not be followed.

Pleuroperitoneal Cavity

Four Branches of the Dorsal Aorta

🦖 Pull viscera to the right to find the **dorsal aorta,** injected with red or pink latex, running along the dorsal midline. Many tiny, paired branches, **intersegmental arteries,** leave the dorsal aorta dorsally to supply axial muscles. But only four major, unpaired branches leave the dorsal aorta ventrally to supply specific regions of viscera. In order, they are the celiac, anterior mesenteric, gastrosplenic, and posterior mesenteric arteries.

Celiac Artery The first of four major branches off the dorsal aorta is the **celiac artery,** departing well forward in the pleuroperitoneal cavity, cranially, above the liver (figures 8.5, 8.6, and 8.7). At this point, it is impractical to clear tissue and viscera to get an unobstructed view of the origin of the celiac artery, but you should try to find the approximate location of this branching point. The celiac artery runs caudally without branching until it emerges dorsal to the stomach along the lesser curvature. Here it divides into the **gastric artery** to the body of the stomach and the **pancreaticomesenteric artery** primarily to the intestine (figures 8.5 and 8.6). After tracing the branches of the pancreaticomesenteric artery, the gastric artery will be followed further.

Near the pyloric sphincter, the pancreaticomesenteric artery sends a small branch to the pylorus and ventral lobe of the pancreas (unnamed), and at about

the same point a larger branch, the **duodenal artery,** to the duodenum. Beyond this point, the artery continues as the **anterior intestinal artery** distributing along the right, anterior side of the valvular intestine.

Return to the base of the gastric artery to find a thin branch, the **hepatic artery,** passing to the liver within the gastrohepatoduodenal ligament and parallel to the common bile duct. Beyond the point of the hepatic artery, the gastric artery divides into **dorsal gastric** and **ventral gastric arteries,** which branch and distribute across dorsal and ventral walls of the stomach, respectively.

Anterior Mesenteric Artery Along the caudal third of the pleuroperitoneal cavity, two adjacent arteries, the anterior mesenteric and gastrosplenic arteries, depart (figures 8.5, 8.6, and 8.7). After leaving the dorsal aorta, they cross and pass in the posterior free edge of the mesentery proper. The first of these, the **anterior mesenteric artery,** divides into branches that course in the spiral creases of the left, middle, and posterior sides of the valvular intestine.

Gastrosplenic Artery Just caudal to the anterior mesenteric artery, the dorsal aorta gives off the **lienogastric** (= gastrosplenic) **artery,** which passes to the spleen, stomach, and dorsal lobe of the pancreas (figures 8.5, 8.6, and 8.7).

Posterior Mesenteric Artery The fourth median branch of the dorsal aorta is the **posterior mesenteric artery,** which runs along the cranial free edge of the mesorectum to supply the rectal gland (figures 8.5, 8.6, and 8.7).

Caudal Branches of the Dorsal Aorta

🦖 Usually, poor injection of latex and obstructing viscera of the urogenital system prevent clear exposure of more posterior vessels. In suitable specimens, or on demonstration, the dorsal aorta caudal to the posterior mesenteric artery can be seen to give off paired **iliac arteries** that pass to the body wall and then into pelvic fin musculature. Beyond the exit of the iliac arteries, the dorsal aorta continues into the tail as the **caudal artery.**

Hepatic Portal System

The hepatic portal system consists of vessels that collect blood from capillaries in the stomach, intestine, pancreas, and spleen and carry it to capillaries in the liver. Because of its unique position between these capillary beds, the hepatic portal system must be injected separately from the arterial and venous systems. This is also true for the *Necturus* and cat circulatory systems. If the hepatic portal system of your animal has been injected with latex, the color will likely be distinct from the color used to inject the veins. Typically, blue and yellow are used for these two venous systems, but variation exists in which of these colors is used for the hepatic portal system. Your instructor will be able to confirm if the hepatic portal system was injected, and what color was used.

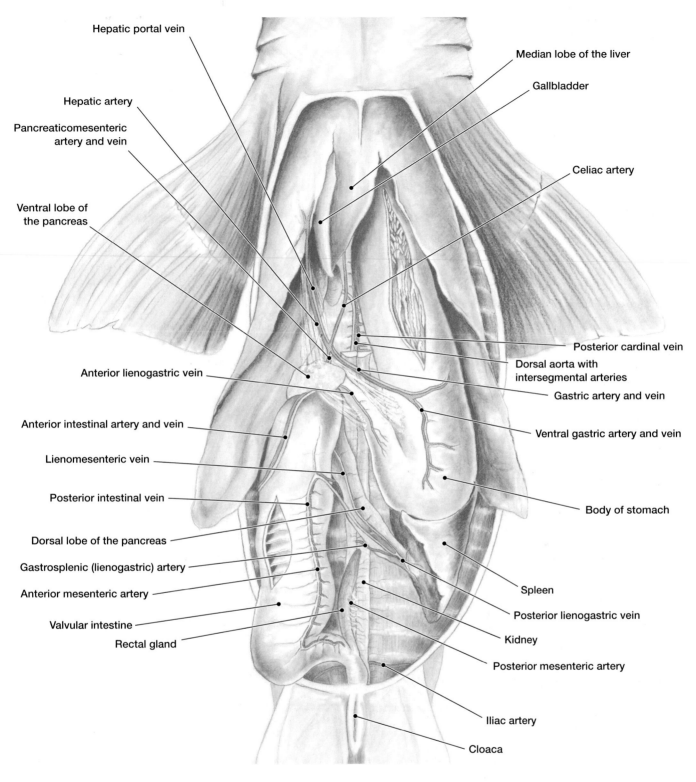

Hepatic portal vein

Hepatic artery

Pancreaticomesenteric
artery and vein

Ventral lobe of
the pancreas

Anterior lienogastric vein

Anterior intestinal artery and vein

Lienomesenteric vein

Posterior intestinal vein

Dorsal lobe of the pancreas

Gastrosplenic (lienogastric) artery

Anterior mesenteric artery

Valvular intestine

Rectal gland

Median lobe of the liver

Gallbladder

Celiac artery

Posterior cardinal vein

Dorsal aorta with
intersegmental arteries

Gastric artery and vein

Ventral gastric artery and vein

Body of stomach

Spleen

Posterior lienogastric vein

Kidney

Posterior mesenteric artery

Iliac artery

Cloaca

FIGURE 8.5 Ventral view of the pleuroperitoneal cavity of a shark.

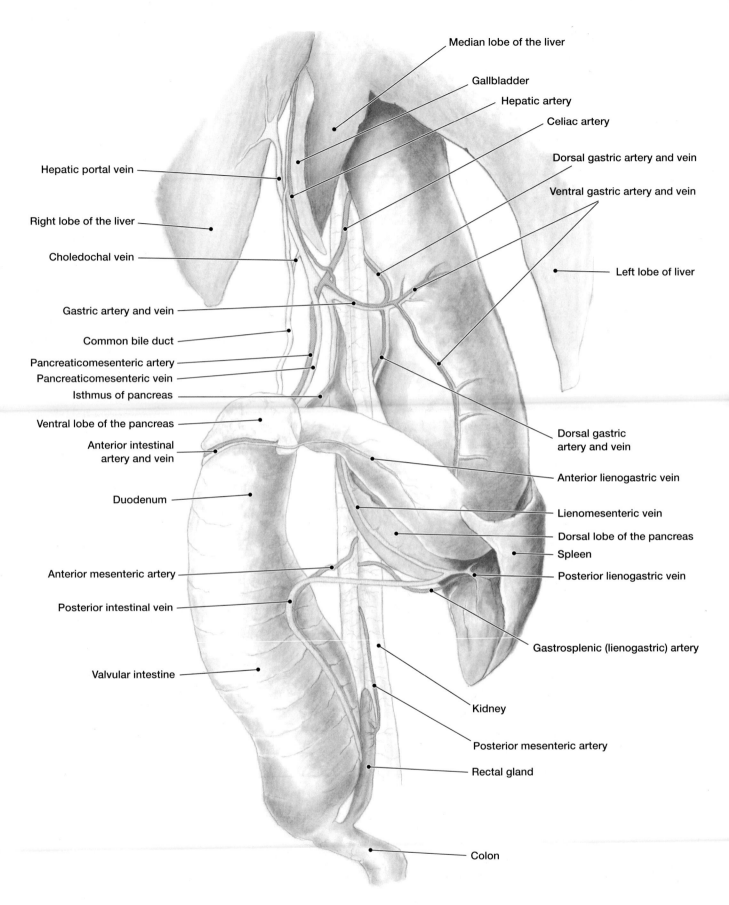

Median lobe of the liver

Gallbladder

Hepatic artery

Celiac artery

Dorsal gastric artery and vein

Ventral gastric artery and vein

Hepatic portal vein

Left lobe of liver

Right lobe of the liver

Choledochal vein

Gastric artery and vein

Common bile duct

Pancreaticomesenteric artery

Pancreaticomesenteric vein

Isthmus of pancreas

Dorsal gastric artery and vein

Ventral lobe of the pancreas

Anterior intestinal artery and vein

Anterior lienogastric vein

Lienomesenteric vein

Dorsal lobe of the pancreas

Spleen

Duodenum

Anterior mesenteric artery

Posterior lienogastric vein

Posterior intestinal vein

Gastrosplenic (lienogastric) artery

Valvular intestine

Kidney

Posterior mesenteric artery

Rectal gland

Colon

FIGURE 8.6 Blood supply to alimentary canal of a shark.

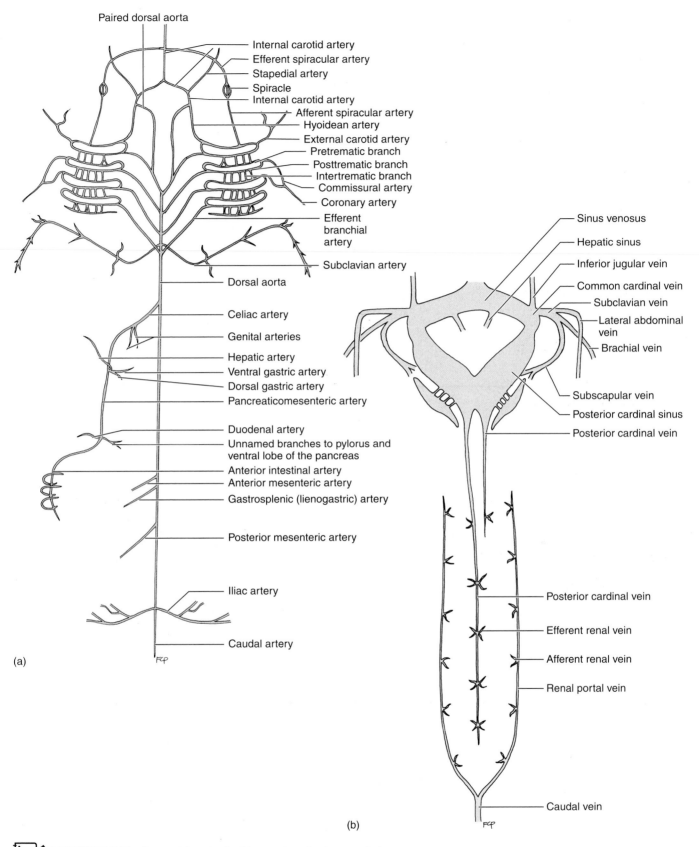

Paired dorsal aorta

Internal carotid artery
Efferent spiracular artery
Stapedial artery
Spiracle
Internal carotid artery
Afferent spiracular artery
Hyoidean artery
External carotid artery
Pretrematic branch
Posttrematic branch
Intertrematic branch
Commissural artery
Coronary artery
Efferent branchial artery

Subclavian artery

Dorsal aorta

Celiac artery

Genital arteries

Hepatic artery
Ventral gastric artery
Dorsal gastric artery
Pancreaticomesenteric artery

Duodenal artery
Unnamed branches to pylorus and ventral lobe of the pancreas
Anterior intestinal artery
Anterior mesenteric artery
Gastrosplenic (lienogastric) artery

Posterior mesenteric artery

Iliac artery

Caudal artery

(a)

Sinus venosus
Hepatic sinus
Inferior jugular vein
Common cardinal vein
Subclavian vein
Lateral abdominal vein
Brachial vein

Subscapular vein
Posterior cardinal sinus
Posterior cardinal vein

Posterior cardinal vein

Efferent renal vein

Afferent renal vein

Renal portal vein

Caudal vein

(b)

FIGURE 8.7   Artery (a) and vein (b) maps in a shark, ventral view.

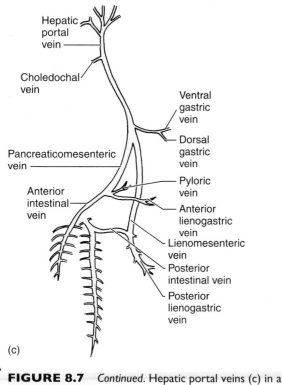

Hepatic portal vein

Choledochal vein

Pancreaticomesenteric vein

Anterior intestinal vein

Ventral gastric vein

Dorsal gastric vein

Pyloric vein

Anterior lienogastric vein

Lienomesenteric vein

Posterior intestinal vein

Posterior lienogastric vein

(c)

FIGURE 8.7 *Continued.* Hepatic portal veins (c) in a shark, ventral view.

🦖 The large vessel running in the hepatoduodenal ligament next to the tiny hepatic artery is the **hepatic portal vein** (figures 8.5, 8.6, and 8.7c). Note that anteriorly it divides and distributes to the liver. Following it posteriorly from the liver, it is joined by a thin vein that runs along the common bile duct, the **choledochal vein** (figures 8.6 and 8.7c). (Keep in mind that some veins may be poorly injected with color, requiring you to search for the plain, tan-colored vessel.) The caudal end of the hepatic portal vein forms by the confluence of three major branches: **gastric, pancreaticomesenteric,** and **lienomesenteric veins.**

🦖 **Gastric Vein** The **gastric vein** comes from the stomach and is itself formed by union of the **dorsal gastric** and **ventral gastric veins** (figures 8.5, 8.6, and 8.7).

🦖 **Pancreaticomesenteric Vein** The second branch joining the hepatic portal, the pancreaticomesenteric vein, begins posteriorly as the **anterior intestinal vein** extending from the right, anterior side of the valvular intestine (figures 8.5, 8.6, and 8.7). Trace the anterior intestinal vein cranially to where it is joined by a longer, thin vein, the **anterior lienogastric,** from the spleen and portions of the pylorus. A small tributary from the ventral side of the pylorus, the **pyloric vein,** also joins at about this same point. Cranially from the juncture of these three veins, the vessel is properly termed the **pancreaticomesenteric vein.**

🦖 **Lienomesenteric Vein** The third branch joining the hepatic portal, the **lienomesenteric vein,** begins posteriorly as the **posterior intestinal vein** draining the left, posterior valvular intestine (figures 8.5, 8.6, and 8.7). It passes away from the intestine and runs freely toward the pancreas, where it is joined by a short vessel, the **posterior lienogastric vein** from the spleen and bend of the stomach. Cranially from this juncture with the lienogastric vein, the vessel is properly the **lienomesenteric vein.** It passes in the surface of the dorsal lobe of the pancreas, from which it receives several small vessels, **pancreatic veins.** Pick away pancreatic tissue to expose several of these pancreatic veins.

(Notice the quirk in naming arterial and venous vessels to the intestine. Anterior and posterior intestinal veins exist, but only an anterior intestinal artery is identified. The expected posterior arterial "counterpart" is termed the anterior mesenteric artery whose counterpart in turn is the posterior mesenteric artery to the rectal gland.)

Veins

🦖 Along each lateral body wall runs a **lateral abdominal vein** returning blood from the fins and adjacent areas (figures 8.2 and 8.7b). Trace it cranially to the region just caudal to the transverse septum. Here, the lateral abdominal vein is joined by the **subscapular vein** running ventrally from the dorsolateral side of the body wall. Clear away the parietal peritoneum covering the subscapular vein and note the deep **brachial vein,** from the fin, joining the subscapular vein. Together, the lateral abdominal and subscapular veins empty into the short **subclavian vein,** which immediately supplies the more medial and expanded **common cardinal vein.** Left and right common cardinals are each joined by a **hepatic sinus,** from the ventral end of the liver, and by the more cranial **inferior jugular veins,** before emptying into the single **sinus venosus,** the first heart chamber.

Turn now to the pleuroperitoneal cavity dorsal to the stomach. Take care not to damage the kidney. Next to the dorsal aorta runs the paired **posterior cardinal veins,** which receive many **efferent renal veins** draining the kidneys. Follow the posterior cardinal veins forward to where they empty into the very large **posterior cardinal sinuses.**

Necturus

The Heart, Branches of the Ventral Aorta, and Gills

🦖 The pericardial cavity has already been entered through the throat musculature. The **left atrium** and **right atrium** lie on the cranial base of the single large muscular **ventricle** (figures 8.8a and 8.9). Lift the apex of the ventricle to find the thin-walled **sinus venosus.** Poor injection usually makes it difficult to identify veins that

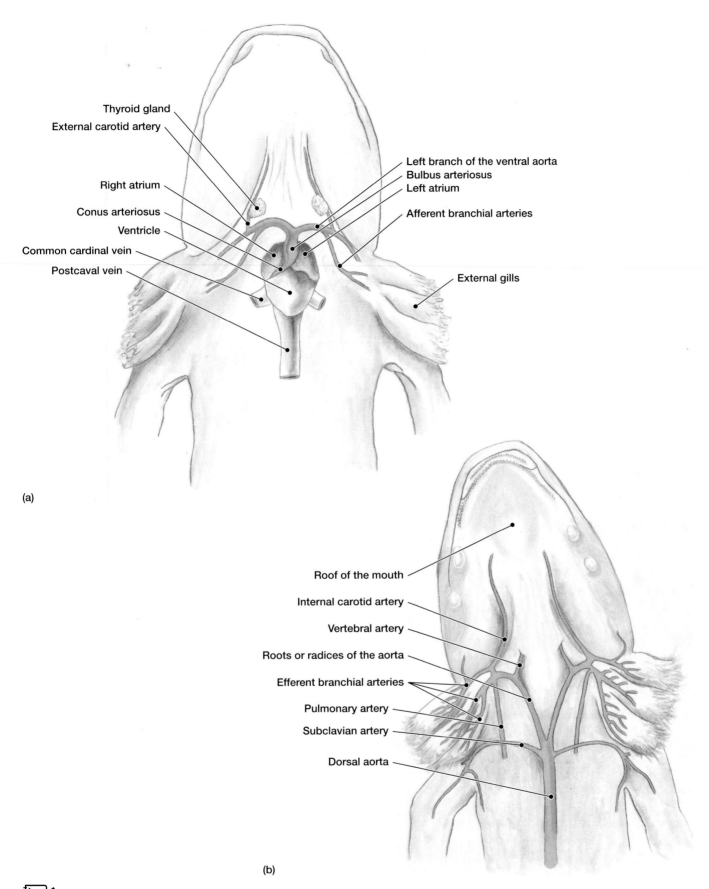

FIGURE 8.8 Heart chambers and associated blood vessels in *Necturus*. (a) Ventral view of heart and associated blood vessels. (b) Ventral view of blood vessels in the roof of the mouth.

Labels in figure (a):
Thyroid gland
External carotid artery
Right atrium
Conus arteriosus
Ventricle
Common cardinal vein
Postcaval vein
Left branch of the ventral aorta
Bulbus arteriosus
Left atrium
Afferent branchial arteries
External gills
(a)

Labels in figure (b):
Roof of the mouth
Internal carotid artery
Vertebral artery
Roots or radices of the aorta
Efferent branchial arteries
Pulmonary artery
Subclavian artery
Dorsal aorta
(b)

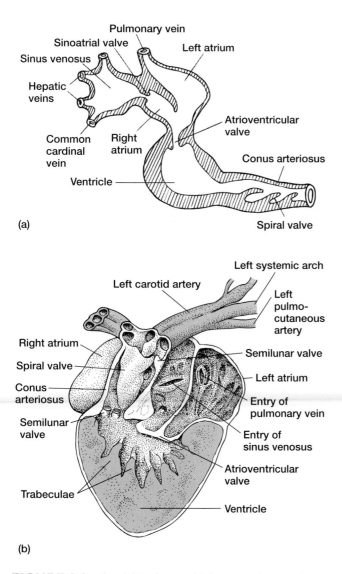

Pulmonary vein
Sinoatrial valve
Left atrium
Sinus venosus
Hepatic
veins
Atrioventricular
valve
Common
cardinal
vein
Right
atrium
Conus arteriosus
Ventricle
(a)
Spiral valve

Left systemic arch
Left carotid artery
Left
pulmo-
cutaneous
artery
Right atrium
Semilunar valve
Spiral valve
Left atrium
Conus
arteriosus
Entry of
pulmonary vein
Semilunar
valve
Entry of
sinus venosus
Atrioventricular
valve
Trabeculae
Ventricle
(b)

FIGURE 8.9 Amphibian hearts. (a) Diagram of a typical amphibian heart. Cranial is to the right. Notice that the atrium is divided into left and right chambers but the ventricle lacks an internal septum. (b) Ventral view of a front section of a bullfrog (*Rana catesbeiana*) heart.

(b) After M. H. Wake, ed., Hyman's Comparative Vertebrate Anatomy, 3d ed., © 1979 by The University of Chicago.

to each side to distribute to the feathery, **external gills** (three on each side; figure 8.8a and figure 8.10). Follow each branch, left and right, by carefully removing all the overlying throat musculature. Each branch soon divides, the anterior fork going to the first gill, the posterior fork branching again to supply the middle and posterior external gills. These three separate supplies to the external gills are the **afferent branchial arteries.** Before reaching the gill, the first afferent branchial artery gives off a thin branch, the **external carotid artery,** which runs forward between gill-arch cartilages and branchiohyoideus muscle. At the fork between external carotid and first afferent branchial artery lies the small, lobed **thyroid gland,** supplied by the external carotid artery.

Clear muscle and connective tissue so that the distribution of afferent branchial arteries on both sides can be followed into the external gills.

Dorsal Aorta and Lungs

To expose the efferent branchial arteries and their convergence into the dorsal aorta, it will be necessary to reach the roof of the mouth. Cut through the angle of the jaws on both sides up to the gill arches. On the **left** side, continue this cut laterally through the bases of the gill arch cartilages by keeping one blade of the scissors in the esophagus, the other external. Remove the mucous membrane from the roof of the mouth.

At the back of the mouth, two large vessels, the **roots,** or **radices of the aorta,** arch inward, pass posteriorly, and unite to form the **dorsal aorta** (figures 8.8b and 8.10). On the right side, locate the source of these roots, the three **efferent branchial arteries** in the external gills. Note that the middle and posterior efferent branchial arteries join immediately upon leaving the external gills. Near their union, the **pulmonary artery** exits, passing posteriorly to supply the right lung. (The lungs were previously identified in the digestive system chapter. They are slender and saclike, running along the dorsolateral sides of the pleuroperitoneal cavities.) The first efferent branchial artery leaves the gill, runs parallel to the root of the aorta joining it by a short, unnamed shunt, then swings forward in the roof of the mouth as the **internal carotid artery,** eventually disappearing

contribute to the sinus venosus. Generally, blood returning from the forelimbs and anterior body enters the sinus venosus laterally through the paired **common cardinal vein;** blood from the posterior body returns from the long, single **postcaval vein,** which passes through the transverse septum and upon entering the pericardial cavity divides into **hepatic sinuses** that join the common cardinal veins in entering the sinus venosus (figure 8.10). From the right, anterior corner of the ventricle exits the narrow **conus arteriosus,** which between the atria expands into the swollen ventral aorta here termed the **bulbus arteriosus** (figures 8.8–8.10).

Immediately upon exiting from the pericardial cavity, the bulbus arteriosus divides, sending a single branch

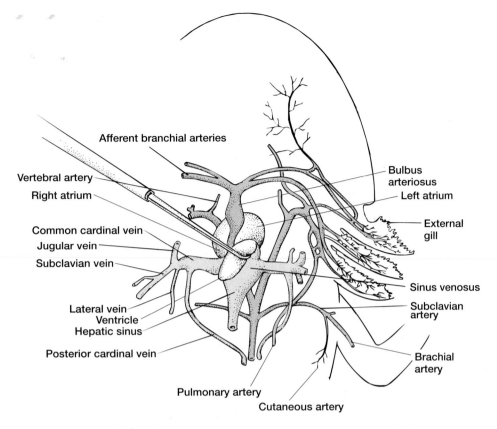

Afferent branchial arteries

Vertebral artery

Right atrium

Common cardinal vein

Jugular vein

Subclavian vein

Lateral vein

Ventricle

Hepatic sinus

Posterior cardinal vein

Pulmonary artery

Cutaneous artery

Bulbus arteriosus

Left atrium

External gill

Sinus venosus

Subclavian artery

Brachial artery

FIGURE 8.10 Aortic arches in *Necturus*. This figure integrates the separate views indicated in figure 8.8a,b.

into the cranium. The thin external carotid artery, identified earlier, should be located again and followed dorsally around the side of the gills beneath the skin to find where it leaves the first afferent branchial artery.

Just as the root of the aorta turns caudally, it gives off a branch running deep into neck muscles, the **vertebral artery** (figures 8.8b and 8.10). Immediately following the union of the roots of the aorta, the dorsal aorta gives off a paired, lateral branch running into the arm, the **subclavian artery.** The subclavian sends one branch along the body wall and several branches into the shoulder and arm.

Pull viscera to the right and locate the dorsal aorta along the dorsal body wall (figures 8.11 and 8.12). Dorsally, it gives off **parietal** (= intercostal) **arteries** that pass segmentally to the axial musculature. After the subclavian, the next major branch to the viscera is the unpaired **gastric artery,** which soon forks into **dorsal** and **ventral gastric arteries** to respective sides of the stomach. The ventral gastric artery also supplies the spleen. In sharks and cats, the gastric artery is a branch off the celiac. Here in *Necturus*, the gastric artery branches from the dorsal aorta, cranial to the celiac.

The next, midventral branch of the dorsal aorta, further posteriorly, is the **celiacomesenteric artery,** followed immediately by numerous **mesenteric arteries** to the intestine (figures 8.11 and 8.12). The celiacomesenteric artery divides into four branches: the **splenic artery** passes cranially to supply the posterior end of the

Patterns & Connections

Like most larval salamanders, *Necturus* simultaneously possesses gills and lungs! In larval salamanders about to undergo metamorphosis, the gills will soon be lost as the lungs become the primary respiratory structures (with the exception of plethodontid salamanders). But why is this adaptive for *Necturus*, an animal that never transforms to a life on land? For *Necturus* living in occasionally oxygen-poor water, lungs can supplement gill respiration. By gulping air at the surface and forcing it into the lungs, *Necturus* obtains supplemental oxygen for this relatively large and occasionally active body. In addition, lungs filled with air also give *Necturus* a means to control its underwater buoyancy.

spleen; the **pancreaticoduodenal artery** supplies and is embedded in the pancreas and eventually reaches the pylorus. The remaining two, best seen on the right side, are the short **mesenteric branch** of the celiacomesenteric artery to the intestine following the duodenum; and the **hepatic artery** passing to the posterior end of the liver.

The numerous **mesenteric arteries** supply the remainder of the small and all of the large intestine (figures 8.11 and 8.12). Thus, *Necturus* lacks the posterior mesenteric arteries typical of other vertebrates. Near the bases of the first few mesenteric arteries are the **genital arteries** (spermatic arteries or ovarian arteries) passing laterally to the gonad. Posterior to this the dorsal aorta

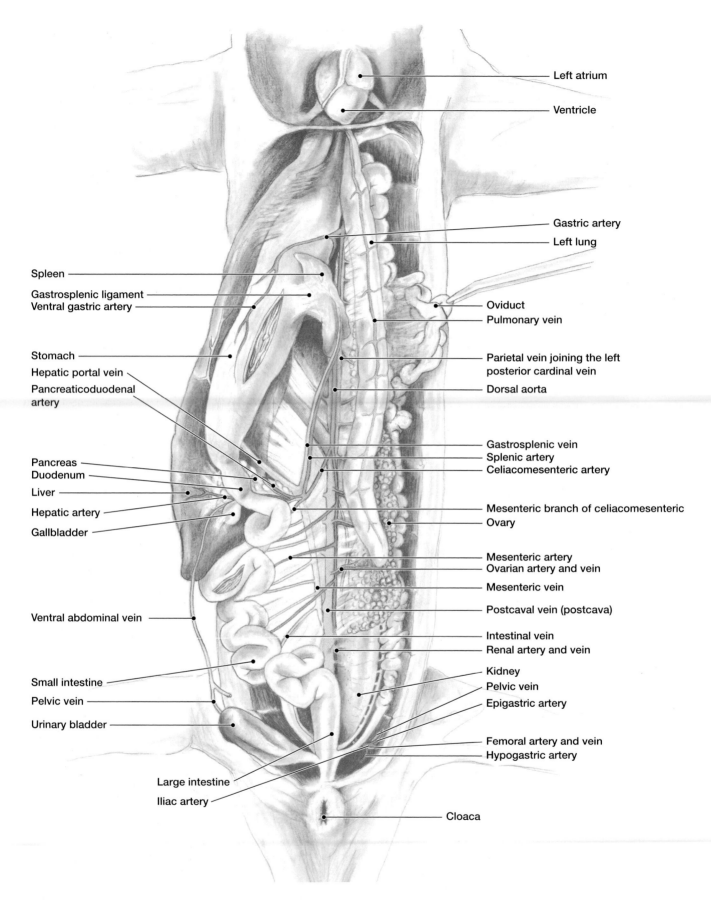

Left atrium

Ventricle

Gastric artery

Left lung

Spleen

Gastrosplenic ligament

Ventral gastric artery

Oviduct

Pulmonary vein

Stomach

Hepatic portal vein

Pancreaticoduodenal
artery

Parietal vein joining the left
posterior cardinal vein

Dorsal aorta

Gastrosplenic vein

Splenic artery

Celiacomesenteric artery

Pancreas

Duodenum

Liver

Hepatic artery

Gallbladder

Mesenteric branch of celiacomesenteric

Ovary

Mesenteric artery

Ovarian artery and vein

Mesenteric vein

Postcaval vein (postcava)

Ventral abdominal vein

Intestinal vein

Renal artery and vein

Kidney

Pelvic vein

Epigastric artery

Small intestine

Pelvic vein

Urinary bladder

Femoral artery and vein

Hypogastric artery

Large intestine

Iliac artery

Cloaca

FIGURE 8.11   Ventral view of the pleuroperitoneal cavity of a female *Necturus*.

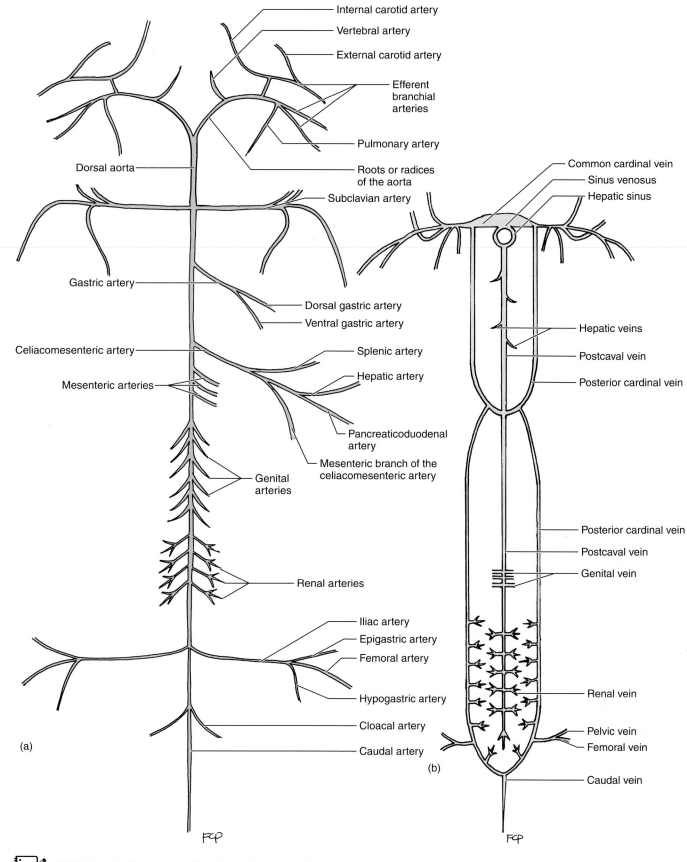

Internal carotid artery
Vertebral artery
External carotid artery
Efferent branchial arteries
Pulmonary artery
Roots or radices of the aorta
Subclavian artery

Dorsal aorta

Gastric artery
Dorsal gastric artery
Ventral gastric artery

Celiacomesenteric artery
Splenic artery
Hepatic artery

Mesenteric arteries
Pancreaticoduodenal artery
Mesenteric branch of the celiacomesenteric artery

Genital arteries

Renal arteries

Iliac artery
Epigastric artery
Femoral artery
Hypogastric artery
Cloacal artery
Caudal artery

Common cardinal vein
Sinus venosus
Hepatic sinus

Hepatic veins
Postcaval vein
Posterior cardinal vein

Posterior cardinal vein
Postcaval vein
Genital vein

Renal vein
Pelvic vein
Femoral vein
Caudal vein

(a)

(b)

FIGURE 8.12 Artery (a) and vein (b) maps in *Necturus,* ventral view.

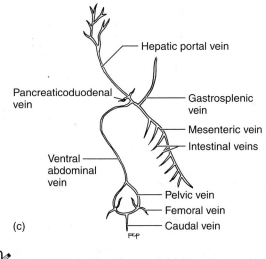

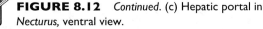

- Hepatic portal vein
- Pancreaticoduodenal vein
- Gastrosplenic vein
- Mesenteric vein
- Intestinal veins
- Ventral abdominal vein
- Pelvic vein
- Femoral vein
- Caudal vein

(c)

FIGURE 8.12 *Continued.* (c) Hepatic portal in *Necturus,* ventral view.

gives off numerous, but very short, **renal arteries** to the kidney that are usually difficult to find because the kidney is so tightly applied to the aorta.

Near the hindlimb, the dorsal aorta gives off the paired **iliac arteries** (figures 8.11 and 8.12). It has three major branches. The first, the **epigastric artery,** passes cranially in the body wall. The second, the **femoral artery,** penetrates the body wall to enter the limb. The third, the **hypogastric artery,** swings medially to the urinary bladder and cloaca.

Beyond the iliacs, the dorsal aorta gives off cloacal arteries and becomes the caudal artery to supply the tail. These two arteries need not be followed.

Hepatic Portal System

Next to the epigastric artery runs the **pelvic vein,** which joins its partner of the opposite side to form the unpaired **ventral abdominal vein,** which passes in the ventral body wall cranial to the liver (figures 8.11 and 8.12c). In most specimens, neither pelvic nor ventral abdominal veins are sufficiently injected to permit clear identification, although uninjected sections can certainly be discovered. Often the cranial part of the ventral abdominal vein can be found injected running past the gallbladder on the most posterior tip of the liver.

The hepatic portal system begins in the **mesenteric vein** from the large intestine (figures 8.11 and 8.12c). Short, numerous **intestinal veins** join the mesenteric vein on its course forward. As the mesenteric vein approaches the liver, it is joined by three veins all at about the same point. Part of the pancreas may need to be removed to expose the junction of these veins. The first is the **pancreaticoduodenal vein,** from duodenum and pancreas. Next, the **gastrosplenic vein,** from spleen and stomach, and the **ventral abdominal vein** join at about the same point. Anterior to this juncture, the collective vein becomes the **hepatic portal vein.** It soon ramifies to distribute across the dorsal surface of the liver, delivering blood to liver sinuses.

Systemic Veins

The **postcaval vein** begins posteriorly between the kidneys, which empty into it through many small renal veins (figures 8.11 and 8.12b). As the postcaval vein journeys forward, **genital veins** (spermatic veins or ovarian veins, specifically) contribute. The postcava leaves the dorsal body wall and passes to the dorsal surface of the liver, coursing as a vessel through the organ and receiving more veins that drain the liver, **hepatic veins** (figure 8.12b). Pick away liver tissue to follow the postcava on its course. Note, for instance, the prominent hepatic veins on the midventral surface of the liver that join the postcava. The postcava is large as it exits from the cranial corner of the liver (now, by the way, in the coronary ligament) to pierce the transverse septum and, as it enters the pericardial cavity, to divide into hepatic sinuses identified previously. Again remember that these hepatic sinuses and the sinus venosus into which they enter are often poorly injected.

The thin, paired **posterior cardinal vein** receives blood from the tail and hindlimb and can be identified where it runs along the lateral edge of the kidney delivering some of its blood through small tributaries to this organ (figures 8.11 and 8.12b). The posterior cardinal vein, through a short shunt, joins the postcava where this large vessel leaves the dorsal body wall and slants toward the liver. The still paired posterior cardinal vein does not end here, but continues forward parallel with the dorsal aorta. Segmental **parietal veins** join the posterior cardinal vein dorsally. Nearing the transverse septum, the posterior cardinal vein diverges laterally, piercing the septum, and enters the **common cardinal vein.**

Pulmonary Vein

Both pulmonary artery and vein are likely injected with pink latex. You have already traced the pulmonary artery from the efferent branchial artery to the surface of the lung opposite its mesenteric attachment. The **pulmonary vein** runs along the lung next to this mesenteric attachment (figure 8.11). It joins its partner from the opposite side and enters the left atrium (figure 8.9).

Cat

We begin with the dissection of the major arteries emanating from the heart. The veins, usually more poorly injected, will be traced later. Most veins accompany respective arteries of the same name; thus, knowing arteries will help with finding veins. However, be careful not to destroy the veins in the process of finding the arteries!

Recall that the *deep* dissection of the muscular system was confined to the left side of the cat to limit damage to the blood vessels on the right side. Therefore, the right member of paired blood vessels will be identified first.

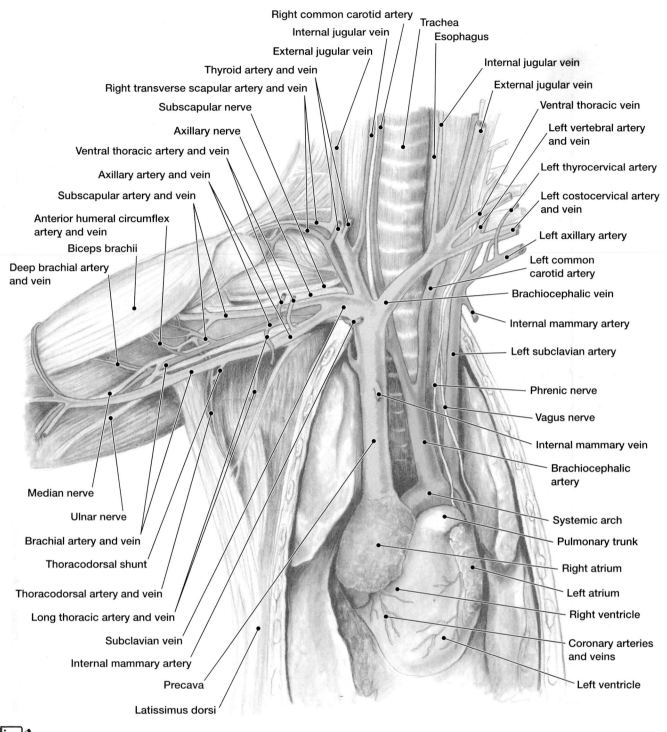

Right common carotid artery
Internal jugular vein
Trachea
Esophagus
External jugular vein
Thyroid artery and vein
Right transverse scapular artery and vein
Subscapular nerve
Axillary nerve
Ventral thoracic artery and vein
Axillary artery and vein
Subscapular artery and vein
Anterior humeral circumflex artery and vein
Biceps brachii
Deep brachial artery and vein

Internal jugular vein
External jugular vein
Ventral thoracic vein
Left vertebral artery and vein
Left thyrocervical artery
Left costocervical artery and vein
Left axillary artery
Left common carotid artery
Brachiocephalic vein
Internal mammary artery
Left subclavian artery
Phrenic nerve
Vagus nerve
Internal mammary vein
Brachiocephalic artery
Systemic arch
Pulmonary trunk
Right atrium
Left atrium
Right ventricle
Coronary arteries and veins
Left ventricle

Median nerve
Ulnar nerve
Brachial artery and vein
Thoracodorsal shunt
Thoracodorsal artery and vein
Long thoracic artery and vein
Subclavian vein
Internal mammary artery
Precava
Latissimus dorsi

 FIGURE 8.13 Ventral view of thoracic components of the circulatory system in a cat.

Heart and Its Major Vessels

🦖 If the heart is not already exposed, continue a midventral incision through the sternum anteriorly, until the entire rib cage can be opened. Be careful of large vessels at the anterior corner of the rib cage. A glandular mass, the thymus, lies just cranial to the heart and may hide some of the major arteries and veins. This thymus and associated connective tissue should be cleared

away. In young, the thymus is important in setting up the immune system. In adults, the thymus regresses but does not entirely disappear. Also try to locate the thin, phrenic nerve (figure 8.13) that passes between the heart and root of the lung. The phrenic innervates the diaphragm and helps regulate breathing. Cut away completely the parietal pericardium (= pericardial sac) covering the heart to near the base of the

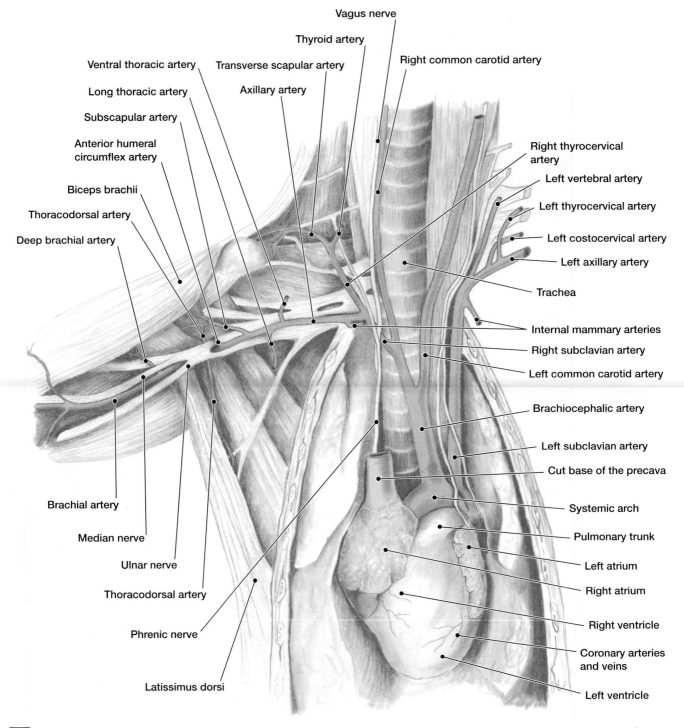

Vagus nerve

Thyroid artery

Ventral thoracic artery

Transverse scapular artery

Long thoracic artery

Axillary artery

Subscapular artery

Anterior humeral
circumflex artery

Biceps brachii

Thoracodorsal artery

Deep brachial artery

Right common carotid artery

Right thyrocervical
artery

Left vertebral artery

Left thyrocervical artery

Left costocervical artery

Left axillary artery

Trachea

Internal mammary arteries

Right subclavian artery

Left common carotid artery

Brachiocephalic artery

Left subclavian artery

Cut base of the precava

Systemic arch

Pulmonary trunk

Left atrium

Right atrium

Right ventricle

Coronary arteries
and veins

Left ventricle

Brachial artery

Median nerve

Ulnar nerve

Thoracodorsal artery

Phrenic nerve

Latissimus dorsi

FIGURE 8.14 Ventral view of systemic arch distribution, precava and its tributaries removed, in a cat.

anterior vessels entering and leaving the heart. The large muscular **ventricles,** left and right, are covered by thin **coronary arteries** and **veins** that supply the tissues. Across the anterior end of the ventricles lie the **left** and **right atria** (figures 8.13–8.15). Lift the apex of the heart to find the single, thin-walled **postcava**

(perhaps injected with blue latex),[1] passing cranially through the diaphragm to the right atrium. The large, blue **precava** also enters here and can be seen coming from the cranial region of the body. The **aorta** exits from between the atria on the cranial side of the heart (figures 8.13–8.15). It then gives off two large branches that pass cranially, and then itself arches caudally along the dorsal wall of the thoracic cavity. To the left of the aorta, near its base, lies the **pulmonary trunk.**

---

[1]Also termed *posterior* and *anterior,* or *inferior* and *superior venae cavae,* respectively.

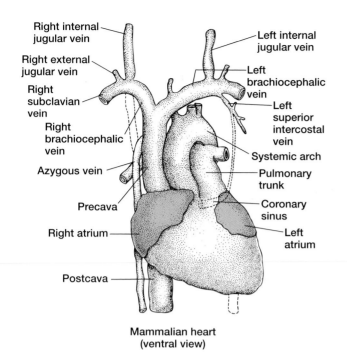

Right internal jugular vein

Right external jugular vein

Right subclavian vein

Right brachiocephalic vein

Azygous vein

Precava

Right atrium

Postcava

Left internal jugular vein

Left brachiocephalic vein

Left superior intercostal vein

Systemic arch

Pulmonary trunk

Coronary sinus

Left atrium

Mammalian heart
(ventral view)

FIGURE 8.15 Mammalian heart, ventral view, to show the complex venous and arterial components surrounding the heart. *After Lawson.*

This large blood vessel passes dorsally from the right ventricle to divide into **left** and **right pulmonary arteries** supplying the respective lungs. Note that this artery, named so because it transports blood away from the heart, carries blood with some of the lowest levels of oxygen anywhere in the body. Clear any obscuring fat to follow these vessels. A short piece of connective tissue, the **ligamentum arteriosum,** is often seen extending between the bases of the pulmonary trunk and aorta. This ligament is a remnant of the fetal shunt, the ductus arteriosus.

Aorta

➤ To aid location of the aorta, remove and discard the left lobes of the lung by cutting cleanly through their bases. Be careful to avoid damaging the nerves that pass through this region. Three regions of the aorta are often recognized: the **aortic trunk,** the base of the vessel where it leaves the heart; the curved **systemic arch;** and the long **dorsal aorta** running posteriorly along the dorsal midline.[2]

In the cat, two vessels project from the systemic arch. The first and slightly larger on the cat's right is the single **brachiocephalic artery.** Next, adjacent to it, is the **left subclavian artery** to the left arm (figure 8.14). Note: In mammals, there is considerable variation in the branching pattern of arteries arising from the systemic arch.

Brachiocephalic Artery Follow the brachiocephalic first. At about the same point, it divides into three

branches. The first two branches to come off are the **left** and then **right common carotid arteries,** which lie next to and parallel the trachea. The third branch, the **right subclavian artery,** passes laterally into the right arm (figures 8.14 and 8.16).

Right Subclavian Artery Four branches exit at about the point where the subclavian passes the first rib (figures 8.14 and 8.16a1). First, the **internal mammary artery** arises from the ventral surface of the subclavian, passes to the chest wall in the company of the corresponding vein, and runs caudally along the ventral side of the thoracic cavity. Second, the **vertebral artery** arises opposite the internal mammary artery, off of the dorsal side of the subclavian. Lift the subclavian to find the vertebral's point of departure and passage deep into the transverse foramen of the seventh cervical vertebra. Third, near the base of the vertebral artery, is the **costocervical artery,** which passes dorsally and forks immediately. One fork runs cranially and deep along the outer wall of the rib cage, reaching epaxial muscles of the neck; the other fork runs caudally along the inner wall of the rib cage. Fourth, the **thyrocervical artery** springs from the anterior side of the subclavian, passing forward for a short time in the company of the external jugular vein. It gives off a thin **thyroid artery** to the thyroid gland and continues as the **transverse scapular artery** to the shoulder.

As the subclavian passes the first rib to enter the axilla, it becomes known as the **axillary artery.** Continue tracing the branches of the axillary artery on the right side (figures 8.14 and 8.16a1). Four branches exit at about this point. The **ventral thoracic artery,** the first to spring from the axillary, passes with its corresponding vein to the ventral pectoralis muscles. Next arises the **long thoracic artery** passing posteriorly with its venous counterpart into pectoralis muscles. The axillary next gives off the large **subscapular artery** extending anteriorly to the upper arm. The **thoracodorsal artery** exits from the posterior side of the subscapular and passes caudally along the medial side of the teres major and latissimus dorsi muscles in the company of its venous counterpart. Just lateral to the base of the subscapular is the smaller **anterior humeral circumflex artery** running dorsally over the biceps muscle near its origin. Beyond the point of departure of the anterior humeral circumflex, the axillary becomes the **brachial artery** passing toward the elbow. About midway along the upper arm, the brachial gives off a lateral branch, the **deep brachial artery,** crossing the dorsal belly of the biceps muscle.

Common Carotid Artery

➤ Trace the two common carotid arteries cranially from their departure from the brachiocephalic artery. Each gives off tiny branches to the trachea and to neck muscles. Near the cranial end, the **superior thyroid artery** extends medially to the thyroid gland, and opposite it, the **muscular artery** extends laterally to the neck

---

[2]These regions are roughly comparable to the ascending aorta, aortic arch, and descending aorta of human anatomy.

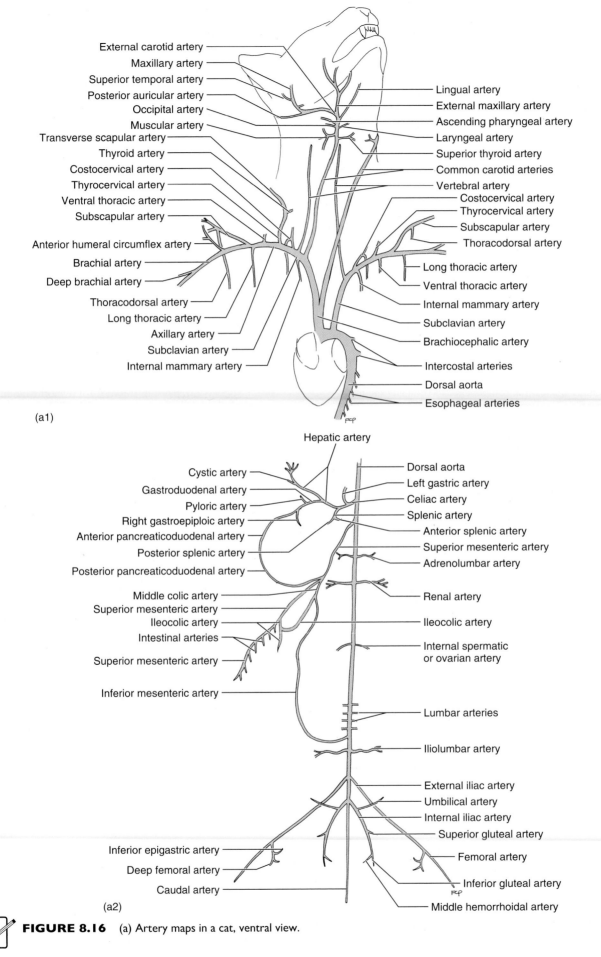

External carotid artery
Maxillary artery
Superior temporal artery
Posterior auricular artery
Occipital artery
Muscular artery
Transverse scapular artery
Thyroid artery
Costocervical artery
Thyrocervical artery
Ventral thoracic artery
Subscapular artery
Anterior humeral circumflex artery
Brachial artery
Deep brachial artery
Thoracodorsal artery
Long thoracic artery
Axillary artery
Subclavian artery
Internal mammary artery

Lingual artery
External maxillary artery
Ascending pharyngeal artery
Laryngeal artery
Superior thyroid artery
Common carotid arteries
Vertebral artery
Costocervical artery
Thyrocervical artery
Subscapular artery
Thoracodorsal artery
Long thoracic artery
Ventral thoracic artery
Internal mammary artery
Subclavian artery
Brachiocephalic artery
Intercostal arteries
Dorsal aorta
Esophageal arteries

(a1)

Hepatic artery

Cystic artery
Gastroduodenal artery
Pyloric artery
Right gastroepiploic artery
Anterior pancreaticoduodenal artery
Posterior splenic artery
Posterior pancreaticoduodenal artery
Middle colic artery
Superior mesenteric artery
Ileocolic artery
Intestinal arteries
Superior mesenteric artery
Inferior mesenteric artery

Dorsal aorta
Left gastric artery
Celiac artery
Splenic artery
Anterior splenic artery
Superior mesenteric artery
Adrenolumbar artery
Renal artery
Ileocolic artery
Internal spermatic
or ovarian artery
Lumbar arteries
Iliolumbar artery
External iliac artery
Umbilical artery
Internal iliac artery
Superior gluteal artery
Femoral artery

Inferior epigastric artery
Deep femoral artery
Caudal artery

Inferior gluteal artery

(a2)
Middle hemorrhoidal artery

FIGURE 8.16   (a) Artery maps in a cat, ventral view.

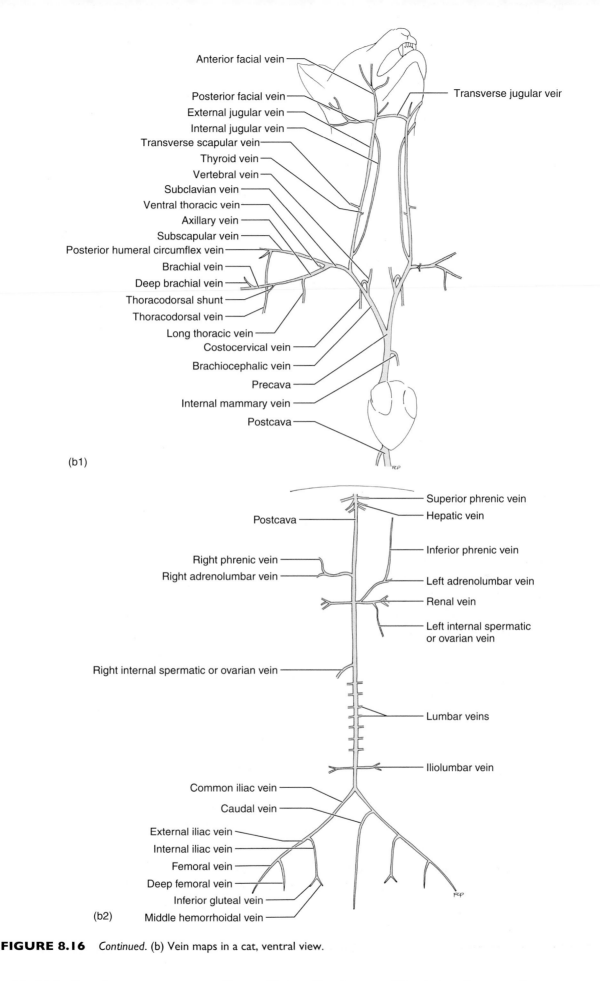

Anterior facial vein —

Posterior facial vein —
External jugular vein —
Internal jugular vein —
Transverse scapular vein —
Thyroid vein —
Vertebral vein —
Subclavian vein —
Ventral thoracic vein —
Axillary vein —
Subscapular vein —
Posterior humeral circumflex vein —
Brachial vein —
Deep brachial vein —
Thoracodorsal shunt —
Thoracodorsal vein —
Long thoracic vein —
Costocervical vein —
Brachiocephalic vein —
Precava —
Internal mammary vein —
Postcava —

— Transverse jugular vein

(b1)

Postcava —

Right phrenic vein —
Right adrenolumbar vein —

Right internal spermatic or ovarian vein —

Common iliac vein —
Caudal vein —

External iliac vein —
Internal iliac vein —
Femoral vein —
Deep femoral vein —
Inferior gluteal vein —
Middle hemorrhoidal vein —

— Superior phrenic vein
— Hepatic vein

— Inferior phrenic vein

— Left adrenolumbar vein
— Renal vein

— Left internal spermatic
or ovarian vein

— Lumbar veins

— Iliolumbar vein

(b2)

FIGURE 8.16 *Continued.* (b) Vein maps in a cat, ventral view.

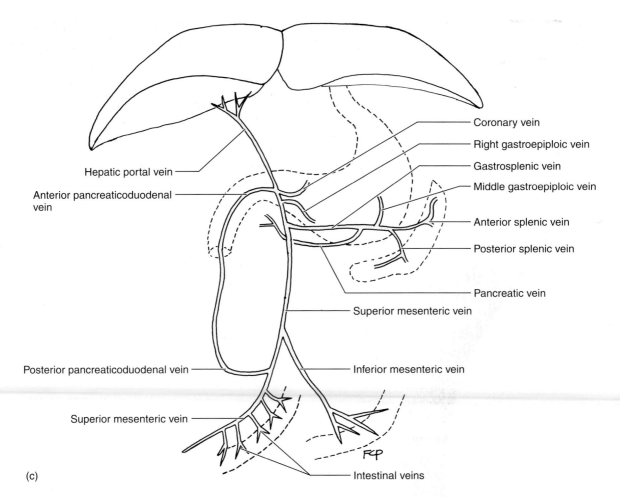

Hepatic portal vein

Anterior pancreaticoduodenal vein

Posterior pancreaticoduodenal vein

Superior mesenteric vein

Coronary vein

Right gastroepiploic vein

Gastrosplenic vein

Middle gastroepiploic vein

Anterior splenic vein

Posterior splenic vein

Pancreatic vein

Superior mesenteric vein

Inferior mesenteric vein

Intestinal veins

(c)

FIGURE 8.16 *Continued.* (c) Hepatic portal map in a cat, ventral view.

musculature (figures 8.14 and 8.16a1). At the level of the larynx, a ventral branch, the **laryngeal artery** (possibly destroyed), supplies the larynx, and at the same point a dorsal branch, the **occipital artery,** passes around the tympanic bulla and into the neck muscles. Just anterior to the occipital artery, the **ascending pharyngeal artery**[3] exits also from the dorsal side of the common carotid. The ascending pharyngeal artery dips deep and eventually enters the skull. Next anteriorly, the common carotid artery forks. The ventral fork, the **lingual artery,** supplies the tongue. The dorsal fork, the **external carotid artery,** passes toward the pinna. The external carotid, near its base, gives off the **external maxillary artery,** which courses along the ventral border of the masseter. The external carotid itself passes in the opposite direction along the posterior border of the masseter, gives off first the **posterior auricular artery** extending dorsally behind the ear, then the **superior temporal artery** extending dorsally in front of the ear, and thereafter the external carotid becomes the **maxillary artery** passing from view into the masseter muscle.

Branches of the Left Subclavian Artery

Names and distribution of arteries from the left subclavian are the same as already discussed for the right subclavian, although the first few branches exit at slightly different points. Near the first rib, the left subclavian gives off four arterial branches: **internal mammary** (ventral, runs to the interior thoracic cavity), **vertebral** (dorsal, runs deep to the vertebrarterial canal), **costocervical** (to inner and outer side of the rib cage), and **thyrocervical** (anteriorly, into shoulder as the **transverse scapular artery**).

Passing into the axilla, the subclavian becomes the **axillary artery,** whose subsequent distribution is similar to that already followed along the right side.

Respiratory System

Anatomy

Most of the structure of the respiratory system of the cat has already been identified at various points in the digestive and circulatory system dissections. Now return to these regions for an integrated look at this system. Trace the flow of air as it passes into the body to the lungs.

---

[3]A true internal carotid artery is absent, being replaced in the cat by the ascending pharyngeal artery, its functional equivalent.

Why does the trachea but not the esophagus have cartilaginous rings? Food is propelled down the esophagus by rhythmic muscular contractions that force the esophagus open as necessary. But air is drawn through the trachea by negative pressure, resulting from a lowering of the diaphragm and an expansion of the chest. This negative pressure in the trachea is similar to the airflow through a vacuum cleaner hose. In both the trachea and the hose, a series of stiff rings (C-shaped cartilages in the trachea) give it flexibility to bend with changes in body posture and to keep the tube from collapsing during breathing.

Why does the thoracodorsal vein drain into both the axillary and subscapular veins? Throughout the circulatory systems of vertebrates, we find many "alternate" blood pathways. Different body postures during sleeping may exert pressure to one area of the body and require rerouting of blood past a compressed region. Junctions like the thoracodorsal shunt provide such useful alternatives. Such "cross connections" between blood vessels are called anastomoses.

Air is inhaled through the **external nares** into the **nasopharynx,** the general region dorsal to the secondary palate. It then passes through the **laryngopharynx** into the **larynx** through the **glottis.** After passing through the larynx, air travels down through the trachea into primary, then secondary, and tertiary bronchioles, which continue to branch until reaching the blind-ended respiratory surfaces called *alveoli.* Gently probe the structure of the trachea to detect the many C-shaped cartilaginous rings that help keep the trachea open. Return to the cut base of the left lobes of the lung. Probe the stumps of the remaining structure to distinguish the hollow bronchioles from the surrounding blood vessels.

Pulmonary Arteries and Veins

The **pulmonary trunk** exits from the right ventricle, next to the aortic trunk, and divides into **left** and **right pulmonary arteries** supplying respective lungs (figure 8.13).

The **left** and **right pulmonary veins** receive blood returning from the lungs and empty into the left atrium. Find this point of entry on the back (dorsal side) of the heart by carefully picking away extraneous tissue and fat.

Precava

Follow the precava cranially to identify tributaries. First, at the base of the precava, enters the **azygous vein** (figure 8.15). Lift the precava to find the base of the azygous vein. Next, dig deep to follow the azygous dorsally to the roof of the thoracic cavity where it turns posteriorly and runs parallel along the right side of the dorsal aorta. Along this length, the azygous receives regular **intercostal veins** from between ribs of both sides. Next along the midventral side of the precava is an unpaired vessel, the **internal mammary vein,** draining the ventral wall of the thoracic cavity.

Anterior to the internal mammary, the **left** and **right** brachiocephalic **veins** join to begin the precava (figure 8.13). Each brachiocephalic has similar, but not symmetrical, tributaries. Thus, dissect and follow each side simultaneously. About halfway along the brachiocephalic, it is joined dorsally by the **vertebral vein** coming from the vertebrarterial canal anteriorly in the neck collecting from the spinal cord. (Frequently, the vertebral vein first joins the precava.) Trace the vertebral vein dorsally from its base, locating the **costocervical vein,** from more posterior muscles of the back, that joins it. The brachiocephalic is first formed near the first rib, by union of the lateral **subclavian vein** from the limb and the more cranial **external jugular vein.**

Right Subclavian Vein

Follow the subclavian first, spreading pectoralis muscles and ribs as needed (figures 8.13 and 8.16b1). The short subclavian itself forms from two major branches: the **subscapular vein** from the upper arm and dorsal side of the humerus, and the more posterior **axillary vein.**

Axillary Vein The base of the axillary vein is joined by the **ventral thoracic vein** from medial regions of the pectoralis muscles (figures 8.13 and 8.16b1). Further laterally, the axillary is joined by a **long thoracic vein** from other portions of the pectoralis and then by a short shunt from the **thoracodorsal vein** draining the latissimus dorsi muscle. The thoracodorsal vein continues cranially beyond this shunt to connect to the subscapular vein. These three tributaries to the axillary vein (ventral thoracic, long thoracic, and thoracodorsal) roughly parallel their arterial counterparts. Beyond juncture with the thoracodorsal vein, the axillary becomes the **brachial vein** passing along the upper arm. It receives a vessel, the **deep brachial vein,** that courses closely along with the deep brachial artery, from above the biceps muscle.

Subscapular Vein The subscapular vein first forms laterally at the junction of the thoracodorsal and lateral posterior humeral circumflex vein (figures 8.13 and 8.16b1). The **posterior humeral circumflex vein** drains the lateral side of the upper arm.

External Jugular Vein

Return to the base of the external jugular vein and from here follow it anteriorly (figures 8.13 and 8.16b1). It is joined medially by the slender **internal jugular vein** (often difficult to identify). Farther forward, a thin

thyroid vein from the thyroid gland enters ventromedially; and at about the same point a larger branch from the shoulder, the **transverse scapular vein,** enters laterally. The transverse scapular vein can be followed over the lateral side of the shoulder, anastomosing with other veins, and through one circuit, connects with the posterior humeral circumflex vein. (Often, on the left side, the brown, chainlike **thoracic lymphatic duct** can be seen to join the venous system near the base of the subscapular, external jugular, or subclavian veins.)

Near the head, the external jugular forms by the confluence of three veins previously identified with the salivary glands (see figures 7.7 and 8.16b). Ventrally, the **transverse jugular vein** passes across the throat to connect with the opposite side. The middle tributary, the **anterior facial vein,** runs along the ventral margin of the masseter muscle. The dorsal tributary, the **posterior facial vein,** comes from beneath the parotid gland and passes superficially across the round submaxillary gland.

Dorsal Aorta

🦖 Three categories of vessels depart from the dorsal aorta: **somatic** branches to the body wall, **median visceral** branches to the digestive tract and its associated glands, and **lateral visceral** branches to the urogenital system.

Somatic Branches

🦖 Pull viscera to the right to permit a left side view of the dorsal aorta. Shortly beyond the departure of the left subclavian artery, tiny paired **intercostal arteries** emerge segmentally from the top of the dorsal aorta to supply the wall of the thoracic cavity. Further caudally, after the dorsal aorta enters the abdominal cavity and passes the level of the kidneys, it gives off several more dorsal pairs of arteries to the adjacent abdominal wall; these are the **lumbar arteries.**

Return now to the area of the aorta in the thoracic cavity so the course of the dorsal aorta can be followed. Note the several, tiny **esophageal arteries** to the esophagus. Also note the two arteries that supply the diaphragm, the **phrenic arteries,** one to the anterior and one to the posterior surface of the diaphragm.

Median Visceral Branches

🦖 Median visceral branches are unpaired. The first, the **celiac artery,** arises as the dorsal aorta enters the abdominal cavity; the second, the **superior mesenteric artery,** arises immediately next to it; and the third, the **inferior mesenteric artery,** arises further posteriorly above the colon, which it also supplies (figures 8.16a2 and 8.17).

Celiac Artery   Follow the celiac artery first. It soon divides into three branches: the **left gastric, hepatic,** and **splenic arteries** (figure 8.16a2).

The **left gastric artery** passes to the lesser curvature of the stomach. The lesser omentum should be selectively removed for a better view.

The **hepatic artery** swings cranially into the liver via the hepatoduodenal ligament. The first branch of the hepatic artery is the **gastroduodenal artery,** which itself almost immediately sends off a branch, the **pyloric artery,** to supply the pyloric region. The gastroduodenal artery next divides. One branch, the **right gastroepiploic artery,** veers to the left along the pylorus and the greater curvature of the stomach. The other branch of the gastroduodenal artery, the **anterior pancreaticoduodenal artery,** continues on to supply the duodenum and adjacent pancreas. Finally, just prior to joining the liver, the hepatic artery sends a short **cystic artery** to the gallbladder.

The **splenic artery,** the third and largest branch of the celiac artery, courses in the greater omentum. Near the spleen, it forks. The first fork, the **anterior splenic artery,** distributes to the anterior end of the spleen. The other fork, the **posterior splenic artery,** passes through and supplies the pancreas, to eventually reach the posterior end of the spleen.

Form & Function

Within the circulatory system of the cat, many arterial loops have been identified. For example, the anterior joins the posterior pancreaticoduodenal artery, the pyloric artery (gastroduodenal) meets the left gastric artery (celiac), and the anterior connects to the posterior humeral circumflex vein. What could be the function of these loops? Like the thoracodorsal shunt to the axillary vein, these loops provide alternate routes for blood to flow to a particular region. When blood flow through a particular route is decreased, perhaps by pressure, alternate pathways can be used.

Superior Mesenteric Artery   Clear away fat to expose this vessel and its branches that lie in mesenteries of the gut (figures 8.16a2 and 8.17). Note the prevalent lymph nodes. The superior mesenteric first gives rise to the **posterior pancreaticoduodenal artery,** which loops forward through the pancreas supplying it and the adjacent duodenum. Follow it forward to confirm that it joins the previously identified anterior pancreaticoduodenal artery, a branch of the hepatic artery. The **middle colic artery** next branches off the superior mesenteric. It passes to the transverse and descending parts of the colon. The superior mesenteric next gives rise to the **ileocolic artery** to the caecum and terminal portion of ileum; some branches loop to the left along the colon, anastomosing with the middle colic artery. Finally, numerous small, looping branches, **intestinal arteries,** leave the superior mesenteric to distribute to the small intestine.

Inferior Mesenteric Artery   The **inferior mesenteric artery** supplies the descending colon and rectum, and joins anteriorly with the middle colic artery (figures 8.16a2 and 8.17).

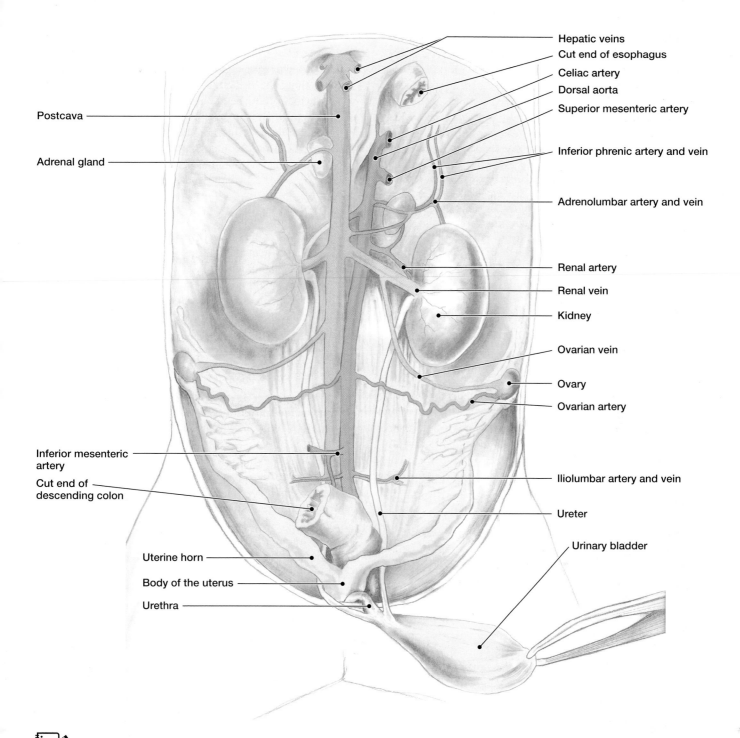

Postcava

Adrenal gland

Inferior mesenteric artery

Cut end of descending colon

Uterine horn

Body of the uterus

Urethra

Hepatic veins

Cut end of esophagus

Celiac artery

Dorsal aorta

Superior mesenteric artery

Inferior phrenic artery and vein

Adrenolumbar artery and vein

Renal artery

Renal vein

Kidney

Ovarian vein

Ovary

Ovarian artery

Iliolumbar artery and vein

Ureter

Urinary bladder

FIGURE 8.17 Ventral view of branches off the dorsal aorta and tributaries to the postcava in the trunk of a female cat.

Lateral Branches of the Dorsal Aorta

Adrenolumbar and Inferior Phrenic Arteries Just caudal to the departure of the superior mesenteric artery, the dorsal aorta sends off laterally, at slightly different levels, the **adrenolumbar arteries,** each supplying a small, hard, oval-shaped adrenal gland (figure 8.17). Before the adrenolumbar reaches the adrenal glands, the thin **inferior phrenic artery** departs cranially to supply the diaphragm.

Renal Arteries The next caudal branches off of the dorsal aorta are the large **renal arteries** extending to the kidneys (figures 8.16a2 and 8.17). A small branch to the adrenal gland may be located leaving the anterior side of each renal artery.

Internal Spermatic or Ovarian Arteries Past the kidneys, the dorsal aorta gives rise to arteries supplying the gonads (figures 8.16a2 and 8.17). In males, the slender **internal spermatic arteries** make the long

Think back to the thin esophageal arteries and the prominent visceral arteries to parts of the gut (e.g., celiac artery). Why do the diameters of these blood vessels vary so greatly? The differences in arterial size reflect differences in size and activity of the organs supplied. The esophagus primarily serves as a mere conduit to the stomach. The stomach, however, churns and mixes food with gastric juice and engages in some absorption, which requires additional blood flow to support these activities. But the intestines, through their many large arteries, receive the greatest blood supply. Like the stomach, the intestines engage in extensive muscular activities, churning and moving food along. But in addition, the intestines are the primary site of absorption of digested molecules into the blood supply. Here, as with the earlier muscular and skeletal systems, differences in function are correlated to differences in form.

The volume of blood flow to parts of the respiratory system shows similar correlations. The trachea and bronchi, serving primarily as gas conduits, have only a minimal vascular supply. But the lungs, the site of gas exchange, are one of the most extensively vascularized regions in the body.

journey posteriorly out of the abdominal cavity and into the scrotum containing the testes, reflecting the descent of the testes from their embryonic origin near the kidneys. In females, the larger, convoluted **ovarian arteries** pass laterally to the ovaries.

Iliolumbar Arteries   Near the inferior mesenteric artery, the dorsal aorta gives off laterally the **iliolumbar arteries** to the adjacent abdominal wall, crossing first the psoas minor muscle, then iliopsoas muscle (figures 8.16a2 and 8.17).

External Iliac, Internal Iliac, and Caudal Arteries Shortly after giving rise to the iliolumbar arteries, the dorsal aorta forks into two **external iliac arteries,** which pass to the hindlimbs (figures 8.16a2 and 8.18). Past this major fork, the dorsal aorta is considerably reduced in size and almost immediately sends off another lateral pair of branches, the **internal iliac arteries** (= hypogastric arteries). Beyond this point, the dorsal aorta is termed the **caudal artery** (= median sacral artery), eventually running into the tail.

Branches of the Internal Iliac Artery   Near its base, the internal iliac gives off the **umbilical artery** to the urinary bladder, the **superior gluteal artery** passing deep to gluteal and other thigh muscles, and finally, the internal iliac ends in a bifurcation (figure 8.16a2). One branch, the **middle hemorrhoidal artery,** swings ventrally to supply the rectum and itself sends, in females, a **uterine artery** to the uterus. The other branch, the **inferior gluteal artery,** dips dorsally passing next to the sciatic nerve.

Branches of the External Iliac Artery   The external iliac passes out of the abdominal cavity and into the thigh (figures 8.16a2 and 8.18). Its first branch is medial—the **deep femoral artery,** which passes the adductor longus

on its way to supply deep muscles of the thigh. At the base of the deep femoral arises, among other branches, the **inferior epigastric artery,** which swings forward to run along the rectus abdominis, eventually joining the superior epigastric artery, a continuation of the internal mammary. Beyond this point of departure of the deep femoral, the external iliac becomes the **femoral artery** coursing along the medial side of the thigh, giving off branches that will not be followed.

Postcava

🦖 Small veins joining the postcava and its large tributaries parallel their arterial counterparts and bear the same names.

The postcava first forms, at the rear of the abdominal cavity, by the union of the two **common iliac veins,** one from each hindlimb (figures 8.16b2 and 8.18). The **caudal vein** (= median sacral vein) joins one common iliac. In turn, the common iliac forms by union of the large **external iliac vein** and smaller, medial **internal iliac vein** (= hypogastric vein). (The external iliac receives the **femoral** and **deep femoral veins.** The internal iliac receives **middle hemorrhoidal** and **inferior gluteal veins.**)

As the postcava passes forward, it receives first the lateral **iliolumbar veins,** then the **right internal spermatic** or the **right ovarian vein,** and finally the large **renal veins** (figures 8.16b2 and 8.17). However, the **left internal spermatic** or **left ovarian vein** usually contributes to the left renal vein before reaching the postcava. Small left and right **adrenolumbar veins** join the postcava at slightly different points just anterior to the entry of the renal veins. Before passing through the diaphragm, the postcava receives several **hepatic veins** from the liver and the paired **superior phrenic vein** from the adjacent diaphragm. The **inferior phrenic vein,** as you might guess from identification of its arterial counterpart earlier, joins the adrenolumbar vein.

The postcava, while in the abdominal cavity, regularly receives dorsally, paired **lumbar veins** from a median groove between muscle masses. Pull the postcava to one side of its course along the roof of the abdominal cavity to see the entry of these vessels.

Hepatic Portal System

🦖 Veins of the hepatic portal system drain the gut and its associated glands. However, the hepatic portal veins empty not into the postcava, which would place the absorbed nutrients into direct circulation, but into the liver sinuses. From the liver, blood is collected into the hepatic veins flowing eventually to the inferior vena cava. The hepatic portal system must be specifically injected because latex added to the arterial or venous systems will not reach these blood vessels through the capillaries of the gut. Latex injected into the hepatic portal system is usually a different color than the latex in the arterial or venous systems. Your instructor will tell you how your cats were injected.

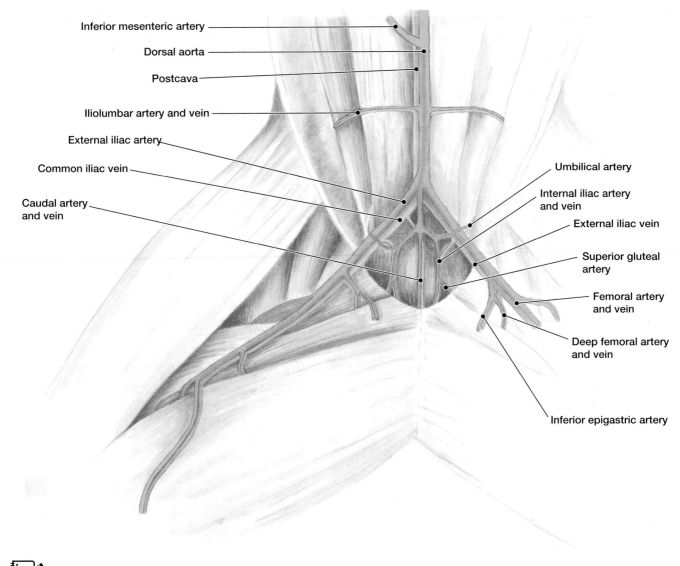

Inferior mesenteric artery

Dorsal aorta

Postcava

Iliolumbar artery and vein

External iliac artery

Common iliac vein

Caudal artery and vein

Umbilical artery

Internal iliac artery and vein

External iliac vein

Superior gluteal artery

Femoral artery and vein

Deep femoral artery and vein

Inferior epigastric artery

FIGURE 8.18 Blood vessels of the medial hindlimb of a cat, ventral view.

Hepatic Portal Vein

🦖 The **hepatic portal vein** is primarily formed along the midline near the lesser curvature of the stomach by the merger of the **superior mesenteric** and **gastrosplenic veins** (figure 8.16c). The tributaries to these two veins are followed in the next sections. Just cranial to this junction, the hepatic portal is joined by three additional veins: the **coronary vein,** draining the lesser curvature of the stomach, the **right gastroepiploic vein,** from the greater curvature of the stomach, and the **anterior pancreaticoduodenal vein,** draining the duodenum and pancreas.

Tributaries of the Gastrosplenic Vein

🦖 Four veins combine to form the **gastrosplenic vein** (figure 8.16c). As their names imply, the **anterior splenic vein** drains the anterior end of the spleen and the **posterior splenic vein** the posterior end. Two smaller veins also join: the **pancreatic vein** from the pancreas and the **middle gastroepiploic vein** from the lesser curvature of the stomach.

Tributaries of the Superior Mesenteric Vein

🦖 The **superior mesenteric vein,** the largest contributor to the hepatic portal vein, primarily forms by the fusion of the **intestinal, inferior mesenteric,** and **posterior pancreaticoduodenal veins** (figure 8.16c). Begin tracing these tributaries by pulling the small intestine away from the body cavity. There, nestled within the mesentery, are many small **intestinal veins** that contribute to the beginning (the most caudal extent) of the superior mesenteric vein. As the superior mesenteric continues forward, it receives the **inferior mesenteric vein** draining the large intestine and the **posterior pancreaticoduodenal vein** from the posterior end of the pancreas and duodenum.

9 Urogenital Systems

Survival of an individual depends upon many factors, including adaptations to escape predators, obtain food, and withstand harsh environmental conditions. But to pass adaptations to descendants, the individual must reproduce successfully, the primary biological role of the genital system.

The reproductive systems of vertebrates must produce and store viable *gametes* (eggs or sperm) prior to mating. In males, the secretions of many glands mix with the sperm to form *seminal plasma,* the fluid used to stabilize and transfer the sperm to the female. Male vertebrates often have *copulatory structures* (e.g., claspers in sharks, a penis in mammals) used to transfer sperm. However, some male fish and most male salamanders (including *Necturus*) transfer sperm using *spermatophores,* gelatinous structures capped with sperm. Cloacal glands in these vertebrates produce components of the spermatophore.

Female reproductive systems vary according to the site of development of the embryos. *Viviparous species* are "live" bearing, giving birth to young not encased in a shell. The dogfish shark and cat are viviparous. Their young develop within a uterus. Their reproductive tracts are therefore adapted to receive sperm and support embryonic development. *Oviparous species,* including *Necturus* and most other amphibians, are egg laying. The embryo emerges from the mother wrapped in a shell or jelly coat secreted by the uterus. Their reproductive tracts include ducts and glands that produce the jelly coats (or shells, in reptiles and birds) around the eggs.

The prominence of reproductive components of the urogenital system varies with maturity and seasonal cycle. Be prepared for differences in size and development of the systems. Also, remember to move around the laboratory to see dissections and appreciate the significant differences in male and female sexual anatomy.

Compared to the reproductive system, the urinary system of vertebrates is devoted to quite different functions, including the elimination of waste products, the regulation of water balance, and the control of electrolyte levels. We treat the urinary and reproductive systems together as the urogenital system because they share so many of the same ducts.

Shark

Male Urogenital System

Pelvic Fin

🦖 The pelvic fins of males possess enlarged and fused radials, the **claspers,** derived from pterygiophores (figure 9.1). The medial side bears a groove that carries

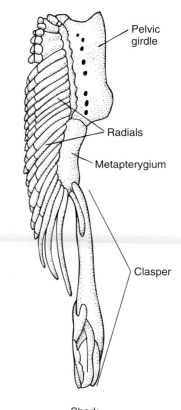

Shark
(Chlamydoselachus)

FIGURE 9.1 Ventral view of the left pelvic girdle, pelvic fin, and intromittent organ of the shark *Chlamydoselachus.*
After vanTienhoven.

seminal fluid during copulation. At the ventral base of each pelvic fin, just under the skin, lies the collapsed, muscular, **siphon,** usually dark brown in color (figure 9.2). Cut open the siphon along its lateral edge to find a medial pore. Fluid secreted by the siphon passes into the groove, travels along the clasper, and thus contributes to seminal fluid. Slide a blunt probe along the groove and into the pore to discover its course.

In the male, remove and discard the liver by cutting the lobes close to their points of attachment. When necessary, you may also want to cut through the puboischiac bar to expose parts of the urogenital system.

Testis and Its Ducts

🦖 Each oval, soft **testis** is suspended from the anterior roof of the pleuroperitoneal cavity by the **mesorchium**

172

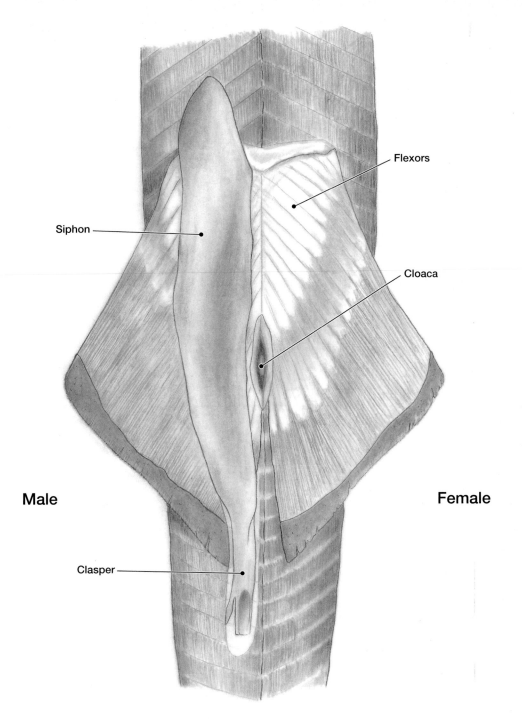

FIGURE 9.2 Ventral view of the female and male shark pelvis.

Siphon

Flexors

Cloaca

Male

Female

Clasper

(figure 9.3). The tightly convoluted, paired duct along the dorsal body wall is the **ductus deferens** (= vas deferens, wolffian duct, mesonephric duct, or archinephric duct). Several threadlike, **efferent ductules** pass in the mesorchium and carry sperm from the testis to the **epididymis** within the anterior end of the kidney. These small efferent ductules may be seen by shining a light through the mesorchium from behind. The epididymis is composed of a knot of many small tubules concentrated just lateral to the mesorchium. From the epididymis, sperm enter the larger but still coiled **ductus deferens** (figure 9.3). As the ductus deferens extends caudally, it enlarges and straightens, forming the **seminal vesicle.** Trace the seminal vesicle caudally. Just where it looks as though the seminal vesicle will enter the cloaca, it instead terminates in a bilobed **sperm sac.**

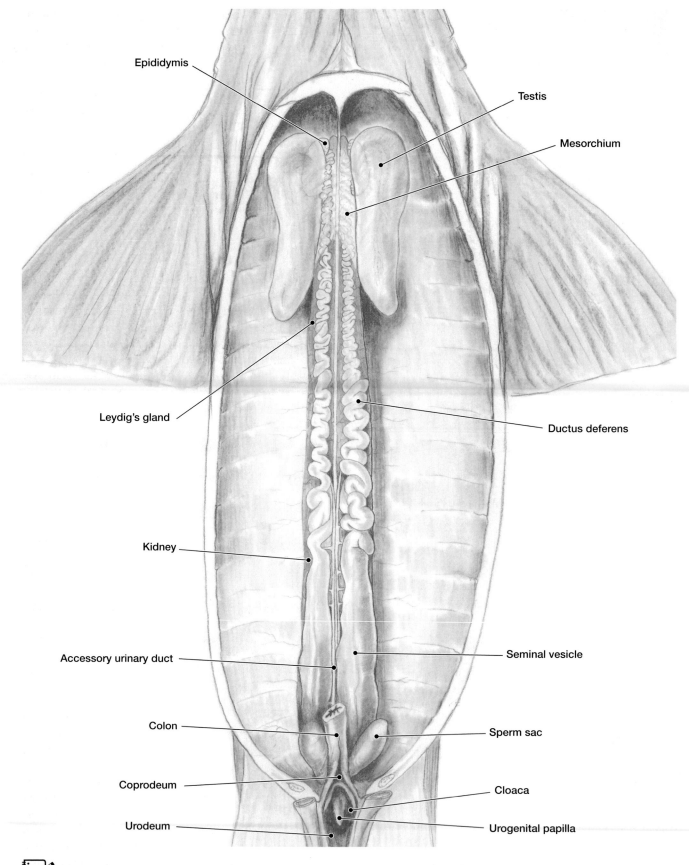

Epididymis

Testis

Mesorchium

Leydig's gland

Ductus deferens

Kidney

Accessory urinary duct

Seminal vesicle

Colon

Sperm sac

Coprodeum

Cloaca

Urodeum

Urogenital papilla

FIGURE 9.3 Ventral view of the urogenital system of a male shark, with the digestive tract removed.

Most of the urogenital system of embryonic vertebrates typically begins as gender neutral. That is, parts are present, but whether they will develop, and how they will develop, will be influenced by the sex of the individual. One example is the müllerian and archinephric ducts (= wolffian duct, pronephric duct, mesonephric duct, opisthonephric duct, ductus deferens, vas deferens).

In males, the müllerian ducts rarely develop. Instead, the archinephric ducts develop into parts of the urogenital system variously associated with sperm and urine transport, depending upon the particular vertebrate.

In females, the archinephric ducts tend to function only within the urinary system. Instead, the müllerian ducts variously form the oviduct, uterus, and vagina, again, depending upon the particular vertebrate. Watch for these names throughout the dissections of the urogenital systems.

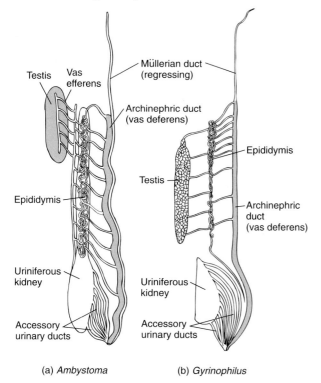

(a) *Ambystoma*   (b) *Gyrinophilus*

Urogenital systems of male salamanders. (a) Salamander *Ambystoma*. (b) Salamander *Gyrinophilus*. In both animals, the left testis and associated ducts have been reflected to the other side. The right testis and ducts are not shown.

(a) After C. L. Baker and W. W. Taylor, Jr., 1964, "The urogenital system of the male Ambystoma," Jrnl. Tenn. Acad. Sci. 39:2. (b) After P. Strickland, 1966, "The male urogenital system of Gyrinophilus danielsi dunni," Jrnl. Tenn. Acad. Sci. 41:1.

Kidney

The shark **kidney,** or **opisthonephros,** is positioned outside the peritoneum within a long, narrow groove along the roof of the pleuroperitoneal cavity. The anterior region of the kidney, called **Leydig's gland,** serves no urinary function, but instead helps in sperm transport.

It is primarily composed of the **epididymis** and **ductus deferens,** identified previously. Glands around the ductus deferens produce a milky fluid that contributes to the seminal fluid. Cut a small slit in the wall of the ductus deferens to reveal these milky secretions.

The remaining posterior region is termed the **kidney,** or **opisthonephros** proper. In males, urine produced by the opisthonephros is collected into the **accessory urinary duct,** running along the medial side of the kidney. Cut through the middle of one seminal vesicle and reflect the ends. This exposes the microscopic accessory urinary duct, which runs caudally to enter the base of the sperm sac.

Cloaca

The **cloaca** receives products of the urogenital and digestive systems. Open the cloaca with a short, medial cut. A partial fold transversely divides the cloaca into ventral **(coprodeum)** and dorsal **(urodeum)** regions. The intestine drains into the **coprodeum** and the urogenital system empties into the **urodeum.** The single, nipplelike process in the urodeum is the **urogenital papilla.** All urinary and genital products passed to the sperm sac empty into the cloaca through this urogenital papilla.

Female Urogenital System

Pelvic Fin

The pelvic fin of females lacks claspers, siphon, or any other copulatory specializations found in males.

Ovary and Its Ducts

Each **ovary** hangs from the anterior roof of the pleuroperitoneal cavity by the **mesovarium** (figure 9.4). In immature females, the ovaries are soft and oval structures. The **oviducts** in immature females are slender tubes running free without mesenteries along the dorsal pleuroperitoneal cavity next to the dorsal aorta.

In sexually mature females, the ovaries are usually lumpy and the oviducts are now suspended by the **mesotubarium** (figure 9.4). Follow the oviducts cranially. They pass dorsal to the liver, arch toward the midline, pass into the falciform ligament, and join in a common opening, the **ostium.** The ostium is slitlike and its rim is usually collapsed. If you have difficulty finding it, again follow the arching oviducts forward to where they unite in the falciform ligament.

As with the male, remove and discard the liver at its base, leaving undisturbed the regions immediately associated with the falciform ligament, the oviducts, and the ostium. When necessary, you may also want to cut through the puboischiac bar to expose parts of the urogenital system.

Beginning at the ostium, trace the oviduct posteriorly. Just dorsal to the ovary in mature specimens, the oviduct slightly enlarges, forming the **shell gland** (= nidamental gland), which secretes a thin membrane

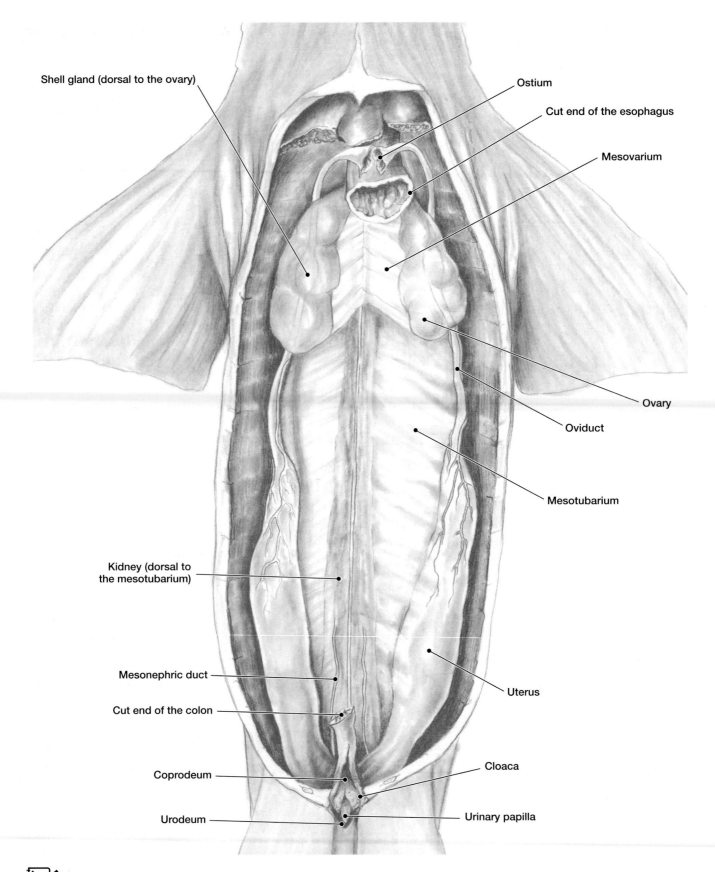

Shell gland (dorsal to the ovary)

Ostium

Cut end of the esophagus

Mesovarium

Ovary

Oviduct

Mesotubarium

Kidney (dorsal to the mesotubarium)

Mesonephric duct

Cut end of the colon

Coprodeum

Urodeum

Uterus

Cloaca

Urinary papilla

FIGURE 9.4   Ventral view of the urogenital system of a female shark, with the digestive tract removed.

around eggs as they pass (figure 9.4). Caudal to this gland, the oviduct again narrows for a short distance, then enlarges greatly into the **uterus,** which contains developing embryos. In immature specimens, the uterus is only slightly wider than the oviduct.

Kidney

🦖 In the female, the anterior part of the kidney has no urinary or reproductive role and fails to develop. What persists is the middle and posterior portions that form the adult kidney (figure 9.4). Transect the oviduct of one side and reflect the ends. Next, cut through the peritoneum to expose the **kidney,** which runs outside the parietal peritoneum next to the dorsal aorta (as in the male). No accessory urinary duct of males exists in females. Instead, the kidney is drained by the **mesonephric duct** (= **wolffian**) lying along the ventral surface of the kidney. Before emptying into the cloaca, the mesonephric duct expands, forming the urinary vesicle (often impractical to find in immature females).

Cloaca

🦖 As in the male, the cloaca is divided horizontally into a ventral **coprodeum,** receiving contents from the intestine, and a dorsal **urodeum,** receiving the products of the urogenital system. Make a short, midventral cut in the cloaca to see these regions. Within the urodeum is the single, nipplelike **urinary papilla** formed by union of mesonephric ducts. Urinary products eventually enter the cloaca through this urinary papilla. The uterus enters the cloaca through pores on either side of the urinary papilla.

Necturus

The kidneys and the arrangement of the associated ducts are similar, but not identical, in the shark and *Necturus.* For example, both the shark and *Necturus* possess *opisthonephric kidneys.* As in the male shark, the *Necturus ductus deferens* (= vas deferens = mesonephric duct) serves in transport of sperm and seminal fluids. But unlike in the shark, *urinary collecting tubules* transport urine from the kidneys into the caudal end of the ductus deferens. Thus, part of the ductus deferens in male *Necturus,* but not in male sharks, transports both urine and sperm. In females, the *mesonephric duct* only drains the kidney. The *oviduct* (= müllerian duct) transports eggs between the ovary and cloaca and deposits jelly coats along the way.

Male Urogenital System

Testis, Kidney, and Related Ducts

🦖 Each elongate **testis** hangs from the middorsal wall of the pleuroperitoneal cavity by the **mesorchium**

Form & Function

The caudal region of the ductus deferens of *Necturus* and most larval salamanders can potentially conduct both urine and sperm in the sexually mature adult. However, that dual role is not typically found in adult salamanders. Instead, the connection between the urinary collecting tubules and ductus deferens shifts caudally, so that urine is passed either separately or only at the most caudal end of the ductus deferens. That *Necturus* males retain this juvenile trait is yet another example of their unusual paedomorphic morphology.

But why do the urinary collecting tubules shift caudally in the typical adult salamander? Consider the roles of the adult ductus deferens. Sperm are stored within the ductus deferens and epididymis of male salamanders just prior to the breeding season. But storing sperm in a region of the ductus deferens that also must transport urine may harm the sperm or the sperm may interfere with the passage of urine. Thus, by shifting the connections between the ductus deferens and urinary collecting tubules more caudally, additional space is available in the ductus deferens to store sperm.

(figure 9.5). Running through the mesorchium are **efferent ductules.** These ducts transport sperm from the testis to the cranial end of the kidney, or **epididymis.** (As in the male shark, the kidney is divisible into a narrower anterior region, the epididymis, involved in sperm transport, and a larger posterior region that produces urine.) These tiny **efferent ductules** may be seen by shining a light through the mesorchium from behind, as was done in the male shark. Numerous **spermatic arteries** and **veins** will be prominent within the mesorchium.

From the epididymis, sperm are transported to the **ductus deferens** (= mesonephric duct = wolffian duct), a larger and somewhat coiled duct running parallel to the kidney along its lateral margin. At the cranial end of the ductus deferens, just beyond the epididymis, a thin, black cord parallels the ductus deferens and then continues cranially on its own. This cord is what remains of the **müllerian duct,** which fails to develop in males (see box page 175).

Follow the ductus deferens posteriorly (figure 9.5). Near the caudal end of the pleuroperitoneal cavity, it straightens and receives numerous **urinary collecting tubules** from the urinary kidney. Beyond the kidney it slants medially to empty into the **cloaca.**

Cloaca

🦖 The collapsed **urinary bladder** empties into the terminus of the large intestine just prior to its entry into the cloaca (figure 9.5). If not already done, make a short, midventral incision through the cloaca to find points of entry of the large intestine and of the paired mesonephric ducts.

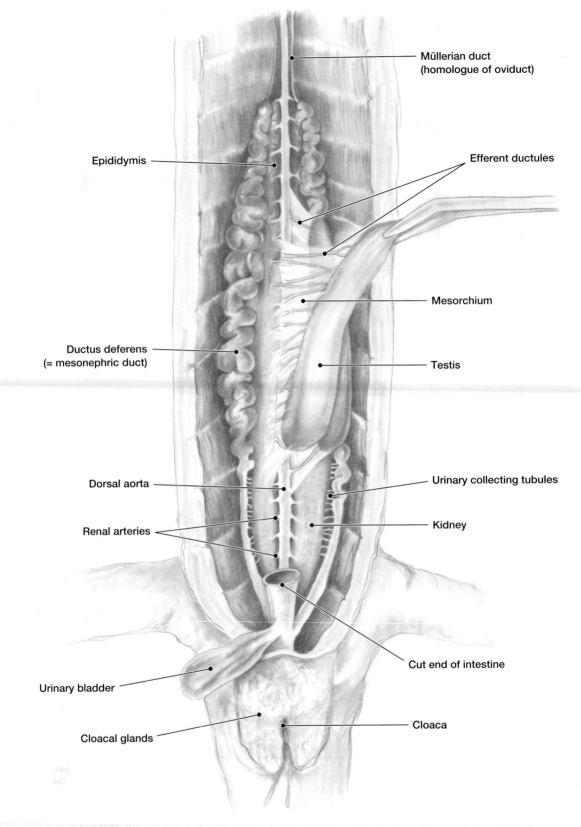

Müllerian duct
(homologue of oviduct)

Epididymis

Efferent ductules

Mesorchium

Ductus deferens
(= mesonephric duct)

Testis

Dorsal aorta

Urinary collecting tubules

Renal arteries

Kidney

Cut end of intestine

Urinary bladder

Cloacal glands

Cloaca

FIGURE 9.5 Ventral view of the urogenital system of a male *Necturus*, with the digestive tract removed.

Female Urogenital System

Ovary and Its Ducts

🦖 Each **ovary,** containing hundreds of beadlike ova, or eggs, hangs by the **mesovarium** from the middorsal wall of the pleuroperitoneal cavity (figure 9.6). **Ovarian arteries** and **veins** pass through the mesovarium. The paired, tortuous **oviducts** (derived from the müllerian ducts) run along the roof of the pleuroperitoneal cavity. Trace one oviduct forward to find its funnel-shaped beginning and the opening into it, the **ostium.** Eggs, released from the ovary into the body cavity, are directed into the ostium to begin their transport down the oviduct. At its posterior terminus, the oviduct empties into the cloaca.

Kidney

🦖 The paired **kidneys** lie in the posterior roof of the pleuroperitoneal cavity. Unlike those of the male, the female kidneys are composed of only the posterior urinary section. Numerous **urinary collecting tubules** transport urine from the kidneys to the **mesonephric duct** (= wolffian), running along the lateral edge of the kidney (figure 9.6). The mesonephric duct drains into the **cloaca.**

Cloaca

🦖 As in the male, the unpaired **urinary bladder** empties into the terminus of the large intestine, which in turn empties into the cloaca (figure 9.6). Make a short, midventral incision through the cloaca to find points of entry of the large intestine, the mesonephric ducts, and finally the oviducts through their small, nipplelike **papillae.**

Cat

Male Urogenital System

Kidney, Urinary Bladder, and Associated Structures

🦖 The general structures of the urinary system will first be explored to provide basic orientation to the pelvic region. Arising from the medial indentation (**hilus**) on each kidney is a thin, white tube, the **ureter,** that passes posteriorly to the base, or **fundus,** of the muscular **urinary bladder** (figure 9.7). The free cranial end of the urinary bladder is its **vertex** (= apex). Pull the vertex toward you and thus away from the descending colon. The pouchlike cavity between the bladder and colon is the **rectovesical pouch.** The fundus of the bladder narrows into the **urethra.** In this region, the descending colon straightens into the final section of the large intestine, the **rectum,** located dorsal to the urethra. The rectum will be better exposed and followed shortly.

The inguinal canals of adult male humans are regions of potential problems. For example, heavy lifting or other activities that generate strong abdominal pressures can cause the intestines to protrude into the inguinal canal, resulting in an inguinal hernia and often great pain.

Inguinal Canal and Associated Structures

🦖 Return now to the dorsal aorta and postcava. On one side, locate the very thin **internal spermatic artery** and **vein,** if proper injection permits. Follow these posteriorly until they pierce the rear wall of the abdominal cavity. This exit portal is the **inguinal canal.** Its openings on inner and outer sides are, respectively, **internal** and **external inguinal rings.** Note that a white cord arches over the ureter and extends laterally to join the internal spermatic vessels as they enter the inguinal canal. This cord is the **ductus deferens** (figure 9.7), carrying sperm from the testis to the **urogenital canal** at the base of the penis. Clear out fat and connective tissue to expose the ureter, ductus deferens, and spermatic vessels.

The three structures—ductus deferens, internal spermatic artery, and vein—become bound in a connective tissue sheath as they pass out of the abdominal cavity. All three tubules so wrapped are collectively known as the **spermatic cord.** Cut through the inguinal canal to locate the spermatic cord. Continue the cut through the skin following the cord posteriorly into the **scrotum,** the saclike fold of skin containing the teardrop-shaped **testis** (figure 9.7). The testis is also wrapped in connective tissue and will be dissected in detail shortly.

Reproductive Glands and Associated Structures

🦖 Now return to the urethra to find the point where the ductus deferens loops medially to join it. Cut through the pubic and ischial symphyses with heavy scissors. Spread the cut by pressing the knees of the cat laterally to the bottom of the dissecting pan. As needed, extend this cut by trimming off more bone adjacent to the symphyses.

The ductus deferens of each side extends cranially from the internal inguinal ring, arches over the corresponding ureter, and then turns posteriorly and medially to join the **urethra,** thus forming the **urogenital canal** (figure 9.7). Follow each ductus deferens to its junction with the urethra. At this point of union resides the **prostate gland,** a small, dorsal swelling (figure 9.7). The urogenital canal runs for about 2 cm before entering the penis. At the base of the penis the urogenital canal is joined by ducts from two small, lateral swellings usually enmeshed within connective tissue, the **bulbourethral** (= Cowper's) **glands** (figure 9.7).

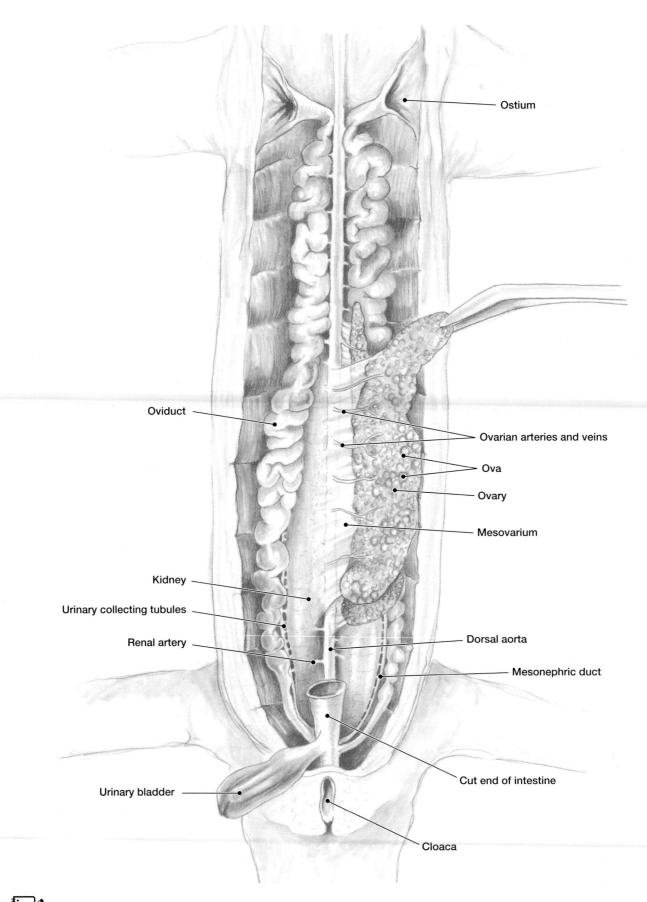

Ostium

Oviduct

Ovarian arteries and veins

Ova

Ovary

Mesovarium

Kidney

Urinary collecting tubules

Renal artery

Dorsal aorta

Mesonephric duct

Cut end of intestine

Urinary bladder

Cloaca

FIGURE 9.6 Ventral view of the urogenital system of a female *Necturus*, with the digestive tract removed.

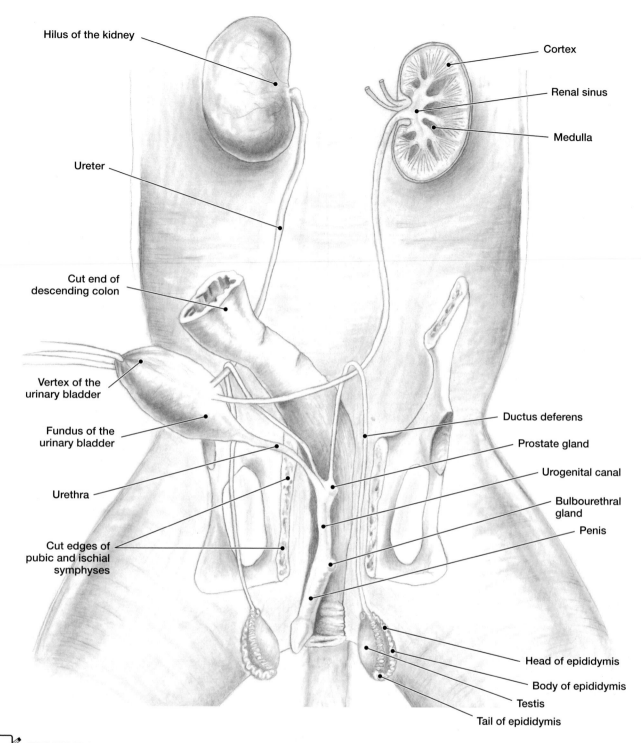

Hilus of the kidney

Ureter

Cut end of
descending colon

Vertex of the
urinary bladder

Fundus of the
urinary bladder

Urethra

Cut edges of
pubic and ischial
symphyses

Cortex

Renal sinus

Medulla

Ductus deferens

Prostate gland

Urogenital canal

Bulbourethral
gland

Penis

Head of epididymis

Body of epididymis

Testis

Tail of epididymis

FIGURE 9.7 Ventral view of the urogenital system of a male cat, with the digestive tract removed.

(Most mammals possess a third gland contributing to the genital ducts, the seminal vesicle, but this is absent in the cat.)

Dorsal to the urethra and urogenital canal runs the rectum. Now, with the symphyses cut and spread, the rectum is better exposed. At its terminus is an **anal sphincter,** a band of smooth muscle. This sphincter controls the posterior opening of the digestive tract, the

anus. A pair of rounded **anal glands** lie along the sides of the rectum near the anus.

Testis

During fetal life, each testis descends from the abdominal cavity, through the then larger inguinal canal, and into a pouch of skin, the **scrotum** (figure 9.8). The fascial sheath, or **cremasteric fascia** (= spermatic fascia

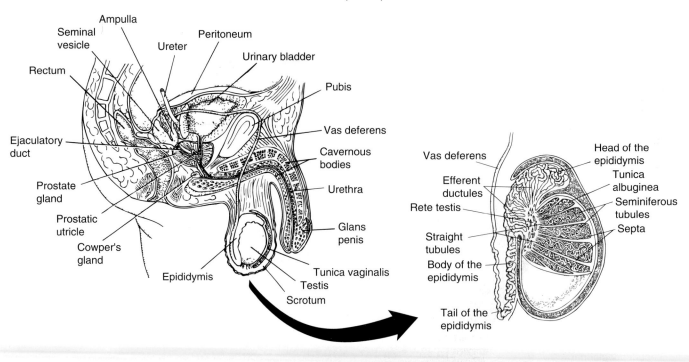

FIGURE 9.8 Human male reproductive system. (Top) This sagittal section of the male pelvis shows the reproductive organs and their relationships to the urinary and digestive systems. (Bottom) This midsagittal section reveals the testis and its complex duct system. Spermatozoa produced in the seminiferous tubules eventually pass through the straight tubules into the rete testis and enter the epididymis. Fluid is added as spermatozoa are moved through the vas deferens by contractions of sheets of smooth muscle in its walls.

or cremasteric pouch), encloses components of the spermatic cord and also wraps the testis. Embryologically it arises as an outpocketing of abdominal wall that projects into the scrotum. It is derived specifically from fascia of the external oblique and from the internal oblique muscle. The latter muscle also gives rise to the cremasteric muscle (atrophied in the cat).

The cremasteric fascia forms a sac around the testis. Cut open this sac along the ventral edge. The cavity opened, between the testis and fascia, is the **tunica vaginalis** (figure 9.8). The posterior end of the testis attaches to the cremasteric fascia by a short, tough ligament, the **gubernaculum** (= caudal ligament of epididymis). The C-shaped epididymis wraps around the ends and back side of the testis. The cranial part of the epididymis is its **head,** the section along the side of the testis its **body,** and the region over the caudal end of the testis the **tail** (figure 9.8). The tail empties into the thin, convoluted ductus deferens. The internal structure of the testis is depicted in figure 9.8. Sperm are produced in the walls of the seminiferous tubules, then pass to the straight tubules (tubuli recti), to the rete testis, then out of the testis into efferent ductules and so enter the head of the epididymis.

Internal Kidney Structure

🦖 Clear away fat, and note the kidney's legendary bean shape. The medial indentation is the **hilus** through which renal blood vessels, lymphatic vessels, and ureter pass.

Horizontally cut one kidney to divide it into approximately equal dorsal and ventral halves. The cavity within the kidney entered when passing through the hilus is the **renal sinus** (figure 9.7). However, because the renal sinus is also occupied by blood vessels, the ureter, and fat, you may not initially find the renal sinus. Note that within the renal sinus, the ureter is expanded, forming the **pelvis** of the ureter. The projection of kidney tissue into the renal pelvis is the **renal papilla.** The outer part of the kidney is the **cortex;** its inner core, including the papilla, is the **medulla** (figure 9.7).

Female Urogenital System

Urinary Structures

🦖 Drainage of the kidney is similar to that in the male. The **ureter,** a white cord, leaves each kidney at its medial indentation, and runs posteriorly to the base or **fundus** of the **urinary bladder** (figure 9.9). The free anterior end of the urinary bladder is the **vertex** (= apex). The fundus narrows into a slender **urethra** passing dorsal to the pubic symphysis and ventral to the uterus (figure 9.9). This will be followed shortly. Now, however, pull the vertex of the bladder toward you to locate the deep pocket between the urethra and uterus, the **vesicouterine pouch.**

Refer to the description of the kidney in the male for identification of internal parts and regions.

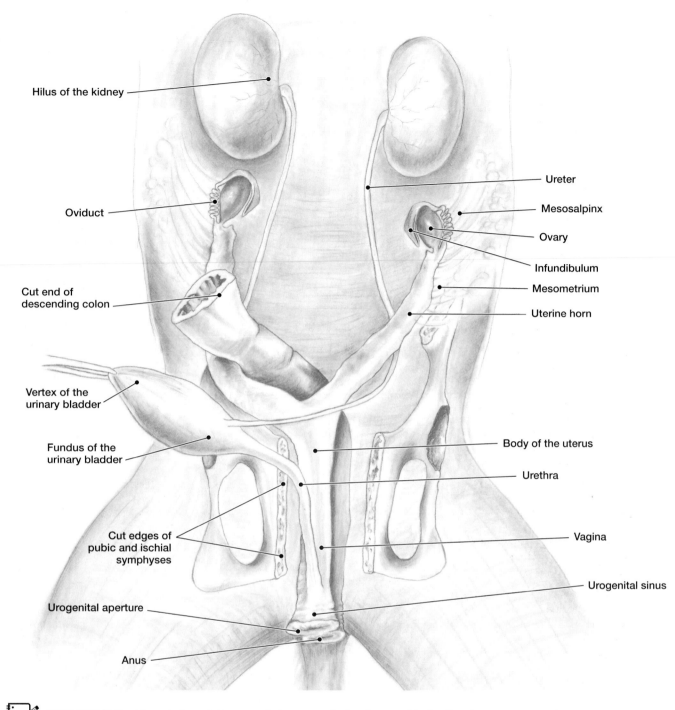

Hilus of the kidney

Oviduct

Cut end of
descending colon

Vertex of the
urinary bladder

Fundus of the
urinary bladder

Cut edges of
pubic and ischial
symphyses

Urogenital aperture

Anus

Ureter

Mesosalpinx

Ovary

Infundibulum

Mesometrium

Uterine horn

Body of the uterus

Urethra

Vagina

Urogenital sinus

FIGURE 9.9   Ventral view of the urogenital system of a female cat, with the digestive tract removed.

Reproductive Structures

The ovarian artery and vein of one side, if properly injected, can be followed laterally to the small, oval **ovary.** The **oviduct** (= fallopian or uterine tube) is the thin, convoluted tube that lies just lateral to the ovary. Ova are ovulated from the ovary and collected into the **infundibulum,** the funnel-shaped cranial end of the oviduct (figure 9.9). The opening into the oviduct is the **ostium,** whose frilly rim is the **fimbriae.** The **oviduct** transports the ova to its associated **uterine horn,** where

development of the embryos occurs. The paired uterine horns, one from each side, join dorsal to the urinary bladder to form the single **body of the uterus.** After identifying the associated mesenteries, the reproductive tract will be followed further.

Mesenteries of the Reproductive Tract   The ovary is suspended from the abdominal roof by the **mesovarium.** The short, cordlike **ovarian ligament** attaches the posterior end of the ovary to the oviduct. The oviduct is supported by the **mesosalpinx,** and each uterine horn is supported by a **mesometrium.** Collectively, the mesovarium, mesosalpinx, and mesometrium form the **broad ligament** of the mammalian uterus.

Deep Dissection of the Reproductive Tract   To follow the reproductive tract further, cut through the pubic and ischial symphyses with heavy scissors. Spread the cut by pressing the knees of the cat laterally to the bottom of the dissecting pan. As needed, extend this cut by trimming off more bone adjacent to the symphyses.

The body of the uterus is short and soon gives way to the more posterior **vagina** (figure 9.9). No external demarcation marks the boundary between these regions. However, internally, a strong muscular sphincter, the **cervix,** controls the movement of young into the vagina during the birthing process (parturition). Even more

caudally, the vagina merges with the urethra to jointly form the **urogenital sinus** (= vaginal vestibule) with external opening the **urogenital aperture** (figure 9.9).

Free the urogenital sinus, vagina, and body of the uterus from the more dorsal rectum. Make a longitudinal cut along the dorsal side to open this part of the reproductive system (figure 9.9). On the ventral side of the urogenital sinus, near its external opening, is a small, unpaired process, the **clitoris.** Anterior to this is the **urethral orifice,** the point of entry of the urethra. The **vagina** is the region between the urethral orifice and the more cranial **cervix,** which should now be located.

The descending colon straightens dorsal to the body of the uterus to form the **rectum.** Smooth muscle, the **anal sphincter,** controls its terminus and external opening, the **anus.** The pair of rounded structures lying next to the rectum near the anus are the **anal glands.**

Form & Function

The uteri of female placental mammals are quite variable. Cats possess a bipartite uterus and humans a simplex uterus. Why are there such variations? One reason is packing. Recall that development in placental mammals occurs within the uterus. Animals with large litters need more room for the many young to attach and grow. The paired uterine horns in these animals provide for more space than the simplex uterus in humans, who typically have only one or two children during a typical pregnancy.

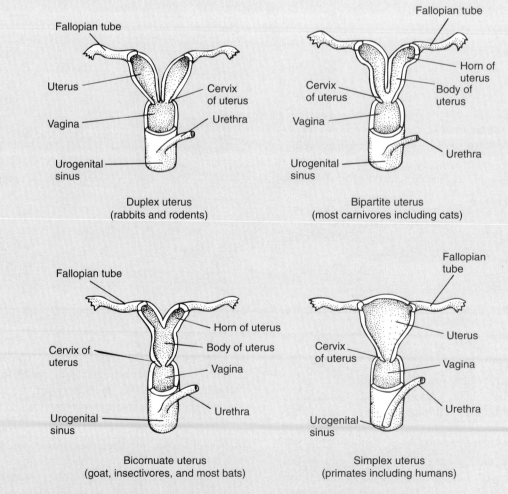

Duplex uterus
(rabbits and rodents)

Bipartite uterus
(most carnivores including cats)

Bicornuate uterus
(goat, insectivores, and most bats)

Simplex uterus
(primates including humans)

Reproductive tracts of female placental mammals. The four types of uteri vary by the degree of fusion of the paired (lateral) uteri.

10

Nervous Systems

Introduction

The nervous system receives stimuli from one or more *receptors,* such as eyes or ears, that monitor the external and internal environment. It then transmits information to one or more *effectors,* such as muscles or glands, that respond appropriately to the signals. Between receptors and effectors the information undergoes extensive processing within the central nervous system. In general, the more information arriving, the larger is the part of the brain responsible for initial sensory input. For example, where eyes gather extensive visual stimuli, the parts of the brain receiving this information are also large. But in blind cave fish, without eyes, the corresponding part of the brain may be quite small. The nervous system further regulates an animal's performance by integrating immediate incoming sensory information with *stored information,* the results of past experience. Past and present information is then translated into action by effectors.

The nervous system is divided into the *central nervous system* (CNS), which includes the brain and spinal cord, and the *peripheral nervous system* (PNS), which includes all nerves, ganglia, and other nervous tissue outside the CNS. Neurons and their processes are often known by different terms, depending on whether they occur in the CNS or the PNS. A collection of nerve fibers in the CNS is a *nerve tract* and a *nerve* in the PNS. A collection of nerve cell bodies is a *nucleus* in the CNS and a *ganglion* in the PNS.

The differentiation and enlargement of the anterior part of the nervous system into a regionalized brain were early events in vertebrate evolution (figure 10.1). This anterior concentration of distinctive nervous structures and formation of a head, called *cephalization,* seems to be built upon a segmental plan, like the trunk. But the segments contributing to the head fuse, enlarge, and in vertebrates are usually indistinguishable from one another.

Descriptive anatomy takes advantage of this ancestral segmentation as the basis for organizing and relating nervous system structures, especially the cranial nerves, among vertebrate groups. In fact, embryology confirms a general retention of relationships among specific nerves, muscles, and skeletal elements. For example, particular nerves tend to supply homologous muscles from one vertebrate group to the next. The practical benefit for the student is that by learning a general scheme of nerve distribution, any particular vertebrate can be examined with some initial idea of what to expect in terms of nerve supply.

Nerve Function

Nerves are classified upon the basis of the information they carry (sensory or motor) and upon the basic type of tissue they serve (visceral or somatic). *Sensory,* or *afferent,* sensations travel along nerve fibers to the central nervous system (figure 10.2). *Motor,* or *efferent,* impulses travel away to peripheral effector organs (e.g., muscles, glands). *Visceral tissues* derive from the embryonic gut and adjacent tissue. *Somatic tissues* are all else, namely derivatives of the embryonic body wall (skin, musculature, skeleton). The terms *general* and *special* recognize that some senses or effectors (e.g., taste, sight, olfaction, hearing) became specialized from more general functions. Thus, a nerve carrying information to skeletal muscle would be a *general somatic motor nerve.* A nerve carrying sensations from the gut would be a *general visceral sensory nerve,* and so on.

Spinal Nerves

Spinal nerves are sequentially arranged and numbered (C-1, T-1, L-1, S-1) according to their association with regions of the vertebral column (cervical, thoracic, lumbar, sacral).

Early anatomists recognized *dorsal* and *ventral roots* of each spinal nerve. Afferent fibers enter the spinal cord via the dorsal root, and efferent fibers leave by way of the ventral root (figure 10.3). The *dorsal root ganglion,* a swelling on the dorsal root, is a collection of neuron bodies whose

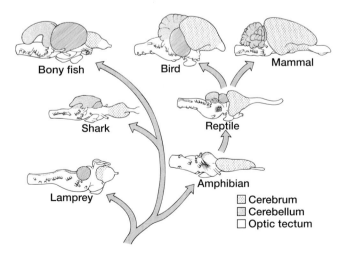

FIGURE 10.1 Evolution of the vertebrate brain. Note the phylogenetic enlargement of the cerebrum.

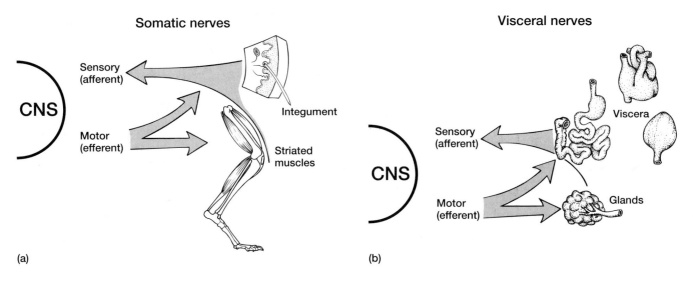

Somatic nerves

Sensory (afferent)

CNS

Motor (efferent)

Integument

Striated muscles

(a)

Visceral nerves

Sensory (afferent)

CNS

Motor (efferent)

Viscera

Glands

(b)

FIGURE 10.2 Functional categories of neurons of the peripheral nervous system. Some neurons supply somatic tissues (a), others visceral tissues (b). They can be sensory and respond to stimuli from these tissues, or they can be motor and deliver stimuli to these tissues. Sensory (afferent) neurons carry action potentials to and motor (efferent) neurons carry action potentials away from the central nervous system (CNS).

axons contribute to the spinal nerve. Parallel to the spinal cord and attached to each spinal nerve through the *ramus communicans* is the *sympathetic chain* of ganglia (paravertebral ganglia), a paired series of linked ganglia adjacent to the vertebral column or notochord. Other peripheral ganglia form the *collateral ganglia* (prevertebral ganglia). The paired *cervical, coeliac,* and *mesenteric ganglia* are examples of the collateral ganglia. The *visceral ganglia* occur within the walls of the visceral effector organs. Thus, there are three types of ganglia: *sympathetic, collateral,* and *visceral.*

Cranial Nerves

Cranial nerves have roots enclosed in the braincase. These are named and numbered by roman numbers from anterior to posterior. The conventional system for numbering these nerves is sometimes inconsistent. For instance, most anamniotes are said to have ten cranial nerves. A few anamniotes and all amniotes are said to have 12. In fact, there is an additional terminal nerve at the beginning of this series. If counted at all, it is numbered 0 to avoid renumbering the conventionally numbered sequence. Further, the second cranial nerve (II) is not a nerve at all, but an extension of the brain. Nevertheless, by convention it is called the optic "nerve." The eleventh cranial nerve (XI) represents the merger of a branch of the tenth cranial nerve (X) with elements of the first two spinal nerves (C-1 and C-2). Despite its composite structure, it is called the spinal accessory nerve and designated roman numeral XI.

Phylogenetically, the cranial nerves are thought to have evolved from dorsal and ventral roots of a few anterior spinal nerves that became incorporated into the braincase.

Like spinal nerves, the cranial nerves supply somatic and visceral tissues and carry general sensory and motor information. Some cranial nerves consist of only *sensory* or only *motor* fibers. Other nerves are *mixed,* containing both types. Cranial nerves concerned with localized senses (e.g., sight, hearing, olfaction, taste) are called *special cranial nerves* to distinguish them from those concerned with the sensory or motor innervation of the more widely distributed viscera, *general cranial nerves.*

Anamniotes

Most anamniotes possess ten cranial nerves (all but cranial nerves XI and XII listed for amniotes in the next section; figure 10.4). The first few spinal nerves caudal to the braincase become housed in the skull of later derived groups. But in anamniotes, these anterior spinal nerves are still partially outside the skull, although they have lost their dorsal roots and differ from other spinal nerves. In cyclostomes, these anterior spinal nerves outside the skull are called *occipitospinal nerves.* In other fishes and amphibians, the anterior spinal nerves become partially incorporated into the braincase. When this occurs, the nerves exit via foramina in the occipital region of the skull and are termed *occipital nerves.* Occipital nerves unite with the next few cervical spinal nerves to form the composite *hypoglossal nerve,* which supplies hypobranchial muscles in the throat.

Amniotes

Amniotes are traditionally said to possess 12 cranial nerves (figure 10.5), although the inaccuracies of this statement have just been explained. The anterior and posterior occipitospinal

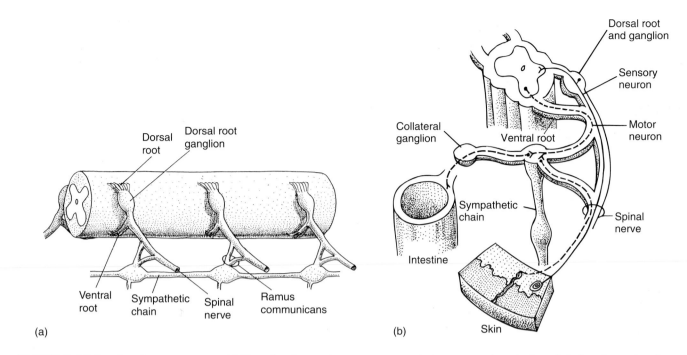

FIGURE 10.3 Spinal nerve anatomy. (a) Dorsal and ventral roots connect spinal nerves to the spinal cord. A dorsal root is enlarged into a dorsal root ganglion. Spinal nerves join with the sympathetic chain through communicating rami. (b) Diagrammatic representation of established afferent and efferent neurons within spinal nerves.

After Tuchmann–Duplessis et al.

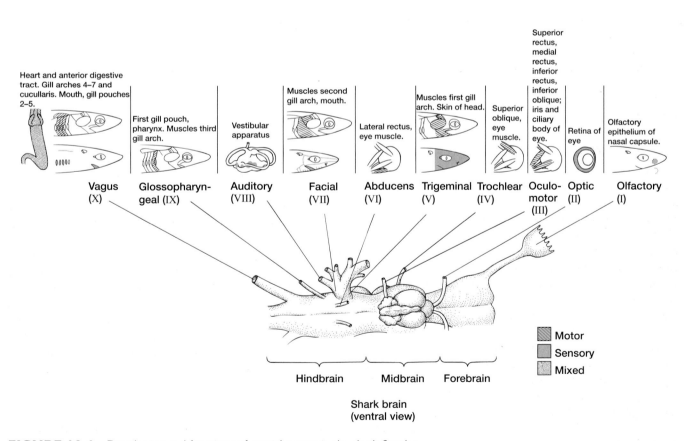

FIGURE 10.4 Distribution and functions of cranial nerves in the shark *Squalus*.

After Gilbert.

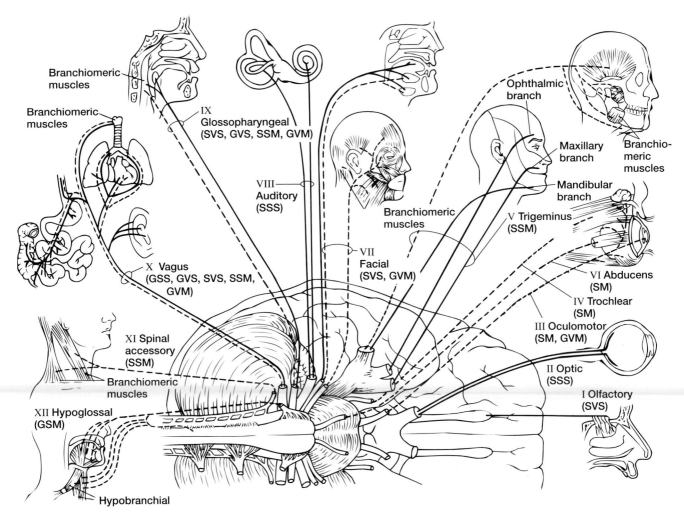

FIGURE 10.5 Distribution of cranial nerves in a mammal, *Homo sapiens*. Sensory (solid lines) and motor (dashed lines) nerve fibers are indicated. Enlarged views of innervated structures of cranial nerves are shown around the human brain in ventral view. Abbreviations: general somatic sensory (GSS), general visceral sensory (GVS), general somatic motor (GSM), general visceral motor (GVM), special somatic motor (SSM), special somatic sensory (SSS), special visceral sensory (SVS), somatic motor (SM).

After H. Smith.

nerves have become incorporated into the skull, modified, and their nuclei within the spinal cord shifted forward so they now actually issue from the medulla oblongata. Thus, by such a phylogenetic shift, amniotes derive the eleventh and twelfth cranial nerves.

Nervus Terminalis (0)—Mixed The *terminal nerve* is not numbered. It may be testimony to an ancient anterior head segment that has been lost. This nerve is present in all classes of gnathostomes except birds. It runs to blood vessels of the olfactory epithelium in the olfactory sac and most likely carries visceral sensory and some motor fibers.

Olfactory Nerve (I)—Sensory The *olfactory nerve* is a sensory nerve associated with the sense of smell. Olfactory cells lie in the mucous membrane of the olfactory sac. A short axon called an *olfactory fiber* leads from each cell to the *olfactory bulb*. Collectively, the olfactory fibers form the short olfactory nerve.

Optic Nerve (II)—Sensory As noted previously, the *optic nerve* is not a nerve but a sensory tract. That is, it is not a collection of peripheral axons; it is a collection of fibers in the CNS. Embryologically, it develops as an outpocketing of the brain. However, once it is differentiated, it lies outside the brain. Its fibers synapse in the thalamus and midbrain.

Oculomotor Nerve (III)—Motor The *oculomotor nerve* primarily supplies extrinsic eye muscles (superior rectus, medial rectus, inferior rectus, and inferior oblique muscles) derived from preotic myotomes. It is a motor nerve that also carries a few visceral motor fibers to the iris and ciliary body of the eye. Fibers arise in the oculomotor nucleus in the floor of the midbrain.

Trochlear Nerve (IV)—Motor The *trochlear nerve* is a motor nerve that supplies the extrinsic, superior oblique eye muscle. Fibers arise in the trochlear nucleus of the midbrain.

Trigeminal Nerve (V)—Mixed The *trigeminal nerve,* or *trigeminus,* is so named because in amniotes it is formed of three branches: *ophthalmic* (V1), *maxillary* (V2), and *mandibular* (V3). The *ophthalmic nerve,* sometimes called the *deep ophthalmic nerve* to distinguish it from a more superficial nerve, usually merges with the other two branches. However, in lower vertebrates, the ophthalmic nerve often emerges from the brain separately, possibly representing an ancient condition in which the ophthalmic nerve supplied an anterior arch that has been lost. The other two branches, the maxillary ramus (V2) to the upper jaw and the mandibular ramus (V3) to the lower jaw, presumably represent rami of a typical branchial nerve, the mandibular arch.

The mixed trigeminus includes sensory fibers from the skin of the head and areas of the mouth, and motor fibers to derivatives of the first branchial arch. Sensory fibers of the trigeminus return to the brain from the skin, teeth, and other areas through each of the three branches. The mandibular branch also contains visceral motor fibers to muscles of the mandibular arch.

Abducens Nerve (VI)—Motor The *abducens* is the third of the three cranial nerves that innervates muscles controlling movements of the eyeball. It is a motor nerve that supplies the extrinsic, lateral rectus eye muscle. Fibers arise in the abducens nucleus located in the medulla oblongata.

Facial Nerve (VII)—Mixed The mixed *facial nerve* includes sensory fibers from the lateral line of the head, ampullae of Lorenzini, and taste buds as well as motor fibers to derivatives of the second (hyoid) arch.

Auditory Nerve (VIII)—Sensory The sensory *auditory nerve* (= acoustic, vestibulocochlear, or statoacoustic) carries sensory fibers from the inner ear, which is concerned with balance and hearing. The nerve synapses in several regions of the medulla oblongata.

Glossopharyngeal Nerve (IX)—Mixed The mixed *glossopharyngeal nerve* supplies the third branchial arch. It contains sensory fibers from taste buds, the first gill pouch, and the adjacent pharyngeal lining and lateral line. Motor fibers innervate muscles of the third branchial arch.

Vagus Nerve (X)—Mixed The term *vagus* is Latin for wandering and aptly applies to this mixed nerve. The vagus meanders widely, serving areas of the mouth, pharynx, and most of the viscera. It is formed by the union of several roots across several head segments.

Spinal Accessory Nerve (XI)—Motor In anamniotes, the *spinal accessory nerve* is probably composed of a branch of the vagus nerve and several occipitospinal nerves. In amniotes, especially in birds and mammals, it is a small but distinct motor nerve that supplies derivatives of the cucullaris muscle (cleidomastoid, sternomastoid, and trapezius). A few of its fibers accompany the vagus nerve to supply part of the pharynx and larynx and perhaps the heart. Fibers arise from several nuclei with the medulla oblongata.

Hypoglossal Nerve (XII)—Motor The *hypoglossal nerve* is a motor nerve that innervates hyoid and tongue muscles. Fibers originate in the hypoglossal nucleus within the medulla oblongata.

Lateral Line Nerves In addition to cranial nerves, fishes and aquatic amphibians possess anterior and posterior *lateral line nerves* that are rooted in the medulla oblongata and supply the lateral line system. They were once thought to be components of the facial and vagal nerves, but they are now recognized as independent cranial nerves. Unfortunately, this late recognition as distinct cranial nerves has left them without an identifying roman numeral!

Shark

Dissection Procedure

☞ To expose the brain and cranial nerves, most of the roof of the chondrocranium will have to be removed. This should be done carefully, taking care not to damage major nerves, white or tan strings passing through the chondrocranium. The removal of the roof of the chondrocranium will be done in steps as follows (a–d):

(a) Remove the skin over the head and snout. Clear away the watery tissue and relocate some landmarks of the chondrocranium: rostrum, filled with a gel-like plug; medial epiphyseal foramen; endolymphatic fossa. Remove the upper eyelids. Between the epiphyseal foramen and orbit emerges the **superficial ophthalmic trunk** (figure 10.6a) from a foramen in the chondrocranium. This nerve trunk runs forward, lateral to the rostrum, branching profusely into many threadlike nerves to the skin.

(b) Pick away the chondrocranium along the dorsal rim starting at the orbit. Note that the superficial ophthalmic trunk runs along the dorsomedial wall of the orbit, over the extrinsic eye muscles. Clear the orbit to expose the extrinsic eye muscles and locate two of the three cranial nerves supplying them (figure 10.6a). The slender **trochlear nerve** emerges from a foramen in the dorsomedial wall of the orbit and passes under the superficial ophthalmic to innervate the superior oblique muscle. The **oculomotor nerve** is best seen at the point of convergence of rectus muscles on the medial wall of the orbit. The slender **deep ophthalmic nerve** can also be seen now passing under the superior rectus muscle, across the surface of the eyeball, under the superior oblique muscle, then out of the orbit.

(c) Remove the skin around the spiracle. The **hyomandibular nerve** passes along the posterior rim of the spiracle.

(d) Remove enough of the axial muscles along their attachment to the chondrocranium to obtain an

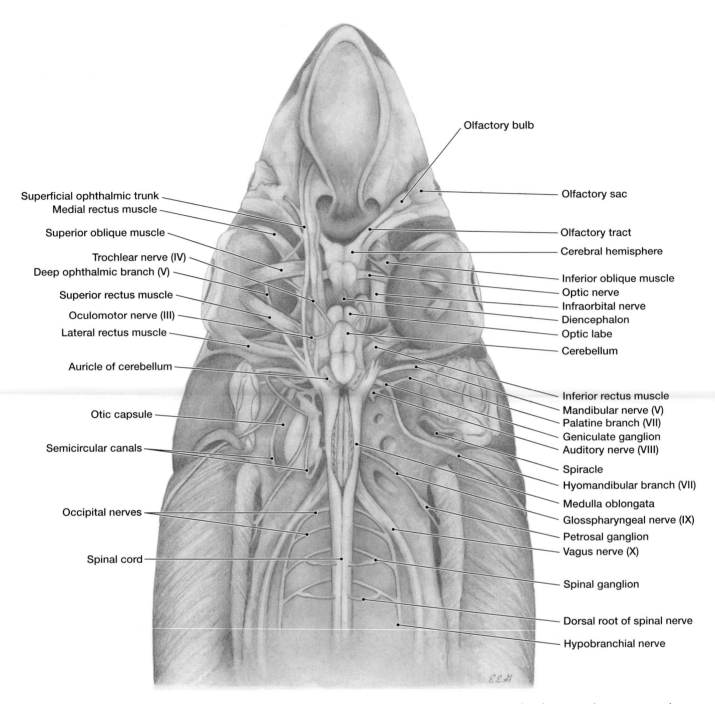

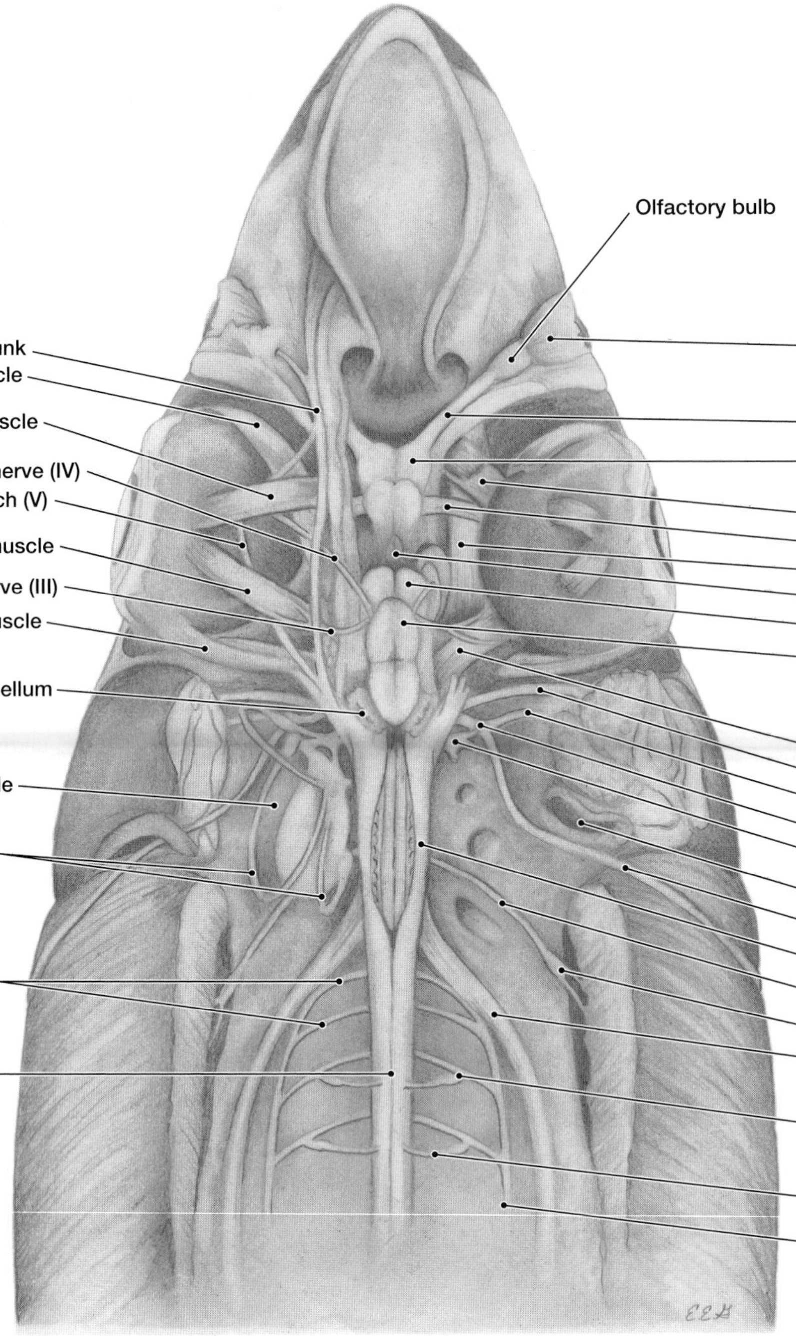
Olfactory bulb
Superficial ophthalmic trunk
Medial rectus muscle
Olfactory sac
Superior oblique muscle
Olfactory tract
Cerebral hemisphere
Trochlear nerve (IV)
Deep ophthalmic branch (V)
Inferior oblique muscle
Optic nerve
Superior rectus muscle
Infraorbital nerve
Oculomotor nerve (III)
Diencephalon
Lateral rectus muscle
Optic labe
Cerebellum
Auricle of cerebellum
Inferior rectus muscle
Mandibular nerve (V)
Palatine branch (VII)
Otic capsule
Geniculate ganglion
Auditory nerve (VIII)
Semicircular canals
Spiracle
Hyomandibular branch (VII)
Medulla oblongata
Occipital nerves
Glosspharyngeal nerve (IX)
Petrosal ganglion
Vagus nerve (X)
Spinal cord
Spinal ganglion
Dorsal root of spinal nerve
Hypobranchial nerve

FIGURE 10.6 Shark brain and cranial nerves. (a) Dorsal view, with the chondrocranium removed and eye muscles cut to reveal deeper structures.

unobstructed view of the posterior chondrocranium and anterior first few vertebrae. The roof of the chondrocranium will next be removed by picking away cartilage pieces. The strategy is to remove first the roof, then the sides of the chondrocranium, until the dorsal side of the brain is exposed and the full extent of exiting cranial nerves lies bare. One approach is to begin by dissecting the olfactory sacs dorsally and then working posteriorly. The other is to begin between the orbits. Whichever method is used, be sure not to damage the soft, even mushy, shark brain within. One further note. Between the spiracles, next to the brain, lie the **semicircular canals** embedded in calcified cartilage of the otic capsules. These semicircular canals can be removed or picked away as needed. Now remove the roof, then sides of the chondrocranium.

Brain (Dorsal Aspect)

In the middle of the brain is the large, unpaired **cerebellum** (figures 10.6a and 10.7a). Adjacent to the posterior end of the cerebellum are the two folded, earlike

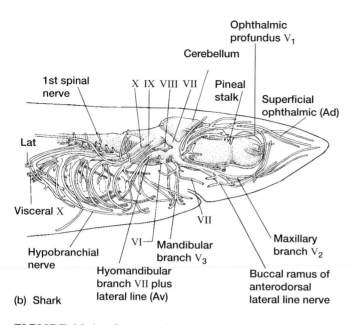

Ophthalmic
profundus V₁ → V_1

Cerebellum

1st spinal
nerve

X IX VIII VII

Pineal
stalk

Superficial
ophthalmic (Ad)

Lat

Visceral X

VI

VII

Hypobranchial
nerve

Mandibular
branch V₃

Hyomandibular
branch VII plus
lateral line (Av)

Maxillary
branch V₂

Buccal ramus of
anterodorsal
lateral line nerve

(b) Shark

FIGURE 10.6 *Continued.* (b) Lateral view.

(b) After Norris and Hughes, 1920.

auricles (figure 10.6a). The region caudal to and continuous with the cerebellum is the **medulla oblongata.** In life, the medulla oblongata is covered dorsally by a thin, pigmented roof, a **tela choroidea,** one area where cerebral spinal fluid is elaborated. Remove the tela choroidea, if it is still present. Where the medulla oblongata narrows posteriorly, it gives way to the **spinal cord.**

Now examine the floor of the open medulla oblongata where longitudinal ridges can be seen. Immediately to either side of the midventral groove lies the **somatic motor column;** the next lateral ridge is the **visceral sensory column;** the groove between these two columns is the sunken **visceral motor column.** The lateral row of very tiny, beadlike swellings is the **somatic sensory column.**

The paired **optic lobes** (tectum) lie immediately cranial to the cerebellum (figure 10.7a). **Cerebral hemispheres** lie cranial to the optic lobes and are the most anterior enlargements of the brain. Extending from each hemisphere, anterolaterally, is the **olfactory tract.** Before reaching the olfactory sac, the tract swells slightly to form the **olfactory bulb.** Olfactory nerves,

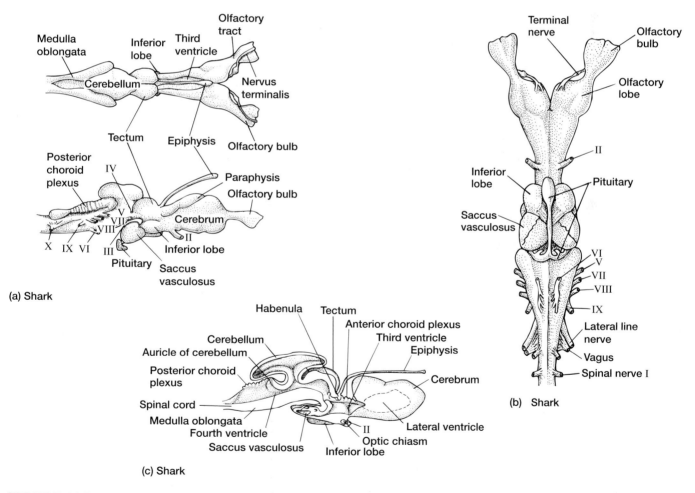

Medulla
oblongata

Inferior
lobe

Third
ventricle

Olfactory
tract

Cerebellum

Nervus
terminalis

Tectum

Epiphysis

Olfactory bulb

Posterior
choroid
plexus

IV

Paraphysis

Olfactory bulb

V

VII

VIII

Cerebrum

X IX VI

III

II

Inferior lobe

Pituitary

Saccus
vasculosus

(a) Shark

Terminal
nerve

Olfactory
bulb

Olfactory
lobe

Inferior
lobe

II

Pituitary

Saccus
vasculosus

VI
V
VII
VIII

IX

Lateral line
nerve

Vagus

Spinal nerve I

(b) Shark

Habenula

Tectum

Anterior choroid plexus

Third ventricle

Epiphysis

Cerebellum

Auricle of cerebellum

Posterior choroid
plexus

Cerebrum

Spinal cord

Lateral ventricle

Medulla oblongata

Fourth ventricle

II

Saccus vasculosus

Optic chiasm

Inferior lobe

(c) Shark

FIGURE 10.7 The brain in the shark *Scymnus*, for comparison. (a) Dorsal (top) and lateral (bottom) views. (b) Ventral view. (c) Sagittal view.

(a) Redrawn from A. S. Romer and T. S. Parsons, 1985, The Vertebrate Body, Saunders College Publishing, after Butschli, Ahlborn; (b) Redrawn from A. S. Romer and T. S. Parsons, 1985, The Vertebrate Body, Saunders College Publishing, after Butschli, Weinstein, Sisson; (c) Redrawn from A. S. Romer and T. S. Parsons, 1985, The Vertebrate Body, Saunders College Publishing.

Brain (Dorsal Aspect)

impractical to see, run from the lining of the sac to the olfactory bulb. With good dissection, and a bit of luck, the very thin, threadlike **nervus terminalis** may be seen on the anterior side, next to and parallel with the olfactory tract (figure 10.7a,b).

The brain narrows between cerebral hemispheres and optic lobes. The small, dorsal dark patch on this narrow region of the brain is a second **tela choroidea.** A delicate membrane, the **primitive meninx** (pl., **meninges**), covers the surface of the brain and carries blood vessels serving the brain. In life, cerebral spinal fluid fills the space between the brain and its cavity within the chondrocranium, and also fills ventricles, hollow chambers inside the brain. These ventricles will be examined later.

Cranial Nerves

Nervus Terminalis (CN 0)

🐾 This thin nerve runs from the olfactory sac caudally in parallel with the olfactory tract and passes into the median fissure (crease) between the two cerebral hemispheres (figure 10.7a). It has no olfactory fibers and its function remains somewhat in doubt. Most likely, however, it carries visceral sensory and visceral motor fibers. Thus, in part (visceral motor) it contributes to the autonomic system.

Olfactory Nerve (CN I)

🐾 It is impractical to see the short **olfactory nerve** fibers within the wall of the **olfactory sac** (figure 10.6a). They synapse in the olfactory bulb on the posterior wall of the olfactory sac. The **olfactory bulb** narrows into the **olfactory tract,** an elongate extension of the brain, which continues posteromedially to a cerebral hemisphere. Fibers carry olfactory sensations.

Optic Nerve (CN II)

🐾 The stout **optic nerve** fibers (figure 10.6a) depart the retina opposite the laterally positioned pupil on the medial wall of the eyeball. The optic nerve can be found somewhat between the superior and inferior oblique extrinsic eye muscles. The nerve runs medially towards the cerebral hemisphere's ventral surface then turns posteriorly, coursing to and entering the optic lobe of the brain. Optic fibers carry only visual information (special somatic sensory).

Oculomotor Nerve (CN III)

🐾 The **oculomotor nerve** (figure 10.6a) springs from the floor of the midbrain. It passes to the origins of the rectus extrinsic eye muscles, then branches to supply the inferior oblique, and the superior, inferior, and medial rectus muscles. Cut through the insertions of the superior oblique and superior rectus muscles and reflect them to aid in tracing out the distributions of this nerve. The branch to the inferior rectus muscle in turn sends off another branch that accompanies a blood vessel into

the wall of the eyeball. This is the ciliary nerve, often impractical to see, part of the autonomic system. The oculomotor nerve carries mostly general somatic motor (and proprioceptive) fibers, but also contains a few general visceral motor fibers as well.

Trochlear Nerve (CN IV)

🐾 Externally, the **trochlear nerve** (figure 10.6a) springs from the roof of the midbrain between optic lobes and cerebellum (in fact, its motor neurons lie on the floor of the midbrain). Upon emergence from the chondrocranium, it passes under the superficial ophthalmic trunk and to the superior oblique muscle that it exclusively supplies. The trochlear nerve is a general somatic motor nerve (with a few proprioceptive fibers).

Abducens Nerve (CN VI)

🐾 To complete the identification of nerves supplying extrinsic eye muscles, skip to the abducens nerve. With the brain exposed only dorsally, it is not yet feasible to see the emergence of the abducens nerve from the floor of the medulla oblongata. However, it can be seen along the origin of the lateral rectus muscle that it exclusively supplies. The abducens is a general somatic motor (and proprioceptive) nerve.

Trigeminal Nerve (CN V)

🐾 Recall that the trigeminal nerve innervates derivatives of the mandibular arch and gathers sensory information from the skin of the head. In sharks, the origin of the trigeminal nerve (V) is merged with the facial (VII) and acoustic (VIII) nerves, giving rise to a large knot of nerves next to the auricle.

The trigeminal nerve of sharks has four main branches: superficial ophthalmic, deep ophthalmic, mandibular, and maxillary branches (figure 10.6b).

Superficial Ophthalmic Branch The **superficial ophthalmic branch,** together with a branch of the **facial nerve** of the same name, form the **superficial ophthalmic trunk.** This trunk was previously identified as it passed next to the rostrum of the chondrocranium on its way to the skin over the top and sides of the snout. The superficial ophthalmic branch carries only general somatic sensory fibers.

Deep Ophthalmic Branch Upon branching from the trigeminus, the **deep ophthalmic branch** (= profundus) passes through the chondrocranium, across the dorsomedial surface of the eyeball, and out the anterior wall of the orbit via the orbitonasal canal. Lift the deep ophthalmic branch slightly as it crosses the eyeball to find a few thin branches, ciliary nerves, that penetrate the wall of the eyeball. Trace the deep ophthalmic branch forward through the canal. It joins, then separates from, the superficial ophthalmic trunk to eventually reach the skin on the top and sides of the snout. The deep ophthalmic branch carries a few general visceral motor (ciliary nerves) and many general somatic sensory (from skin of snout) fibers.

Mandibular Branch The **mandibular branch** departs at about the same point as the superficial and deep ophthalmic branches, but lies ventral to them, passing ventral to the lateral rectus muscle. The mandibular branch skirts the rear side of the orbit, supplying various gill arch muscles on the floor of the orbit, then arches ventrally into muscles of the lower jaw. Follow it for a distance through this muscle mass that it supplies. The mandibular branch carries general visceral motor (mandibular muscles) and a few general somatic sensory (skin over lower jaw) fibers.

Maxillary Branch The **maxillary branch** of the trigeminus, together with the **buccal branch of the facial nerve,** are merged into a prominent broad ribbon-like trunk running along the floor of the orbit, the **infraorbital trunk** (figure 10.6a). Remove the eyeball on one side by cutting further insertions of extrinsic muscles, nerves close to the wall of the eyeball, and by pulling out the optic pedicle. The infraorbital trunk crosses the medial wall of the orbit, then out the anterior wall. The maxillary branch returns fibers from the skin over the upper jaw and underside of the rostrum. The maxillary branch thus carries general somatic sensory fibers.

Facial Nerve (CN VII)

Recall that the **facial nerve** innervates derivatives of the hyoid arch and returns sensory information from the lateral line of the head, ampullae of Lorenzini, and areas of the mouth. The facial nerve also has four main branches: superficial ophthalmic, buccal, hyomandibular, and palatine branches (figure 10.6b). As just described, these first two branches join with branches of the trigeminus to form composite trunks.

Superficial Ophthalmic Branch The **superficial ophthalmic branch** joins with the branch of the trigeminus of the same name to form the **superficial ophthalmic trunk.** The superficial ophthalmic branch of the facial nerve returns sensory fibers from the lateral line over the top and sides of the head and from the ampullae of Lorenzini. Thus, the superficial ophthalmic nerve carries special somatic sensory fibers.

Buccal Branch The **buccal branch** joins with the maxillary branch of the trigeminus to form the infraorbital trunk across the medial wall of the orbit. The buccal branch returns sensory fibers from the lateral line and ampullae of Lorenzini over the upper jaw and underside of the rostrum. The buccal branch carries special somatic sensory fibers.

Hyomandibular Branch The **hyomandibular branch** of the facial nerve skirts the rear edge of the spiracle. Near the base of the hyomandibular, where it

first branches from other nerves, it bears a small swelling, the **geniculate ganglion.** Follow the nerve peripherally around the side of the head. The hyomandibular branch contains special visceral motor (hyoid muscles) fibers, special somatic sensory (lateral line) fibers, general visceral sensory (mouth lining) fibers, and special visceral sensory (taste buds) fibers.

Palatine Branch The **palatine branch** departs at the geniculate ganglion, then runs forward and ventrally to the lining of the mouth, from which it returns sensory fibers. The palatine branch carries general and special sensory fibers (mouth lining and taste buds, respectively).

Auditory Nerve (CN VIII)

The stout **auditory nerve** arises between the base of the hyomandibular nerve and medulla oblongata. It returns fibers from the inner ear. Pick away the otic capsule to better follow this nerve. The auditory nerve carries special somatic sensory fibers.

Glossopharyngeal Nerve (CN IX)

The **glossopharyngeal nerve,** slender at its base, is easily confused for part of the auditory nerve. The glossopharyngeal nerve supplies the first complete gill pouch. To find where it springs from the medulla oblongata, slightly lift this side of the brain. The glossopharyngeal nerve passes posterolaterally through the otic capsule, generally in parallel with the hyomandibular nerve. Pick away the otic capsule. Just before emerging from the capsule, the glossopharyngeal nerve swells, forming the **petrosal ganglion** (figure 10.6a,b). At or just distal to this ganglion, the nerve divides into **pretrematic** and **posttrematic branches** to the anterior and posterior walls of the gill pouch, respectively (figure 10.6b). A still smaller pharyngeal branch to the wall of the pharynx and a dorsal branch to the lateral line can occasionally be discovered. The glossopharyngeal nerve mostly carries visceral sensory fibers (general and special), but also a few somatic sensory (special) fibers from the lateral line.

Vagus Nerve (CN X)

Basically, the **vagus nerve** supplies the remaining gill pouches, lateral line, and viscera (figure 10.6). Remove skin, muscles, and deeper cardinal sinus as needed to follow this nerve. It has a fan-shaped origin from the posterior end of the medulla oblongata. It sweeps posteriorly dorsal to the gill arches, giving off four branchial branches, each to separate gill pouches. Like the glossopharyngeal nerve previously examined, each of the four branchial branches of the vagus nerve has in turn **pretrematic, posttrematic,** and **pharyngeal branches.**

Formally, the four branches arise themselves from the **visceral branch** of the vagus. The visceral branch continues posteriorly, entering pericardial and pleuroperitoneal cavities to supply viscera within.

The medial coursing **lateral branch** of the vagus does not become anatomically separate until near the end of the row of gill pouches. The lateral branch carries somatic sensory fibers from the lateral line.

The vagus nerve carries all functional types of fibers, except somatic motor (and proprioceptive).

Occipital, Hypobranchial, and Spinal Nerves

🦖 At about the level of the last gill pouch, the **hypobranchial nerve** can be seen as it crosses the visceral branch of the vagus. Follow the hypobranchial nerve proximally toward the spinal cord. The first few spinal nerves merge to form the hypobranchial nerve. In amniotes, the hypobranchial nerve is incorporated into the cranium as the hypoglossal nerve.

Expose the **spinal cord** caudal to the medulla oblongata by picking away the neural spines and neural arches of the first few vertebrae. Next to the spinal column, serially between vertebrae, are small swellings, the spinal **ganglia** of the dorsal root (figure 10.6a). From the side of the medulla oblongata, between the first such ganglion and the root of the vagus, spring two **spinal roots** without ganglia. These are the **occipital nerves.** The occipital nerves, like the next several spinal nerves that follow, can now be traced peripherally to where they contribute to the hypoglossal nerve.

The hypobranchial nerve swings around the gill region to supply hypobranchial muscles in the throat (figure 10.6b). The hypobranchial nerve carries general somatic motor (and proprioceptive) fibers.

Brain (Ventral Aspect)

Of all sharks dissected by members of the class, only the brains of one or two sharks should be removed. Check with your instructor before proceeding with the removal of the brain.

Removal of the Brain

🦖 To free the brain from the chondrocranium, first cut through the olfactory tracts and the spinal cord posterior to the medulla oblongata. Lift the anterior end of the brain, noting the optic nerves. Cut both optic nerves distally, leaving generous portions still attached to the ventral surface of the brain. Continue to lift the brain and cut remaining cranial nerves, leaving generous lengths of nerve attached. Continue until the brain is completely removed. Observation is sometimes aided by floating the brain in a shallow dissection pan of water.

Regions

🦖 Now in ventral view, refamiliarize yourself with the general brain regions already seen from the dorsal aspect (figure 10.7a,b).

Where the optic nerves converge and enter the brain is seen an unpaired medial swelling, the **optic chiasm** (figure 10.7b,c). Posterior to the optic chiasm

are a pair of larger swellings, the **inferior lobes of the infundibulum** (figure 10.7a–c). Between the inferior lobes lies a median projection, sometimes torn off when the brain is removed, the **hypophysis** (= pituitary body) (figure 10.7a,b). To either side and slightly dorsally above the posterior tip of the hypophysis lie the heavily vascularized (and hence slightly discolored) **vascular sacs** (saccus vasculosus), also part of the infundibulum (figure 10.7a–c).

Oculomotor nerves (CN III) spring from the wall of the midbrain just dorsal to the vascular sacs (figure 10.6a). Approximately between trigeminal nerves, on the floor of the medulla oblongata, lies the paired abducens nerve (CN VI; figure 10.7b). Lateral to the abducens is the glossopharyngeal nerve (IX).

Brain (Sagittal Aspect)

🦖 Use one of the brains removed previously for viewing ventral brain structures. With a razor blade or very sharp scalpel, make a midsagittal cut bisecting the brain into two equal, longitudinal halves.

The brain and spinal cord are hollow. This cavity in the spinal cord is the **central canal.** In the brain, the cavity is irregularly enlarged, forming four **ventricles,** continuous with each other and with the central canal (figure 10.7c). In life, the ventricles and central canal are filled with cerebral spinal fluid formed by the tela choroidea and probably reabsorbed in the primitive meninx. The first two ventricles (lateral ventricles) lie in the two cerebral hemispheres. The third ventricle resides in the optic lobes (here it is properly termed an optic ventricle) and extends into the hypophysis and cranially to the first two ventricles. The passage between the third and the first two ventricles is the **interventricular foramen** (= foramen of Monro), not an obvious portal in the shark, but more so in mammals. The fourth ventricle occupies the medulla oblongata and extends up into the cerebellum, forming the **cerebellar ventricle.** The slender canal connecting the fourth and third ventricles is the **cerebral aqueduct** (of Sylvius).

Ventral to the anterior end of the optic lobes lies a small swelling, the **habenula** (figure 10.7c). The slender epiphysis (**pineal body**) arises from the habenula and projects dorsally, but is often lost when removing the brain (figure 10.8c). Directly across the ventricle

Patterns & Connections

Based largely on early embryology, the brain is divided into three regions: forebrain (prosencephalon), midbrain (mesencephalon), and hindbrain (rhombencephalon). These in turn differentiate into five regions composed of specific structures indicated in figure 10.8. Reexamine the shark brain to identify roughly the boundaries of these five regions (telencephalon, diencephalon, mesencephalon, metencephalon, and myelencephalon).

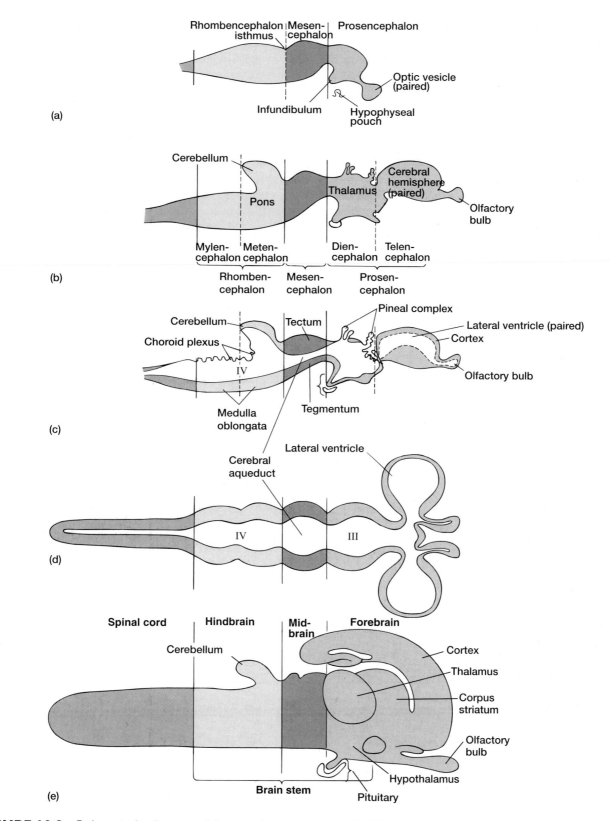

FIGURE 10.8 Embryonic development of the central nervous system. (a–c) Embryonic development. (d) Fluid-filled ventricles within the central nervous system. (e) Anatomical regions of the adult brain. Note that embryonic terms are generally replaced by descriptive terms in the adult.

from the habenula (sagittal view) is the optic chiasm (figure 10.8b,c). The walls of the brain between cerebral hemispheres and optic lobes represent the **thalamus** (figure 10.8b). The roof of the brain above the thalamus constitutes the **epithalamus;** the floor below and its derivatives constitute the **hypothalamus** (figure 10.8b).

Sheep Brain

Four preparations of the sheep brain will be studied: whole brains, sagittal sections, brain stems, and cross sections. Each will provide a different view of previously identified parts, and expose new structures for identification.

As with vertebrate brains generally, the mammalian brain early in ontogeny differentiates into three regions: prosencephalon, mesencephalon, and rhombencephalon. In turn, these differentiate into specific regions of the adult nervous system (figure 10.8e).

Overview—Whole Brain

Meninges

The brain is wrapped by three layered meninges: from outside to inside, *dura, arachnoid,* and *pia mater* (figure 10.9). The outer two layers (dura mater and arachnoid) are usually left behind in the cranium when the skull is removed. The

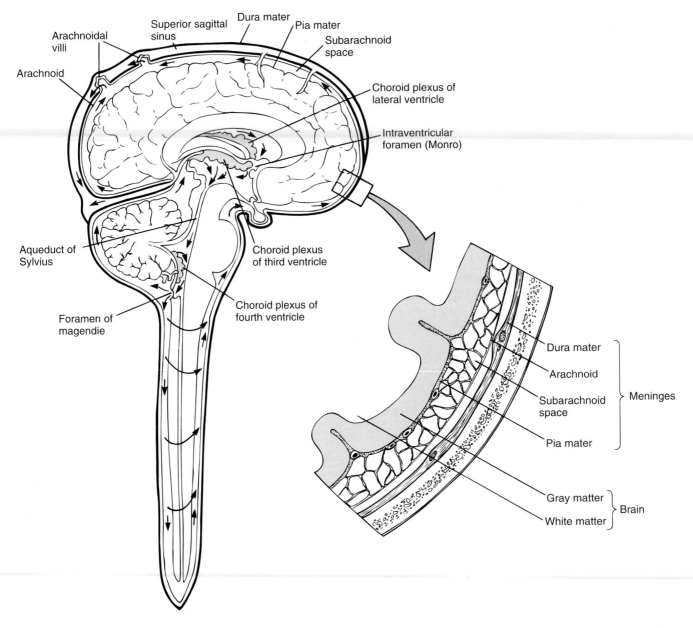

FIGURE 10.9 Cerebrospinal fluid and meninges. Arrows trace the circulation of cerebrospinal fluid through the brain and spinal cord of a mammal. The triple-layered meninges are enlarged to the right.

After H. Smith, 1960, Evolution of Chordate Structure, *Holt, Rinehart and Winston, NY.*

A bacterial or viral infection may cause an inflammation of the meninges of the brain, a condition called meningitis. If the infectious agents spread to the central nervous system, significant damage may result. Meningitis is typically diagnosed by sampling cerebrospinal fluid that surrounds the spinal cord to look for signs of immunological response to infection (lymphocytes) or for infectious agents (bacteria or viruses).

Patterns & Connections

In many mammals, the cerebral cortex is folded in a complicated fashion, increasing the volume of this region of the brain. But not all mammals show such folding. In the duckbill platypus, opossum, and many rodents, the cerebral cortex is smooth. In the echidna, kangaroo, and most primates, the degree of folding is variable. In all groups of mammals, the extent of folding seems to be more pronounced in larger species.

inner most meningeal lining, the pia mater, carries blood vessels and tightly follows the contours of the brain.

Dorsal Aspect (Figure 10.10)

In dorsal view, the large, greatly expanded **cerebral hemispheres** constitute the most anterior part of the brain (figure 10.10). The convoluted ridges, termed **gyri** (sing., **gyrus**) are partially separated from one another by furrows, termed **sulci** (sing., **sulcus**). With a blunt probe, break through the **pia mater** over a sulcus and gently spread the adjacent gyri to see that the furrow continues down a few millimeters. Along the dorsal midline between hemispheres lies the **superior sagittal sinus,** which receives blood exiting from the brain. In preserved specimens, it is usually not distinct. With a blunt probe, slit the superior sagittal sinus and gently spread the two hemispheres. This cleft between hemispheres is the **median sulcus** (figure 10.10). Along the ventral side of the sulcus lies the band joining the bases of the cerebral hemispheres, the **corpus callosum.**

The large, unpaired lobe of the brain in contact with the posterior ends of the cerebral hemispheres is the **cerebellum** (figure 10.10). Posterior to the cerebellum runs the **spinal cord.** The surface of the cerebellum is formed of many irregular folds termed **folia,** separated by sulci. Because it looks like a segmental worm tightly coiled into a circle, the median sagittal part of the cerebellum is termed the **vermis;** the lateral parts are termed the **hemispheres.** The most lateral superficial tip of each cerebellar hemisphere, in contact with the medulla oblongata, is the **flocculus.** It receives vestibular input and is thought to be homologous with the auricular lobes of lower vertebrates.

Gently spread the cerebral hemispheres and the cerebellum, clearing connective tissue between these two parts of the brain (see figure 10.13). Revealed are four swellings termed **colliculi** (= corpora quadrigemina), dorsal parts of the midbrain. The anterior pair, **superior colliculi,** looks like two, small embedded balls; the posterior pair forms a transverse ridge with raised corners, the **inferior colliculi.**

Ventral Aspect (Figure 10.11)

In ventral view, the flattened, oval-shaped **olfactory bulbs** lie at the anterior tip of the cerebral hemispheres. The broad **olfactory tract** begins on the posterior corner of the olfactory bulb, but almost immediately divides into medial and lateral bands. The medial division, the **medial olfactory stria,** is a broad, short band vaguely discernible. Clearing some pia mater may bring it into relief as it slants inward to the median sulcus. The lateral division of the olfactory tract is the **lateral olfactory stria,** which passes posteriorly, eventually expanding into a prominent lobe on the base of the cerebral hemisphere, the **pyriform** (= piriform) **lobe.** The pyriform lobe lacks sulci within, but is set off along its dorsal edge from the rest of the cerebral hemisphere lateral to the **rhinal fissure.**

The most prominent nerves seen on the base of the brain are the X-shaped **optic nerves.** The **optic chiasm** is formed at their point of crossing. The prominent **pituitary gland** resides centrally on the base of the brain, but is almost invariably torn away. Instead, the remains of its hollow stalk of attachment to the brain, the **infundibulum,** can be found just posterior to the optic chiasma. Following the infundibulum is a small, medial swelling, the **mammillary body,** continuous with two low ridges, **tuber cinerei** (sing., **cinereum**), that pass anteriorly to either side of the infundibulum.

The **circle of Willis** is the blood vessel encircling the base of the optic nerves, mammillary body, and tuber cinerei. The **internal carotid artery** empties into the circle laterally, adjacent to each tuber cinereum. If the blood vessels are preserved, note how branches off the circle of Willis supply regions of the cerebrum.

The **cerebral peduncle** is the region lateral to and extending posteriorly from the mammillary body. The ribbonlike **oculomotor nerve** springs from this peduncle. The posterior part of this peduncle is crossed by a broad band, the **pons,** appearing as swellings on either side of the ventral midline. The pons continues laterally up into the cerebellum as a stout stalk, the **cerebellar peduncle.** The **medulla oblongata** follows posterior to the pons, continuing until it constricts to give way to the **spinal cord.**

Cranial Nerves—Whole Brains (Figure 10.11)

The cranial nerves seem especially susceptible to destruction when the sheep brain is removed. You may thus need to inspect several whole brains to see them all.

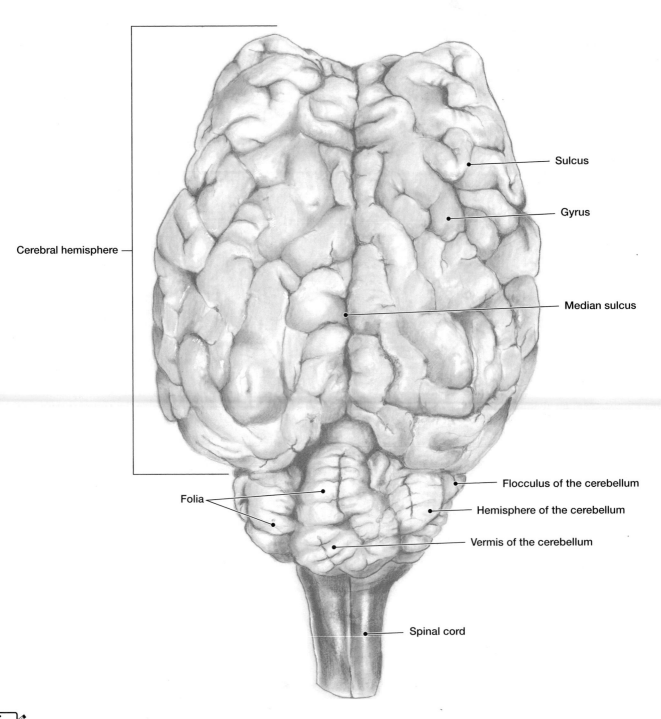

Cerebral hemisphere

Sulcus

Gyrus

Median sulcus

Flocculus of the cerebellum

Hemisphere of the cerebellum

Folia

Vermis of the cerebellum

Spinal cord

FIGURE 10.10 Dorsal view of the sheep brain.

Olfactory Nerve (I)

🐾 Fibers contributing to the olfactory nerve are short. They exit from the mucous membrane lining the nostrils, pass to and end in the olfactory bulb, and are impractical to see in gross dissection.

Optic Nerve (II)

🐾 The large optic nerves cross, forming the **optic chiasm,** before entering the base of the brain. In the chiasma, some fibers of one nerve turn and proceed with fibers of

the other crossing nerve. Thus, visual information coming from one eye distributes to both sides of the brain.

Oculomotor Nerve (III)

🐾 The **oculomotor nerve** is broad and ribbonlike. It springs from the base of the cerebral peduncle.

Trochlear Nerve (IV)

🐾 The **trochlear nerve** springs from the dorsal surface of the brain stem next to the inferior colliculus. It passes

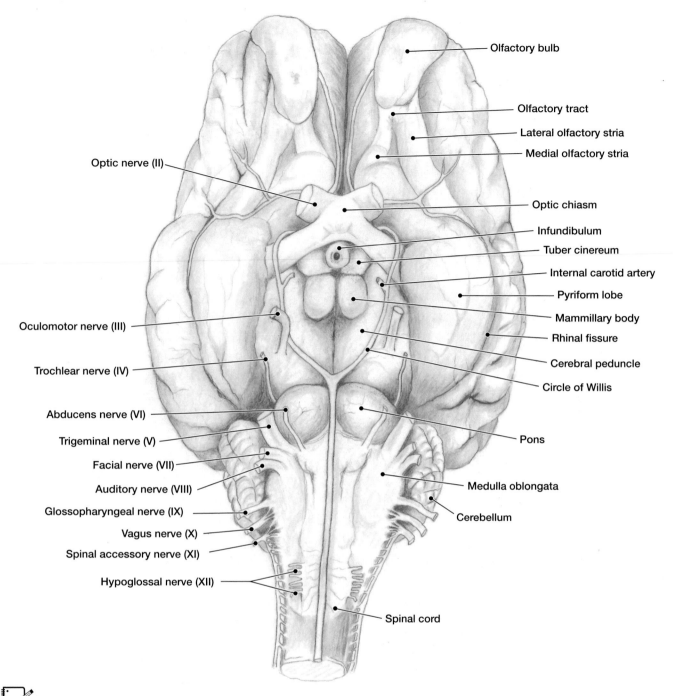

Olfactory bulb

Olfactory tract

Lateral olfactory stria

Medial olfactory stria

Optic nerve (II)

Optic chiasm

Infundibulum

Tuber cinereum

Internal carotid artery

Pyriform lobe

Mammillary body

Oculomotor nerve (III)

Rhinal fissure

Cerebral peduncle

Trochlear nerve (IV)

Circle of Willis

Abducens nerve (VI)

Trigeminal nerve (V)

Pons

Facial nerve (VII)

Auditory nerve (VIII)

Glossopharyngeal nerve (IX)

Medulla oblongata

Vagus nerve (X)

Cerebellum

Spinal accessory nerve (XI)

Hypoglossal nerve (XII)

Spinal cord

FIGURE 10.11   Ventral view of the sheep brain.

around the sides of the brain stem where it can usually be found next to the cerebral peduncle.

Trigeminal Nerve (V)

🐾 The **trigeminal nerve** is the largest nerve of the hindbrain. It emerges from the posterior and lateral side of the pons. If the nerve has been cut, the remaining large stump can usually be located at this point.

Abducens Nerve (VI)

🐾 The slender **abducens nerve** arises from the medulla oblongata next to the ventral midline posterior to the

pons. It extends anteriorly, and in sheep brain preparations, its free cut end usually lies across the pons.

Facial Nerve (VII)

🐾 The cut stump of the **facial nerve** can be located in a low depression immediately posterior to the root of the trigeminus.

Auditory Nerve (VIII)

🐾 The root of the **auditory nerve** can be found just posterior to the facial nerve and ventral to the flocculus of the cerebellum. Note that the root of the auditory nerve rests on a low ridge, the **trapezoid body,** which runs downward and around the sides and ventral surface of the medulla oblongata. The facial nerve arises from the slight depression between trapezoid body and cerebellar peduncle. The abducens nerve arises from a more medial part of the trapezoid body.

Glossopharyngeal (IX), Vagus (X), Spinal Accessory (XI), and Hypoglossal (XII) Nerves

These four nerves arise from the medulla oblongata, but their close proximity to each other, and damaged condition resulting from the removal of the brain, can make their identification challenging.

The roots of the first three (IX, X, XI) are found in order (anterior to posterior) along the dorsolateral side of the medulla oblongata. The **glossopharyngeal**

nerve, and **vagus** posterior to it, have small but defined stumplike roots. The **spinal accessory,** posterior to these two, arises as numerous tiny, threadlike rootlets that combine to form the nerve. Rootlets of the last cranial nerve, **hypoglossal** (XII), are also tiny and threadlike. This final cranial nerve arises more medially and slightly caudal to the spinal accessory nerve, from the ventrolateral side of the medulla oblongata.

Brain Regions—Sagittal Sections (Figure 10.12)

This part of the dissection provides another view of structures already identified and reveals deep structures not yet described. However, it is equally intended to introduce the three brain regions, their subdivisions, and specific structures derived from each. Refer again to figure 10.8 to recall the five embryonic brain regions, discussed here from posterior to anterior.

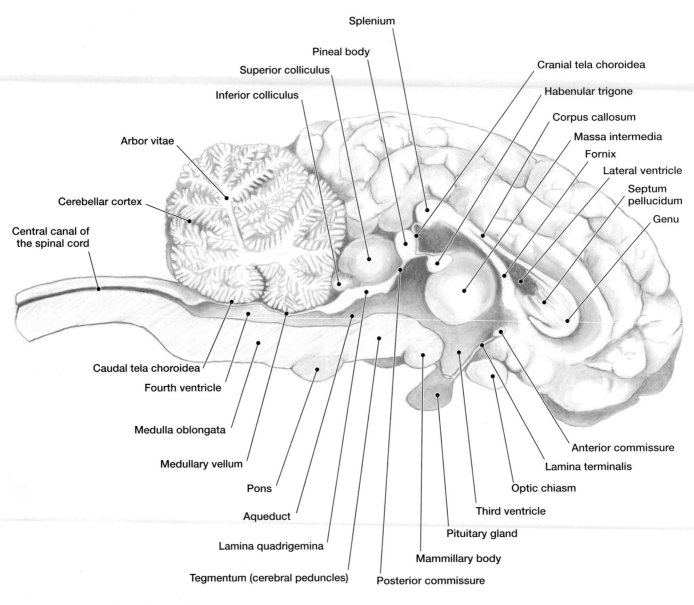

FIGURE 10.12 Sagittal view of the sheep brain.

Rhombencephalon (Hindbrain)

🦕 **Myelencephalon** (= *medulla oblongata*) The myelencephalon is composed of the **medulla oblongata,** the anatomical connection between the brain and spinal cord. It serves as a major reflex center and forms the base for cranial nerves (VII through XII).

Metencephalon Two parts compose the metencephalon, the **pons** and **cerebellum.** The outer surface of the cerebellum, the **cerebellar cortex,** is primarily highly folded gray matter. The finely branched white matter within is the **arbor vitae,** or "tree of life."

The **fourth ventricle** (figures 10.9 and 10.12) lies beneath the cerebellum and is continuous posteriorly with the **central canal** of the spinal cord and anteriorly with the **aqueduct** of the midbrain. The thin membrane between the anterior base of the cerebellum and fourth ventricle is the **medullary vellum.** A similar membrane, the **caudal tela choroidea,** occurs between the posterior base of the cerebellum and the roof of the fourth ventricle (figure 10.12).

Mesencephalon (Midbrain)

🦕 **Corpora Quadrigemina** The roof, or **tectum,** of the midbrain over the **aqueduct,** differentiates into the **lamina quadrigemina,** which supports two paired swellings, the **superior** and **inferior colliculi** (collectively the **corpora quadrigemina**).

🦕 **Cerebral Peduncle** The thick floor of the midbrain, just anterior to the pons, is the **tegmentum,** or **cerebral peduncles.**

Prosencephalon (Forebrain)

🦕 **Diencephalon** The diencephalon encloses the **third ventricle.** The roof, sides, and floor compose the three basic regions of the diencephalon: **epithalamus, thalamus,** and **hypothalamus,** respectively. The most prominent part of the epithalamus is the small, oval **pineal body** resting on the superior colliculus. The base of the pineal continues posteriorly into the **posterior commissure,** which in turn continues posteriorly into the **lamina quadrigemina,** a thicker strip just anterior to the cerebellum and ventral to the superior colliculus. The base of the pineal continues anteriorly into the **habenular trigone.** Dorsal to the pineal lies another thin, vascular membrane, the **cranial tela choroidea.** The largest part of the diencephalon, the **thalamus,** bears a prominent, circular, and medial enlargement, the **massa intermedia** (= intermediate mass), partially cut in making the sagittal section of brain.

🦕 The **optic chiasm** passes just ventral to a thin membrane, the **lamina terminalis,** the anterior boundary of the third ventricle. The tiny **anterior commissure** lies at the dorsal end of the lamina terminalis.

The ventral projection from the hypothalamus is the **pituitary gland** (= hypophysis), whose narrow stalk connecting it to the brain is the **infundibulum.** The **mammillary body** is a swelling found just posterior to the infundibulum.

🦕 **Telencephalon** (= *cerebrum*) The band of the **corpus callosum** connecting both hemispheres is seen in cross section. Both ends are curled and enlarged, forming a **genu** (anterior) and a **splenium** (posterior). The **septum pellucidum** is a thin membrane between the genu and splenium. It closes the entrance into the lateral ventricle of the cerebral hemisphere. Free the lower border of this septum from the **fornix,** a band of white matter seen in cross section and lying dorsal to the massa intermedia. The fornix begins anteriorly in the mammillary body (impractical to see this beginning in sagittal section alone). Pull the brain stem away from the hemisphere enough to allow you to follow the fornix posteriorly. The fornix sweeps around the inside of the cerebral hemisphere to end in the pyriform lobe. The entire free edge of this part of the fornix within the hemisphere is termed the **fimbria.** Fold out the edge of the fimbria to discover that its inner margin is greatly expanded. This expanded region is the **hippocampus.**

---

[651] **Form & Function**

In fishes and amphibians, the midbrain is often the most prominent region of the brain. The tectum (roof) receives direct input from the eyes. In addition, information from the acousticolateralis system, the cerebellum, the olfactory epithelium, and the cutaneous sensors may be transmitted indirectly to the tectum. The tegmentum (floor) is also prominent in these vertebrates. In fact, in some fishes, the tegmentum seems to be an important area associated with learned behaviors.

[603] **Form & Function**

The pineal organ participates in photoreception among more basal vertebrates but tends to become an endocrine gland (thus, pineal "gland") in more derived vertebrates. In more basal vertebrates (including reptiles), it affects skin pigmentation and plays a role in regulating photoperiod. In birds and mammals, it helps to regulate biological rhythms.

Form & Function **[651]**

The hypothalamus houses a collection of nuclei that regulate homeostasis to maintain the body's internal physiological balance. Homeostatic mechanisms adjusted by these nuclei pertain to temperature, water balance, appetite, metabolism, blood pressure, sexual behavior, alertness, and some aspects of emotional behavior. The hypothalamus stimulates the pituitary gland beneath it to regulate these homeostatic functions.

Brain Stem—Parts (Figure 10.13)

🦖 The brain regions, minus the cerebral hemispheres and cerebellum, collectively constitute the **brain stem.** Relocate structures already identified in whole and sagittal section, region by region.

Follow the optic tract as it sweeps upward along the side of the diencephalon. Most fibers reach the large, rounded region on top of the thalamus. This swollen part of the thalamus is specifically termed the **pulvinar nucleus.** Optic fibers also synapse in the **lateral geniculate body,** a small lateral swelling at the base of the colliculi. The **medial geniculate body** lies immediately posterior to the lateral geniculate, but the lateral geniculate is not well defined here on the surface of the thalamus.

If still present, remove the cerebellum at its base. The pedicle by which the cerebellum attaches is formed by three peduncles: **middle cerebellar peduncle** (from pons along the lateral side of the myelencephalon), **anterior cerebellar peduncle** (parallel with and medial to the middle peduncle), and **posterior cerebellar peduncle,** or **restiform** (a short section completing

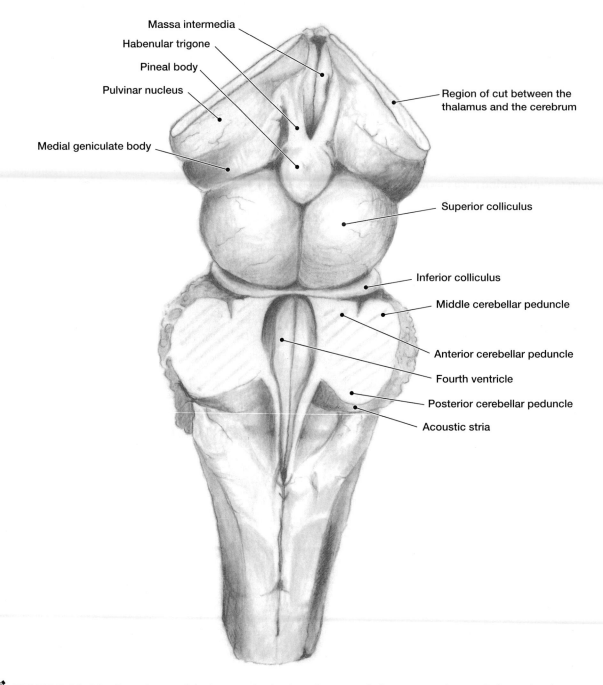

Massa intermedia
Habenular trigone
Pineal body
Pulvinar nucleus
Medial geniculate body
Region of cut between the thalamus and the cerebrum
Superior colliculus
Inferior colliculus
Middle cerebellar peduncle
Anterior cerebellar peduncle
Fourth ventricle
Posterior cerebellar peduncle
Acoustic stria

FIGURE 10.13   Dorsal view of the brain stem of a sheep brain, cerebellum removed at cerebellar peduncles.

the posterior side of the collective peduncle). The small, dorsal transverse swelling tucked in behind the posterior peduncle is the **acoustic stria.**

Brain—Optional Cross Sections

🦖 You may find it helpful to use cross sections to see structures from another view. Try following the caudate nucleus, fornix, hippocampus, corpus callosum, and other identified regions in the cerebral hemispheres through sequential cross sections. If you get lost, return to the earlier description to begin again.

In sections through the cerebral hemisphere, note the outer folded cortex.

Spinal Nerves

The dissection of the previous animal systems has made a thorough analysis of the spinal nerves difficult. However, in the cat, many cervical and thoracic spinal nerves should still be available for study. Spinal nerves often unite to form a complex network, or *plexus,* serving a broad region of the body. The last cervical and first thoracic spinal nerves form the *brachial plexus.* Similarly, the last lumbar and sacral spinal nerves form the *lumbosacral plexus.*

🦖 In the axillary region, identify the last **cervical** and first few **thoracic spinal nerves,** which parallel the axillary artery to form the **brachial plexus.** These nerves will appear fibrous and have a tan or off-white color. In the lumbar region of the peritoneal cavity, identify several **lumbar** and **sacral spinal nerves** paralleling the femoral artery on the ventral side of the groin and thigh. These spinal nerves contribute to the **lumbosacral plexus.** The specific names of spinal nerves will not be followed.

Glossary

A

adaptation A phenotypic feature of an individual that contributes to that individual's survival; a feature's form/function and associated biological role with respect to a particular environment.

advanced *See* derived. *Compare* primitive.

afferent Bringing to, as sensory afferent fibers bringing impulses to the central nervous system. *Compare* efferent.

agnathan Vertebrate lacking jaws.

akinetic skull A skull lacking cranial kinesis.

amnion A saclike membrane that holds the developing embryo in a compartment of water.

amniote Vertebrate whose embryo is wrapped in an amnion.

amphystyly Jaw suspension via two major attachments, the hyomandibula and the palatoquadrate.

analogy Features in two or more organisms performing a similar function; common function.

anamniote Vertebrate whose embryo lacks an amnion.

antagonist A muscle with an opposite action to other muscles. *Compare* synergist, fixator.

aponeurosis Broad, flat tendon.

archetype Fundamental type; basic underlying blueprint or model upon which the definitive animal or part of an animal is thought to be based.

archinephric duct General term for the urogenital duct that may go by alternative names (wolffian duct) at different embryonic stages (pronephric duct, mesonephric duct, opisthonephric duct) or in different functional roles (vas deferens).

artery Blood vessel carrying blood away from the heart; blood carried may be high or low in oxygen tension. *Compare* vein.

atavism Evolutionary throwback; reappearance of a lost ancestral trait. *Compare* vestigial.

atrophy Decrease in size or density.

auditory Pertaining to the perception of sound.

autostyly Jaw suspension in which the jaws articulate directly with the braincase.

axon Nerve fiber of a neuron carrying an impulse away from the cell body.

B

baleen Keratinized straining plates arising from the integument in the mouth of some species of whales.

benthic Bottom dwelling.

bilateral symmetry A body in which left and right halves are mirror images of each other.

biological role How the form and function of a part performs in an environmental context contributing to survival. *Compare* function.

biomechanics The study of how physical forces affect and are incorporated into animal designs.

bipedal Walking or running by means of only the two hindlegs. *Compare* quadrupedal.

bolus Soft mass of food in the mouth or stomach. *Compare* chyme.

brachyodont Low crown teeth. *Compare* hypsodont.

braincase That part of the skull that contains the cranial cavities; part of the skull housing the brain.

brain stem Posterior part of brain composed of midbrain, pons, medulla.

branchial basket The expanded chordate pharynx serving suspension feeding.

bunodont Teeth with peak-shaped cusps. *Compare* lophodont, selenodont.

C

calcification The process of calcium deposition in tissue. *Compare* ossification.

carapace Dorsal, dome-shaped bony part of a turtle shell. *Compare* plastron.

carnassials Sectorial teeth of carnivores including upper premolar and lower molar. *See* sectorial teeth.

caudal Toward the tail or back end of the body; posterior.

cecum Blind-ended outpocketing from the intestines.

cementum Cellular and acellular layer on teeth, usually located on the roots but in some herbivores may contribute to the occlusal surface. *See* enamel, dentin.

central nervous system The brain and spinal cord.

centrum The body or base of a vertebra.

ceratotrichia Fan-shaped array of keratinized rods internally supporting the elasmobranch fin. *Compare* lepidotrichia.

chemoreceptor

chemoreceptor Sense organ responsive to chemical molecules. *Compare* radiation receptor, mechanoreceptor.

chiasm Crossing of fibers.

choana The internal naris; the openings of the nasal passage into the mouth.

chondrocranium That part of the skull formed by endochondral bone or cartilage that underlies and supports the brain, plus the fused or associated nasal capsules.

chromatophore General term for a pigment cell.

chyme The liquefied bolus of partially digested food after it leaves the stomach and enters the intestine; digesta. *Compare* bolus.

claws Sharp, curved laterally compressed nail at the end of a digit; talon.

coelom Fluid-filled body cavity formed in mesoderm.

copulation Coitus involving an intromittent organ.

coracoid Posterior coracoid; first evolved in early synapsids or their immediate ancestors. *Compare* procoracoid.

cornified Having a layer of keratin; keratinized.

cortex The outer portion or rim of an organ.

cosmine Older term for a derivative of dentin that covers some fish scales; cosmoid scale.

cranial Toward the head or front end of the body; anterior, rostral.

cranial kinesis Movement between the upper jaw and braincase about joints between them; in restricted sense, skulls with a movable joint across the roofing bones. *Compare* akinetic skull, prokinesis, mesokinesis, metakinesis.

cranial nerve Any nerve entering or departing from the brain.

crop Baglike expansion of the esophagus.

crown group The smallest clade encompassing the living members of a group and the extinct taxa nested within. *Compare* stem group, total group.

cutaneous respiration Gas exchange directly between blood and the environment through the skin.

D

deglutition The act of swallowing.

dendrite Nerve fiber of a neuron carrying impulses toward the cell body.

dendrograms A branching diagram representing relationships or history of a group of organisms.

dental formula Shorthand expression of the characteristic number of each type (incisor, canine, premolar, molar) of upper and lower teeth in a mammalian species.

dentin Forms the bulk of the tooth; similar in structure to bone, but harder; yellowish in color; composed of inorganic hydroxyapatite crystals and of collagen; secreted by odontoblasts of neural crest origin. *See* enamel, cementum.

dentition A set of teeth.

derived Denoting that an organism or species evolved late within its phylogenetic lineage; advanced; opposed to primitive.

dermal papilla The part of the tooth-forming primordium that is derived from neural crest cells, becomes associated with the enamel organ, and differentiates into odontoblasts that secrete dentin. *See* enamel organ.

dermatocranium That part of the skull formed from dermal bones.

dermis Skin layer beneath the epidermis derived from mesoderm.

design The structural and functional organization of a part related to its biological role.

diffusion The movement of molecules from an area of high to an area of low concentration; if the movement is random and unaided, it is passive diffusion.

digestion The mechanical and chemical breakdown of foods into their basic end products that can be absorbed, usually simple carbohydrates, proteins, fatty acids.

digitigrade Foot posture in which the balls of the feet, middle of the digits, support the weight (as in cats). *Compare* plantigrade, unguligrade.

dikinetic skull Kinetic skull with two joints passing transversely through the braincase. *Compare* monokinetic.

dioecious Female and male gonads in separate individuals.

diphyodont Pattern of tooth replacement involving only two sets of teeth, usually milk teeth and permanent teeth.

diplospondyly The condition in which a vertebral segment is composed of two centra. *Compare* monospondyly.

dissection The careful exposure of anatomical parts for the purpose of allowing students to discover and master for themselves the extraordinary morphological organization of an animal, in order to understand the processes these parts perform and the remarkable evolutionary history out of which they come. Pronounced *dis*-section, as opposed to *di*-section, which is chopping into two halves.

distal Toward the free end of an attached part such as the limb. *Compare* proximal.

dorsal Toward the back or upper surface of the body; opposite of ventral.

E

ecomorphology The study of the relationship between the form and function of a part and how it is actually used in a natural environmental setting; the basis for determining biological role.

effector An organ responding to nervous stimulation; muscle or gland.

efferent Carrying away, as motor neurons carry impulses away from the central nervous system.

electric organ Specialized block of muscle producing electrical fields and often high jolts of voltage.

emargination Large notches in the bony braincase. *Compare* fenestra.

emulsify To break up fats into smaller droplets. *Compare* digestion.

enamel Forms the occlusal cap on most teeth. Hardest substance in vertebrate body, consisting almost entirely of calcium salts as apatite crystals; secreted by ameloblasts of epidermal origin. *See* dentin, cementum.

enamel organ The part of the tooth-forming primordium that is derived from epidermis, becomes associated with the dermal papilla, and differentiates into the ameloblasts that secrete the enamel. *See* dermal papilla.

endochondral bone formation Embryonic formation of bone preceded by a cartilage precursor that is subsequently ossified; cartilage bone, replacement bone. *Compare* intramembranous bone formation.

endocrine Denoting a gland that releases its product directly into blood vessels. *Compare* exocrine.

endoskeleton Supportive or protective framework beneath the integument; within the body. *Compare* exoskeleton.

endothelium Single-celled lining to vascular channels.

ependyma The layer of cells lining the central canal of the chordate spinal cord.

epidermis Skin layer over the dermis derived from ectoderm.

epigenetics Above the genes; the study of developmental events above the level of the genes; the embryonic processes not directly arising from the genes that contribute to the developing phenotype.

epiphysis 1. Secondary center of ossification on the end of a bone; the end of a bone. 2. Pineal gland.

evolutionary morphology The study of the relationship between change in anatomical design through time and the processes responsible.

excretion Removal from the body of wastes and substances in excess.

exocrine Denoting release of secretion into ducts. *Compare* endocrine.

exoskeleton Supportive or protective framework lying on the outside of the body. *Compare* endoskeleton.

extant Living.

extinct Dead.

extrinsic Originating outside the part on which it acts. *Compare* intrinsic.

F

fascicle Within a muscle organ, a bundle of muscle fibers defined by a connective tissue coat.

fenestra An opening within the bony braincase.

fermentation Anaerobic extraction of energy from food in vertebrates by microorganisms that release cellulases to enzymatically break down plant material.

fin An external, projecting plate or membrane from the body of an aquatic animal (as in fish).

fixator A muscle functioning to stabilize a joint. *Compare* synergist, antagonist.

foramen A perforation or hole through bone or cartilage.

foramen of ovale One-way connection between right and left atria in an embryonic mammal; closes at birth.

foramen of Panizza Connecting vessel between the bases of left and right aortic arches in crocodilians.

foregut Anterior embryonic gut that gives rise to pharynx, esophagus, stomach, and anterior intestine. *Compare* hindgut.

foregut fermentation *See* gastric fermentation.

frontal plane A plane passing from one side to the other of an organism so as to divide the body into dorsal and ventral parts. *Compare* transverse plane.

function How a part performs within the organism. *Compare* biological role.

functional morphology The study of the relationship between the anatomical design of a structure and the function or functions it performs.

G

gametogenesis The production of gametes.

ganglion Gathering of nerve cell bodies within the peripheral nervous system.

ganoin Older term for a derivative of enamel that covers some fish scales; ganoid scale.

gas bladder Gas-filled bag in fish derived from the gut. Because the composition of the gas may vary, the term "air bladder" is less appropriate. *Compare* swim bladder.

gastralia Rib-shaped dermal bones located in the abdominal region.

gastric fermentation Digestion involving microbial fermentation centered near specialized stomach; foregut fermentation. *See* intestinal fermentation.

gill Aquatic respiratory organ.

gill slit Pharyngeal slit associated with a gill.

gizzard An especially well-muscularized region of the stomach used to grind hard foods.

gnathostome A vertebrate with jaws.

H

heterochrony Within an evolutionary lineage, the change in time of embryonic appearance of an organ relative to its time of embryonic appearance in a related member of the phylogeny. Usually heterochrony is concerned with the time of onset of sexual maturity relative to somatic development. *Compare* paedomorphosis.

heterodont Dentition in which the teeth are different in general appearance.

hindgut Posterior embryonic gut that gives rise to the posterior intestines. *Compare* foregut.

hindgut fermentation *See* intestinal fermentation.

holospondyly The condition in which vertebral elements, centra and spines, are anatomically fused into a single piece.

homeostasis The maintaining of a relatively stable internal environment within an organism.

homodont Dentition in which the teeth are similar in general appearance.

homology Features in two or more organisms resulting from the derivation of those features from common ancestors; common ancestry. *Compare* serial homology.

homoplasy Features in two or more organisms that look alike; similar appearance.

hoof Enlarged cornified plate on the end of an ungulate digit.

horn Unbranched keratinized sheath with a bony core located on the head; horns usually occur in both males and females and are retained year round.

hydrofoil Any object that produces lift when placed in a moving stream of water (as a shark pectoral fin).

hydrostatic organ A structure whose mechanical integrity depends upon a fluid-filled core enclosed by defining walls of connective tissue.

hydrostatic pressure Fluid force, as in blood resulting from heart contraction.

hyostyly Jaw suspension primarily through attachment with the hyomandibula.

hypsodont High crown teeth. *Compare* brachyodont.

I

incus Mammalian, middle ear bone, derived phylogenetically from the quadrate.

insertion For a muscle, its relatively movable point of attachment.

integument The skin covering the body.

intervertebral body A pad of cartilage or sometime fibrous connective tissue between articular ends of successive vertebral centra.

intervertebral disk In the adult mammal, a pad of fibrocartilage with a gel-like core derived from the notochord and located between articular ends of successive vertebral centra. *Compare* intervertebral body.

intestinal fermentation Digestion involving microbial fermentation centered in the intestines; hindgut fermentation, cecal fermentation. *See* gastric fermentation.

intramembranous bone formation Embryonic formation of bone directly from mesenchyme without a cartilage precursor; dermal bone. *Compare* endochondral bone formation.

intratarsal joint An archosaur ankle type in which the line of flexion passes *between* calcaneus and astragalus. *Compare* mesotarsal joint.

intrinsic Belonging entirely to the part. *Compare* extrinsic.

intromittent organ Male reproductive organ that is inserted into the female reproductive tract to deliver sperm; penis or phallus.

invagination Indentation or infolding of the surface.

ischemia When blood flow to a region is insufficient to meet metabolic demands of the tissue.

isometry Geometric similarity in which proportions remain constant with a change in size.

J

jaws Skeletal elements of bone or cartilage that reinforce the lower borders of the mouth.

K

keratin Fibrous protein.

keratinization The process by which the proteins, especially keratin, are formed in the skin.

kinesis Denotes movement, usually in reference to relative movement of skull bones. *See* cranial kinesis.

L

lactation Release of milk from mammary glands to suckling young.

lamina Thin sheet, layer, or plate—e.g., gill lamella.

larva An immature (nonreproductive) stage morphologically different from the adult.

lateral Toward or on the side of the body.

lepidotrichia Fan-shaped array of ossified or chondrified dermal rods internally supporting the fin of bony fishes.

lepospondyly A holospondylous vertebra with a husk-shaped centrum usually pierced by a notochordal canal. *Compare* ceratotrichia.

lift The force produced by an airfoil perpendicular to its surface.

lingual feeding Prey are captured by use of the tongue.

load In mechanics, the forces to which a structure is subjected.

lophodont Teeth with broad, ridged cusps useful in grinding plant material. *Compare* bunodont, selenodont.

lymph Clear fluid carried in lymphatic vessels.

lymphoid tissue Blood-forming tissue outside of bone cavities—in the spleen, lymph nodes, etc.

M

malleus In mammals, one of the three middle ear bones phylogenetically derived from the articular.

mastication Chewing of food; mechanical breakdown of a large bolus of food into smaller pieces usually by use of teeth.

meatus A canal or channel.

mechanoreceptor Sense organ responsive to small changes in mechanical force. *Compare* chemoreceptor, photoreceptor.

medulla The inner portion or core of an organ.

merycism Remastication together with microbial fermentation in nonruminants. *Compare* rumination.

mesenchyme Loosely associated cells of mesodermal origin.

mesokinesis Skull with a transverse joint passing through the dermatocranium posterior to the orbit. *Compare* prokinesis, metakinesis.

mesonephros Kidney formed of nephric tubules arising in the middle of the nephric ridge; usually a transient embryonic stage replacing the pronephros, but itself replaced by the adult opisthonephros or metanephros. *Compare* pronephros, opisthonephros, metanephros.

mesotarsal joint An archosaur ankle type in which the calcaneous and astragalus fuse and the line of flexion passes between them and the distal tarsals. *Compare* intratarsal joint.

metakinesis Skull with a transverse hinge posteriorly between the deep neurocranium and outer dermatocranium. *Compare* prokinesis, mesokinesis.

metamorphosis Abrupt transformation from one anatomically distinct stage (juvenile) to another (adult).

metanephric duct Ureter; distinct from pronephric and mesonephric ducts.

metanephros Kidney formed of nephric tubules arising in the posterior region of the nephric ridge drained by a ureter; usually replaces embryonic pronephros and mesonephros. *Compare* pronephros, mesonephros, opisthonephros.

metapterygial fin Basic fin type in which the axis (metapterygial stem) is located posteriorly in the fin.

metapterygial stem The chain of endoskeletal elements within the fish fin that define the major internal supportive axis.

microcirculation The capillary beds plus the arterioles that supply them and the venules that drain them.

midsagittal plane Median sagittal plane passing through the central long axis of the body.

molariform Cheek teeth; general term for similar appearing premolar and molar teeth.

molt The shedding of parts or all of the cornified layer of the epidermis; loss of feathers or hair, usually on an annual rhythm; ecdysis.

monokinetic skull Kinetic skull with a single transverse joint passing through the braincase.

monospondyly The condition in which a vertebral segment is composed of one centrum. *Compare* diplospondyly.

morph A term referring to the general form or design of an animal—for example, juvenile morph (tadpole) and adult morph (sexually mature stage) of a frog.

morphological cross section The cross-sectional area of a structure perpendicular to its long axis, as in muscles. *Compare* physiological cross section.

morphology The study of anatomy and its significance.

motor neuron A nerve cell carrying impulses to an effector organ. *Compare* sensory neuron.

motor pattern A defined, local pattern of activity produced by muscles that shows little variation when repeated.

motor unit One motor neuron and the subset of muscle fibers it supplies; important in producing graded muscle force.

mucous gland An organ secreting a protein-rich mucin that is usually a thick fluid. *Compare* serous gland.

müllerian duct Paired embryonic ducts that variously form the oviduct, uterus, and vagina in females but rarely yield adult structures in males.

muscle fiber A muscle cell; a contractile part of a muscle organ.

muscle organ Muscle cells together with noncontractile tissues that support them (connective tissue, blood vessels, nerves).

myeloid tissue Blood-forming tissue housed inside bones.

myofibril Contractile unit in a muscle cell; a chain of repeating sarcomeres, composed of myofilaments.

myofilament Composed of myosin and actin.

myomeres Differentiated segments of adult body muscle.

myotomes Undifferentiated, embryonic blocks of presumptive muscle.

N

naris A nostril.

natural selection The process by which organisms with poorly suited features, on average, fare less well in a particular environment and tend to perish, thereby leaving (preserving) those individuals with more favorable adaptations; survival of the fittest.

neomorph Denoting in a derived species a new morphological structure that has no equivalent evolutionary antecedent.

neoteny Paedomorphosis produced by delaying the onset of somatic development.

nephridium Tubular excretory organ.

nephron The portion of the uriniferous tubule concerned with the formation of urine, composed of proximal, intermediate, and distal regions; nephric tubule.

nephrotome Segmental, early embryonic urinary structure that is the forerunner of a nephron.

nerve Collection of nerve fibers coursing together in the peripheral nervous system.

network Anything reticulated or decussated at equal distances, with interstices between the intersections.

neural crest A paired strip of tissue that pinches off from the dorsal edges of the neural groove as it forms into the neural tube.

neurocranium That part of the braincase that contains cavities for the brain and associated sensory capsules (nasal, optic, otic).

neuroglia Nonnervous supportive cells of the nervous system.

neurotransmitter Chemical released into the synapse at the end of a nerve fiber, usually an axon.

notochord Long, axial rod composed of a fibrous connective tissue wall around cells and/or a fluid-filled space.

nucleus 1. Membrane-bound organelle within the body of a cell. 2. Gathering of nerve cell bodies within the central nervous system.

O

occlusion Closure; the meeting of opposite tooth rows.

olfaction The act of smelling.

ontogeny The sequences of events during embryonic development.

operculum Lid or cover, as over the gills of fishes.

opisthonephros The adult kidney formed from the mesonephros together with additional tubules from the posterior region of the nephric ridge. *Compare* pronephros, mesonephros, metanephros.

origin For a muscle, its relatively fixed point of attachment. *Compare* insertion.

osmoregulation Active maintenance of water and solute levels.

ossification The process of bone formation; the appearance of bone cells and their surrounding matrix. *Compare* calcification.

osteoderm A dermal bone located under and in support of an epidermal scale.

osteon Highly ordered arrangement of bone cells into concentric rings and bone matrix around a central canal carrying blood vessels and nerves; haversian system.

otolith Single calcareous mass in the cupula of hair cells.

oviduct Urogenital duct transporting ova and often involved in protection and nourishment of the embryo; müllerian duct.

oviparity Egg laying.

oviposition The act of laying eggs.

ovulation Release of the ovum from the ovary.

P

paedomorphosis Juvenile-like; the general retention of juvenile features of ancestors in the late developmental stages of descendants. *See* neoteny.

parallel muscle A muscle organ in which all its muscle fibers lie in the same direction and are aligned with its long axis. *Compare* pinnate muscle.

parasagittal plane A sagittal plane parallel with the midsagittal plane.

parturition The act of giving birth via viviparity. *Compare* oviposition.

patterning The process of establishing the main topographical regions and body axes in an embryo; dorsoventral, anteroposterior for instance.

pentidactyl Five digits per limb; thought to be the basic pattern characteristic of tetrapods, but modified by functional demands.

peripheral nervous system The cranial and spinal nerves and their associated ganglia; the part of the nervous system outside the central nervous system.

peristalsis Muscular contractions in the walls of a tubular structure, as in the digestive tract.

pharyngeal slit An elongate opening in the lateral wall of the pharynx.

photoreceptor Radiation receptor responsive to visible light.

phylogeny The course of evolutionary change within a related group of organisms.

physiological cross section In muscles, the total cross-sectional area of all muscle fibers perpendicular to their long axes. *Compare* morphological cross section.

pinnate muscle A muscle organ in which all of its muscle fibers are aligned oblique to its long axis. *Compare* parallel muscle.

plantigrade Foot posture in which the entire sole is in contact with the ground. *Compare* digitigrade, unguligrade.

plastron Ventral bony part of turtle shell. *Compare* carapace.

platysma Unspecialized muscle derived from hyoid arch musculature that spreads as a thin subcutaneous sheet into the neck and over the face.

plexus A network of intermingling parts, as in blood vessels or nerves.

polydactyly An increase in the number of digits over the basic pendidactylous number. *Compare* polyphalangy.

polyphalangy An increase in the number of phalanges in each digit. *Compare* polydactyly.

polyphyodont Pattern of continuous tooth replacement. *Compare* diphyodont.

polyspondyly The condition in which a vertebral segment is composed of two or more centra.

portal system Set of venous vessels beginning and ending in capillary beds.

preadaptation The concept that features possess the necessary properties of form and function to meet the demands of a particular environment before the organism experiences that particular environment.

prehension Prey capture characterized by rapid grasping of the prey, usually with the jaws or with claws.

preservationism (embryology) The observation that embryonic development is conservative in that some features (e.g., gill slits) of ancestral embryos are retained in embryos of descendants.

primitive Denoting that an organism or species appeared early within its phylogenetic lineage; opposite of derived.

procoracoid Anterior coracoid (= precoracoid); first evolved in fishes. *Compare* coracoid.

proctodeum The embryonic invagination of surface ectoderm that contributes to the hindgut, usually giving rise to the cloaca.

project In the nervous system, to send neural impulses to.

prokinesis Skull with a transverse joint passing through the dermatocranium anterior to the orbit. *Compare* mesokinesis, metakinesis.

pronephros Kidney formed of nephric tubules arising in the anterior region of the nephric ridge; usually transient embryonic stage. *Compare* mesonephros, opisthonephros, metanephros.

proximal Toward the base of an attached part where it joins the body. *Compare* distal.

pterylae Feather tracks.

Q

quadrupedal Walking or running by means of four legs. *Compare* bipedal.

R

radial symmetry A regular arrangement of the body about a central axis.

radiation receptor Sense organ responsive to light and other forms of electromagnetic radiation.

reflex Involuntary action mediated by the nervous system.

reticulum The second of four chambers in the complex ruminant stomach; a specialized region of the esophagus. *Compare* rumen.

rhachitomous vertebra An aspidospondylous vertebra characteristic of some crossopterygians and some early amphibians.

rumen The first of four chambers in the complex ruminant stomach; an expanded specialization of the esophagus. *Compare* reticulum.

ruminant Placental mammals with a specialized expansion of the digestive tract, the rumen, for processing of plant material; Ruminatia.

rumination Remastication together with microbial fermentation in ruminants. *Compare* merycism.

S

sagittal plane Any plane parallel with the long axis of the animal's body, oriented dorsoventrally.

secondary cartilage Cartilage that forms after initial bone ossification is complete; formed usually in response to mechanical stress, especially that cartilage that forms on the margins of intramembranous bone.

sectorial teeth Teeth with opposing sharp ridges specialized for cutting.

segmentation A body made up of repeating sections or parts; metamerism.

selection force or **pressure** The biological or physical demands arising from the environment that affect the survival of the individual living there.

selenodont Teeth with crescent-shaped cusps, as in artiodactyls. *Compare* bunodont, lophodont.

sensory neuron A nerve cell carrying responses from a sensory organ. *Compare* motor neuron.

serial homology Similarity between successively repeated features in the same individual.

serial theory The hypothesis that jaws evolved from one of the anterior branchial arches.

serous gland An organ secreting a thin, watery fluid.

sesamoid bone A bone that develops directly in tendons.

sinusoids Tiny vascular channels slightly larger than capillaries lined or partially lined only by endothelium.

somatic Pertaining to the body, usually the skeleton, its muscles, and the skin, but not the viscera.

spermatophore A reproductive structure capped by sperm, used to deliver sperm to a female.

sphincter A band of muscle around a tube or opening, functioning to constrict or close it.

spiracle A reduced gill slit, first in series.

splanchnocranium That part of the skull arising first in support of the pharyngeal slits and later contributing to the jaws and other structures of the head; branchial arches and derivatives; visceral cranium.

stapes In tetrapods, a derivative of the fish hyomandibula. Initially a brace to the back of the skull, eventually becoming slenderized and incorporated into the specialized sound detection system. In mammals, one of three middle ear bones.

stem group A paraphyletic assemblage of extinct taxa related to but not part of the crown group. *Compare* crown group, total group.

stereospondylous vertebra A monospondylous vertebra in which the single centrum (an intercentrum) is separate (aspidospondylous).

stigmata Extensively subdivided pharyngeal slit.

stolon Rootlike process of ascidians and other invertebrates that may fragment into pieces that asexually grow into more individuals.

stratified Formed of layers.

streptostyly The condition in which the quadrate is movable relative to the braincase.

suspension feeding Feeding based upon filtering suspended food from water, usually involving cilia and secreted mucus; filter feeding, cilia-mucous feeding.

swim bladder Gas bladder functioning primarily in buoyancy control.

synapse The region of contact between one nerve fiber and another, a nerve fiber and a neuron, or a nerve fiber and an effector.

synapticules Cross-linking connections between pharyngeal bars in amphioxus.

synergist Two or more muscles cooperating to produce motion in the same direction. *Compare* antagonist, fixator.

T

talons Specialized bird claws used in striking or catching live prey.

tendon Noncontractile, fibrous connective tissue band connecting a muscle organ to bone. *Compare* aponeurosis.

thermoreceptor Radiation receptor sensitive to infrared energy.

total group The monophyletic clade formed of the stem group plus the crown group. *Compare* crown group, stem group.

tract Collection of nerve fibers coursing together in the brain or spinal cord. *Compare* nerve.

transverse plane A plane passing from one side to the other of an organism so as to divide the body into anterior and posterior parts.

transverse process A general term for any bony or cartilaginous projection from the centrum or neural arch.

tusk Specialized, long teeth protruding from the mouth; elongate incisors (elephants), left upper incisor (narwhal), canines (walruses).

tympanum Ear drum; tympanic membrane.

U

ungulate Hoofed placental mammals belonging to the orders Perissodactyla (horses) and Artiodactyla (cattle, deer, pigs, etc.).

unguligrade Foot posture in which the weight is carried on the tips of the toes (as in horses).

ureter Metanephric duct; urinary duct arising as a ureteric diverticulum and draining the metanephros.

V

vasoconstriction Narrowing a blood vessel; usually resulting from active smooth muscle contraction. *Compare* vasodilation.

vasodilation Widening a blood vessel; may be active or passive enlargement. *Compare* vasoconstriction.

vein Blood vessel carrying blood toward the heart; blood carried may be low or high in oxygen tension. *Compare* artery.

ventilation The active movement of water or air across respiratory exchange surfaces.

ventral Toward the belly or bottom of an animal; opposite of dorsal.

vertebra One of several bone or cartilage blocks firmly joined into a backbone, defining the major body axis of vertebrates.

vestigial Evolutionary decline; reduction of a trait in descendants. *Compare* atavism.

villus A fingerlike projection of a tissue layer, as in the small intestine.

viviparity Live birth.

W

wolffian duct Mesonephric duct.

Z

zygapophysis Projection of a neural arch that establishes an articulation with the adjacent neural arch.

Credits

Chapter 2

Fig. 2.5A–F: Cephalochordate, amphioxus. © Kenneth V. Kardong and Edward J. Zalisko. Figure 2.7A: Urochordate, Class Ascidiacea. R. A. Cloney, "Ascidian larvae and the events of metamorphosis," *American Zoologist* 22:817–826, 1982, American Society of Zoologists. Fig. 2.7C: Adult Urochordate. Copyright © Kenneth V. Kardong and Edward J. Zalisko.

Chapter 3

Fig. 3.1: From P.C.J. Donoghue, P.L. Forey, and R.J. Aldridge, "Condont Affinity and chordate Phylogeny" in Biological Reviews of the Cambridge Philosophical Society, Vol. 75 (2000): 191–251. Reprinted with permission of Wiley-Blackwell. Fig. 3.4A–C: The adult lamprey, Petromyzon. © Kenneth V. Kardong and Edward J. Zalisko. Fig. 3.6, 3.8A–C: Cross sections of the adult lamprey; Ammocoetes larva. © Kenneth V. Kardong and Edward J. Zalisko.

Chapter 4

Fig. 4.3: Skin derivatives. Redrawn from M. H. Wake, ed., *Hyman's Comparative Vertebrate Anatomy*, 3d ed., © 1979 by The University of Chicago. All rights reserved. Reprinted by permission.

Chapter 5

Fig. 5.9A–C; Fig. 5.11A–D: Skeleton of the mudpuppy; Skeleton of an alligator. © Kenneth V. Kardong and Edward J. Zalisko. Fig. 5.11D: Cervical vertebrae of an alligator. © Kenneth V. Kardong and Edward J. Zalisko. Fig. 5.12; 5.13; 5.14; 5.16; 5.17; 5.18; 5.19; 5.20: Skeleton of a painted turtle; skeleton of a pigeon; Axial column of a cat; Right pelvic and right pectoral girdles of a shark; Right pelvic and right pectoral girdles of *Necturus;* Right pelvic and right pectoral girdles of an alligator; Right pelvic and right pectoral girdles of a turtle; Right pelvic and right pectoral girdles of a pigeon. © Kenneth V. Kardong and Edward J. Zalisko. Fig. 5.21: The development of the pelvic girdle in birds… © Kenneth V. Kardong and Edward J. Zalisko; Redrawn after Alexis L. Romanoff, *The Avian Embryo,* figure 37.1, p. 1010, the MacMillan Company; after Lebedinsky, 1913,

Fig. 5.22; 5.24; 5.29; 5.30; 5.31; 5.32; 5.33; 5.34, 5.35, 5.39: Right pelvic girdle and right hindlimb of a cat; Right scapula, clavicle, and forelimb skeleton of a cat; Chondrocranium and splanchnocranium of the shark; Skull of *Necturus;* Skull of an alligator; Skull of a turtle; Skull of a chicken; Skull of a cat. © Kenneth V. Kardong and Edward J. Zalisko. Box Figure 5.3: After M. Hildebrand et al., eds., 1985, Functional Vertebrate Morphology, Belknap Press of Harvard University Press.

Chapter 6

Fig. 6.6; 6.7; 6.8; 6.9; 6.10; 6.11; 6.12; 6.13; 6.14; 6.15; 6.16; 6.17; 6.18; 6.19; 6.20; 6.21; 6.22; 6.23: Lateral view of the trunk, branchial, and cephalic musculature of a shark; Ventral view of the pelvic fins of a male…; Branchiomeric musculature; Trunk, branchial, and cephalic musculature of a shark; Dorsal view of extrinsic eye musculature of the shark; Ventral trunk, branchial, and cephalic musculature of a *Necturus;* Trunk, branchial, and cephalic musculature of *Necturus;* Lateral trunk and hindlimb musculature of *Necturus;* Ventral trunk and hindlimb musculature of *Necturus;* Superficial ventral trunk, brachial, and cephalic musculature of a cat; Head and neck musculature of a cat; Superficial neck and shoulder musculature of a cat; Musculature of the dorsal left forelimb of a cat; Lateral thoracic muscles of a cat; Musculature of the dorsal left forelimb of a cat; Musculature of the ventral left forelimb of a cat; Lateral musculature of cat hindlimb; Medial musculature of cat hindlimb. © Kenneth V. Kardong and Edward J. Zalisko.

Chapter 7

Fig. 7.3; 7.4; 7.6; 7.7; 7.8A–B; 7.9A, B; 7.10: Ventral view of the pleuroperitoneal cavity…; Blood supply to alimentary canal of a shark; Ventral view of the pleuroperitoneal cavity of *Necturus;* Lateral view of the neck and face of a cat; Oral and pharyngeal structures of a cat; Ventral view of the thoracic cavity…; Abdominal structures in a cat… © Kenneth V. Kardong and Edward J. Zalisko.

Chapter 8

Fig. 8.2; 8.3; 8.5; 8.6; 8.7; 8.8; 8.11; 8.12; 8.13; 8.14; 8.16; 8.17; 8.18: Ventral view of the shark heart…; Ventral view of the branchial structure and vascular supply…; Ventral view of the

pleuroperitoneal cavity of a shark; Blood supply to alimentary canal of a shark; no legend; Heart chambers and associated blood vessels in *Necturus;* Ventral view of the pleuroperitoneal cavity of *Necturus;* Artery, vein, and hepatic portal vein…; Ventral view of thoracic components of the circulatory system…; Ventral view of aortic arch distribution…; Artery, vein, and hepatic portal vein…; Ventral view of branches off the dorsal aorta…; Blood vessels of the medial hindlimb… © Kenneth V. Kardong and Edward J. Zalisko.

Chapter 9

Fig. 9.1: Ventral view of the left pelvic girdle… Redrawn from Ari van Tienhoven, *Reproductive Physiology of Vertebrates,* 2d ed. Copyright © 1983 by Cornell University Press. Fig. 9.2; 9.3; 9.4; 9.5; 9.6; 9.7; 9.9: Ventral view of the female and male shark pelvis; Ventral view of the urogenital system of a male shark…; Ventral view of the urogenital system of a female shark…; Ventral view of the urogenital system of a male *Necturus;* Ventral view of the urogenital system of a female *Necturus;* Ventral view of the urogenital system of a male cat; Ventral view of the urogenital system of a female cat. © Kenneth V. Kardong and Edward J. Zalisko.

Chapter 10

Fig. 10.6A; 10.10; 10.11; 10.12; 10.13: Dorsal view of shark brain; Dorsal view of the sheep brain; Ventral view of the sheep brain; Sagittal view of the sheep brain; Dorsal view of the brain stem of a sheep brain… © Kenneth V. Kardong and Edward J. Zalisko.

ILLUSTRATORS:

Hazen Audel: 5.31A–F, 5.32, 5.33.
Kathleen Bodley: 3.4, 3.6B–C, 5.11D, 5.20D, 5.29, 5.32A–C, 5.34, 5.35, 6.6, 6.7, 6.9, 6.11–6.23, 7.3, 7.4, 7.6–7.10, 8.2, 8.3, 8.5, 8.6, 8.8, 8.11, 8.13, 8.14, 8.17, 8.18, 9.2–9.7, 9.9, 10.10–10.13.
Julie Jordon Brown: 5.11A–C, 5.14A, 5.19A–C, 5.22A, 5.24A.
Gavin Lawson: 2.5A–F, 3.6, 3.8.
Laszlo Meszoly: 5.9B–C, 5.12, 5.14B–H, 5.18.
Faye Prevedell: 5.9A, 5.13A–C, 5.16A–C, 5.17A–C, 5.20A–C, 5.22B–I, 5.24B–H, 5.35A-F, 7.6 8.7A–C, 8.11, 8.12A–C, 8.13, 8.14, 8.17, 8.18.
Emily Glenn: 6.10, 10.6.

210

Index

A

Abdomen, cat, 143, 144*f*, 145
Abdomen muscles, cat, 105, 106, 106*t*, 108
Abdominal pores, shark, 131
Abducens nerve (VI), 189
 shark, 192
 sheep, 199, 199*f*
Abductor pollicis longus, cat, 116*f*, 120*f*, 121
Accessory lobe, cat lung, 141
Accessory pancreatic duct, cat, 145
Accessory process, cat, 51, 52*f*
Accessory urinary duct, shark, 174*f*, 175
Acetabulum
 alligator, 55
 bird, 57, 60*f*
 cat, 59, 61*f*
 Necturus, 53
Acorn worms, 9*f*, 10–11
Acoustic stria, sheep brain, 202*f*, 203
Acromiodeltoid, cat, 114*f*, 115*f*, 118
Acromion
 cat, 64*f*, 66
 turtle, 56*f*, 63
Acromiotrapezius, cat, 114*f*, 115*f*, 116, 117*b*
Actinopterygian, vertebrae, 46*f*
Adductor, shark, 95*f*
Adductor femoris, cat, 126*f*, 127, 127*f*, 128, 128*f*
Adductor longus, cat, 127, 127*f*, 128*f*
Adductor mandibulae
 Necturus, 102, 103*f*, 117*b*, 118*t*
 shark, 93, 93*f*, 96*f*, 97*f*, 117*b*, 118*t*
Adductor muscle, shark, 148, 148*f*
Adductor process, shark, 69*f*, 70
Adhesive papillae, sea squirts, 16*f*, 17
Adipose tissue, as connective tissue, 29*f*, 40, 40*f*
Adrenal gland, cat, 169*f*
Adrenolumbar arteries, cat, 169, 169*f*
Adrenolumbar vein, cat, 165*f*, 169*f*, 170
Adult forms, sea squirts, 17
Afferent branchial arteries
 lamprey, 24
 Necturus, 155*f*, 156, 157*f*
 shark, 146, 148, 148*f*, 149*f*
Afferent renal vein, shark, 153*f*

Afferent spiracular artery, shark, 148*f*, 150, 153*f*
Agnathans, 18–21. *See also* Lampreys (petromyzoniformes)
 hagfishes, 18, 20*f*
 lampreys, 18, 22–27, 23*f*–25*f*
 relationships to other cephalochordates, 19*f*
 shift in feeding mechanisms in, 18–21
 vertebrae and ribs, 44
Alimentary canal, 131
 bird, 132*f*
 shark, 135*f*, 152*f*
Alisphenoid, cat, 81*f*, 82
Alligator
 cranial skeleton, 72, 73–74*f*, 74–75
 pelvic girdle and posterior appendages, 55, 57*f*
 teeth, 84
 vertebrae, ribs, skeleton of, 47–49, 48*f*
Alveolar process, cat, 79, 81*f*
Alveoli, cat, 167
Amia. See Bowfin *(Amia)*
Ammocoete larva, 18, 27–28
Amniotes
 cranial nerves of, 186, 188–89, 188*f*
 ribs of, 43*f*
Amphibians
 cranial skeleton, 71–72, 72*f*
 heart, 156*f*
 integument, 33, 33*f*
 midbrain and tectum, 201*b*
 pectoral girdle, sternum, and appendages, 62
 pelvic girdle and posterior appendages, 53–54, 55*f*
 teeth, 84
 vertebrae and ribs, 46–47, 47*f*
Amphioxus, 1*f*, 2, 2*t*, 11–15
 circulatory system, 13–14, 14*f*
 digestive system, 11–12
 examination of structure of, 12*f*, 13*f*, 14–15, 14*f*
 filter feeding, 11, 12*f*
 nervous system and muscles, 12, 12*f*
 notochord, 12
Ampullae of Lorenzini, shark, 90, 90*f*
Anal glands, cat, 181, 184
Anal sphincter, cat, 181, 184

Anamniotes, cranial nerves of, 186
Anatomical terminology, 7*t*
Anatomy
 terminology and directions of, 42*t*
Anatomy, external, 89
 cat, 105
 Necturus, 99–100
 shark, 89–90
Angle, rib, 51, 52*f*
Angular
 alligator, 74*f*, 75
 bowfin *(Amia),* 71, 71*f*
 turtle, 76*f*, 77
Angular process, cat, 81*f*, 83
Annular cartilage, lamprey, 21*f*, 26
Annulus fibrosis, cat, 50–51
Antebrachium musculature, cat, 116*f*, 120–21, 120*f*, 121–24
Anterior border, cat, 64*f*, 66
Anterior cardinal vein, lamprey, 24, 25*f*
Anterior cerebellar peduncle, sheep brain, 202, 202*f*
Anterior choroid plexus, shark, 191*f*
Anterior commissure, sheep, 200*f*, 201
Anterior facial vein, cat, 114*f*, 139*f*, 165*f*, 168
Anterior horns, of hyoid apparatus, 83
Anterior humeral circumflex artery, cat, 161*f*, 162*f*, 163, 164*f*
Anterior humeral circumflex vein, cat, 161*f*
Anterior intestinal artery, shark, 150, 151*f*–153*f*
Anterior intestinal vein, shark, 151*f*, 152*f*, 154, 154*f*
Anterior lienogastric vein, shark, 151*f*, 152*f*, 154, 154*f*
Anterior lobe
 cat liver, 144*f*
 cat lung, 141, 142*f*, 144*f*
Anterior mesenteric artery, shark, 135*f*, 150, 151*f*–153*f*
Anterior nostrils, bowfin *(Amia),* 71, 71*f*
Anterior orbital shelf, shark, 69*f*, 70
Anterior palatine vacuity, turtle, 77
Anterior pancreaticoduodenal artery, cat, 164*f*, 168
Anterior pancreaticoduodenal vein, cat, 166*f*, 171
Anterior part, cat liver, 143

Anterior splenic artery, cat, 164*f*, 168
Anterior splenic vein, cat, 166*f*, 171
Anterior xiphisternal process, bird, 58*f*, 63
Antlers, 38–39, 38*f*
Antorbital process
 Necturus, 72, 72*f*
 shark, 68, 69*f*, 70
Antorbitals, bowfin *(Amia),* 71, 71*f*
Anura, vertebrae and ribs, 46–47, 47*f*
Anus
 acorn worms, 9*f*, 10
 amphioxus, 14
 cat, 105, 181, 183*f*, 184
 fish, 133*f*
 lamprey, 27, 28*f*
 sea squirts, 17
Aorta
 cat, 161*f*, 162–63
 Necturus, 154, 155*f*, 156
 shark, 146, 147*f*, 148, 148*f*, 149*f*
Aortic arches
 cat, 161*f*, 162*f*, 163
 Necturus, 157*f*
 shark, 149*f*
Aortic trunk, cat, 163
Apical foramen (tooth), 83, 83*f*
Appendages. *See also* Limbs
 pectoral girdle, sternum, and, 62–66
 pelvic girdle and posterior, 53–62
Aqueduct, sheep brain, 200*f*, 201
Aqueduct of Sylvius, 196*f*
Arachnoid layer of meninges, 196, 196*f*
Arbor vitae, sheep brain, 200*f*, 201
Arches, teleost fish, 45, 46*f*
Archinephric duct, 26, 173, 175*b*
Arm muscles, *Necturus,* 100*t*
Arrector pili muscle, 36*f*, 37
Arterial loops, 168*b*
Arteries
 arterial size, 170*b*
 cat, 161–62*f*, 161*f*–162*f*, 162, 164*f*, 166
 digestive system and, 131
 Necturus, 155–56*f*, 155*f*–156*f*, 156, 157, 157*f*, 158–59*f*, 160
 shark, 146, 147–48*f*, 147*f*–148*f*, 148, 149–50, 149*f*, 151–53*f*, 151*f*–153*f*

Articular, 67, 68f
   alligator, 74f, 75
   bird, 78f, 79
   turtle, 76f, 77
Articular facets, cat, 51
Articular scar, cat, 59, 61f
Arytenoid cartilages, cat, 140f, 141
Ascending colon, cat, 143, 144f
Ascending pharyngeal artery, cat,
   164f, 166, 166n
Ascidiacea, 15–17
Ascidian tadpole, 15, 16f
Associated branchial arteries,
   shark, 149–50
Astragalus
   alligator, 57f
   cat, 60, 61f
   turtle, 54
Atlas
   alligator, 47, 48f
   bird, 49
   cat, 51, 52f
Atriopore, amphioxus, 11, 12f, 14
Atrioventricular valve,
   amphibian, 156f
Atrium
   amphibian, 156f
   amphioxus, 14, 14f
   cat, 144f
   lamprey, 24
   Necturus, 137f, 154, 155f
   sea squirts, 16f, 17
   shark, 146, 146f–148f
Atrophy, of bone, 41
Auditory nerve (VIII), 189
   shark, 98f, 190f, 193
   sheep, 199–200, 199f
Auditory (otic) capsule, shark, 70
Auditory tubes, cat, 140f, 141
Auricle of cerebellum, 98f,
   190f, 191f
Aves (bird)
   cranial skeleton, 78–79, 78f
   girdles and limbs, 56–59
   vertebrae and ribs, 49–50, 50f
Axial musculature in fishes,
   24, 26
   shark, 91–99
Axial skeleton
   bowfin (Amia calva), 46f
   placoderm, 44f
   shark, 45f
Axilla, cat, 105
Axillary artery, cat, 161f, 162f,
   163, 164f, 166
Axillary border, cat, 64f, 66
Axillary nerve, cat, 161f
Axillary vein, cat, 165f, 167
Axis
   alligator, 47, 48f
   bird, 49
   cat, 51, 52f
Azygous vein, 163f, 167

B

Baleen, 38, 39, 39f
Barb, feather, 34f, 35
Barbules, feather, 34f, 35, 35f

Basal plate, 67f
   Necturus, 72
   shark, 69f, 70
Basals, shark, 53
Basapophyses
   shark, 44, 45f
   teleost fish, 45, 46f
Basement membrane, 29, 29f
Base (tooth), 83
Basibranchial, 67f, 72f
   shark, 69f, 70
Basihyal
   chicken, 78f
   shark, 69f, 70
Basioccipital, 67f, 74f, 75
   bird, 78f, 79
   turtle, 76f, 77
Basisphenoid, 67f, 74f, 75, 80f
   bird, 78f, 79
   cat, 81f, 82
   turtle, 76f, 77
Basitrabecular process, shark, 69f, 70
Beak, turtle, 77
Beam, antler, 38
Benthic organisms, 9
Biceps brachii, cat, 110f, 120,
   161f, 162f
Biceps femoris, cat, 124, 125f, 126f
Bicipital ribs
   alligator, 48
   bird, 49
   cat, 51, 52f
   Necturus, 46
Bicipital tuberosity, cat, 64f, 66
Bilateral body symmetry, 6f
Bilateria, 2f, 4f
Biliverdin, 22
Birds
   alimentary canal, 132f
   cranial skeleton, 78–79, 78f
   embryonic development of
     pelvic girdle, 59, 60f
   integument, 34–35
   pectoral girdle, sternum, and
     appendages, 63
   pelvic girdle and posterior
     appendages, 56–59, 58f
   vertebrae and ribs, 49–50, 50f
Blood
   alternate pathways of, 167b
   as connective tissue, 40, 40f
   to shark alimentary canal, 135f
Blood vessels
   acorn worms, 11
   cat, 161–66
   mammalian integument, 37
   Necturus, 154–60
   shark, 146–54
   size of, 170b
Body
   of dentary-baring teeth in cats,
     81f, 83
   of epididymis in cats, 181f, 182
   of epididymis in humans, 182f
   of hyoid, 75
   of hyoid apparatus, 83
   of ilium in cats, 59, 61f
   of rib, 51, 52f

of sea squirts, 16f, 17
of sternum, 66
of stomach, cat, 143, 144f
of stomach, shark, 131, 134f,
   169f, 183f
of uterus, cat, 183
Body plans, chordates, 4f, 5
Body symmetries, explained, 6f
Body symmetry, 6f
Bone
   architecture of, 42f
   as connective tissue, 41–42
   osteocytes and chondrocytes
     of, 40, 42f
Bone mass and gravity, 41b
Bony fishes, integument, 30f,
   31–32, 32f
Bowfin (Amia)
   axial skeleton, 46f
   cranial skeleton, 71, 71f
   pectoral girdle, sternum, and
     appendages, 62, 62f
   pectoral girdle evolution, 65b
   vertebrae and ribs, 45
Brachial artery
   cat, 161f, 162f, 163, 164f
   Necturus, 157f
Brachialis, cat, 116f, 119
Brachial plexus, sheep brain, 203
Brachial vein
   cat, 165f, 167
   shark, 147f, 153f, 154
Brachiocephalic artery, cat,
   161f–164f, 163
Brachiocephalic vein, cat, 165f
Brachioradialis, cat, 115f, 120f,
   121, 123f
Brain
   adult vertebrate, 185f
   lamprey, 23f, 27
   regions of, 194
   shark, 189, 190–92, 191f,
     192f, 194
   sheep, 196–203
   skull and size of, 77b
Brain stem, sheep, 202–3, 202f
Branchial arches, 15, 67f
   lamprey, 24f
   shark, 70
Branchial arteries, shark, 146
Branchial basket
   lamprey, 21f, 26
   sea squirts, 16f, 17
Branchial chamber, fish, 133f
Branchial musculature, 89, 92t,
   93–95
Branchial pore, 9f, 11
Branchial pouch
   acorn worms, 9f, 11
   fish, 133f
   lamprey, 24f–26f
Branchials, shark, 95f
Branchial tube, lamprey, 22
Branchiohyoideus, Necturus, 99f,
   102, 103f
Branchiomeric musculature, 89
   homologies of, 118t
   shark, 95f

Branchiostegal, bowfin (Amia),
   71, 71f
Branchiostoma, 11n. See also
   Amphioxus
Broad ligament of the mammalian
   uterus, cat, 184
Buccal branch of facial nerve,
   shark, 193
Buccal cavity
   cat, 138, 140f
   lamprey, 22, 23f
   Necturus, 99
Buccal cirri, amphioxus, 11, 12f
Buccal funnel, lamprey, 22, 23f
Budding, reproduction by, 17
Bulbourethral (Cowper's) glands,
   cat, 179, 181, 181f
Bulbus arteriosus, Necturus, 155f,
   156, 157f
Burrowing organisms, 9, 10f

C

Calamus, feather, 34f, 35, 35f
Calcaneum
   alligator, 57f
   cat, 61f
   turtle, 54
Calcium, 40
Callus, 30f
Cancellous bone, 41
Canines, 84f, 85
Canine tooth, 81f
Capitate, cat, 64f, 66
Capitulum, 43f
   alligator, 48, 48f
   bird, 49
   cat, 51, 52f, 64f, 66
   Necturus, 46, 47f
   turtle, 49, 49f
Carapace, 33, 49, 49f
Cardia, cat stomach, 143
Carina, bird, 63
Carnassial teeth, 85, 86f
Carotid artery
   amphibian, 156f
   cat, 161f, 162f, 163, 164f, 166
   lamprey, 24, 25f
   Necturus, 155f, 156–57, 159f
   shark, 148f, 149–50, 149f, 153f
   sheep, 197, 199f
Carotid canal, shark, 69f, 70
Carpalia, turtle, 63
Carpals
   alligator, 57f, 63
   Necturus, 55f, 62
   turtle, 56f
Carpometacarpus, bird, 58f, 63
Cartilage, 40–41
   as connective tissue, 40
   types of, 40–41, 41f
Cartilaginous capsules, 66
Cat
   axial column, 52f
   circulatory and respiratory
     system, 160–71
   classification of, 2t
   cranial skeleton, 79, 80–81f,
     82–83

determining sex of, 105
digestive system, 138–45
dissection approach, 105
external anatomy, 105
muscle homologies with other vertebrates, 117b
musculature, 105–30, 106t–108t
pectoral girdle, sternum, and appendages, 63, 64f, 66
pelvic girdle and posterior appendages, 59–62, 61f
as representative form, 1
skinning, for dissection, 105
urogenital system, 179–84
vertebrae and ribs, 50–51, 52f
Caudal artery
cat, 170, 171f
lamprey, 24, 25f
Necturus, 159f
shark, 150, 153f
Caudal fin
amphioxus, 12f, 14
lamprey, 22, 23f, 27, 28f
Caudal region
alligator, 48
Necturus, 46
turtle, 49
Caudal vein
cat, 165f, 170, 171f
lamprey, 24, 25f
Necturus, 160f
shark, 153f
Caudal vertebrae, 44
alligator, 47, 48, 48f
aves (bird), 49, 50, 50f
cat, 51, 52f
frog, 47, 47f
turtle, 49f
Caudate lobe, cat liver, 143
Caudocruralis, Necturus, 102, 104f
Caudofemoralis
cat, 124, 125f, 126f
Necturus, 102, 104f
Ceca, 131
Cecum, cat, 143, 144f
Celiac artery
cat, 164f, 168, 168b, 169f
shark, 135f, 149f, 150, 151f–153f
Celiacomesenteric artery, Necturus, 157, 158f, 159f
Cementum, 83f
Center of ossification, bird pelvic girdle, 59
Centra (centrum), 43, 43f
cat, 51, 52f
shapes, 44f
shark, 44–45, 45f
structure of, 44t
teleost, 45, 46f
Central canal of spinal cord
shark, 194
sheep, 200f, 201
Centralia
cat, 60, 61f
turtle, 63
Central nervous system, 185

Central tendon, cat, 141
Cephalization, nervous system, 185
Cephalochordates, 2, 4, 11–15. See also Amphioxus
examination of, 14–15
relationships among, 19f
systems, 11–13
Ceratobranchial, 67f, 72f
chicken, 78f
shark, 69f, 70
Ceratohyal
Necturus, 72f
shark, 69f, 70
Ceratotrichia, shark, 53, 54f, 62
Cerebellar cortex, sheep, 200f, 201
Cerebellar fossa, cat, 81f, 83
Cerebellar peduncle, sheep, 197, 202, 202f
Cerebellar ventricle, shark, 194
Cerebellum
shark, 190–91, 190f
sheep, 197, 198f
Cerebellum cat, 140f
lamprey, 27
shark, 191f
sheep, 199f, 201
Cerebral aqueduct, shark, 194
Cerebral cortex, folding of, 197b
Cerebral fossa, cat, 81f, 83
Cerebral ganglion, sea squirts, 16f, 17
Cerebral hemispheres
shark, 191
sheep, 197, 198f
Cerebral peduncle, sheep brain, 197, 199f, 200f, 201
Cerebrospinal fluid, 196f
Cerebrum cat, 140f
shark, 191f
sheep, 201
Cervical ganglia, 186
Cervical region
bird, 49, 50f
cat, 50, 52f
turtle, 49, 49f
Cervical region alligator, 47
frog, 46, 47f
Necturus, 46, 47f
Cervical spinal nerves, sheep, 203
Cervical vertebrae, cat, 51, 52f
Cervical vertebrae alligator, 48f
Cervix, cat, 184
Chemoreception, shark, 90
Chest musculature, cat, 106t, 109, 111
Chicken
cranial skeleton, 78–79, 78f
pectoral girdle, sternum, and appendages, 63
pelvic girdle and posterior appendages, 56–57, 58f, 59
vertebrae and ribs, 49–50
Chimaera, digestive tracts, 133f
Choledochal vein, shark, 152f, 154, 154f

Chondrichthyes (cartilaginous fishes)
classification of, 2t
integument, 31
vertebrae and ribs, 45, 46f
Chondrocranium, 66, 68f
embryonic development of, 67f
Necturus, 71–72, 72f
shark, 68, 69f
Chondrocyte, 40, 41, 41f
Chondrogenesis, bird pelvic girdle, 59
Chondrostei, vertebrae and ribs, 45, 46f
Chordates, 2–5
body plan, 4f, 5
characteristics of, 2–5
evolution of, 8f
possible evolutionary sequence, 1f
protostomes and deuterostomes divisions in, 2, 3f, 4f, 4t
traditional classification of, 2t
Choroid plexus of fourth ventricle, 196f
Choroid plexus of lateral ventricle, 196f
Chromatophores, 29f, 31, 31f
amphibians, 33, 33f
in bird integument, 34, 35f
Cilia, amphioxus, 15
Ciliary mucous feeding, 10f
Ciliary tracts, amphioxus, 11
Ciliated band, sea squirts, 16f, 17
Circle of Willis, sheep, 197, 199f
Circulatory system, 146–66
acorn worms, 11
amphioxus, 13–14, 14f
cat, 160–66
lamprey, 23f, 24, 25f
Necturus, 154–60
respiratory system and, 146
shark, 146–54
Circuli, 32
Clasper, shark, 91, 94f, 172, 172f, 173f
Classification, 1
of chordates, 1f
Linnaean, of representative forms, 2t
traditional, 2t
Clavicle
bird, 58f, 63
bowfin (Amia), 62f
cat, 52f, 63, 64f, 111, 112f
Clavobrachialis, cat, 109f, 111, 112f, 114f, 115f, 116, 120f, 123f
Clavotrapezius
cat, 106t, 109f, 111, 112f, 114, 114f, 115f, 116, 117b, 139f
meaning of term, 88
Claws, 37, 37f
alligator, 55, 57f
cat, 62, 64f, 66
turtle, 54
Cleavage
radial, in deuterostomes, 4f
spiral, in protostomes, 4f

Cleidomastoid, cat, 109f, 112, 112f
Cleithrum, bowfin (Amia), 62, 62f, 71f
Clitoris, cat, 184
Cloaca
bird, 132f
fish, 133f
Necturus, 100, 137f, 138, 158f, 177, 178f, 179, 180f
shark, 91, 94f, 132, 134f, 151f, 173f, 174f, 175, 176f, 177
Cloacal artery, Necturus, 159f
Cloacal bursa, bird, 132f
Cloacal glands, Necturus, 100, 104f, 136, 137f, 178f
Cloacal opening, Necturus, 104f
Coccosteus, axial skeleton of, 44f
Coccyx, cat, 51
Coeliac ganglia, 186
Coelom, 2
lamprey, 22, 24
origins of, 136f
Coelomates, 2
Collagen fibers, 29, 40, 41, 41f, 84f
Collagen in integument, 29
Collar, acorn worms, 9f, 10, 10f
Collar nerve cord, acorn worms, 11
Collateral ganglia, 186, 187f
Collector loops, shark, 149–50
Colliculi, sheep brain, 197
Colon
cat, 143, 181f, 183f
fish, 133f
shark, 132, 134f, 135f, 152f, 174f
Colonial organisms, 9
Commissural artery, shark, 148f, 149f, 153f
Common bile duct
cat, 145
Necturus, 136
shark, 131, 135f, 152f
Common cardinal vein
lamprey, 24
Necturus, 155f, 156, 157f, 159f, 160
shark, 146f, 147f, 153f, 154
Common carotid artery, cat, 161f, 162f, 163, 166
Common coracoarcuals, shark, 94, 96f
Common iliac vein, cat, 165f
Common iliac veins, cat, 170, 171f
Compact bone, 41, 42f
Composite skull, 66–68
Compression teeth, 86f
Condyle of quadrate, alligator, 73f
Condyles
alligator, 74
bird, 58f, 59
Condyloid process, cat, 81f, 83
Connective tissue, categories of, 40f. See also Blood vessels; Bone; Cartilage
Constrictor muscles, shark, 95f
Constrictor series, shark musculature, 92t, 93, 93f, 95f, 96f

Contour feather, 34f
Conus arteriosus
    lamprey, 24
    Necturus, 155f, 156, 156f
    shark, 146, 146f–149f
Coprodeum, shark, 174f, 175,
    176f, 177
Copulatory structures, 172
Coracobrachialis
    cat, 110f, 111
    Necturus, 99f, 101
Coracobranchials, shark, 146, 148f
Coracohyoid, shark, 94, 96f, 98
Coracoid
    alligator, 57f, 63
    bird, 58f, 63
    turtle, 56f, 63
Coracoid bar, shark, 54f, 62, 96f,
    148f
Coracoid cartilage, *Necturus,*
    55f, 62
Coracoid process, cat, 64f, 66
Coracomandibularis, shark, 94, 96f
Cornified layer of epidermis, 29
Coronary artery
    cat, 144f, 161f, 162, 162f
    shark, 146, 147f–149f, 153f
Coronary ligament
    cat, 143
    Necturus, 138
    shark, 133
Coronary sinus, 163f
Coronary vein, cat, 161f, 162, 162f,
    166f, 171
Coronoid
    alligator, 74f, 75
    turtle, 76f, 78
Coronoid process, cat, 81f, 83
Corpora quadrigemina, sheep brain,
    197, 201
Corpus callosum, sheep brain, 197,
    200f, 201
Corpus of stomach, cat, 143, 144f
Cortex of kidney, cat, 181f, 182
Cortical bone, 41
Cosmine, 32
Cosmoid scale, 30f, 32, 32f
Costal cartilage, cat, 51
Costal cartilage alligator, 48
Costal process, bird, 58f, 63
Costocervical artery, cat, 161f,
    162f, 163, 164f, 166
Costocervical vein, cat, 165f, 167
Countershading, shark, 90
Cranial cartilage, lamprey, 23f,
    25f, 26
Cranial cavity, shark, 68
Cranial muscles, homologies of,
    118t
Cranial nerves, 186–89
    lamprey, 27
    shark, 187f, 190f, 192–94
    sheep, 197, 199f
Cranial region, cat, 79
Cranial skeleton, 42, 68–83
    amphibian *(Necturus),*
        71–72, 72f
    aves (bird), 78–79, 78f

elasmobranchii (shark), 68–70
    mammal (cat), 79, 80–81f,
        82–83
    osteichthyes bowfin *(Amia),*
        71, 71f
    reptiles, 72–78
Cremasteric fascia, cat, 181–82
Crests
    bird, 59
    of ilium in cats, 59
Cribriform plate, cat, 81f, 83
Cricoid cartilage, cat, 141, 142f
Cricothyroid, cat, 113, 113f
Crop, bird, 132f
Crown (tooth), 83, 83f
Ctenoid scales, 32, 32f
Cucullaris
    Necturus, 102, 103f
    shark, 93, 93f, 95f, 97f
Cusps (tooth), 83, 83f, 84f
Cutaneous artery, *Necturus,* 157f
Cutaneous maximus, cat, 105
Cutaneous respiration, 33, 156b
Cycloid scale, 32, 32f
Cyclostomes, 18, 20f, 44.
    See also Lampreys
        (petromyzoniformes)
Cystic artery, cat, 164f, 168
Cystic duct
    cat, 145
    Necturus, 136

D

Deciduous dentition, 83, 84f
Deep brachial artery, cat, 161f,
    162f, 163, 164f
Deep brachial vein, cat, 161f,
    165f, 167
Deep fascia, cat, 105
Deep femoral artery, cat, 164f,
    170, 171f
Deep femoral vein, cat, 165f,
    170, 171f
Deep ophthalmic branch of
    trigeminal nerve, shark,
    192–93
Deep ophthalmic nerve, shark,
    189, 190f
Deltoid muscle group, cat, 116
Deltoid ridge
    bird, 58f, 63
    cat, 64f, 66
Demifacet, cat, 51, 52f
Dental formula, 85
Dentary/ies
    alligator, 74f, 75
    bird, 78f, 79
    bowfin *(Amia),* 71, 71f
    cat, 83
    Necturus, 71, 72f
    turtle, 76f, 77
Dentin, 31, 31f, 32, 32f, 83f
Depression, muscle action, 88f
Depressor mandibulae, *Necturus,*
    102, 103f
Dermal bones, 29, 41, 42t, 49,
    67–68
Dermal muscle, cat, 105

Dermal papillae, 36f, 37, 39
Dermal scale, 30, 30f, 32, 37
Dermatocranium, 42t, 66, 68f
    in bowfin *(Amia),* 71
    major bones of, 67f
    Necturus, 71, 72f
    six series of bones forming, 68t
Dermis, 29, 29f
    amphibian, 30f, 33, 33f
    fish, 31
    mammalian, 36f, 37
Descending colon, cat, 143, 144f,
    169f, 181f, 183f
Deuterostomes, 2, 3f, 4f, 4t
Diaphragm, cat, 141, 142f
Diapophysis, 43f
    alligator, 48, 48f
    bird, 49
    as common type
        of apophyses, 44t
    Necturus, 46, 47f
Diapsid skull, alligator, 72
Diastema, 85
Diencephalon
    shark brain, 98f, 190f
    sheep brain, 201
Digastric, cat, 109f, 110f, 112f,
    113, 117b, 139f
Digesta, 131
Digestive systems, 131–45
    bird, 132f
    cat, 138–45
    fish, 133f
    introduction, 131
    Necturus, 136–38, 137f
    shark, 131–33, 134f, 135f
    sturgeon, 133f
Digits, 124b
Dilatator laryngis, *Necturus,*
    102, 103f
Dioecious organisms, 9, 26
Diphyodont teeth, 83, 84–85
Dissection
    cat, 105–30
    defined, 89
    Necturus, 99–105
    process, 89
    safety precautions, 89
    shark, 89–99
Diverticulum, fish, 133f
Dog, teeth of, 84f
Dorsal aorta
    cat, 163, 164f, 168–70,
        169f, 171f
    lamprey, 23f, 24, 25f, 28f
    Necturus, 137f, 155f, 156–57,
        158f, 159f, 160, 178f, 180f
    shark, 148f, 149, 149f, 150,
        151f, 153f
Dorsal blood vessel, 11
Dorsal constrictors, shark, 93f, 97f
Dorsal-epaxial trunk muscles,
    Necturus, 101t, 102, 103f
Dorsal fin, 14, 14f, 15, 22, 23f,
    28f, 90
Dorsal gastric artery
    Necturus, 157, 159f
    shark, 150, 152f, 153f

Dorsal gastric vein, shark, 152f, 154
Dorsalis scapulae, *Necturus,*
    102, 103f
Dorsalis trunci, *Necturus,* 102,
    103f, 104f
Dorsal lobe of pancreas, shark,
    131, 134f, 135f, 152f
Dorsal longitudinal bundles, shark,
    92, 92t, 93f, 97f
Dorsal mesentery
    Necturus, 138
    shark, 132
Dorsal muscles, *Necturus,* 100t
Dorsal nerve cord, 4–5, 5f, 14
Dorsal projection of lateral process,
    cat, 60, 61f
Dorsal ribs, 43, 43f
Dorsal root ganglion, 186, 187f
Dorsal root of spinal nerve,
    shark, 190f
Dorsal shoulder musculature, cat,
    114–16, 118–20
Dorsal skeletogenous septum,
    shark, 45
Down feather, 34f
Ductus deferens
    cat, 179, 181f
    Necturus, 177, 177b, 178f
    shark, 173, 174f, 175
Duodenal artery, shark, 150, 153f
Duodenum
    cat, 143, 144f
    Necturus, 136, 137f, 158f
    shark, 131, 134f, 135f, 152f
Dura layer of meninges, 196

E

Ectopterygoid
    alligator, 73f, 75
    bowfin *(Amia),* 71
Effectors, nervous system, 185
Efferent branchial artery
    Necturus, 155f, 156, 159f
    shark, 148f, 149, 149f, 153f
Efferent collector loop, shark,
    148f, 149
Efferent ductules
    human, 182f
    Necturus, 177, 178f
    shark, 173
Efferent renal veins, shark,
    153f, 154
Efferent spiracular artery, shark,
    148f, 150, 153f
Elasmobranchii. *See also* Shark
    cranial skeleton, 68–70
    pectoral girdle, sternum, and
        appendages, 62
    pelvic girdle and posterior
        appendages, 53, 54f
    vertebrae and ribs, 44–45, 45f
Elastic cartilage, 41, 41f
Elastin fibers, 40, 41f
Elbow
    bird, 63
    cat, 66
Electroreceptors, shark, 90

Elevation, muscle action, 88*f*
Embryonic development
 of bone, 41
 of central nervous system, 195*f*
 of chondrocranium, 67*f*
 myotomes and, 91*f*
 of pelvic girdle in birds,
  59–62, 60*f*
Enamel, 31, 31*f*, 32*f*, 83, 83*f*
Endochondral bone, 41, 42*t*
Endolymphatic ducts, shark, 90
Endolymphatic foramina, shark,
 69*f*, 70
Endolymphatic fossa, shark,
 69*f*, 70
Endolymphatic pores
 amphioxus, 11–12
 shark, 90
Endostyle, 4, 11, 14
Ensiform cartilage, cat, 66
Enteropneusta, 9, 9*f*, 10–11
Entoplastron, turtle, 63
Epaxial musculature, shark, 45, 92
Epaxial myotomes, shark, 95*f*
Epibranchial, 67*f*, 72*f*
 chicken, 78*f*
 shark, 69*f*, 70
Epibranchial cartilage, shark, 97*f*,
 148, 148*f*
Epibranchial groove
 amphioxus, 11, 14, 14*f*, 15
 lamprey, 28*f*
Epidermal scale, 30, 30*f*, 34
Epidermis, 29–30
 amphibian, 29*f*, 33, 33*f*
 derivatives of, 30*f*, 34*f*
 fish, 31
 mammalian, 36–37, 36*f*
Epididymis
 cat, 181*f*
 human, 182*f*
 Necturus, 177, 177*b,* 178*f*
 salamander, 175*b*
 shark, 173, 174*f*, 175
Epigastric artery, *Necturus,* 158*f*,
 159*f*, 160
Epiglottis, cat, 140*f*, 141, 142*f*
Epiphyseal foramen, shark, 68, 69*f*
Epiphysis, shark, 68, 191*f*
Epiplastra, turtle, 63
Epiploic foramen, cat, 145
Epipterygoid, 67, 68*f*
Epipubic cartilage, turtle, 54, 56*f*
Epithalamus
 shark, 196
 sheep, 201
Epitrochlearis, cat, 109*f*, 120, 123*f*
Esophageal arteries, cat, 168
Esophagus
 amphioxus, 14
 bird, 132*f*
 cat, 140*f*, 141, 142, 161*f*, 169*f*
 fish, 133*f*
 lamprey, 22, 23*f*, 25*f*, 26*f*,
  27, 28*f*
 shark, 131, 134*f*
Ethmoid plate, 67*f*, 72, 72*f*
Eustachian canal, cat, 80*f*, 82

Evolution
 pattern of change and, 85, 87
 of pectoral girdle, 65*b*
 of vertebrate brain, 185*f*
Excurrent siphons, sea squirts, 17
Exoccipitals, 67*f*
 alligator, 74*f*, 75
 bird, 78*f*, 79
 Necturus, 72, 72*f*
 turtle, 76*f*, 77
Extension, muscle action, 88*f*, 116*f*,
 120*f*, 121, cat
Extensor carpi radialis longus, cat,
 115*f*, 116*f*, 120*f*, 121, 123*f*
Extensor carpi ulnaris, cat, 115*f*,
 116*f*, 120*f*, 121
Extensor digiti secundi, cat, 121
Extensor digitorum communis, cat,
 115*f*, 116*f*, 120*f*, 121
Extensor digitorum lateralis, cat,
 115*f*, 116*f*, 120*f*, 121
Extensor digitorum longus, cat,
 125*f*, 126*f*, 129
Extensor pollicis, cat, 121
Extensors
 Necturus, 102, 103*f*
 shark, 92, 93*f*, 97*f*
External auditory meatus
 alligator, 73*f*, 75
 cat, 79, 80*f*, 81
 turtle, 77
External carotid artery
 cat, 164*f*, 166
 Necturus, 155*f*, 156, 159*f*
 shark, 148*f*, 149*f*, 150, 153*f*
External gills, *Necturus,* 99, 155*f*,
 156, 157*f*
External gill slits
 lamprey, 22, 23*f*, 27
 shark, 148, 148*f*
External iliac arteries, cat, 170, 171*f*
External iliac vein, cat, 165*f*,
 170, 171*f*
External inguinal rings, cat, 179
External intercostals, cat, 108, 110*f*
External jugular vein, cat, 112*f*,
 114*f*, 138, 139*f*, 161*f*, 165*f*,
 167–68
External mandibular fenestra,
 alligator, 74*f*, 75
External maxillary artery, cat, 166
External nares
 alligator, 72, 73*f*
 cat, 79, 80*f*, 167
 Necturus, 99
 turtle, 75, 76*f*
External oblique
 cat, 108, 110*f*, 127*f*, 128*f*
 Necturus, 102, 103–4*f*
Extrascapulars, bowfin *(Amia),*
 71, 71*f*
Extrinsic eye muscles, shark, 92*t*,
 93*f*, 98–99, 98*f*
Eyes
 lamprey, 22, 23*f*
 musculature of shark, 92*t*, 93*f*,
  97*f*, 98–99, 98*f*
 Necturus, 99
Eye spot, lamprey, 27, 28*f*

F

Facet, cat, 51
Facial nerve (VII), 189
 cat, 139*f*
 shark, 193
 sheep brain, 199, 199*f*
Facial region, cat, 79
Facial series (bones), 67*f*, 68*f*, 68*t*
Falciform ligament
 cat, 143
 Necturus, 136, 138
 shark, 133
False rib, cat, 51
False vocal cords, cat, 141
Fascia
 cat, 105
 shark, 91
Fascia lata, cat, 124, 125*f*, 128*f*
Fauces, cat, 141
Feathers, 34–35
 contour, 34*f*
 down, 34*f*
 flight, 34*f*
 follicles, 35, 35*f*
Feeding mechanisms, vertebrate
 evolution and shifts in,
 18–21
Female urogenital system
 cat, 182–84
 ducts in, 175*b*
 mammal, 184*b*
 Necturus, 180*f* reproductive
  tracts of placental, 182–84
 shark, 175–77
Femoral artery
 cat, 164*f*, 170, 171*f*
 Necturus, 158*f*, 159*f*, 160
Femoral vein
 cat, 165*f*, 170, 171*f*
 Necturus, 158*f*, 159*f*, 160, 160*f*
Femur
 alligator, 55, 57*f*
 bird, 58*f*, 59
 cat, 59
 Necturus, 53, 55*f*
 turtle, 54, 56*f*
Fenestra, *Necturus,* 72*f*
Fibrocartilage, 41, 41*f*
Fibrous connective tissue, 40, 40*f*
Fibula
 alligator, 55, 57*f*
 bird, 58*f*, 59
 cat, 60, 61*f*
 Necturus, 53–54, 55*f*
 turtle, 54, 56*f*
Fibulare
 alligator, 55, 57*f*
 cat, 60, 61*f*
 turtle, 54, 56*f*
5th afferent branchial arteries,
 shark, 146, 148*f*
Filiform papillae, cat, 140*f*, 141
Filoplume, 34*f*
Filter feeding, 11, 12*f*, 15
Fimbriae, cat, 183
Fin musculature, shark, 92–93, 92*t*
Fin rays, lamprey, 22, 25*f*
Fins

amphioxus, 14
basic components of, 53*f*
lamprey, 18, 21*f*
1st afferent branchial artery, shark,
 148, 148*f*
Fishes
 digestive tract, 133*f*
 integuments, 31–32, 31*f*, 32*f*
 midbrain and tectum in brain
  of, 201*b*
 vertebrae and ribs, 43–45
Flexion, muscle action, 88*f*
Flexor carpi radialis, cat, 123, 123*f*
Flexor carpi ulnaris, cat, 123, 123*f*
Flexor digitorum longus, cat, 127*f*,
 128*f*, 129
Flexor digitorum profundus, cat,
 120*f*, 123, 123*f*
Flexor digitorum superficialis, cat,
 123, 123*f*
Flexor hallucis longus, cat,
 127*f*, 129
Flexors, shark, 93, 94*f*, 96*f*, 173*f*
Flight feather, 34*f*
Flocculus, sheep brain, 197, 198*f*
Folia, sheep brain, 197, 198*f*
Foliate papillae, cat, 140*f*, 141
Follicles, feather, 35, 35*f*
Foptic nerve (II), shark, 190*f*, 192
Foramen ischiadicum, bird, 60*f*
Foramen magnum
 alligator, 74*f*, 75
 bird, 79
 cat, 79, 80*f*
 Necturus, 72, 72*f*
 shark, 69*f*, 70
 turtle, 77
Foramen of magendie, 196*f*
Foramen ovale, 80*f*
 alligator, 75
 cat, 82
Foramen rotundum, cat, 80*f*, 82
Foramina
 shark, 70
 turtle, 76*f*, 77
Forearm. *See* Forelimbs
Forebrain (prosencephalon), 194*b*
 lamprey, 27
 sheep, 201
Forelimbs, 102. *See also* Limbs
 diversity and underlying
  patterns, 87*f*
 muscles of cat, 107*t*, 114–24
 muscles of *Necturus,* 100*t*,
  101–2
Form and function
 alternate blood pathways, 167*b*
 arterial loops, 168*b*
 bone mass and gravity, 41*b*
 cartilaginous rings in
  trachea, 167*b*
 cutaneous respiration, 156*b*
 ductus deferens in *Necturus* and
  salamanders, 177*b*
 elastic cartilage *vs.* rigid
  bone, 40*b*
 forearm and digit
  musculature, 124*b*

Form and function—*Cont.*
    forearms of digger *vs.*
        runner, 121*b*
    gill lamellae, 148*b*
    hindlimb musculature, 122*b*
    hypothalamus and
        homeostasis, 201*b*
    limb tendons, 130*b*
    mammals, 184*b*
    mastication and tooth
        occlusion, 87*b*
    mesenteries, 133*b*
    midbrain and tectum of fishes
        and amphibians, 201*b*
    palatine processes, 77*b*
    palpebral bone, 75*b*
    parallel and pinnate
        muscles, 129*b*
    pineal organ, 201*b*
    receptors in neck for
        positioning, 90*b*
    rectal gland of shark, 132*b*
    spiral valve in intestine,
        131*b*, 132*b*
    of teeth, 86*b*
    uropygial gland secretions, 35*b*
    vertebral column of birds, 49*b*
    vertebrate jaws, 70*b* vestibular
        apparatus and stretch
        vessels, 170*b*
Fornix, sheep, 200*f*, 201
4th afferent branchial arteries,
    shark, 146, 148*f*
Fourth ventricle
    shark, 191*f*
    sheep brain, 200*f*, 201, 202*f*
Fovea capitis, cat, 59, 61*f*
Free (floating) rib, cat, 51
Fringes, 39, 39*f*
Frog, vertebrae and ribs, 46–47, 47*f*
Frontal process, cat, 79, 80*f*, 81*f*
Frontals
    alligator, 72, 73*f*
    bird, 78*f*, 79
    cat, 79, 80*f*
    Necturus, 71, 72*f*
    turtle, 76*f*, 77
Frontal sinus, cat, 81*f*, 82
Full facet, cat, 51, 52*f*
Fundus
    stomach, of cat, 143
    urinary bladder, of cat, 179,
        181*f*, 182, 183*f*
Fungiform papillae, cat, 140*f*, 141
Furcula, bird, 58*f*, 63

G

Gallbladder
    cat, 142*f*, 143, 144*f*
    lamprey, 27, 28*f*
    Necturus, 136, 137*f*, 158*f*
    shark, 131, 134*f*, 135*f*, 151*f*,
        152*f*
Gametes, 26, 172
Ganglia of dorsal root, shark, 194
Ganoid scale, 32, 32*f*
Ganoin, 32

Gastralia, 34
    alligator, 49
    turtle, 63
Gastric artery
    cat, 164*f*, 168
    Necturus, 157, 158*f*, 159*f*
    shark, 135*f*, 150, 151*f*, 152*f*
Gastric vein, shark, 135*f*, 151*f*,
    152*f*, 154, 154*f*
Gastrocnemius
    cat, 108*t*, 125*f*–128*f*, 129
    defined, 88
    orientation of fibers in, 129*b*
Gastrocolic ligament, cat, 145
Gastroduodenal artery, cat, 168
Gastrohepatic ligament
    cat, 144*f*, 145
    Necturus, 137*f*, 138
    shark, 133, 134*f*
Gastrohepatoduodenal ligament,
    shark, 133
Gastrosplenic artery, shark, 150,
    151*f*–153*f*
Gastrosplenic ligament
    cat, 144*f*, 145
    Necturus, 137*f*, 138, 158*f*
    shark, 132
Gastrosplenic vein
    cat, 166*f*, 171
    Necturus, 158*f*, 160, 160*f*
General cranial nerves, 186
Generalized connective tissues,
    40, 40*f*
General somatic motor nerve, 185
General visceral sensory nerve, 185
Geniculate ganglion, shark, 98*f*,
    190*f*, 193
Geniohyoid, 113*f*
    cat, 106*t* 109*f*, 110*f*, 112*f*, 113,
        113*f*, 117*b*, 118*t*
    Necturus, 99*f*, 100*t*, 102, 118*t*
Genital arteries
    Necturus, 157, 159*f*
    shark, 153*f*
Genital veins, *Necturus,* 159*f*, 160
Genu, sheep brain, 200*f*, 201
Gill adductor muscles, lamprey, 24*f*
Gill arches
    bowfin *(Amia),* 71
    shark, 70
Gill capillaries, 24
Gill filament, lamprey, 24*f*, 27, 28*f*
Gill lamellae
    lamprey, 22, 25*f*, 28
    shark, 148, 148*f*
Gill pouch
    lamprey, 22
    shark, 148, 148*f*
Gill raker, shark, 148, 148*f*
Gill rays
    bowfin *(Amia),* 71
    shark, 69*f*, 70
Gills, 20
    Necturus, 156
    shark, 70, 148, 148*f*, 149
Gill slits
    fish, 133*f*
    lamprey, 27, 28*f*

    shark, 91, 93*f*, 97*f*
Gingiva, 83, 139
Giraffe
    horns of, 39, 39*f*
    limb tendons and muscles
        of, 130*b*
Girdles and limbs, 53–66
    basic components of the fin and
        limb, 53, 53*f*
    introduction, 53
    pectoral girdle, sternum, and
        appendages, 62–66
    pelvic girdle and posterior
        appendages, 53–62
Gizzard, 131, 132*f*
Glenoid fossa
    alligator, 63
    bird, 63
    cat, 63, 64*f*
    Necturus, 62
    turtle, 63
Glomerulus, 11
Glossopharyngeal nerve (IX), 189
    shark, 69*f*, 70, 98*f*, 190*f*, 193
    sheep, 199*f*, 200
Glottis, cat, 141, 167
Gluteus maximus
    cat, 108*t*, 124, 126*f*
    defined, 88
Gluteus medius, cat, 124, 126*f*
Gnathostomes, 19*f*, 21, 23*f*, 70*b*
Gonad
    amphioxus, 14, 14*f*, 15
    lamprey, 22, 23*f*, 26
    Necturus, 138
    sea squirt, 16*f*
    shark, 132, 134*f*
Gracilis, cat, 125, 127*f*, 128*f*
Gravity, effects of, on bone
    mass, 41*b*
Gray matter, 196*f*
Greater curvature of stomach, 143
    cat, 144*f*
    shark, 131, 134*f*
Greater multangular, cat, 64*f*, 66
Greater omentum, cat, 142*f*, 145
Greater trochanter, 58*f*
    bird, 59
    cat, 59–60, 61*f*
Greater tuberosity
    bird, 58*f*, 63
    cat, 64*f*, 66
Grinding teeth, 86*f*
Ground substance, cartilage, 40, 41*f*
Gubernaculum, cat, 182
Gular, bowfin *(Amia),* 71, 71*f*
Gular fold, *Necturus,* 99–100, 99*f*
Gyri (gyrus), sheep brain,
    197, 198*f*

H

Habenula, shark, 191*f*, 194
Habenular trigone, sheep, 200*f*,
    201, 202*f*
Hagfishes (myxini), 18, 20*f*
Haikouella, 18, 19*f*
Haikouichthyes, 18, 19*f*

Hair
    cuticle, 36*f*, 37
    mammalian, 36, 36*f*
    root, 37
    shaft, 37
Hallux
    bird, 58*f*, 59
    cat, 62
Hamate, cat, 64*f*, 66
Hamstring muscles, 128*b*
Hamulus, cat, 80–81*f*, 82
Hard palate, cat, 82, 138, 140*f*
Hatschek's pit (Hatschek's groove),
    amphioxus, 11, 12*f*
Head
    lamprey, 22
    nervous system and, 185
    shark, 89
Head of epididymis, cat, 181*f*
Head of femur
    alligator, 55
    cat, 59, 61*f*
Head of fibula, cat, 60, 61*f*
Head of humerus
    bird, 63
    cat, 64*f*, 66, 116*f*
Head of pancreas, cat, 143, 144*f*
Head of radius, cat, 64*f*, 66
Heart
    amphibian, 156*f*
    cat, 161–66
    lamprey, 23*f*, 24, 27, 28*f*
    Necturus, 155*f*, 156, 157*f*
    shark, 146*f*, 148*f*
Hemal arch, 43, 43*f*
    alligator, 48, 48*f*
    cat, 51, 52*f*
    holostei, 46*f*
    Necturus, 46, 47*f*
    shark, 44, 45*f*
    teleostei, 46*f*
    turtle, 49
Hemal spine, 43, 43*f*, 44, 45*f*, 47*f*
Hemibranch, shark, 148
Hemichordata
    examination of enteropneusta,
        9–11
    generalized acorn worm, 9*f*
    pterobranchs, 9, 10*f*
Hemispheres, sheep brain, 197, 198*f*
Hemopoietic tissue, 28*f*
Hepatic artery
    cat, 164*f*, 168
    Necturus, 157, 158*f*, 159*f*
    shark, 135*f*, 150, 151*f*–153*f*
Hepatic ducts
    cat, 145
    Necturus, 136
Hepatic portal system
    cat, 166*f*, 170–71
    lamprey, 24
    Necturus, 160, 160*f*
    shark, 150, 151–52*f*, 151*f*–152*f*,
        154, 154*f*
Hepatic portal vein
    cat, 166*f*, 171
    Necturus, 158*f*, 160, 160*f*
    shark, 135*f*, 151*f*, 152*f*, 154, 154*f*

Hepatic sinus
 Necturus, 156, 157f, 159f
 shark, 147f, 153f, 154
Hepatic vein
 amphibian, 156f
 amphioxus, 13
 cat, 165f, 169f, 170
 Necturus, 159f, 160
 shark, 146f
Hepatocavopulmonary ligament,
 Necturus, 138
Hepatoduodenal ligament
 cat, 144f, 145
 Necturus, 138
 shark, 133, 134f
Hepatopancreatic ampulla, cat, 145
Hepatorenal ligament, cat, 145
Hermaphrodites, ascidians as, 17
Heterocercal tail, shark, 90
Heterocoelous vertebrae, bird,
 49, 50f
Heterodont teeth, 83, 84–85, 84f
Hilus, cat, 179, 181f, 182, 183f
Hindbrain (rhombencephalon), 194b
 lamprey, 27
 sheep, 201
Hindgut, amphioxus, 12f, 13f, 14
Hindlimbs
 blood vessels of cat, 171f
 muscles of cat, 108t, 124–30
 muscles of *Necturus,* 101t, 102,
  104f, 105
Hinge, reptile skin, 33, 33f
Hippocampus, sheep, 201
Holobranch, shark, 148
Holostei
 pectoral girdle, sternum,
  and appendages, 62
 vertebrae and ribs, 46f
Homeostasis, hypothalamus
 and, 201b
Homodont teeth, 83, 84
Homologies in musculature cranial
 muscles, 117b, 118t
Hooves, 37, 37f
Horizontal septum
 Necturus, 102, 103f, 104f
 shark, 91, 92, 93f, 97f
Horizontal skeletogenous septum,
 43, 45
Horns, 38–39, 38f, 75
 of hyoid apparatus, 83
Humans
 buccal structures in mouth
  of, 139b
 inguinal canals and adult
  male, 179b
 male reproductive
  system, 182f
Humeroantebrachialis, *Necturus,*
 99f, 101, 103f
Humerus
 alligator, 57f, 63
 bird, 58f, 63
 cat, 66
 Necturus, 55f, 62
 turtle, 56f, 63
Hyaline cartilage, 41, 41f

Hyoglossus, cat, 110f, 112f,
 113, 113f
Hyoid, cat, 52f, 112f, 113f
Hyoid apparatus, 66
 alligator, 75
 cat, 83
Hyoid arch, 67f, 68f, 70
Hyoid arches shark, 70
Hyoid bone, 67
 bird, 79
 cat, 140f
Hyoidean artery, shark, 148f,
 149, 153f
Hyomandibula, 67, 67f, 68f, 71, 71f
Hyomandibular, shark, 69f, 70
Hyomandibular branch of facial
 nerve, shark, 98f, 191f, 193
Hyomandibular nerve (VII) shark,
 93f, 189, 190f
Hypapophyses, bird, 49, 50f
Hypapophyses alligator, 48
Hypaxial musculature, shark,
 45, 92
Hypertrophy of bone, 41
Hypobranchial, 67f, 69f, 70, 94,
 95f, 98
Hypobranchial musculature, 89,
 92t, 94, 118t
Hypobranchial nerve, 98f
Hypobranchial nerve, shark, 190f
Hypodermis, 29, 29f
Hypogastric artery, *Necturus,* 158f,
 159f, 160
Hypoglossal nerve (XII), 186, 189
 sheep, 199f, 200
Hypohyal, *Necturus,* 72f
Hypophyseal pouch, lamprey, 23f,
 25f, 27
Hypophysis, shark, 194
Hypothalamus
 adult brain, 195f, 196
 homeostasis and, 201b
 sheep, 201

I

Ileocolic artery, cat, 164f, 168
Ileum, cat, 143
Iliac arteries
 Necturus, 158f, 159f, 160
 shark, 150, 151f, 153f
Iliac process, shark, 53, 54f
Ilioextensorius, *Necturus,* 104f, 105
Iliofibularis, *Necturus,* 105
Ilio-ischiac foramen, bird, 57,
 59, 60f
Iliolumbar artery, cat, 164f, 169f,
 170, 171f
Iliolumbar vein, cat, 165f, 171f
Iliopectineal eminence, cat, 59, 61f
Iliopsoas, cat, 127, 127f, 128f
Iliotibialis, *Necturus,* 104f, 105
Ilium
 alligator, 55, 57f
 bird, 50f, 56–57, 59, 60f
 cat, 59
 frog, 47, 47f
 Necturus, 53, 55f
 Ornithischian hip, 59b

Saurischian hip, 59b
 turtle, 54, 56f
Incisive foramen
 alligator, 73f, 74
 cat, 80f, 82
 turtle, 76f, 77
Incisors, 81f, 84f, 85
Incurrent siphon, sea squirts, 17
Incus, origins of, 82b
Inferior colliculi, sheep brain, 197,
 200f, 201, 202f
Inferior epigastric artery, cat, 164f,
 170, 171f
Inferior gluteal artery, cat, 164f, 170
Inferior gluteal veins, cat, 165f, 170
Inferior jugular vein
 lamprey, 23f, 24, 25f
 shark, 147f, 153f, 154
Inferior lobes of infundibulum,
 shark, 194
Inferior mesenteric artery, cat,
 164f, 168, 169f, 171f
Inferior mesenteric vein, cat,
 166f, 171
Inferior oblique, shark, 93f, 96f, 98,
 98f, 190f
Inferior phrenic artery, cat,
 169, 169f
Inferior phrenic vein, cat, 165f,
 169f, 170
Inferior rectus, shark, 93f, 96f, 98
Inferior rectus muscle, shark, 190f
Infraorbital canal, cat, 79, 80f, 81f
Infraorbital gland, cat, 138
Infraorbital nerve, shark, 99, 190f
Infraorbital trunk, shark, 193
Infraspinatus, cat, 116f, 118
Infraspinous fossa, cat, 64f, 66
Infratemporal arcade, alligator, 72
Infratemporal fenestra, alligator,
 72, 73f
Infundibulum
 adult brain, 195f
 cat, 183, 183f
 shark, 194
 sheep brain, 197, 199f, 201
Inguinal canals
 of adult male humans, 179b
 cat, 179, 179b
 human, 179b
Inguinal region, cat, 105
Inner ear cavity, cat, 83
Innominate, bird, 56
Innominate bone, 50f
 bird pelvic girdle, 59
 cat, 59
Insertion, origin, and function of
 musculature
 Necturus, 100, 100t–101t, 101
 shark, 91–99, 92t
Integuments, 29–39
 cutaneous respiration, 156b
 dermis of, 29
 epidermis of, 29–30
 of fishes, 31–32, 31f, 32f
 skin derivatives, 30f
 specializations of, 37–39
 of tetrapods, 33–37

Interarcual muscle series, shark,
 94, 97f, 98
Interbranchial septum, shark,
 148, 148f
Intercentrum
 alligator, 47
 bowfin (*Amia*), 45
Interclavicle
 alligator, 57f, 63
 bird, 58f, 63
Intercondyloid fossa, cat, 60, 61f
Intercostal arteries, cat, 164f, 168
Intercostal veins, cat, 167
Interhyoideus
 Necturus, 99f, 102, 103f
 shark, 93, 96f
Intermandibularis
 Necturus, 99f, 101, 103f
 shark, 93, 93f, 96f
Intermediate rib, alligator, 48, 48f
Intermedium, turtle, 56f, 63
Internal auditory meatus, cat, 81f, 83
Internal carotid artery
 Necturus, 155f, 156, 159f
 shark, 148f, 149, 149f, 153f
 sheep brain, 197, 199f
Internal gill slit
 lamprey, 22, 23f
 shark, 148, 148f
Internal iliac arteries, cat, 170
Internal iliac vein, cat, 165f, 170
Internal inguinal rings, cat, 179
Internal intercostals, cat, 108
Internal jugular vein, 161f, 163f,
 165f, 167, 168
Internal mammary artery, cat, 163
Internal mammary vein, cat, 161f,
 165f, 167
Internal nares
 alligator, 73f, 74
 cat, 80f, 140f, 141
 turtle, 76f, 77
Internal oblique
 cat, 108, 110f
 Necturus, 102
Internal spermatic artery, cat,
 169–70, 179
Internal spermatic vein, cat, 179
Interneural arches, shark, 44, 45f
Interopercular, bowfin (*Amia*),
 71, 71f
Interorbital septum, bird, 78f, 79
Interosseous crest, cat, 64f, 66
Interparietal, cat, 79, 80f
Intersegmental arteries, shark,
 150, 151f
Intertemporal, bowfin (*Amia*),
 71, 71f
Intertrematic branch, shark, 148f,
 149, 153f
Intertrochanteric line, cat, 60, 61f
Interventricular foramen, shark, 194
Intervertebral cartilages, 50
Intervertebral disks, 50
Intervertebral foramen, cat,
 51, 52f
Intervertebral notch, cat, 51
Intestinal arteries, cat, 164f, 168

Intestinal vein
cat, 166f, 171
Necturus, 158f, 160, 160f
posterior, shark, 135f, 151f, 152f, 154, 154f
Intestine
bird, 132f
cat, 142f, 143
fish, 133f
lamprey, 22, 23f, 24, 25f, 26f, 27, 28, 28f
sea squirts, 16f, 17
shark, 132
Intraventricular foramen (Monro), 196f
Invertebrates, 3
Ischiac cartilage, *Necturus,* 53, 55f
Ischiac symphysis
alligator, 55
cat, 59
turtle, 54, 56f
Ischial spine, cat, 59, 61f
Ischial symphyses, cat, 179, 181f, 183f
Ischial tuberosity, cat, 59, 61f
Ischiocaudalis, *Necturus,* 102, 104f
Ischioflexorius, *Necturus,* 102, 104f
Ischium
alligator, 55, 57f
bird, 50f, 56–57, 59, 60f
cat, 59, 61f
Necturus, 53, 55f
Ornithischian hip, 59b
Saurischian hip, 59b
turtle, 54, 56f
Isthmus
cat, 141
shark, 131, 135f

J

Jawed fishes (Gnathosomes), 21, 70b
Jawless fishes. See Agnathans
Jaw muscles
cat, 106t, 113–14
shark, 95f
Jaws, 70b
alligator, 74f, 75
cat, 81f, 83
turtle, 76f, 77–78
Jejunum, cat, 143
Jugal
alligator, 72, 73f
bird, 78f, 79
bowfin (*Amia*), 71, 71f
cat, 79, 80f, 81f, 82
turtle, 76f, 77
Jugal bar (arch), bird, 78f, 79
Jugular process, cat, 79, 80f, 81f
Jugular vein
cat, 139f
mammal, 163f
Necturus, 157f

K

Keel, bird, 58f, 63
Keel fringes, 39, 39f
Keratin, 29
Keratinization, 29, 30f, 33, 36

Keratinized layer of epidermis, 29
Keratinocytes, 36
Kidney
cat, 144f, 169f, 179, 181f, 182, 183f
Necturus, 137f, 138, 158f, 177, 178f, 179, 180f
opisthonephric, 26, 177
shark, 134f, 135f, 151f, 152f, 174f, 175, 176f, 177
Knotting behavior, 20f

L

Labia, cat, 138, 140f
Labial cartilage, shark, 69f, 70
Labial frenulum, cat, 139
Lacrimal, 82
alligator, 72, 73f
bird, 78f, 79
bowfin (*Amia*), 71, 71f
Lacrimal canal, alligator, 72, 73f
Lamellar bone, 32, 32f
Lamina, cat, 51
Lamina quadrigemina, sheep brain, 200f, 201
Lamina terminalis, sheep, 200f, 201
Lampreys (petromyzoniformes), 18, 22–27, 23f–25f
circulatory system, adult, 23f, 24, 25f
digestive system, adult, 22, 24, 25f, 26f
external adult anatomy, 22, 23f
larva, anatomy of, 27–28
muscular system, adult, 21f, 24, 25f, 26
nervous system and special senses, 23f, 25f, 26–27
reproductive and urinary systems, adult, 23f, 25f, 26
respiratory system, adult, 22, 23f–25f
sagittal and cross sections, 22–27
sea squirts, 15, 16f
skeletal system, adult, 21f, 23f, 25f, 26
skin, 31
tail, 22, 23f
vertebrae and ribs, 44
Lancelot. See Amphioxus
Large intestine
cat, 143
Necturus, 136, 137f, 158f
Larva
ammocoete, 17, 27–27
ascidian, (sea squirt), 15, 16f
Necturus, 99
Laryngeal artery, cat, 164f, 166
Laryngopharynx, cat, 140f, 141, 167
Larynx, cat, 113, 141, 167
Lateral abdominal vein, shark, 147f, 153f, 154
Lateral branch of vagus nerve, shark, 194
Lateral condyles, cat, 60, 61f

Lateral epicondyles, cat, 60, 61f, 64f, 66
Lateral fringes, 39, 39f
Lateral geniculate body, sheep brain, 202
Lateral head muscles, *Necturus,* 100t, 102
Lateral head of triceps brachii, cat, 114f, 115f, 119, 120f
Lateral ligaments, cat bladder, 145
Lateral line nerves, 189
Lateral line system
cat, 142f, 143, 144f
shark, 90, 90f
Lateral longitudinal bundle, shark, 92, 93f, 97f
Lateral malleolus, cat, 60, 61f
Lateral muscles, *Necturus,* 100t, sheep brain
Lateral olfactory stria, 197
Lateral pectoral muscles, *Necturus,* 100t, 102
Lateral rectus muscle, shark, 97f, 98, 98f, 190f
Lateral shoulder musculature, cat, 111, 114–16, 118
Lateral vein, *Necturus,* 157f
Lateral ventricle
adult brain, 195f
shark, 191f
sheep, 200f
Lateral visceral branches of dorsal aorta, cat, 168, 169–70
Laterosphenoid, alligator, 73f, 75
Latissimus dorsi
cat, 109f, 110f, 111, 114, 114f, 115f
Necturus, 102, 103f
Left adrenolumbar vein, cat, 165f
Left atrium
amphibian, 156f
cat, 144f, 161f, 162f
mammal, 163f
Necturus, 137f, 155f, 157f, 158f
Left axillary artery, cat, 161f, 162f, cat
Left brachiocephalic vein, cat, 167
Left common carotid artery, cat, 162f, 163
Left costocervical artery, cat, 161f, 162f
Left costocervical vein, cat, 161f
Left gastric artery, cat, 164f, 168
Left innominate vein, cat, 167
Left internal spermatic/ovarian vein, cat, 165f
Left internal spermatic vein, cat, 170
Left lateral lobe, cat liver, 143
Left lobe of liver, shark, 131, 134f, 135f, 147f
Left lung, *Necturus,* 137f, 158f
Left medial lobe, cat liver, 143
Left ovarian vein, cat, 170
Left pleural cavity, cat, 141
Left pulmonary artery, cat, 163, 167
Left pulmonary vein, cat, 167

Left subclavian artery, cat, 163, 164f, 166
Left thyrocervical artery, cat, 161f, 162f
Left ventricle, cat, 161f, 162f
Left vertebral artery, cat, 161f, 162f
Left vertebral vein, cat, 161f
Lesser curvature of stomach, 143
cat, 144f
shark, 131, 134f
Lesser multangular, cat, 64f, 66
Lesser omentum, cat, 145
Lesser trochanter, cat, 60, 61f
Lesser tuberosity
bird, 58f, 63
cat, 64f, 66
Levatores arcuum, *Necturus,* 102, 103f
Levator hyomandibulae, shark, 93, 93f, 97f
Levator mandibulae anterior, *Necturus,* 102
Levator mandibulae externus, *Necturus,* 102, 103f
Levator palatoquadrati, shark, 93, 93f, 97f
Levator scapulae ventralis, 110f, 114f, 115f, 116
Levator series, shark musculature, 92t, 93, 93f, 97f
Leydig cell, 33, 33f
Leydig's gland, shark, 174f, 175
Lienogastric artery, shark, 150, 151f–153f
Lienomesenteric vein, shark, 151f, 152f, 154, 154f
Ligaments, 40. See also specific names
Ligamentum arteriosum, cat, 163
Limbs. See also Forelimbs; Hindlimbs
basic components of, 53, 53f
pectoral girdle, sternum, and, 62–66
pelvic girdle and posterior, 53–62
Linea alba
cat, 105
Necturus, 99f, 101, 102
shark, 92, 96f
Linea aspera, cat, 60, 61f
Lingual artery, cat, 164f, 166
Lingual cartilage, lamprey, 23f, 26
Lingual frenulum, cat, 139, 140f
Lingual muscles, lamprey, 26
Lips, *Necturus,* 99
Liver
cat, 143
lamprey, 22, 23f, 24, 25f, 27, 28f
lobes of, shark, 131
Necturus, 136, 137f, 158f
shark, 131b
Long head of triceps brachii, cat, 110f, 114f–116f, 119, 120f
Long thoracic artery, cat, 161f, 162f, 163, 164f
Long thoracic vein, cat, 161f, 165f, 167

Lumbar, alligator, 47
Lumbar arteries, cat, 168
Lumbar region
    alligator, 48
    cat, 50, 52f
Lumbar spinal nerves, sheep, 203
Lumbar vein, cat, 165f, 170
Lumbar vertebrae
    alligator, 48, 48f
    bird, 49, 50f
    cat, 51, 52f
Lumbodorsal fascia, cat, 105, 114, 115f, 124
Lumbosacral plexus, sheep, 203
Lumbricales, cat, 124
Lungfish, digestive tracts, 133f
Lungs
    cat, 141, 142f
    Necturus, 136, 137f, 156–57
Lymph node, cat, 109f, 112f, 114f, 145

M

Main pancreatic duct, cat, 145
Male urogenital system
    cat, 179–82
    ducts in, 177b
    human, 182f
    Necturus, 177, 178f
    shark, 172–75
Malleoli, bird, 58f, 59
Malleus bone, origins of, 82b
Mammals. See also specific mammal
    cranial skeleton, 79, 80–81f, 82–83
    heart, 163f
    integument, 36–37, 36f
    pectoral girdle, sternum, and appendages, 63, 66
    pelvic girdle and posterior appendages, 59–62, 61f
    reproductive tracts of female placental, 184b
    teeth, 83–84
    vertebrae and ribs, 50–51, 52f
Mammary glands, cat, 105
Mammillary body, sheep, 197, 199f, 200f, 201
Mammillary process, cat, 51, 52f
Mandibles
    cat, 139f
    turtle, 77
Mandibular adductor fossa, alligator, 75
Mandibular arch, 67, 67f, 68f, 70
Mandibular branch of trigeminal nerve, shark, 193
Mandibular foramen
    alligator, 74f, 75
    cat, 81f, 83
Mandibular fossa, cat, 80f, 82
Mandibular nerve, shark, 190f
Mandibular series (bones), 68, 68f, 68t
Mandibular symphysis, cat, 81f, 83
Manubrium, cat, 52f, 66

Massa intermedia
    cat, 140f
    sheep, 201
    sheep brain, 200f, 202f
Masseter, cat, 109f, 112f, 113–14, 114f, 119f, 139f
Mastication, 83
Mastoid process, cat, 79, 80f, 81f
Matrix, cartilage, 40, 41
Maxillae
    alligator, 72, 73f
    bird, 78f, 79
    bowfin (Amia), 71, 71f
    cat, 79, 80f
    turtle, 76f, 77
Maxillary arch, alligator, 74n
Maxillary artery, cat, 164f, 166
Maxillary branch of trigeminal nerve, shark, 193
Maxillary process, bird, 78f
Maxillary process of the premaxilla, 79
Mechanical digestion, 83
Meckelian fenestra, alligator, 75
Meckelian foramen, turtle, 76f, 78
Meckelian groove, turtle, 76f, 78
Meckel's cartilage, 67, 67f–69f, 70, 96f
Medial condyles, cat, 60, 61f
Medial epicondyles, cat, 60, 61f, 64f, 66
Medial geniculate body, sheep brain, 202, 202f
Medial head of triceps brachii, cat, 110f, 116f, 119
Medial lobe of liver, cat, 142f, 144f
Medial malleolus, 197
    cat, 60, 61f
    sheep brain, 199f
Medial olfactory stria, 197
Medial rectus, shark, 98, 98f, 190f
Median ligament of bladder
    cat, 145
    Necturus, 138
Median lobe of liver, shark, 131, 134f, 135f, 151f, 152f
Median nerve, cat, 161f, 162f
Median raphe
    cat, 113
    Necturus, 99f, 101
    shark, 93
Median sulcus, sheep brain, 197, 198f
Median visceral branches of dorsa aorta, cat, 168
Mediastinal septum, cat, 141
Mediastinum, cat, 141–42
Medulla oblongata, 195f
    lamprey, 27
    shark, 98f, 190f, 191, 191f
    sheep brain, 197, 199f, 200f, 201
Medulla of kidney, cat, 181f, 182
Medullary bone, 41
Medullary vellum, sheep brain, 200f, 201
Melanin, 36
Melanophores, 36

Meninges
    primitive, shark, 192
    sheep brain, 196, 196f
Meningitis, 197b
Mental foramina, cat, 81f, 83
Mesencephalon (midbrain), 194b, 195f
    in fishes and amphibians, 201b
    lamprey, 27
    sheep, 201
Mesenteric arteries, Necturus, 157, 158f, 159f
Mesenteric branch of the celiacomesenteric, Necturus, 157, 158f, 159f
Mesenteric ganglia, 186
Mesenteric vein, Necturus, 158f, 160, 160f
Mesenteries
    cat, 144f, 145, 183–84
    Necturus, 137f, 138
    shark, 132–33
Mesethmoid, 66, 79
Mesocolon, cat, 145
Mesoduodenum, cat, 145
Mesogaster
    Necturus, 138
    shark, 132
Mesometrium, cat, 183f, 184
Mesonephric duct, Necturus, 180f
Mesonephric (wolffian) duct
    Necturus, 177, 178f, 179
    shark, 176f, 177
Mesopterygium, shark, 62
Mesorchium
    Necturus, 138, 177, 178f
    shark, 133, 172–73, 174f
Mesorectum
    Necturus, 138
    shark, 132, 134f
Mesosalpinx, cat, 183–84, 183f
Mesotubarium
    Necturus, 137f, 138
    shark, 133, 175, 176f
Mesovarium
    cat, 183
    Necturus, 137f, 138, 179, 180f
    shark, 133, 175, 176f
Metacarpals
    alligator, 57f, 63
    bird, 58f
    cat, 64f, 66
    Necturus, 55f, 62
    turtle, 56f, 63
Metacromion, cat, 64f, 66
Metamorphosis, 99
    of Necturus larva, 99
    of sea squirts, 16f, 17
Metapleural folds, amphioxus, 15
Metapterygium, shark, 53, 54f, 62, 172f
Metatarsals
    alligator, 55, 57f
    bird, 59
    cat, 60, 61f
    Necturus, 54, 55f
    turtle, 54, 55, 56f
Metencephalon, sheep, 201

Midbrain (mesencephalon), 194b, 201b
    lamprey, 27
    sheep, 201
Middle cerebellar peduncle, sheep brain, 202, 202f
Middle colic artery, cat, 164f, 168
Middle ear, cat, 79
Middle gastroepiploic vein, cat, 166f, 171
Middle hemorrhoidal artery, cat, 164f, 170
Middle hemorrhoidal vein, cat, 165f, 170
Middle lobe, cat lung, 141, 142f, 144f
Midgut cecum, amphioxus, 11, 12f, 13f, 14, 15
Mixed nerves, 186
Molar gland, cat, 138, 139f
Molars, 83f
Molar teeth, cat, 81f
Monoecious organisms, 9
Motor (efferent) nerves, 185, 188, 189
Motor neuron, 187f
Mouth
    fish, 133f
    lamprey, 22, 23f, 27
Mucous cuticle, 29f
Mucous glands, amphibian, 33, 33f
Mudpuppy. See Salamander
Mule deer, dental formula, 85, 85f
Müllerian (wolffian) ducts
    male vertebrates, 175b
    Necturus, 177, 178f
Muscle groups, 89
Muscles
    amphioxus, 12, 12f
    parallel and pinnate orientation of, 129b
    reflecting, 105
    terminology related to, 88t
Muscular artery, cat, 163, 164f
Muscular pump, 20
Muscular systems, 88–130
    cat, 105–30
    cranial musculature homologies, 118t
    general muscle groups, 89
    introduction, 88–89
    lamprey, 21f, 24, 25f, 26
    muscle actions, 88f
    Necturus, 99–105
    shark, 89–99
    terminology, 88, 88t
Myelencephalon, sheep, 201
Mylohyoid, cat, 109f, 112f, 113, 117b
Myoepithelial cells, amphioxus, 13
Myomeres, 5, 45
    amphioxus, 12, 12f, 13f, 14, 15
    lamprey, 22, 23f, 25f, 26, 28, 28f
    shark, 91, 93f
Myosepta, 43
    amphioxus, 14
    lamprey, 24, 25f, 26
    shark, 91, 93f

Myotomes
  axial musculature arising from, 91, 91f, 92
  segmented, 27
Myxini (hagfishes), 18, 20f

N

Nail matrix, 37, 37f
Nails, 37, 37f
Nares, shark, 68, 69f, 90
Nasal aperture, bird, 78f, 79
Nasal bones, bird, 79
Nasal canal, lamprey, 27
Nasal capsule, 66, 68, 69f, 70, 93f, 96f, 99
Nasal cavity, cat, 83, 141
Nasal opening, fish, 133f
Nasal process, bird, 79
Nasal(s), 79, 81f
  alligator, 72, 73f
  bird, 78f, 79
  bowfin (Amia), 71, 71f
Nasal sac
  lamprey, 23f, 25f, 27
  shark, 90f
Nasohypophyseal duct, lamprey, 23f, 27
Nasohypophyseal opening, lamprey, 22, 23f, 27
Nasolacrimal canal, 80f, 81f
Nasopalatine ducts, cat, 141
Nasopharynx, cat, 140f, 141, 167
Neck
  of femur, cat, 59, 61f
  of humerus, cat, 64f
  of rib, cat, 51, 52f
Neck muscles, cat, 11–13
Nerve cord, 43
  acorn worm, 11
  amphioxus, 12f, 15
  dorsal and tubular, as chordate characteristic, 4–5, 5f
Nerve network, 11
Nerve network, acorn worms, 11
Nerves
  cranial, 186–89
  function of, 185
  somatic and visceral, 185, 186f
  spinal, 185–86, 187f
Nervous system
  acorn worms, 11
  amphioxus, 12, 12f
  cranial nerves, 186–89
  introduction, 185–89
  lamprey, 23f, 25f, 26–27
  nerve function in, 185
  shark, 189–96
  sheep brain, 196–203
  spinal nerves, 185–86, 187f
Nervus terminalis, shark, 191f, 192
Neural arch, 43, 43f
  alligator, 47, 48, 48f
  cat, 51, 52f
  shark, 44, 45f
  teleost fish, 45, 46f
  turtle, 49

Neural canal
  cat, 52f
  shark, 43f, 44, 44f, 45f
Neural spine, 43, 43f, 48, 48f
  cat, 51, 52f
  frog, 46, 47f
  Necturus, 46, 47f
  shark, 44, 45f
  sturgeon, 45, 46f
  teleost fish, 46f
  turtle, 49
Neurocranium, 66
Notochord
  amphioxus, 12, 13f, 14, 14f, 15
  as chordate characteristic, 4, 5f
  lamprey, 18, 21f, 23f, 24, 25f, 26–28, 28f
  sea squirts, 16f, 17
  shark, 69f, 70
  teleost fish, 45, 46f
  vertebral column and, 43
Notochordal canal, shark, 44–45
Nuchal line, cat, 79, 81f
Nucleus pulposus, cat, 50–51

O

Obturator foramen
  alligator, 55
  Necturus, 53, 55f
  turtle, 54, 56f
Occipital arch, 67f
Occipital artery, cat, 164f, 166
Occipital bone, cat, 79, 80f, 81f
Occipital condyle
  alligator, 74f, 75
  bird, 78f, 79
  cat, 79, 80f, 81f
  shark, 69f, 70
  turtle, 76f, 77
Occipital nerves, 186
  shark, 98f, 190f, 194
Occipitals, 66, 67f
Occipitospinal nerves, 186
Occlusal surface (tooth), 83
Ocellus, sea squirts, 17
Oculomotor nerve (III), 98f, 188
  shark, 189, 190f, 192
  sheep brain, 197, 198, 199f
Odontoid process
  alligator, 47–48, 48f
  bird, 49
  cat, 51, 52f
Olecranon fossa, cat, 64f, 66
Olecranon process
  bird, 58f, 63
  cat, 64f, 66
Olfactory bulb, 188
  shark, 190f, 191, 191f, 192
  sheep brain, 197, 199f
Olfactory capsules, lamprey, 26
Olfactory fiber, 188
Olfactory fossa, cat, 81f, 83
Olfactory nerve (I), 188
  shark, 192
  sheep brain, 198
Olfactory sac
  lamprey, 27
  shark, 90, 190f, 192

Olfactory tract
  shark, 98f, 190f, 191, 191f, 192
  sheep brain, 197, 199f
Omental bursa, cat, 145
Omoarcual, Necturus, 99f, 101
Opercular, bowfin (Amia), 71, 71f
Ophthalmic nerve, 189
Opisthonephric kidneys
  lamprey, 26
  Necturus, 177
Opisthonephros, shark, 175
Opisthotic
  Necturus, 72, 72f
  turtle, 76f, 77
Optic capsule, 66, 67f
Optic chiasm
  shark, 191f, 194
  sheep, 197, 198, 199f, 200f, 201
Optic foramen
  cat, 82
  shark, 69f, 70
Optic labe, shark, 190f
Optic lobes, shark, 191
Optic nerve (II), 188
  shark, 98f, 99, 190f, 192
  sheep brain, 197, 198, 199f
Optic pedicel, shark, 69f, 70, 99
Oral cavity
  cat, 138–41, 139, 140f, 141
  lamprey, 27
Oral hood, amphioxus, 11, 12f, 14
Oral tentacles, lamprey, 27, 28f
Orbit
  bird, 79
  cat, 79, 80f
  shark, 70
  turtle, 75, 76f
Orbital fissure, cat, 82
Orbital process
  cat, 79, 80f, 81f
  shark, 69f, 70
Orbital series, 68t
Orbitosphenoid, 78f, 79
Oropharynx, cat, 140f, 141
Osteichthyes
  cranial skeleton, 71, 71f
  pectoral girdle, sternum, and appendages, 62
  vertebrae and ribs, 45
Osteoblasts, 40
Osteoclasts, 40
Osteocyte, 40
Osteon, 41, 42f
Ostium
  cat, 183
  Necturus, 179, 180f
  shark, 133, 175, 176f
Ostracoderms, 18, 19f, 31–32
  vertebrae and ribs, 44
Otic capsule, 66, 67f
  lamprey, 21f, 26
  Necturus, 72
  shark, 69f, 70, 90, 190f
  turtle, 77
Otic notch, turtle, 77
Otolith, sea squirts, 16f, 17
Ova, Necturus, 179, 180f

Ovarian arteries
  cat, 164f, 169–70, 169f
  Necturus, 158f, 179, 180f
Ovarian ligament, cat, 183
Ovarian veins
  cat, 164f, 169f
  Necturus, 158f, 179, 180f
Ovaries
  cat, 169f, 183–84
  lamprey, 26
  Necturus, 137f, 138, 158f, 179, 180f
  shark, 132, 134f, 175, 176f
Oviduct
  cat, 183, 183f
  Necturus, 137f, 158f, 177, 178f, 179, 180f
  shark, 175, 176f
Oviparous species, 172

P

Paedomorphosis, 99
Paired dorsal aorta, shark, 148f, 149, 149f, 153f
Palatal rugae, cat, 138
Palatal series (bones), 67f, 68, 68f, 68t
Palatine, 78f
  bird, 79
  bowfin (Amia), 71
  cat, 79, 80f
  turtle, 76f, 77
Palatine branch (VII), shark, 190f, 193
Palatine process
  alligator, 73f, 74
  cat, 79, 80f, 82
Palatine processes of palatines
  alligator, 73f, 74
  cat, 82
Palatine processes of premaxillae
  alligator, 73f, 74
  cat, 80f
Palatine processes of pterygoids, alligator, 73f, 74
Palatine process of maxilla, 82
  alligator, 73f
  cat, 80f
Palatine process of premaxillae, 82
Palatine tonsils, cat, 140f, 141
Palatoglossal arch, cat, 141
Palatopharyngeal arch, cat, 141
Palatoquadrates shark, 69f, 70
Palmate, 38
Palpebral, alligator, 72, 73f, 74
Pancreas, 143
  bird, 132f
  cat, 143b
  lamprey, 28
  Necturus, 136, 137f, 158f
  shark, 131
Pancreaticoduodenal artery, Necturus, 157, 158f, 159f
Pancreaticoduodenal vein, Necturus, 159f, 160, 160f
Pancreaticomesenteric artery, shark, 150, 151f, 152f, 153

Pancreaticomesenteric vein, shark, 152f, 154, 154f
Pancreatic veins
    cat, 166f, 171
    shark, 154
Papillae
    lamprey, 22, 23f
    Necturus, 179
    shark, 131, 134f
Papillary layer, 36f, 37
Parachordal, 66, 67f
Paraglossal, 78f
Paraphysis, shark, 191f
Parapophysis, 43f
    alligator, 48, 48f
    bird, 49
    as common type
        of apophyses, 44t
    Necturus, 46, 47f
Parasphenoid
    bird, 78f, 79
    Necturus, 71, 72f
Parietal artery, *Necturus,* 157
Parietal musculature, 89
Parietal pericardium
    cat, 142
    Necturus, 138
    shark, 133
Parietal peritoneum
    cat, 143
    Necturus, 138
    shark, 132
Parietal pleura, cat, 141
Parietals
    alligator, 72, 73f
    bird, 78f, 79
    bowfin *(Amia),* 71, 71f
    cat, 79, 80f
    Necturus, 71, 72f
    turtle, 76f, 77
Parietal vein, *Necturus,* 158f, 159f, 160
Parotid duct, cat, 138, 139f
Parotid gland, cat, 112, 112f–114f, 138, 139f
Patella
    bird, 58f, 59
    cat, 60, 61f
Patellar surface, cat, 60, 61f
Patterns and connections
    brain regions, 194b
    buccal structures in human mouth, 139b
    cerebral cortex, folding of, 197b
    ducts in urogenital systems, 175b
    elongation of snout in alligators, bone remodeling, 75b
    gill and lungs in *Necturus,* 157b
    hamstring muscles, 128b
    iliac, pubic, and ischiac bones of pelvic girdle, 53b
    incus, malleus, and stapes bones, origins of, 82b
    inguinal canals in male humans, 179b

lamprey, current methods to control, 21b
larval and adult protochordates, 15b
meaning of "branchial" and "brachial," 149b
meningitis (brain inflammation), 197b
number of centra per body segment, 46b
pectoral girdle evolution, 65b
pelvic structure of dinosaurs, 59b
quadriceps muscles, 127b
relationships of duodenum, gallbladder, and pancreas, 143b
similarities in vertebrate skulls, 82b
skull and size of brain, 77b
Pectineal process, bird, 60f
Pectineus, cat, 127, 127f, 128f
Pectoantebrachialis, cat, 109f, 110f, 111, 112f, 123f
Pectoral fins, shark, 89
Pectoral girdle, 42, 42t, 62–66
    evolution of, 65f
    musculature of, *Necturus,* 101–2
Pectoralis, *Necturus,* 101
Pectoralis major, cat, 109f, 110f, 111, 112f, 123f
Pectoralis minor, cat, 109f, 110f, 111
Pectoralis muscle
    cat, 111
    Necturus, 99f
Pectoral ridge, 64f, 66
Pectoriscapularis, *Necturus,* 102, 103f
Pedicle, cat, 51
Pelagic organisms, 8
Pelvic fins, sharks, 89–90, 94f, 172
Pelvic girdle, 53–62
    embryonic development of bird, 59–60, 60f
    iliac, pubic, and ischiac bones of, 53b
    musculature, *Necturus,* 100–101t, 102, 104f, 105
    shark, 53, 54f, 172f
Pelvic vein, *Necturus,* 158f, 159f, 160, 160f
Pelvis of ureter, cat, 182
Penis, cat, 105, 179, 180f
Perch
    digestive tracts, 62, 133f, and appendages, sternum vertebrae and ribs, 45–46, 47f
Pericardial cavity
    cat, 141, 142
    lamprey, 23f, 24
    Necturus, 138
    shark, 133, 136f, 146
Perichondrium, 40
Perilymphatic foramina, shark, 69f, 70
Periosteum, 40, 41, 42f
Peripheral nervous system, 185, 186f

Peritoneal cavity, cat, 143
Permanent dentition, 84, 84f
Peroneus muscles, cat, 125f, 126f, 129
Petromastoid, cat, 83
Petromyzoniformes. *See* Lampreys (Petromyzoniformes)
Petrosal ganglion, shark, 190f, 193
Petrous, cat, 83
Petrous part of petromastoid, 81f
Phalangeal series, bird, 59
Phalanges (phalanx)
    alligator, 55, 57f, 63
    bird, 58f
    cat, 61f, 62, 64f, 66
    Necturus, 54, 55f, 62
    turtle, 54, 55, 56f, 63
Pharyngeal bar, amphioxus, 15
Pharyngeal branch of vagus nerve, shark, 193
Pharyngeal region, cat, 139–41, 140f
Pharyngeal slits, 67f
    acorn worms, 9f, 11
    amphioxus, 11, 14–15
    as chordate characteristics, 3–4, 5f
    lamprey, 22
    sea squirts, 16f, 17
Pharyngobranchial, 67f, 69f, 70
Pharyngobranchial cartilage, shark, 97f
Pharynx
    amphioxus, 11, 12f, 13, 13f, 14, 14f, 15
    cat, 138
    lamprey, 20, 22, 23f, 27, 28, 28f
    shift to muscular pump, 20
Phrenic arteries, cat, 168
Phrenic nerve, cat, 161f, 162f
Pia mater layer of meninges, 196f, 197
Pigeon, vertebrae and ribs, 49–50, 50f
Pigment granules, bird integument, 34
Pineal body
    shark, 194
    sheep, 200f, 201, 202f
Pineal gland
    lamprey, 22, 23f, 25f, 27
    shark, 68
Pineal organ, 201b
Pinnae, cat, 105, 113
Pisiform
    cat, 64f, 66
    turtle, 56f
Piston cartilage, lamprey, 21f, 26
Pituitary gland
    amphioxus, 11
    shark, 197
    sheep, 200f, 201
Placoderms
    axial skeleton of, 44f
    integument, 31
    lamprey, 21f
    vertebrae and ribs, 44
Placoid scales, 31, 31f

Planktonic organisms, 9
Plantaris, cat, 127f, 128f, 129
Plastron, turtle, 33, 63
Platysma, cat, 105
Pleural cavities, cat, 141
Pleurocentrum, bowfin *(Amia),* 45
Pleuroperitoneal cavity, 131
    lamprey, 22, 24
    Necturus, 136–38, 137f, 138, 158f
    shark, 131–33, 134f, 150, 154
Plexus, sheep, 203
Plicae, *Necturus,* 136, 137f
Plies, dermal, 29
Pneumatic foramen, bird, 58f, 63
Points, horn, 38
Poison glands, amphibian, 33, 33f
Polar cartilage, 67f
Polyphyodont teeth, 83, 84, 84f
Polysaccharides, cartilage, 40
Pons, sheep brain, 197, 199f, 200f, 201
Popliteal notch, cat, 60, 61f
Positioning role of stretch receptors and vestibular apparatus for body and head, 90b
Postanal tail, as chordate characteristic, 4, 5, 5f
Postcava
    cat, 162, 165f, 169f, 170, 171f
    mammal, 163f
Postcaval vein, *Necturus,* 137f, 138, 155f, 156, 158f, 159f, 160
Postcleithrum, bowfin *(Amia),* 62, 62f, 71f
Postcranial skeleton, 42
    girdles and limbs, 53–66
    vertebrae and ribs, 43–52
Posterior auriclular artery, cat, 164f, 166
Posterior cardinal sinus, shark, 153f, 154
Posterior cardinal vein, 28f
    lamprey, 24, 25f
    Necturus, 157f–159f, 160
    shark, 151f, 153f, 154
Posterior cerebellar peduncle, sheep brain, 202, 202f
Posterior choroid plexus, shark, 191f
Posterior commissure, sheep, 200f, 201
Posterior facial vein, cat, 114f, 138, 139f, 165f, 168
Posterior horns, of hyoid apparatus, 83
Posterior humeral circumflex artery, cat, 161f, 162f, 164f, 167
Posterior humeral circumflex vein, cat, 165f, 167
Posterior inferior spine, cat, 59, 61f
Posterior intestinal vein, shark, 151f, 152f, 154, 154f
Posterior lienogastric vein, shark, 151f, 152f, 154, 154f
Posterior lobe, cat lung, 141, 142f, 144f

Posterior mesenteric artery, shark, 135f, 150, 151f–153f
Posterior nostrils, bowfin (Amia), 71, 71f
Posterior palatine canal, cat, 82
Posterior palatine foramina, cat, 80f, 82
Posterior palatine vacuities
    alligator, 73f, 74
    turtle, 76f, 77
Posterior pancreaticoduodenal artery, cat, 164f, 168, 168b
Posterior pancreaticoduodenal vein, cat, 166f, 171
Posterior part, cat liver, 143
Posterior splenic artery, cat, 164f, 168
Posterior splenic vein, cat, 166f, 171
Posterior xiphisternal process, bird, 58f, 63
Postorbital bar, alligator, 72, 73f
Postorbital process
    cat, 79, 80f
    shark, 68, 69f, 70
Postorbitals
    alligator, 72, 73f
    bowfin (Amia), 71, 71f
    turtle, 76f, 77
Postparietals, bowfin (Amia), 71, 71f
Posttemporal, bowfin (Amia), 62, 62f, 71f
Posttrematic branch
    of branchial arteries, shark, 148f, 149, 149f
    of glossopharyngeal nerve, shark, 193
    of vagus nerve, shark, 193
Postzygapophyses
    alligator, 48
    bird, 50f
    cat, 51
    frog, 46
    Necturus, 46
Prearticular, turtle, 76f, 78
Precardial cartilage, lamprey, 21f, 26
Precava
    cat, 161f, 162, 162f, 165f, 167–168
    mammal, 163f
Precerebral cavity, shark, 68, 69f
Precerebral fenestra, shark, 68
Prefrontal
    alligator, 72, 73f
    turtle, 76f, 77
Premaxillae
    alligator, 72, 73f
    bird, 78f, 79
    bowfin (Amia), 71, 71f
    cat, 79, 80f
    Necturus, 71, 72f
    turtle, 75, 76f, 77
Premolars, 84f, 85
Preopercular, bowfin (Amia), 71, 71f
Preorbitalis, shark, 93, 93f

Prepubic (pectineal) process, turtle, 54, 56f
Presphenoid, cat, 80f, 81f, 82
Pretrematic branch
    of branchial arteries, shark, 148f, 149, 149f
    of glossopharyngeal nerve, shark, 193
    of vagus nerve, shark, 193
Prezygapophyses
    alligator, 48
    bird, 50f
    cat, 51
    frog, 46
    Necturus, 46
Primitve meninx (meninges), shark, 192
Proatlas, alligator, 47, 48f
Proboscis, acorn worms, 9f, 10–11, 10f
Procoelous centra, alligator, 48, 48f
Procoracohumeralis, Neeturus, 99f, 101, 103f
Procoracoid
    alligator, 63
    bird, 63
    turtle, 56f, 63
Procoracoid cartilage, Necturus, 55f, 62
Procoracoid process, bird, 58f, 63
Pronator teres, cat, 123, 123f
Pronephric kidney, lamprey, 27, 28f
Pronghorn, horns of, 38, 39f
Prootic
    Necturus, 72, 72f
    turtle, 76f, 77
Propterygium, shark, 53, 54f, 62
Prosencephalon (forebrain), 194b
    lamprey, 27
    sheep, 201
Prostate gland, cat, 179, 181f
Protein fibers, cartilage, 40
Protochordates, 8–17
    cephalochordates, 11–15
    hemichordates, 9–11
    shift in feeding mechanisms in, 18–21
    urochordates, 15–17
Protostomes, 2, 3f, 4f, 4t
Proventriculus, bird, 132f
Pterygiophores, shark, 53, 62
Pterygoid
    alligator, 73f
    bird, 78f, 79
    cat, 82
    Necturus, 71, 72f
    turtle, 76f, 77
Pterygoid process, cat, 80f, 81f, 82
Pterylae, 34f, 35
Pubic cartilage, Necturus, 53, 55f
Pubic symphysis
    alligator, 55
    cat, 59, 61f
    turtle, 54, 56f
Pubis
    alligator, 55, 57f
    bird, 56–57

cat, 59, 60f, 61f
    turtle, 54, 56f
Pubofemoralis, meaning of term, 88
Puboischiac bar, shark, 53, 54f
Puboischiac plate, Necturus, 53
Pubo-ischio-femoralis externus, Necturus, 102, 104f
Pubo-ischio-femoralis internus, Necturus, 104f, 105
Pubo-ischio-tibialis, Necturus, 102, 104f
Pubotibialis, Necturus, 102, 104f
Pulmonary artery
    cat, 167
    Necturus, 155f, 156, 157f, 159f
Pulmonary ligament, Necturus, 137f, 138
Pulmonary trunk, cat, 161f, 162f, 163, 167
Pulmonary vein
    amphibian, 156f
    cat, 167
    Necturus, 158f, 160
Pulp cavity, 83f
Pulp (tooth), 83
Pulvinar nucleus, sheep brain, 202
Pygostyle, bird, 50, 50f
Pyloric artery, cat, 168, 168b
Pyloric cecum, fish, 133f
Pyloric region, cat stomach, 143, 144f
Pyloric sphincter
    cat, 143, 144f
    Necturus, 136, 137f
    shark, 131
Pyloric vein, shark, 154, 154f
Pylorus, shark, 131, 134f
Pyriform lobe, 197, 199f

Q
Quadrate, 67, 68f, 73f, 74
    bird, 78f, 79
    bowfin (Amia), 71, 71f
    Necturus, 71, 72f
    turtle, 76f, 77
Quadrate bone, Necturus, 72, 72f
Quadrate cartilage, Necturus, 72, 72f
Quadrate lobe, cat liver, 142f, 143, 144f
Quadrate process, shark, 70
Quadratojugal
    alligator, 73f, 74
    bird, 78f, 79
    turtle, 76f, 77
Quadriceps muscles, 127b

R
Rachis, feather, 34f, 35, 35f
Radial body symmetry, 6f
Radiale
    bird, 58f, 63
    turtle, 56f, 63
Radials, shark, 53, 54f, 62, 172f
Radices of aorta, Necturus, 155f, 156, 159f

Radius
    alligator, 57f, 63
    bird, 58f, 63
    cat, 64f, 66
    Necturus, 55f, 62
    turtle, 56f, 63
Radix, cat, 142
Rami, cat, 59, 61f
Ramus, 83
    alligator, 75
    cat, 81f
Ramus communicans, 186, 187f
Raphe, shark, 93, 93f
Rays, Necturus, 54
Receptors, nervous system, 185
Rectal gland
    fish, 133f
    shark, 132, 134f, 135f, 151f, 152f
Rectovesical pouch, cat, 179
Rectum, 143, 144f, 179, 184
Rectus abdominis
    cat, 106, 108, 119f, 127f, 128f
    Necturus, 99f, 101, 102, 104f
Rectus cervicis, Necturus, 99f, 101, 103f
Rectus femoris, cat, 125, 126, 128f
Renal arteries
    cat, 169, 169f
    Necturus, 157, 158f, 159f, 160, 178f, 180f
Renal papilla, cat, 182
Renal portal vein, shark, 153f
Renal sinus, cat, 181f, 182
Renal vein
    cat, 165f, 169f, 170
    Necturus, 158f, 159f
Reproductive system. See Urogenital system
Reptiles
    cranial skeleton, 72–78
    girdles and limbs, 54–55, 56f, 57f, 62–66
    integument, 33–34, 33f
    teeth, 84
    vertebrae and ribs, 47–49, 48f, 49f
Respiratory system
    cat, 166–67
    circulatory system and, 146, 148
    cutaneous respiration and, 33, 156b
    lamprey, 21f, 22, 23f
    Necturus, 154–60
    shark, 90, 93f, 146–54
Restiform, sheep brain, 202–3
Reticular layer, mammalian dermis, 36f, 37
Retinaculum (retinacula) cat, 120, 120f, 123f, 125f, 126f
Retroarticular process, alligator, 74f, 75
Rhabdopleura, 10f
Rhinal fissure, sheep brain, 197, 199f
Rhinoceros, horns of, 38, 39f
Rhombencephalon (hindbrain), 194b
    lamprey, 27
    sheep, 201

Rhomboideus, cat, 118, 119f
Rhomboideus capitis, cat, 116f, 118, 119f
Rib facets, bird, 63
Rib muscles, cat, 106, 106t, 108, 111
Ribs. See also Vertebrae and ribs
    bicipital, 46, 48, 49, 51, 52f
    fish and amniotes, 43, 43f
Ribs of trunk vertebrae, turtle, 49
Right adrenolumbar veain, 165f
Right atrium
    amphibian, 156f
    cat, 144f, 161f, 162f
    mammal, 163f
    Necturus, 137f, 154, 155f, 157f, 158f
Right axillary artery, cat, 161f, 162f
Right brachiocephalic vein, cat, 167
Right common carotid artery, cat, 161f, 162f, 163
Right costocervical artery, cat, 161f, 162f
Right costocervical vein, cat, 161f
Right gastric artery, cat, 164f, 168
Right gastroepiploic vein, cat, 166f
Right innominate vein, cat, 167
Right internal spermatic/ovarian vein, cat, 165f
Right internal spermatic vein, cat, 170
Right lateral lobe, cat liver, 143
Right lobe of liver, shark, 131, 134f, 135f
Right lung, Necturus, 137f, 158f
Right medial lobe, cat liver, 143
Right ovarian vein, cat, 170
Right phrenic vein, cat, 165f
Right pleural cavity, cat, 141
Right pulmonary artery, cat, 163, 167
Right pulmonary vein, cat, 167
Right subclavian artery, cat, 163, 164f, 166
Right thyrocervical artery, cat, 161f, 162f
Right ventricle, cat, 161f, 162f
Right vertebral artery, cat, 161f, 162f
Right vertebral vein, cat, 161f
Roof of mouth, Necturus, 155f
Roof of skull, 79
Root canal (tooth), 83, 83f
Root of lung, cat, 142
Roots of aorta, Necturus, 155f, 156, 159f
Root (tooth), 83, 83f
Rostral, bowfin (Amia), 71, 71f
Rostral carina, shark, 69f, 70
Rostral fenestrae, shark, 69f, 70
Rostrum
    alligator, 73f, 75
    bird, 63
    shark, 68, 69f
Round ligament, cat, 143
Rugae
    cat, 143
    Necturus, 136, 137f
    shark, 131, 134f

S
Saccus vasculosus, shark, 191f
Sacral region
    alligator, 47, 48
    cat, 50, 52f
    frog, 46–47
    Necturus, 46
    turtle, 49
Sacral spinal nerves, sheep, 203
Sacral vertebrae, bird, 49, 50f
Sacrum
    cat, 51, 52f
    turtle, 49
Safety precautions, dissection and, 89
Salamander
    classification of, 2t
    cutaneous respiration in, 156b
    muscle homologies with other vertebrates, 117b
    as representative form, 1
    teeth, 84, 84f
Salivary glands, cat, 109f, 138
Sartorius, cat, 124, 125, 125f–128f
Scalenus, cat, 110f, 111
Scalenus anterior, cat, 111, 119f
Scalenus medius, cat, 111, 119f
Scalenus posterior, cat, 111, 119f
Scapholunar, cat, 64f, 66
Scapula
    alligator, 57f, 63
    bird, 58f, 63
    cat, 63, 64f
    Necturus, 55f, 62
    turtle, 56f, 63
Scapular cartilage, shark, 54f, 62
Scapular process, shark, 54f, 62, 93f, 97f
Scapulocoracoid, bowfin (Amia), 62, 62f
Sciatic nerve, cat, 126f
Sclerotic ring, bird, 79
Scrotum, cat, 105, 179, 181
Sea squirts (tunicates), 3, 15–17
    adult, 16f, 17
    larva, 15, 16f, 17
    metamorphosis of, 16f, 17
Sebaceous glands, 37
Secondary palate, alligator, 74
2nd afferent branchial arteries, shark, 148, 148f
Second dorsal gill constrictor, shark, 93, 93f
Secretory cells, 31
Segmented myotomes, lamprey, 27
Sella turcica, cat, 81f, 83
Semicircular canals, shark, 90, 190f
Semilunar notch, cat, 64f, 66
Semilunar valves, lamprey, 24
Semimembranosus, cat, 126f, 128, 128f
Seminal plasma, 172
Seminal vesicle, shark, 173, 174f
Semitendinosus, cat, 125f–127f, 128, 128f
Sensory (afferent) nerves, 185, 186f, 187f
Sensory fibers, 188f, 189

Sensory neurons, 187f
Sensory systems, shark, 90–91
Septum pellucidum, sheep, 200f, 201
Serratus ventralis, cat, 110f, 111, 116f, 119f
Sesamoid bones, 41
Sessile organisms, 9
Sex, determining of, in cats, 105
Shaft, rib, 51, 52f
Shank muscles, cat, 108t, 128–29
Shark
    axial skeleton, 45f
    brain, 189, 190–92, 191f, 192f, 194
    circulatory and respiratory system, 146–54
    classification of, 2t
    cranial nerves, 187f, 190f, 192–94
    cranial skeleton, 68–70
    digestive system, 131–33, 134f, 135f
    dissection of, 89–99
    external anatomy, 89–91
    heart, 146f, 147f
    musculature, 89–99, 117b
    pectoral girdle, sternum, and appendages, 62
    pelvic girdle and posterior appendages, 53, 54f
    rectal gland, 132, 134f, 135f
    as representative form, 1
    skin, 31f
    teeth, 84, 84f
    urogenital system, 172–77
    vertebrae and ribs, 43–45, 45f
Sheep brain
    brain stem, 202–3, 202f
    cranial nerves, 197, 199f
    regions, 200–201
    spinal nerves, 203
    whole brian overview, 196–98
Shell gland, shark, 175, 176, 176f
Short velar tentacles, ampbioxus, 14
Shoulder musculature
    cat, 106t–107t, 111
    homologies in vertebrate, 117b
Sinus venosus
    amphioxus, 13
    lamprey, 24
    Necturus, 154, 156, 156f, 157f
    shark, 146, 147f, 149f, 152f, 154
Siphons
    sea squirts, 16f, 17
    shark, 93, 94f, 172, 173f
Skeletal systems, 40–87
    cranial skeleton, 68–83
    divisions of, 42
    embryonic origins, 42t
    lamprey, 21f, 23f, 25f, 26
    postcranial, girdles and limbs, 53–62
    postcranial, vertebrae and ribs, 43–52
    skull, 66–68
    teeth, 83–87
    tissues of, 40–42

Skin. See Integuments
Skinning procedures
    cat, 105
    shark, 91
Skull, 66–68. See also Cranial skeleton
    chondrocranium, 66
    composite, 66–68
    contributions to, 68f
    dermatocranium, 66, 68t
    splanchnocranium, 66
Small intestines
    cat, 142f, 143, 144f
    Necturus, 136, 137f, 158f
Smooth muscle, mammalian integument, 37
Socket (tooth), 83, 84
Soft palate, cat, 138, 140f
Soleus, cat, 125f, 126f, 129
Solitary organisms, 9
Somatic branches of dorsal aorta, cat, 168
Somatic motor column, shark, 191
Somatic musculature, 89
Somatic sensory column, shark, 191
Somatic skeleton, 42
Somatic tissues, 185
Spathe, 35, 35f
"Special," term applied to sensory effectors, 185
Special cranial nerves, 186
Specialized connective tissues, 40, 40f
Spermatic arteries, Necturus, 177
Spermatic cord, cat, 179
Spermatic veins, Necturus, 177
Spermatophores, 172
Sperm sac, shark, 173, 174f
Sphenethnoid, 67f
Sphenoids, 66
Sphenoid sinus, cat, 81f, 83
Sphenopalatine foramen, cat, 82
Spinal accessory nerve (XI), 189
    sheep, 199f, 200
Spinal cord
    cat, 140f
    lamprey, 23f, 25f, 26–28, 28f
    shark, 190f, 191, 194
    sheep, 197, 198f, 199f
Spinal ganglion, shark, 190f
Spinal nerves
    anatomy of, 187f
    dorsal and ventral roots of, 185–86
    sheep brain, 202f
Spinal roots, shark, 194
Spines, 31
    cat, 63, 66
    cat scapula, 116f
    teleost fish, 45, 46f
Spinodeltoid, cat, 114f, 115f, 118
Spinotrapezius, cat, 114f, 115f, 116
Spinotrapezius, shark, 153f
Spiracle, shark, 91, 190f
Spiracularis, shark, 93f, 97f
Spiral valve, 131, 133f
    lamprey, 24, 26f
    shark, 132, 133f, 134f

Splanchnocranium, 66, 68f
  Necturus, 72, 72f
  primitive, 67f
  shark, 68, 69f, 70
Spleen
  cat, 143, 144f
  Necturus, 136, 137f, 158f
  shark, 131, 134f, 135f,
    151f, 152f
Splenial
  alligator, 74f, 75
  turtle, 76f, 78
Splenic artery
  cat, 168
  Necturus, 157, 158f, 159f
Splenium, sheep, 201, 201f
Splenius, cat, 116f, 119, 119f
Spongy bone, 41, 42f
Squalene in shark liver, 131b
Squamosal
  alligator, 72, 73f
  bird, 78f, 79
  Necturus, 72f
  turtle, 76f, 77
Squamous portion of temporal, cat,
    81f, 82
Stapedial artery, shark, 148f, 149,
    149f, 153f
Stapes, 67, 68f, 75
  origins of, 82b
  turtle, 76f, 77
Sternal cartilage, alligator, 48
Sternal section, bird, 49, 50f
Sternebrae, cat, 52f, 66
Sternohyoid
  cat, 109f, 110f, 112–13,
    112f, 113f
  orientation of muscle fibers
    in, 129b
Sternomastoid, cat, 109f, 110f, 112,
    112f, 114f, 117b
Sternothyroid, cat, 110f, 112f,
    113, 113f
Sternum, 62–66
  alligator, 57f, 63
  bird, 58f, 63
Stigmata, sea squirts, 16f, 17
Stolons, 17
Stomach
  cat, 143
  fish, 133f
  Necturus, 136, 137f, 158f
  sea squirts, 16f, 17
  shark, 134f, 147f, 151f
Stomochord, acorn worms, 9f, 11
Stratified epidermal cells, fish,
    32, 32f
Stratum basale, 33f, 34, 35f, 36, 36f
Stratum corneum, 29, 30f, 33f, 34,
    35f, 36, 36f
Stratum granulosum, 36, 36f
Stratum lucidum, 36, 36f
Stratum spinosum, 36, 36f
Study advice, 5–7
Study strategies, 5–7
Sturgeon
  digestive tracts, 133f
  vertebrae and ribs, 45, 46f

Styloid process, cat, 64f, 66
Stylomastoid foramen, cat, 79,
    80f, 81f
Subarachnoid space, 196f
Subarcual, *Necturus*, 99f, 102
Subclavian artery
  cat, 161f, 162f, 163, 166
  Necturus, 155f, 157, 157f, 159f
  shark, 148f, 149f, 150
Subclavian vein
  cat, 161f, 165f, 167
  Necturus, 157f
  shark, 147f, 153f, 154
Sublingual gland, cat, 138, 139f
Sublingual papilla, cat, 139, 140f
Submandibular glands,
    human, 139b
Submaxillary duct, cat, 138
Submaxillary gland, cat, 112, 112f,
    114f, 138, 139f
Subopercular, bowfin (*Amia*),
    71, 71f
Subpharyngeal gland, lamprey, 28f
Subscapular artery, cat, 161f,
    162f, 163
Subscapular fossa, cat, 64f, 66
Subscapularis, cat, 110f, 111
Subscapular nerve, cat, 161f
Subscapular vein
  cat, 161f, 165f, 167
  shark, 147f, 153f, 154
Sulci (sulcus), sheep brain, 197
Superficial constrictor musculature,
    shark, 92t, 93
Superficial fascia, 29, 29f, 105
Superficial ophthalmic foramina,
    shark, 69f, 70
Superficial opthalmic nerve,
    shark, 191f
Superficial opthalmic trunk, shark,
    189, 190f, 192, 193
Superior colliculi, sheep brain, 197,
    200f, 201, 202f
Superior gluteal artery, cat,
    170, 171f
Superior mesenteric artery, cat,
    168, 169f
Superior mesenteric vein, cat,
    166f, 171
Superior oblique, shark, 93f, 97f,
    98, 98f, 190f
Superior ophthalmic, shark, 98–99
Superior phrenic vein, cat, 165f, 170
Superior rectus, shark, 93f, 97f, 98,
    98f, 190f
Superior sagittal sinus, 196f, 197
Superior temporal artery, cat,
    164f, 166
Superior thyroid artery, cat,
    163, 164f
Supinator, cat, 116f, 120f, 121
Supracleithrum, bowfin (*Amia*), 62,
    62f, 71f
Supracondyloid foramen, cat,
    64f, 66
Supracondyloid ridge, cat, 64f, 66
Supracoracoideus, *Necturus*,
    99f, 101

Supramaxillae, bowfin (*Amia*),
    71, 71f
Supraoccipital, 67f, 74f, 75
  bird, 78f, 79
  turtle, 76f, 77
Supraorbital crest, shark, 68,
    69f, 70
Suprascapular cartilage
  alligator, 57f, 63
  Necturus, 55f, 62
Supraspinatus, cat, 116f, 118
Supraspinous fossa, cat, 64f, 66
Supratemporal, bowfin (*Amia*),
    71, 71f
Supratemporal arcade, alligator, 72
Supratemporal fenestrae, alligator,
    72, 73f
Surangular
  alligator, 74f, 75
  bird, 78f, 79
  bowfin (*Amia*), 71, 71f
  turtle, 76f, 77
Suspension feeders, 4, 10
Suspensor, shark, 70
Sweat glands, 36f, 37
Sympathetic chain of ganglia,
    186, 187f
Symphysis
  alligator, 75
  turtle, 77
Synotic tectum, *Necturus*, 72, 72f
Synsacrum, bird, 49–50, 50f, 56
Systemic arch, cat, 163
Systemic veins, *Necturus*, 158f,
    159f, 160

T

Tail
  (caudal) vertebrae, 44
  musculature, *Necturus*, 102
  postanal, as chordate
    characteristic, 4, 5, 5f
  shark, 89
Tail of epididymis, cat, 181f, 182
Tail of pancreas, cat, 143
Talons, 37, 37f
Tarsale, cat, 60, 61f, 62
Tarsalia
  fourth and fifth in cat, 60
  turtle, 55
Tarsals
  alligator, 55
  Necturus, 54, 55f
  turtle, 54, 56f
Tarsometatarsus, bird, 58f, 59
Tectum, shark, 191f
Teeth, 83–87
  amphibian, 84
  anatomy, 83–84
  diet and type, 86b
  evolutionary change pattern,
    85, 87
  form and function, 85, 86b, 87
  lamprey, 21f, 22
  mammal, 85
  mastication, 86b, 87
  molar anatomy, 85f
  Necturus, 99

  occlusion, 87b
  reptile, 84
  shark, 84, 84f
  specializations of, 84f
  structure, 83f
Tegmentum, sheep brain, 200f, 201
Tela choroidea
  shark, 191, 192
  sheep, 200f, 201
Telencephalon, sheep, 201
Teleost fishes
  pectoral girdle, sternum, and
    appendages, 62
  vertebrae and ribs, 45, 46f
Teleost scales, 32, 32f
Temporal bone, cat, 82
Temporal emarginations, turtle,
    76f, 77
Temporal fenestra, cat, 73f, 79, 80f
Temporal fossa, cat, 80f
Temporalis, cat, 114, 114f, 115f,
    117b, 119f
Temporal series (bones), 67f, 68,
    68f, 68t
Tendons, 40, 130b
Tensor fasciae latae, cat, 124, 125f,
    126f, 128f
Tentorium, cat, 81f, 83
Tenuissimus, cat, 124, 126f
Teres major, cat, 110f, 111, 116f
Teres minor, cat, 118
Terminal nerve (O), 188
Terminal nerve (O), shark, 191f
Test, 15
Testes
  cat, 179, 181–82, 181f
  lamprey, 25f, 26
  Necturus, 138, 177, 178f
  shark, 132, 134f, 172, 174f
Tetrapods, integuments, 33–37
Thalamus
  shark, 196
  sheep, 201, 202
Thigh muscles, cat, 108t, 124–28
3rd afferent branchial arteries,
    shark, 148, 148f
Third ventricle
  shark, 191f
  sheep brain, 200f, 201
Thoracic lymphatic duct, cat, 168
Thoracic region
  alligator, 47
  bird, 49, 50f
  cat, 50, 52f, 141–43
Thoracic spinal nerves, sheep, 203
Thoracic vertebrae
  alligator, 48, 48f
  bird, 49, 50f
  cat, 51, 52f
Thoracodorsal artery, cat, 161f,
    162f, 163
Thoracodorsal shunt, cat, 165f
Thoracodorsa vein, cat, 165f, 167
Throat muscles
  cat, 111–13
  Necturus, 100t, 101–2
Thymus, cat, 142, 142f
Thyrocervical artery, cat, 163, 166

Thyrohyoid, cat, 110f, 112f, 113, 113f
Thyroid artery, cat, 161f, 162f, 163
Thyroid cartilage, cat, 112f, 113f, 141, 142f
Thyroid gland, 4
    cat, 113, 141
    Necturus, 155f, 156
    shark, 96f, 98
Thyroid vein, cat, 165f, 168
Tibia
    alligator, 55, 57f
    cat, 60, 61f
    Necturus, 53, 55f
    turtle, 54, 56f
Tibial crest, 60
Tibiale, alligator, 55, 57f
Tibialis anterior, cat, 125f–128f, 129
Tibialis posterior, cat, 127f, 128f, 129
Tibial tuberosity, cat, 60, 61f
Tibiotarsus, bird, 58f, 59
Tines, horn, 38
Tissues, connective, of skeletal system, 40–42
Tongue
    cat, 140f, 141
    fish, 133f
    lamprey, 22, 23f
Tonsillar fossa, cat, 141
Tooth. See Teeth
Tori, cat, 105
Tornaria larva, 9, 9f
Trabeculae, 66, 67f, 72
Trabecular cartilage, Necturus, 72f
Trabecular horns, Necturus, 72, 72f
Trachea
    cartilaginous rings on, 167b
    cat, 161f, 162f
Trachea, cat, 113, 140f, 141, 142, 142f
Transverse colon, cat, 143, 144f
Transverse foramen
    alligator, 48
    bird, 49
    cat, 51, 52f
Transverse jugular vein, cat, 112f, 138, 139f, 165f, 168
Transverse processes
    alligator, 48, 48f
    bird, 49, 50f
    cat, 51
    frog, 46, 47, 47f
    Necturus, 46
    turtle, 49
Transverse scapular artery, cat, 162f, 163, 166
Transverse scapular vein, cat, 165f, 168
Transverse septum
    Necturus, 137f, 138
    shark, 133, 134f
Transversis ventralis, Necturus, 102
Transversus
    cat, 108, 110f
    Necturus, 102

Transversus costarum, cat, 110f, 111, 119f
Trapezoid body, sheep, 200
Triceps brachii
    cat, 110f, 115f, 116f, 119, 120f
    Necturus, 102, 103f
Trigeminal nerve (V), 189
    shark, 192
    sheep, 199, 199f
Trochanteric fossa, cat, 60, 61f
Trochlea, cat, 64f, 66
Trochlear nerve (IV), 98f, 188, 192
    shark, 189, 190f
    sheep, 198–99, 199f
True antlers, 38
True horns, 38
True rib, cat, 51
Trunk
    acorn worms, 9f, 10, 10f
    lamprey, 22, 23f
    Necturus, 99
    shark, 89
Trunk muscles, Necturus, 101t, 102
Trunk region frog, 46
    Necturus, 46
    turtle, 49
Trunk vertebrae, 44, 46
    bullfrog, 47f
    shark, 45f
    turtle, 49, 49f
Tuber cinereum, sheep, 197, 199f
Tuberculum, 43f
    alligator, 48, 48f
    bird, 49
    cat, 51, 52f
    frog, 47f
    Necturus, 46, 47f
Tuberosity, cat, 64f, 66
Tubular nerve cord, 4–5, 5f
    amphioxus, 14
    sea squirts, 16f, 17
Tunic, 15, 16f, 17
Tunica vaginalis, cat, 182
Tunicin, 17
Turbinate bones, 67, 81f, 83
Turkey, pectoral girdle, sternum, and appendages, 63
Turtles
    cranial skeleton, 75, 76f, 77–78
    pectoral girdle, sternum, and appendages, 63
    pelvic girdle and posterior appendages, 54–55, 56f
    shell of, 33
    vertebrae, ribs, skeleton of, 49f
    vertebrae and ribs, 49
Tympanic bullae
    alligator, 74f, 75
    cat, 79, 80f, 81f

U

Ulna
    alligator, 57f
    bird, 58f, 63
    cat, 64f, 66
    Necturus, 55f, 62
    turtle, 56f, 63

Ulnare
    bird, 58f, 63
    cat, 64f, 66
    turtle, 56f, 63
Ulnar nerve, cat, 161f, 162f
Umbilical artery, cat, 170, 171f
Uncinate process, 49, 50f
Unicellular glands, 31, 32, 32f
Upper arm musculature, cat, 119
Ureter, cat, 169f, 179, 181f, 182, 183f
Urethra, cat, 169f, 179, 181f, 182, 183f
Urethral orifice, cat, 184
Urinary bladder
    cat, 142f, 143, 144f, 169f, 179, 181f, 183
    Necturus, 136, 137f, 138, 158f, 177, 178f, 179, 180f
Urinary collecting tubules, Necturus, 177, 177b, 178f, 179, 180f
Urinary papilla, shark, 91, 176f, 177
Urinary system
    cat, 179
    lamprey, 26
Urochordata, 2, 15–17
Urodela, 46, 47f
Urodeum, shark, 174f, 175, 176f, 177
Urogenital aperture, cat, 183f, 184
Urogenital canal, cat, 179, 181f
Urogenital papilla
    lamprey, 22, 23f, 26
    shark, 91, 174f, 175
Urogenital sinus, cat, 105, 183f, 184
Urogenital system
    cat, 179–84
    human, 182f
    lamprey, 23f, 26
    Necturus, 177–79
    salamander, 175f
    shark, 172–77
Urohyal, chicken, 78f
Uropygial gland, birds, 35b
Urostyle
    cat, 51
    frog, 47, 47f
Uterine artery, cat, 170
Uterine horn, cat, 169f, 183, 183f, 184
Uterus
    cat, 169f
    of placental mammals, 184b
    shark, 176f, 177

V

Vagina, cat, 183f, 184
Vagus nerve (X), 189
    cat, 161f, 162f
    shark, 69f, 70, 98f, 190f, 193
    sheep brain, 199f
Vallate papillae, cat, 140f, 141
Valvular intestine, shark, 132, 134f, 135f, 151f, 152f
Vane, feather, 34f, 35, 35f
Vascular bone, 32f

Vascular sacs, shark, 194
Vastus intermedius, cat, 126
Vastus lateralis, cat, 125, 126, 126f
Vastus medialis, cat, 125–128f
Vault series (bones), 67f, 68, 68f, 68t
Veins. See also specific veins
    cat, 165f
    digestive system and, 131
    Necturus, 157f, 158f, 159–60f, 160
    shark, 152–53f, 154
Velar tentacles
    amphioxus, 11, 12f, 14
    lamprey, 22, 23f, 24f
Velum
    amphioxus, 11, 12f, 14
    lamprey, 27, 28f
Ventilation, lamprey, 24f
Ventral abdominal vein, Necturus, 158f, 160, 160f
Ventral aorta
    amphioxus, 13
    lamprey, 23f, 24, 25f
    Necturus, 154, 155f, 156
    shark, 146, 147f–149f
Ventral blood vessel, 11
Ventral constrictors, shark, 93f, 96f
Ventral fin, amphioxus, 15
Ventral gastric artery
    Necturus, 157, 158f, 159f
    shark, 151f–153f
Ventral gastric vein, shark, 154, 154f
Ventral-hypaxial trunk muscles, 101–2, 101t, 104f, 131, 134f, 135f, 152f
Ventral lobe of the pancreas, 131
Ventral longitudinal bundle, shark, 92, 93f, 96f, 97f
Ventral mesentery
    Necturus, 138
    shark, 133
Ventral musculature of vertebrate chest, 117b
Ventral pectoral muscles, Necturus, 100t, 101–2
Ventral root, 187f
Ventral skeletogenous septum, shark, 45
Ventral (subperitoneal) ribs, 43, 43f
Ventral thoracic artery, cat, 161f, 162f, 163, 164f
Ventral thoracic vein, cat, 161f, 165f, 167
Ventral throat muscles, Necturus, 100t, 101–2
Ventricle
    amphibian, 156f
    cat, 161f, 162
    lamprey, 24
    Necturus, 137f, 154, 155f, 157f, 158f
    shark, 146, 146f, 147f, 149f, 194
    sheep, 200f
Vermiform appendix, cat, 143
Vermis, sheep brain, 197, 198f

Vertebra
  structure, 43, 43*f*
  trunk and tail, 44
Vertebrae and ribs, 43–52
  agnatha, 44
  amphibia, 46–47, 47*f*
  aves (birds), 49–50, 50*f*
  elasmobranchii (sharks),
    44–45, 45*f*
  mammalia (cat), 50–51, 52*f*
  osteichthyes, 45, 46*f*
  placoderms, 44, 44*f*
  reptilia, 47–49, 48*f*, 49*f*
  sturgeon, 45, 46*f*
Vertebral artery
  cat, 163, 164*f*, 166
  Necturus, 157, 157*f*, 159*f*
  shark, 155*f*
Vertebral border, cat, 64*f*, 66
Vertebral column, 43
Vertebral rib
  alligator, 48, 48*f*
  cat, 51
Vertebral section, bird, 49, 50*f*
Vertebral vein, cat, 165*f*, 167
Vertebrarterial canal
  alligator, 48
  bird, 49

Vertebrate(s), 2
  brain of adult, 185*f*
  chordates and subphylum
    of, 2–5
  classification and comparison
    of, 1, 2*t*
  design features of this manual
    on, 7
  integuments, 29–39
  laboratory strategy advice
    of, 5–7
  shifts in feeding mechanisms
    and origins of, 18–21
Vertex, urinary bladder of cat, 179,
  181*f*, 182, 183*f*
Vesicouterine pouch, 182
Vestibular apparatus, shark,
  90–91, 90*f*
Vestibule, cat, 138
Vibrissae, cat, 105
Villi, cat, 143
Viscera
  Necturus, 136, 138
  shark, 131–32
Visceral branch of vagus nerve,
  shark, 193
Visceral ganglia, 186
Visceral motor column, shark, 191

Visceral pericardium
  cat, 142
  Necturus, 138
  shark, 133
Visceral peritoneum
  cat, 143
  Necturus, 138
  shark, 132
Visceral pleura, cat, 141
Visceral sensory column,
  shark, 191
Visceral skeleton, 42
Visceral tissues, 185
Viviparous species, 172
Vocal cords, cat, 141
Volkmann's canal, 42*f*
Vomer
  bowfin *(Amia),* 71
  cat, 80*f*, 82
  Necturus, 71, 72*f*
  turtle, 76*f*, 77

W

Whale, baleen of, 38–39, 39*f*
Wheel organ, amphioxus, 11,
  12*f*, 14
White matter, 196*f*
Wing, of ilium in cats, 59, 61*f*

X

Xiphihumeralis, cat, 109*f*, 110*f*, 111
Xiphisternum, cat, 52*f*, 66
Xiphoid, cat, 66

Z

Zygomatic arch, cat, 79, 80*f*
Zygomatic process, cat, 79, 82
Zygomatic process of jugal, cat,
  81*f*, 82
Zygomatic process of the maxilla,
  cat, 80*f*, 81*f*
Zygomatic process of the temporal,
  cat, 80*f*, 81*f*
Zygomatic process of the temporal
  bone, cat, 82

Comparative Vertebrate Anatomy

Student Art Notebook

Figure 2.2(a)

1. Trunk
2. Collar
3. Proboscis
4. Proboscis coelom
5. Stomochord
6. Mouth
7. Collar coelom
8. Trunk coelom
9. Pharyngeal slit in pharynx
10. Branchial pore
11. Branchial pouch
12. Anus

Figure 2.2(b)

1. Trunk dorsal nerve cord
2. Trunk ventral nerve cord
3. Collar cord
4. Mouth
5. Nerve ring

Figure 2.2(c)

1. Mouth
2. Preoral hood
3. Gut
4. Circumoral band
5. Telotroch
6. Pulsatile vesicle
7. Nephridial pore
8. Nephridial duct
9. Nephridium
10. Coelomic compartment
11. Blastocoel

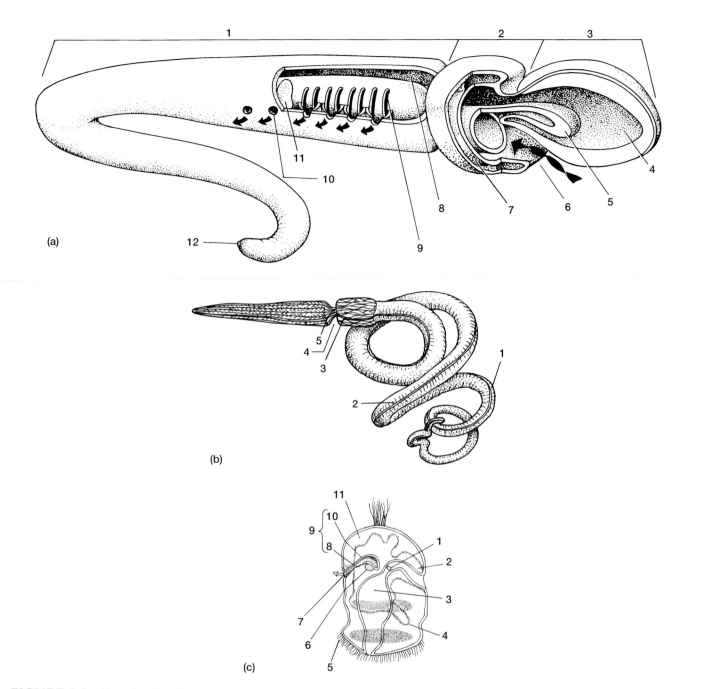

FIGURE 2.2 Hemichordata, Enteropneusta. (a) Generalized acorn worm. Proboscis, collar, and trunk regions are shown in partial cutaway view, revealing the coelom in each region and the associated internal anatomy. Within the proboscis is the stomochord, an extension of the gut. The food-laden cord of mucus (spiral arrow) enters the mouth together with water. The food is directed through the pharynx into the gut. Excess water exits via the pharyngeal slits. Several slits open into a common compartment, the branchial pouch, that in turn opens to the environment by a branchial pore. (b) General structure of the acorn worm, *Saccoglossus.* (c) Hemichordate, generalized tornaria larva.

(b) Source: G. Stiasny, 1910, "Zur kenntnis der lebenweise von Balanoglossus clavigerus," Zoolisches Anzeiger 35, Gustav Fischer Verlag.

Figure 2.5(a)

1. Dorsal and tubular nerve cord
2. Dorsal fin
3. Oral hood
4. Pharyngeal slit
5. Midgut cecum
6. Esophagus
7. Midgut
8. Ileocolic ring
9. Atriopore
10. Hindgut
11. Ventral fin
12. Anus
13. Postanal tail
14. Caudal fin
15. Notochord
16. Myomere

Figure 2.5(b)

1. Ocellus
2. Notochord
3. Velar tentacle
4. Velum
5. Wheel organ
6. Hatschek's pit or groove
7. Oral hood
8. Frontal eye pigment
9. Buccal cirri
10. Pharyngeal slit

Figure 2.5(c)

1. Wheel organ
2. Hatschek's pit or Hatschek's groove
3. Notochord
4. Dorsal and tubular nerve cord

Figure 2.5(d)

1. Dorsal, tubular nerve cord
2. Notochord
3. Myomere
4. Epibranchial groove
5. Pharynx
6. Midgut cecum
7. Endostyle
8. Ovary
9. Metapleural fold

Figure 2.5(e)

1. Dorsal, tubular nerve cord
2. Notochord
3. Hindgut

Figure 2.5(f)

1. Dorsal fin
2. Dorsal, tubular nerve cord
3. Notochord
4. Ventral fin

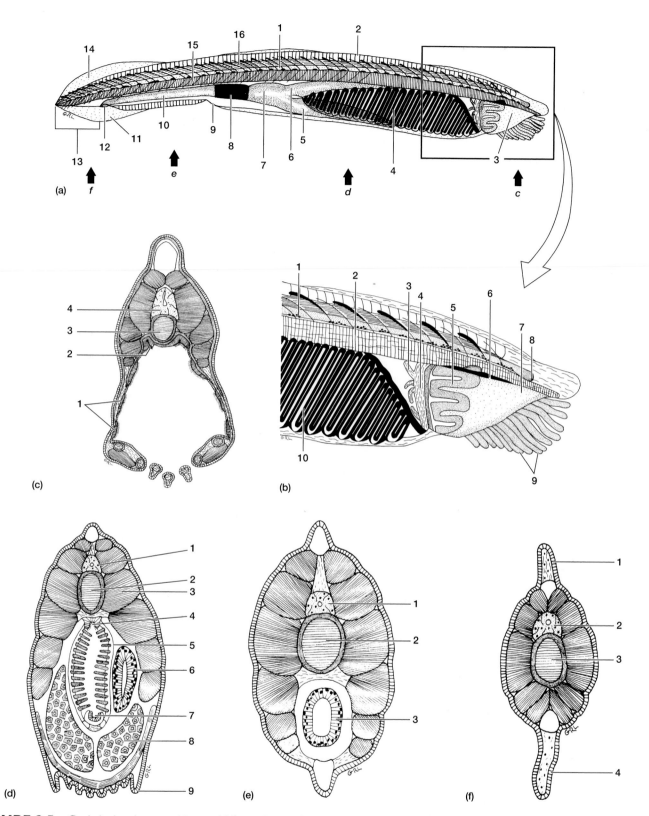

FIGURE 2.5 Cephalochordate, amphioxus. (a) Lateral view. (b) Enlargement of oral and pharyngeal regions. (c) Cross section through oral hood.

Figure 2.6

1. Dorsal nerve cord
2. Notochord
3. Epibranchial groove
4. Ciliary tracts
5. Coelom
6. Skeletal rod
7. Blood vessel
8. Pharyngeal bars
9. Food-moving cilia
10. Water-moving cilia
11. Water
12. Secondary (tongue bar)
13. Primary (septa)
14. Water
15. Accessory artery
16. Endostylar artery
17. Coelom
18. Endostyle
19. Metapleural fold
20. Endostyle
21. Ovary
22. Pharynx
23. Atrium
24. Segmental muscle (myomere)
25. Coelom
26. Epibranchial groove
27. Dorsal aorta
28. Dorsal fin

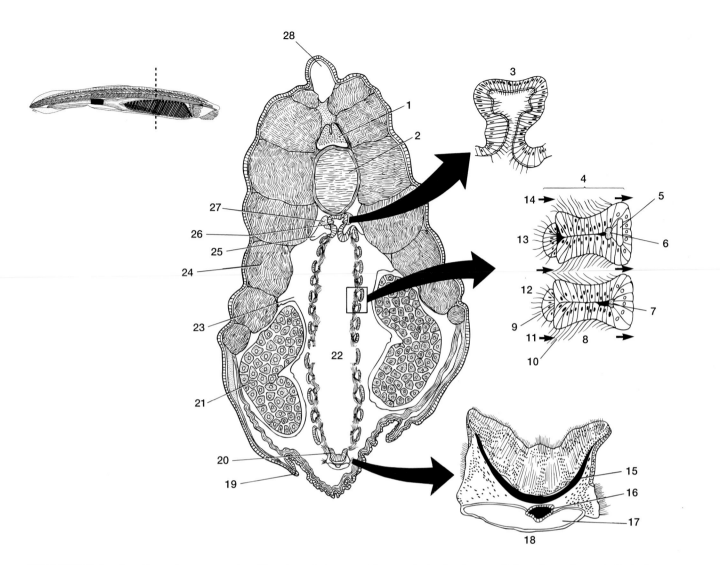

FIGURE 2.6 Amphioxus in cross section. The slanted pharyngeal bars encircle the pharynx. Enlarged at right are individual pharyngeal bars cut transversely at right angles to their long axis. The coelom continues into some of the branchial bars.

After Smith; Moller and Philpott; Baskin and Detmers.

Figure 2.7(a)

1. Postanal tail
2. Notochord
3. Dorsal nerve cord
4. Epidermis
5. Tunic
6. Adhesive papilla
7. Branchial basket with pharyngeal slits
8. Visceral ganglion
9. Incurrent branchial siphon
10. Cerebral ganglion
11. Ocellus
12. Otolith
13. Excurrent atrial siphon

Figure 2.7(b)

1. Outer tunic cuticle
2. Feces
3. Branchial siphon
4. Excurrent siphon
5. Atrium
6. Adhesive papillae

Figure 2.7(c)

1. Oral tentacle
2. Incurrent siphon (mouth)
3. Cerebral ganglion
4. Excurrent (atrial) siphon
5. Atrium
6. Anus
7. Intestine
8. Gonad
9. Stomach
10. Stigmata
11. Ciliated band
12. Heart
13. Stigmata
14. Endostyle
15. Tunic

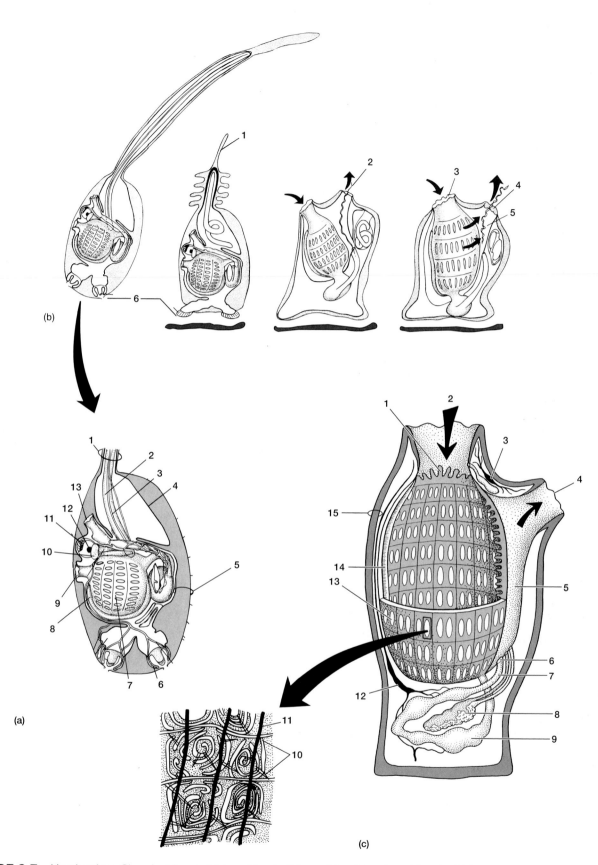

FIGURE 2.7 Urochordate, Class Ascidiacea. (a) Larval head of the ascidian *Distaplia occidentalis*. (b) Metamorphosis of ascidian larva *Distaplia*. Left to right, the planktonic, nonfeeding larva settles and attaches to a substrate. Adhesive papillae hold the larva, the tail contracts, and the outer cuticle of the tunic is shed. (c) Generalized adult urochordate. The flow of water is shown as it enters the incurrent siphon and passes through modified pharyngeal slits (stigmata) into the atrium and out the excurrent siphon. At the right, several of the highly subdivided pharyngeal slits are shown.

(a) Based on the research of R. A. Cloney.

Figure 3.4(a)

1. Caudal fin
2. Dorsal fins
3. Eye
4. Location of pineal gland
5. Nasohypophyseal opening
6. Buccal funnel
7. External gill slit
8. Trunk
9. Region of urogenital papilla
10. Tail

Figure 3.4(b)

1. Teeth on tongue
2. Teeth
3. Papillae

Figure 3.4(c)

1. Notochord
2. Myomere
3. Spinal cord
4. Heart
5. Pericardial cavity
6. Ventral aorta
7. Dorsal aorta
8. Esophagus
9. Hypophyseal pouch
10. Brain
11. Pineal gland
12. Nasal sac
13. Nasohypophyseal duct
14. Cranial cartilages
15. Tooth
16. Papillae
17. Buccal cavity
18. Tongue
19. Pharynx
20. Velar tentacle
21. Lingual cartilage
22. Lingual muscle
23. Branchial tube
24. Internal gill slit
25. Inferior jugular vein
26. Liver
27. Gonad
28. Intestine

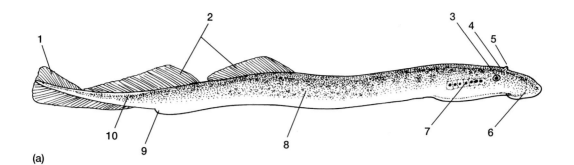

(a)

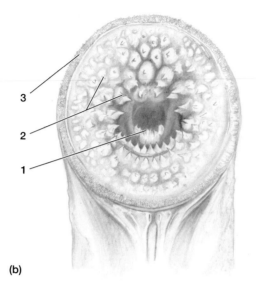

(b)

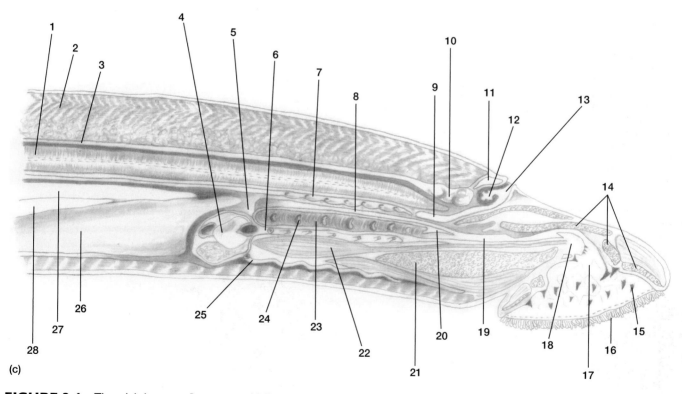

(c)

FIGURE 3.4   The adult lamprey, *Petromyzon*. (a) External morphology of the adult lamprey. (b) Ventral view of the structure of the mouth. (c) Midsagittal section of the anterior half of the body.

Figure 3.6(a)

1. Pineal gland
2. Nasal sac
3. Hypophyseal pouch
4. Cranial cartilage
5. Carotid artery
6. Lens of eye
7. Eye muscle

Figure 3.6(b)

1. Dorsal aorta
2. Esophagus
3. Branchial tube
4. Ventral aorta
5. Inferior jugular vein
6. Lingual muscle
7. Branchial pouch
8. External gill opening
9. Gill lamellae
10. Anterior cardinal vein
11. Notochord
12. Spinal cord

Figure 3.6(c)

1. Spinal cord
2. Notochord
3. Dorsal aorta
4. Posterior cardinal vein
5. Testis
6. Liver
7. Intestine
8. Myosepta
9. Myomere

Figure 3.6(d)

1. Fin rays
2. Spinal cord
3. Notochord
4. Caudal artery
5. Caudal vein

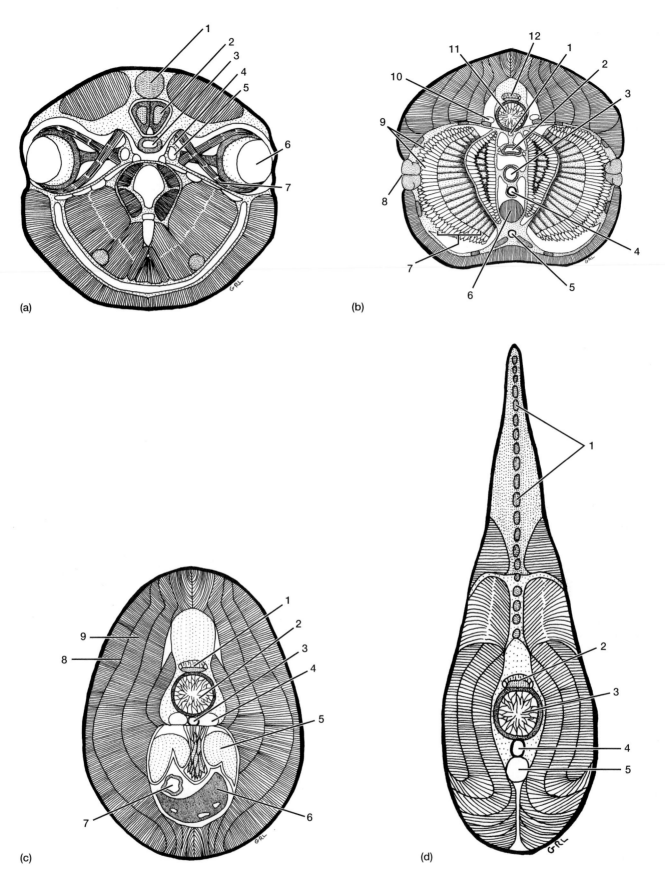

FIGURE 3.6 Cross sections of the adult lamprey. (a) Section through the pineal gland and eyes. (b) Section through the 3rd or 4th branchial slit. Two sets of gill lamellae are seen. The section passes through the most lateral lamellae in the region of an external gill slit. The medial lamellae are slanted laterally and caudally to communicate with a more posterior gill slit. (c) Section through the liver and intestine. (d) Section through the tail region.

Figure 3.8(a)

1. Caudal fin
2. Dorsal fin
3. Spinal cord
4. Notochord
5. Intestine
6. Liver
7. Gallbladder
8. Esophagus
9. Heart
10. Ear vesicle
11. Eyespot
12. Prosencephalon
13. Oral tentacles
14. Velum
15. Gill filament
16. Gill slits
17. Subpharyngeal gland
18. Anus

Figure 3.8(b)

1. Myomere
2. Spinal cord
3. Notochord
4. Dorsal aorta
5. Epibranchial groove
6. Gill filament
7. Pharynx
8. Subpharyngeal gland

Figure 3.8(c)

1. Spinal cord
2. Notochord
3. Posterior cardinal vein
4. Pronephric kidney
5. Intestine
6. Hemopoietic tissue

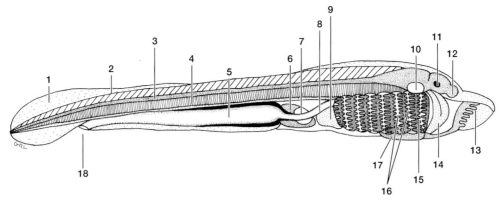

(a)

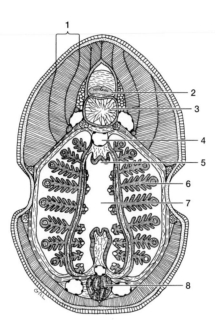

(b)

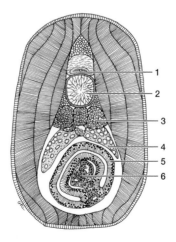

(c)

FIGURE 3.8 Ammocoetes larva. (a) External morphology. (b) Cross section through the pharynx. (c) Cross section through the intestine.

Figure 5.4(a,b)

1. Neural spine
2. Neural arch
3. Neural canal
4. Centrum
5. Notochord
6. Hemal arch
7. Hemal canal
8. Hemal spine
9. Ventral rib
10. Basapophysis
11. Dorsal rib

Figure 5.4(c)

1. Horizontal septum
2. Myoseptum
3. Dorsal rib
4. Centrum
5. Neural arch
6. Neural tube
7. Notochord
8. Blood vessels
9. Skin
10. Mesentery
11. Intestine
12. Median ventral septum
13. Ventral rib
14. Outer wall of coelomic cavity
15. Body wall

Figure 5.4(d)

1. Postzygapophysis
2. Prezygapophysis
3. Diapophysis
4. Tuberculum
5. Parapophysis
6. Capitulum
7. Pleurocentrum
8. Intercentrum
9. Sternum
10. Sternal segment
11. Costal segment
12. False rib
13. Floating rib

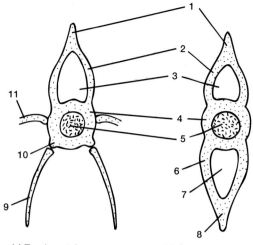

(a) Trunk vertebra (b) Caudal vertebra

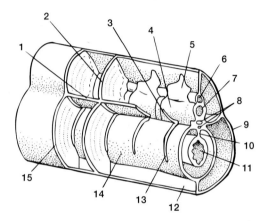

(c) Fish ribs

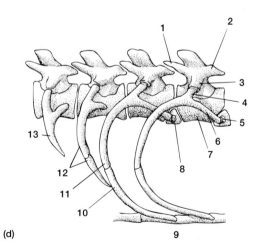

(d)

FIGURE 5.4 Ribs. (a) Cross section of trunk vertebra of a fish. (b) Cross section of caudal vertebra of a fish. (c) In fishes, dorsal ribs develop where myosepta intersect with the horizontal septum, and ventral ribs develop where myosepta meet the wall of the coelomic cavity. (d) Amniote ribs. Ribs are named on the basis of their articulation with the sternum (true ribs), with each other (false ribs), or with nothing ventrally (floating ribs). Primitively, ribs are bicipital, having two heads, a capitulum and a tuberculum, that articulate respectively with the parapophysis on the intercentrum or the diapophysis on the neural arch. The body of the rib may differentiate into a dorsal part, the vertebral rib or costal segment, and a ventral part, the sternal rib or segment that articulates with the sternum.

Figure 5.7

1. Neural arch
2. Interneural arch
3. Nerve foramen
4. Neural canal
5. Notochord
6. Centrum
7. Basapophysis
8. Rib
9. Hemal spine
10. Hemal canal
11. Hemal arch
12. Notochord
13. Centrum
14. Neural canal
15. Neural arch
16. Neural spine

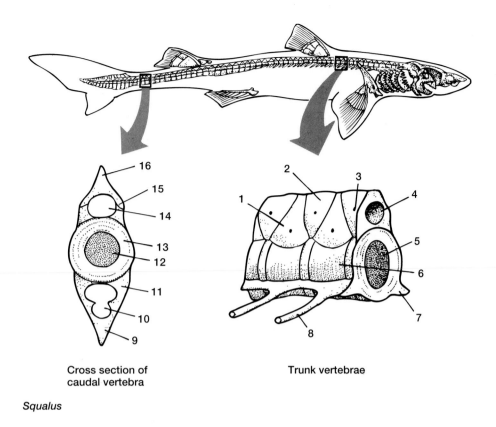

Cross section of
caudal vertebra

Trunk vertebrae

Squalus

FIGURE 5.7 The axial skeleton in the shark *Squalus*. The vertebral elements tend to enlarge in elasmobranchs, surpassing the notochord as the major mechanical support for the body in modern sharks.

Figure 5.8(a)

1. Neural spine
2. Ligament
3. Notochord
4. Ventral arch
5. Rib
6. Interneural
7. Neural arch

Figure 5.8(c,d)

1. Two part neural spine
2. Neural arch
3. Dorsal rib
4. Ligament
5. Basapophysis
6. Ventral rib
7. Dorsal rib
8. Ligament

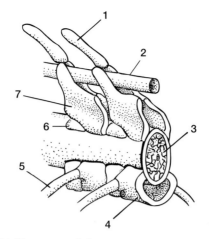

(a) Sturgeon vertebrae
(lateral)

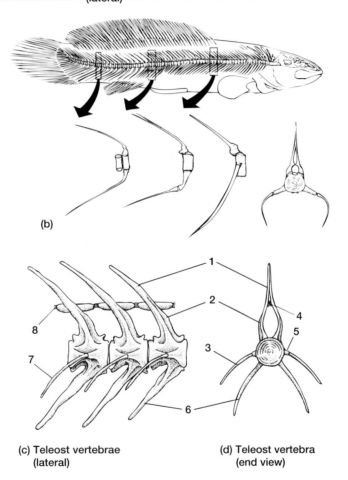

(b)

(c) Teleost vertebrae
(lateral)

(d) Teleost vertebra
(end view)

FIGURE 5.8 Actinopterygian vertebrae. (a) Sturgeon vertebrae, lateral view. (b) Axial skeleton of the bowfin *Amia calva*. (c) Teleost vertebrae, lateral view. (d) Teleost vertebra, end view.

(a, c) After Jollie; (b) after Jarvik.

Figure 5.9(b)

1. Neural spine
2. Prezygapophysis
3. Hemal spine
4. Transverse process
5. Postzygapophysis

Figure 5.9(c)

1. Transverse process
2. Prezygapophysis
3. Parapophysis
4. Capitulum
5. Rib
6. Tuberculum
7. Diapophysis
8. Postzygapophysis
9. Neural spine

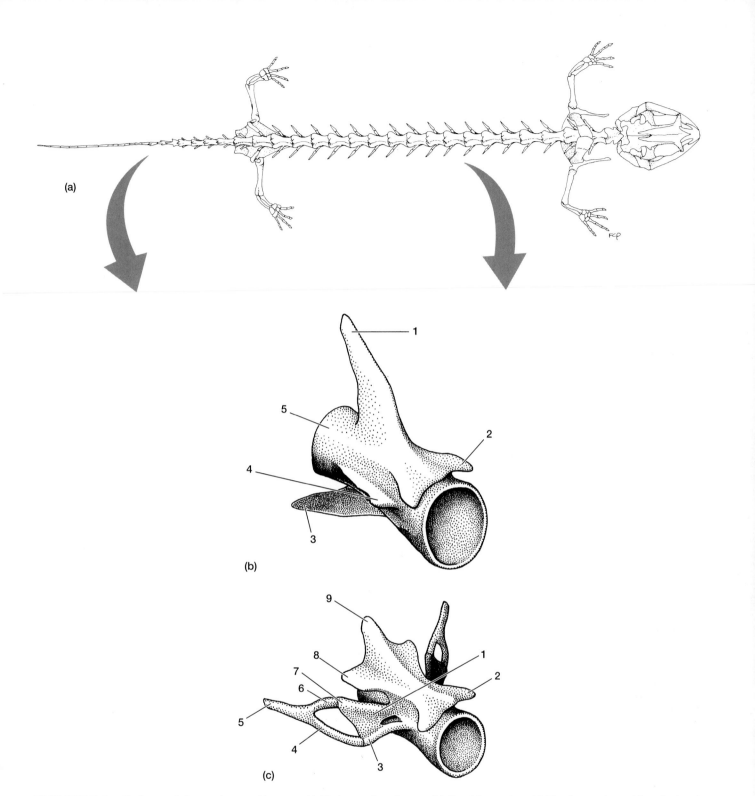

FIGURE 5.9 Skeleton of the mudpuppy, *Necturus*. (a) Skeleton, dorsal view. (b) Caudal vertebra. (c) Trunk vertebra with articulated ribs. Cranial is to the right in both (b) and (c).

Figure 5.10

1. Transverse process of sacral vertebra
2. Ilium
3. Urostyle
4. Transverse process of trunk vertebra

Figure 5.11(b)

1. Hemal arch
2. Transverse process
3. Neural arch
4. Neural spine

Figure 5.11(c)

1. Prezygapophysis
2. Parapophysis
3. Capitulum
4. Sternal or costal rib
5. Intermediate rib
6. Vertebral rib
7. Tuberculum
8. Diapophysis
9. Transverse process
10. Postzygapophysis

Figure 5.11(d)

1. Neural spine
2. Neural arch of axis
3. Neural arch of atlas
4. Proatlas
5. Intercentrum of atlas
6. Odontoid process of axis
7. Ribs
8. Procoelous centra

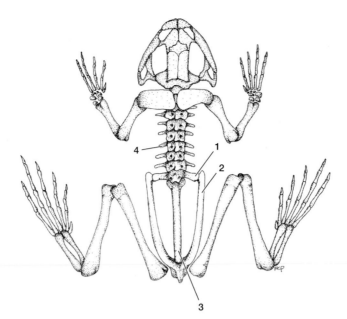

FIGURE 5.10 Skeleton of a bullfrog; dorsal view.

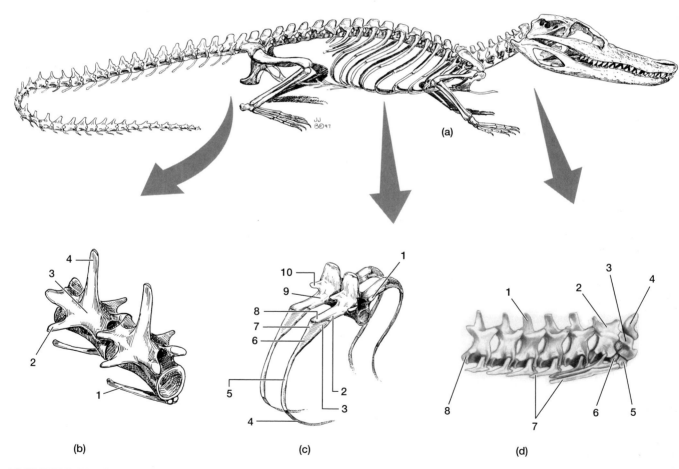

FIGURE 5.11 Skeleton of an alligator. (a) Lateral view of complete skeleton. (b) Two caudal vertebrae. (c) Oblique view of two thoracic vertebrae. (d) Lateral view of cervical vertebrae. Cranial is to the right in (b–d).

(d) Redrawn from E. S. Goodrich, Studies on the Structure and Development of Vertebrates.

Figure 5.12

1. Caudal vertebra
2. Carapace
3. Capitulum of rib
4. Trunk vertebra
5. Cervical vertebra

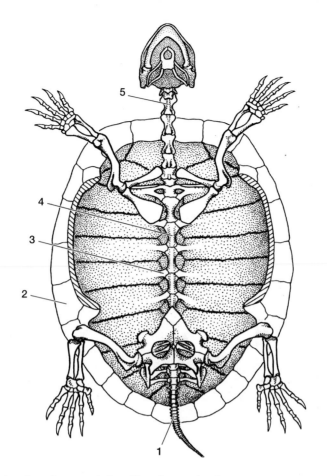

FIGURE 5.12 Skeleton of a painted turtle, ventral view. The plastron has been cut away to reveal the internal skeletal detail.

Figure 5.13(a)

1. Sternal section of rib
2. Uncinate process
3. Vertebral section of rib

Figure 5.13(b)

1. Prezygapophysis
2. Hypapophysis
3. Cervical rib
4. Heterocoelous centra
5. Postzygapophysis

Figure 5.13(c)

1. Thoracic vertebra
2. Lumbar vertebrae
3. Transverse processes of sacral vertebrae
4. Caudal vertebra
5. Synsacrum
6. Pygostyle
7. Free caudal vertebra
8. Pubis
9. Ischium
10. Ilium
11. Innominate bone
12. Free thoracic vertebrae

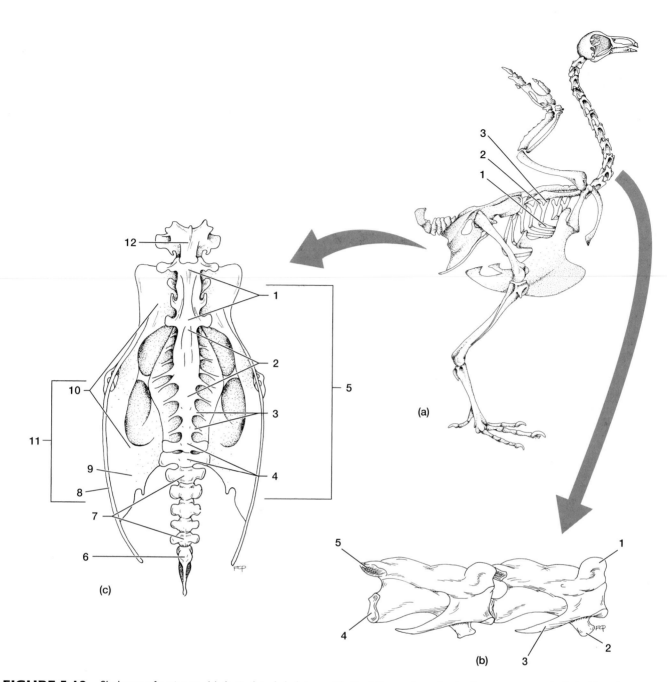

FIGURE 5.13 Skeleton of a pigeon. (a) Articulated skeleton with the left member of paired structures not included for clarity. (b) Cervical vertebrae, lateral view. (c) Synsacrum, ventral view; cranial is up.

Figure 5.14(a)

1. Hyoid
2. Manubrium
3. Clavicle
4. Sternebrae
5. Xiphisternum
6. Intervertebral foramen

Figure 5.14(b)

1. Neural arch
2. Neural canal
3. Prezygapophysis
4. Neural spine
5. Neural canal
6. Centrum
7. Odontoid process
8. Prezygapophysis
9. Transverse foramen
10. Transverse process
11. Postzygapophysis
12. Neural arch
13. Transverse process

Figure 5.14(c)

1. Neural canal
2. Prezygapophysis
3. Transverse foramen
4. Centrum
5. Transverse process
6. Postzygapophysis
7. Neural arch
8. Neural spine

Figure 5.14(d)

1. Neural spine
2. Prezygapophysis
3. Transverse process
4. Neural canal
5. Centrum
6. Demifacet
7. Full facet
8. Neural arch
9. Postzygapophysis

Figure 5.14(e)

1. Tuberculum
2. Angle
3. Body or shaft
4. Capitulum
5. Neck

Figure 5.14(f)

1. Neural spine
2. Mammillary process
3. Prezygapophysis
4. Centrum
5. Transverse process
6. Accessory process
7. Postzygapophysis

Figure 5.14(g)

1. Neural spine
2. Prezygapophysis
3. Centrum
4. Articular surface with ilium
5. Intervertebral foramen
6. Postzygapophysis

Figure 5.14(h)

1. Hemal arch
2. Centrum

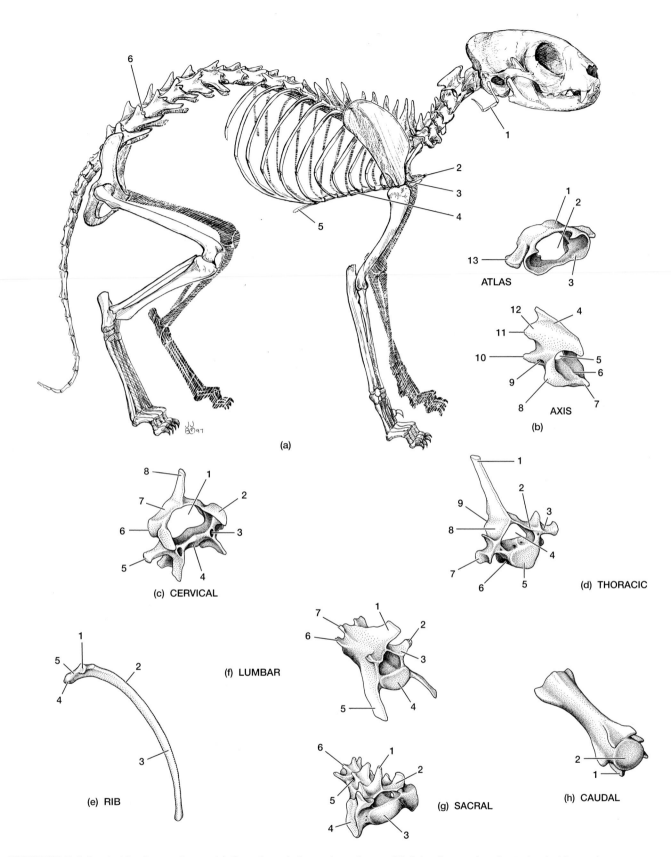

FIGURE 5.14 Axial column of a cat. (a) Complete skeleton, lateral view. (b) Atlas (bottom) and axis (top). All vertebrae (b–d, f–h) are in oblique view, with the cranial end to the right. (c) More caudal cervical vertebra. (d) Thoracic vertebra. (e) Rib. (f) Lumbar vertebra. (g) Sacral vertebra. (h) Caudal vertebra.

Figure 5.16(b)

1. Iliac process
2. Puboischiac bar
3. Clasper
4. Ceratotrichia
5. Radials
6. Metapterygium
7. Propterygium
8. Iliac process
9. Puboischiac bar

Figure 5.16(c)

1. Scapular process
2. Scapular cartilage
3. Ceratotrichia
4. Radials
5. Metapterygium
6. Mesopterygium
7. Propterygium
8. Coracoid bar

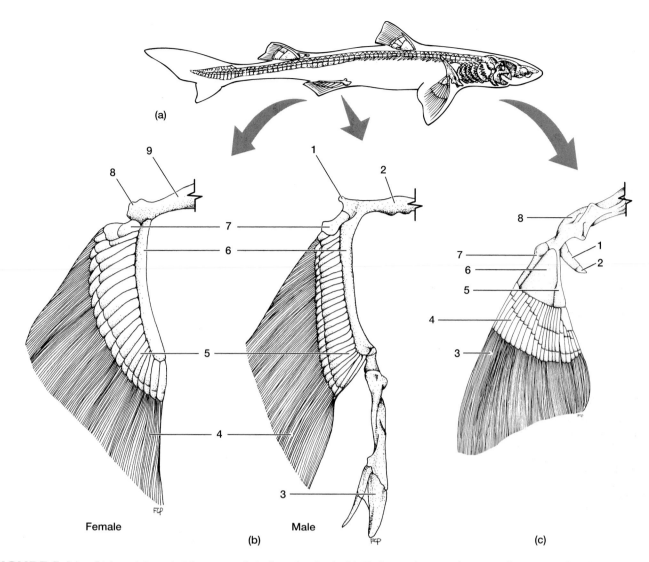

FIGURE 5.16   Right pelvic and right pectoral girdles of a shark. (a) Skeleton showing the general position of each girdle. (b) Female (left) and male (right) pelvic girdles, ventral view. Note the elongate clasper of the male used for sperm transfer. (c) Pectoral girdle (similar in males and females), ventral view.

Figure 5.17(b)

1. Ilium
2. Ischiac cartilage
3. Obturator foramen
4. Pubic cartilage
5. Phalanges
6. Metatarsal
7. Tarsals
8. Fibula
9. Tibia
10. Femur
11. Ischium

Figure 5.17(c)

1. Scapula
2. Suprascapular cartilage
3. Coracoid cartilage
4. Procoracoid cartilage
5. Phalanges
6. Metacarpal
7. Carpals
8. Ulna
9. Radius
10. Humerus

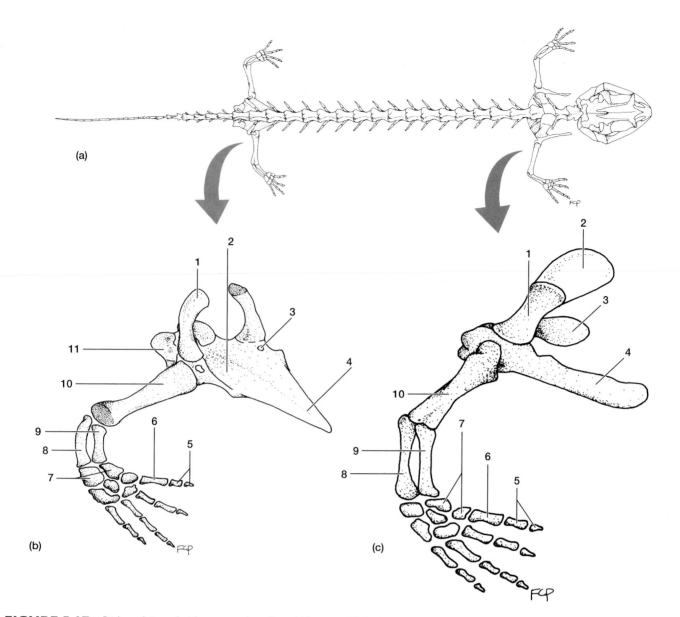

(a)

(b)

(c)

FIGURE 5.17 Right pelvic and right pectoral girdles of *Necturus*. (a) Skeleton showing the general position of each girdle. (b) Pelvic girdle, oblique view. (c) Pectoral girdle, oblique view.

Figure 5.18(b)

1. Prepubic process
2. Epipubic cartilage
3. Pubis
4. Pubic symphysis
5. Obturator foramen
6. Ischium
7. Ischiac symphysis
8. Ilium
9. Fibulare
10. Metatarsal
11. Phalanges
12. Tarsal
13. Tibiale
14. Fibula
15. Tibia
16. Femur

Figure 5.18(c)

1. Phalanges
2. Metacarpal
3. Carpals
4. Radiale
5. Intermedium
6. Radius
7. Humerus
8. Procoracoid (Acromion)
9. Scapula
10. Coracoid
11. Ulna
12. Pisiform
13. Ulnare

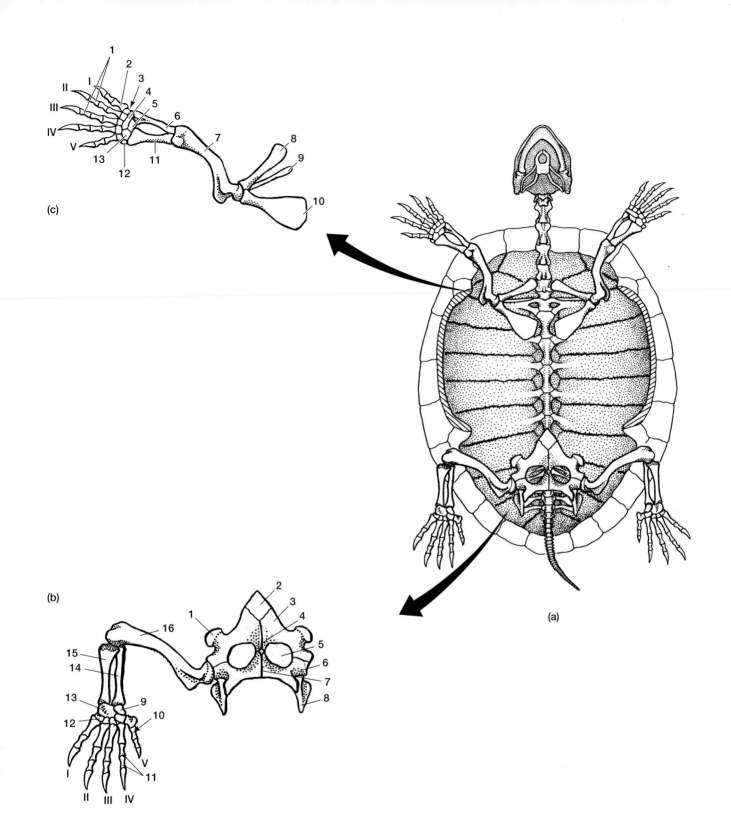

FIGURE 5.18 Right pelvic and right pectoral girdles of a turtle. (a) Skeleton with plastron removed, showing the general position of each girdle. (b) Pelvic girdle, ventral view. (c) Pectoral girdle, ventral view.

Figure 5.19(b)

1. Ilium
2. Head
3. Ischium
4. Femur
5. Pubis
6. Claws
7. Phalanges
8. Metatarsals
9. Astragalus (tibiale)
10. Calcaneum (fibulare)
11. Tibia
12. Fibula

Figure 5.19(c)

1. Suprascapular cartilage
2. Scapula
3. Coracoid
4. Interclavicle
5. Radius
6. Ulna
7. Carpals
8. Metacarpals
9. Phalanges
10. Sternum
11. Humerus

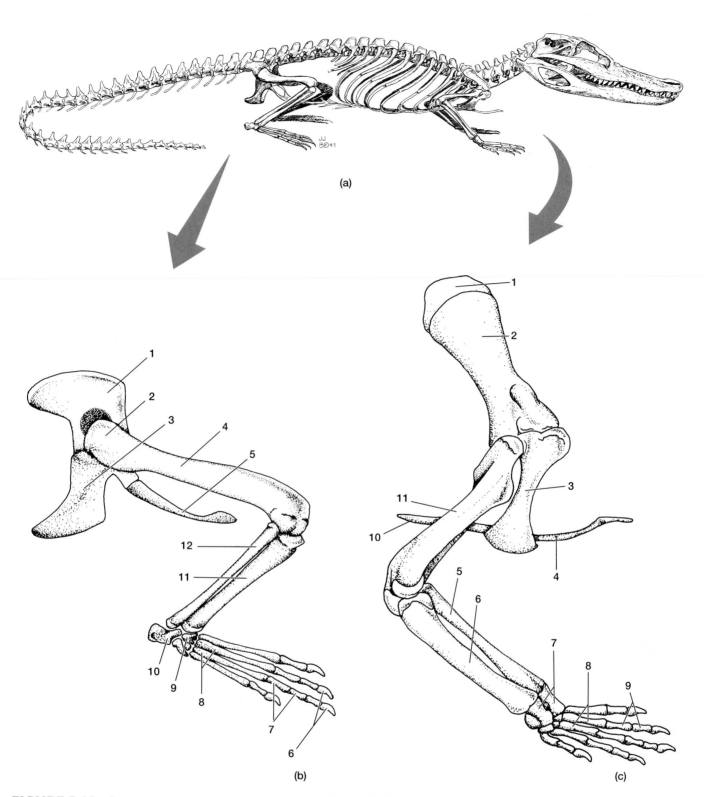

(a)

(b)

(c)

FIGURE 5.19 Right pelvic and right pectoral girdles of an alligator. (a) Skeleton showing the general position of each girdle. (b) Pelvic girdle, lateral view. (c) Pectoral girdle, lateral view.

Figure 5.20(a)

1. Procoracoid process
2. Furcula
3. Costal process
4. Anterior xiphisternal process
5. Sternum
6. Posterior xiphisternal process
7. Keel

Figure 5.20(b)

1. Greater trochanter
2. Femur
3. Condyle
4. Patella
5. Fibula
6. Tibiotarsus
7. Malleolus
8. Tarsometatarsus
9. Phalanges
10. Hallux

Figure 5.20(c)

1. Phalanx
2. Phalanx
3. Metacarpal II
4. Carpometacarpus
5. Radiale
6. Ulnare
7. Ulna
8. Radius
9. Olecranon process
10. Humerus
11. Lesser tuberosity
12. Pneumatic foramen
13. Deltoid ridge
14. Procoracoid process
15. Greater tuberosity
16. Coracoid
17. Clavicle
18. Interclavicle
19. Scapula
20. Metacarpal IV
21. Metacarpal III
22. Phalanx

Figure 5.20(d)

1. Clavicle
2. Interclavicle
3. Furcula

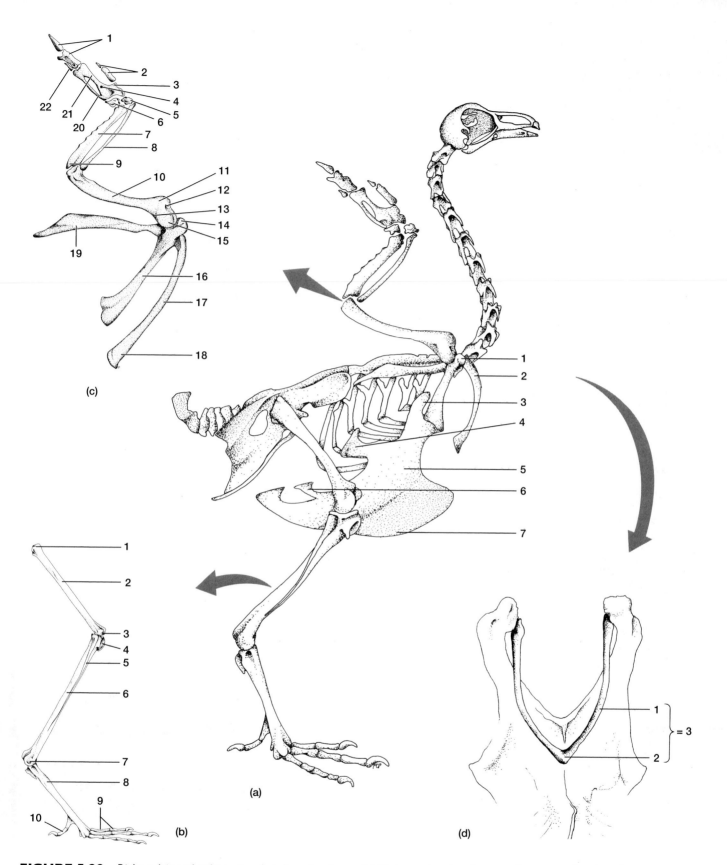

FIGURE 5.20 Right pelvic and right pectoral girdles of a pigeon. (a) Skeleton showing the general position of each girdle. The left element of paired structures is not shown for clarity. (b) Hindlimb, lateral view. (c) Partial pectoral girdle and forelimb, lateral view.

Figure 5.22(b,c)

1. Articular scar
2. Iliopectineal eminence
3. Body of pubis
4. Rami of pubis
5. Rami of ischium
6. Ischial tuberosity
7. Body of ischium
8. Obturator foramen
9. Ischial spine
10. Acetabulum
11. Body of ilium
12. Posterior inferior spine
13. Wing of ilium
14. Articular scar
15. Iliopectineal eminence
16. Pubic symphysis

Figure 5.22(d)

1. Greater trochanter
2. Trochanteric fossa
3. Fovea capitis
4. Head
5. Neck
6. Medial epicondyle
7. Patellar surface
8. Lateral epicondyle

Figure 5.22(e)

1. Greater trochanter
2. Intertrochanteric line
3. Lesser trochanter
4. Linea aspera
5. Intercondyloid fossa
6. Lateral condyle
7. Medial condyle

Figure 5.22(g,h)

1. Medial condyle
2. Popliteal notch
3. Lateral condyle
4. Head of fibula
5. Fibula
6. Lateral malleolus
7. Dorsal projection
8. Medial malleolus
9. Tibia
10. Tibial tuberosity
11. Tibial crest
12. Fibula

Figure 5.22(i)

1. Fibulare (calcaneum)
2. Tibiale (astragalus)
3. First tarsale
4. Centralia 1 & 2
5. First metatarsal
6. Second tarsale
7. Third tarsale
8. Second metatarsal
9. Phalanges
10. Tarsalia 4 & 5 (cuboid)

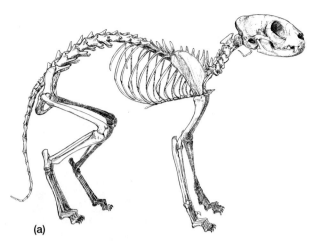

(a)

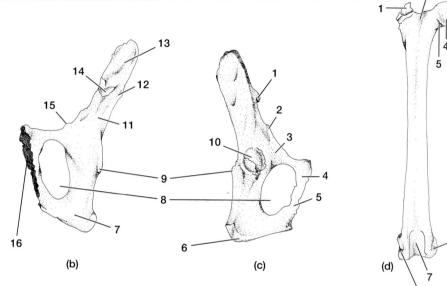

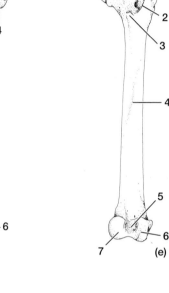

(b)

(c)

(d)

(e)

Patella

(f)

(g)

(h)

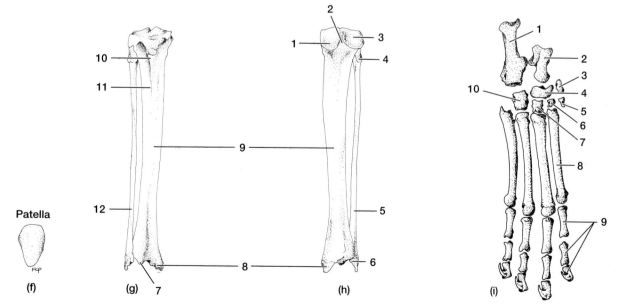

(i)

FIGURE 5.22 Right pelvic girdle and right hindlimb of a cat. (a) Lateral view of cat skeleton. (b) Innominate bone, dorsal view. (c) Innominate bone, ventral view. (d) Femur, anterior view. (e) Femur, posterior view. (f) Patella. (g) Tibia and fibula, anterior view. (h) Tibia and fibula, posterior view. (i) Hindfoot, ventral view.

Figure 5.24(b)

1. Vertebral border
2. Anterior border
3. Supraspinous fossa
4. Acromion
5. Clavicle
6. Glenoid fossa
7. Metacromion
8. Axillary border
9. Tuberosity of spine
10. Infraspinous fossa

Figure 5.24(c)

1. Subscapular fossa
2. Metacromion
3. Glenoid fossa
4. Clavicle
5. Coracoid process

Figure 5.24(d,e)

1. Greater tuberosity
2. Head
3. Supracondyloid ridge
4. Olecranon fossa
5. Supracondyloid foramen
6. Medial epicondyle
7. Trochlea
8. Capitulum
9. Lateral epicondyle
10. Deltoid ridge
11. Pectoral ridge
12. Greater tuberosity
13. Lesser tuberosity

Figure 5.24(f,g)

1. Head
2. Bicipital tuberosity
3. Styloid process
4. Interosseous crest
5. Neck
6. Head
7. Semilunar notch
8. Olecranon process
9. Bicipital tuberosity
10. Styloid process

Figure 5.24(h)

1. Metacarpals
2. Claw
3. Phalanges
4. Capitate
5. Hamate
6. Pisiform
7. Ulnare
8. Scapholunar
9. Greater multangular
10. Lesser multangular

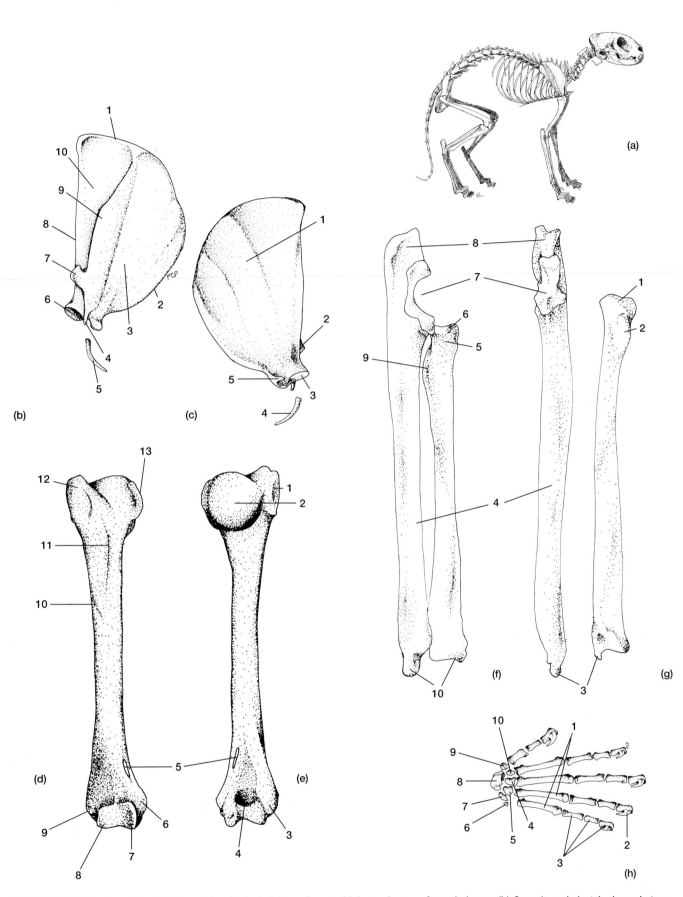

FIGURE 5.24 Scapula, clavicle, and forelimb skeleton of a cat. (a) Lateral view of cat skeleton. (b) Scapula and clavicle, lateral view. (c) Right scapula and clavicle, medial view. (d) Right humerus, anterior view. (e) Right humerus, posterior view. (f) Articulated left radius and ulna, lateral view. (g) Left radius and ulna, anterior view. (h) Right forefoot, dorsal view.

Figure 5.29(a)

1. Perilymphatic foramen
2. Endolymphatic foramen
3. Superficial ophthalmic foramina
4. Epiphyseal foramen
5. Precerebral cavity
6. Rostrum
7. Nasal capsule
8. Antorbital process
9. Supraorbital crest
10. Postorbital process
11. Endolymphatic fossa

Figure 5.29(b)

1. Otic capsule
2. Carotid canal
3. Basitrabecular process
4. Rostral carina
5. Nares
6. Nasal capsule
7. Rostral fenestra
8. Antorbital process
9. Anterior orbital shelf
10. Optic pedicel
11. Basal plate
12. Notochord

Figure 5.29(c)

1. Occipital condyle
2. Foramen magnum
3. Exit of vagus (X) nerve
4. Exit of glossopharyngeal (IX) nerve

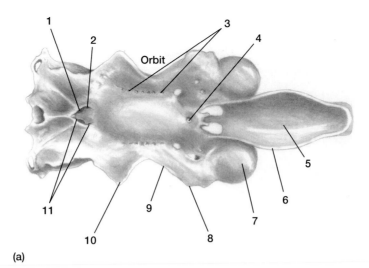

(a)

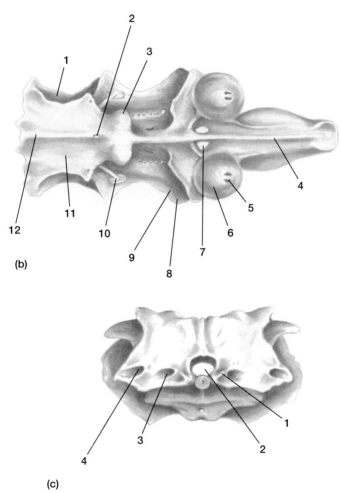

(b)

(c)

FIGURE 5.29 Chondrocranium and splanchnocranium of the shark. (a) Chondrocranium, dorsal view. (b) Chondrocranium, ventral view. (c) Chondrocranium, posterior view.

Figure 5.29(d)

1. Hyomandibular
2. Adductor process
3. Otic capsule
4. Postorbital process
5. Optic pedicel
6. Orbital process
7. Supraorbital crest
8. Antorbital process
9. Nares
10. Nasal capsule
11. Optic foramen
12. Palatoquadrate
13. Labial cartilage
14. Meckel's cartilage
15. Basihyal
16. Ceratohyal
17. Gill rays
18. Hypobranchial
19. Basibranchial
20. Ceratobranchial
21. Epibranchial
22. Pharyngobranchial

Figure 5.29(e)

1. Palatoquadrate
2. Meckel's cartilage
3. Labial cartilage
4. Ceratohyal
5. Basihyal
6. Basibranchials
7. Hypobranchials
8. Pharyngobranchial
9. Ceratobranchial

Figure 5.30(a)

1. Supratemporal
2. Hyomandibula
3. Infraorbito-suborbitals ("postorbitals")
4. Intertemporal
5. Lacrimal
6. Premaxilla
7. Jugal
8. Maxilla
9. Supramaxilla
10. Dentary
11. Surangular
12. Gular
13. Quadrate
14. Angular
15. Preopercular
16. Interopercular
17. Branchiostegals
18. Subopercular
19. Cleithrum
20. Postcleithrum
21. Opercular
22. Supracleithrum
23. Posttemporal
24. Extrascapular
25. Postparietal

Figure 5.30(b)

1. Premaxilla
2. Rostral
3. Nasal
4. Antorbital
5. Lacrimal
6. Jugal
7. Parietal
8. Intertemporal
9. Postparietal
10. Supratemporal
11. Extrascapular
12. Opercular
13. Posttemporal
14. Posterior nostril
15. Anterior nostril

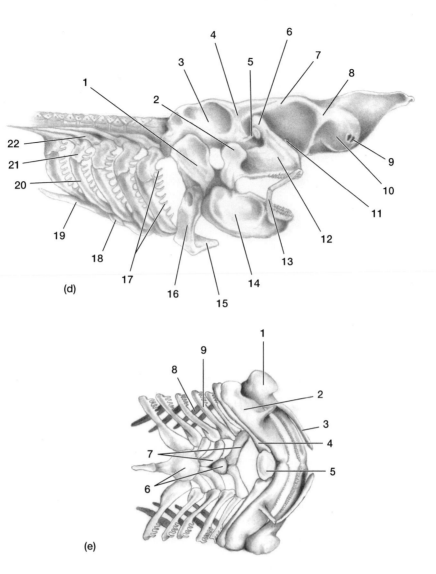

(d)

(e)

FIGURE 5.29 *Continued,* (d) Chondrocranium with attached splanchnocranium, lateral view. (e) Splanchnocranium, ventral view.

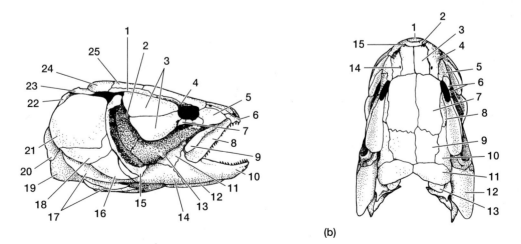

(a)

(b)

FIGURE 5.30 Skull of the bowfin, *Amia,* a chondrostean. Lateral (a) and dorsal (b) views.

Figure 5.31(a)

1. Trabecular horn
2. Ethmoid plate
3. Fenestra
4. Antorbital process
5. Trabecular cartilage
6. Quadrate bone
7. Quadrate cartilage
8. Prootic
9. Opisthotic
10. Synotic tectum
11. Exoccipital
12. Basal plate
13. Foramen magnum
14. Squamosal
15. Parietal
16. Pterygoid
17. Frontal
18. Vomer
19. Premaxilla

Figure 5.31(b)

1. Parasphenoid
2. Opisthotic
3. Squamosal
4. Prootic
5. Quadrate
6. Pterygoid
7. Vomer
8. Premaxilla

Figure 5.31(c)

1. Frontal
2. Vomer
3. Premaxilla
4. Quadrate cartilage
5. Quadrate bone
6. Dentary
7. Ceratohyal
8. Hypohyal
9. Basibranchial
10. Ceratobranchial
11. Basibranchial
12. Epibranchials
13. Parietal
14. Pterygoid

Figure 5.31(d)

1. Hypohyal
2. Ceratohyal
3. Basibranchials
4. Ceratobranchial
5. Epibranchials

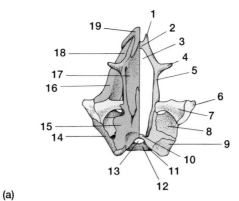

(a)

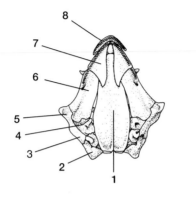

(b)

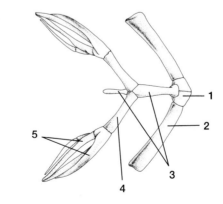

(c)

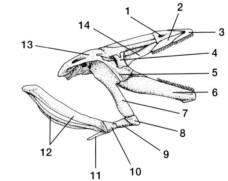

(d)

FIGURE 5.31 Skull of *Necturus*. (a) Dorsal view. Dermatocranium (top, shaded) and chondrocranium (lower, unshaded). (b) Ventral view of skull with jaw removed. (c) Lateral view of skull with splanchnocranium attached. (d) Splanchnocranium, ventral view.

Figure 5.32(a)

1. Postorbital bar
2. External nares
3. Nasal
4. Premaxilla
5. Maxilla
6. Prefrontal
7. Lacrimal
8. Palpebral
9. Jugal
10. Frontal
11. Condyle of quadrate
12. Infratemporal fenestra
13. Postorbital
14. Squamosal
15. Supratemporal fenestra
16. Parietal
17. Quadrate
18. Quadratojugal

Figure 5.32(b)

1. Internal nares
2. Palatine process of pterygoids
3. Ectopterygoid
4. Palatine process of palatines
5. Posterior palatine vacuities
6. Palatine process of maxilla
7. Palatine process of premaxilla
8. Incisive foramen
9. Palpebral
10. Quadrate
11. Quadratojugal
12. Jugal

Figure 5.32(c)

1. External auditory meatus
2. Location of interorbital septum
3. Palpebral
4. Lacrimal canal
5. Nasal
6. Premaxilla
7. Maxilla
8. Jugal
9. Ectopterygoid
10. Pterygoid
11. Rostrum to the basisphenoid
12. Laterosphenoid
13. Foramen ovale
14. Quadratojugal
15. Quadrate

(a)

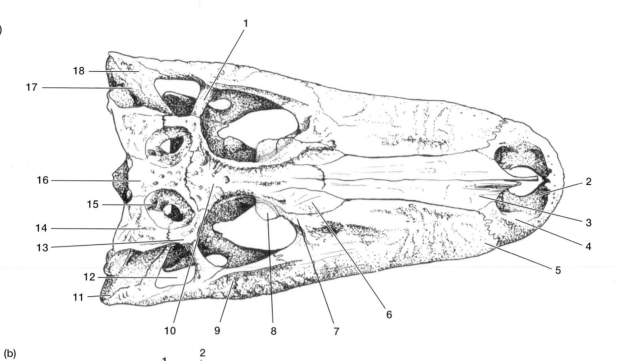

(b)

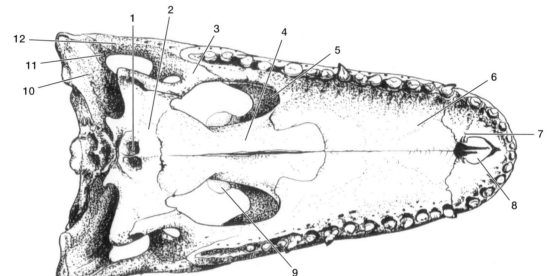

(c)

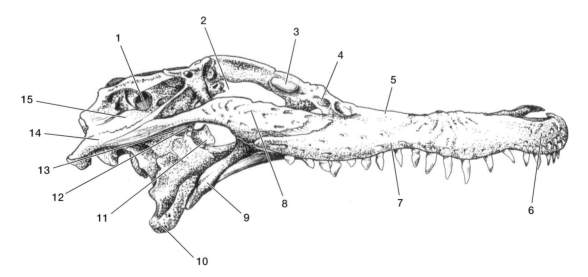

FIGURE 5.32 Skull of alligator. (a) Dorsal view. (b) Ventral view. (c) Lateral view.

Figure 5.32(d)

1. Supraoccipital
2. Exoccipital
3. Foramen magnum
4. Tympanic bulla
5. Occipital condyle
6. Basioccipital
7. Basisphenoid
8. Pterygoid
9. Quadrate
10. Squamosal

Figure 5.32(e)

1. Articular
2. External mandibular fenestra
3. Surangular
4. Dentary
5. Angular
6. Retroarticular process

Figure 5.32(f)

1. Articular
2. Angular
3. Coronoid
4. Mandibular foramen
5. Splenial

(d)

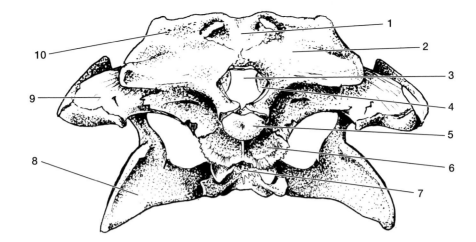

(e)

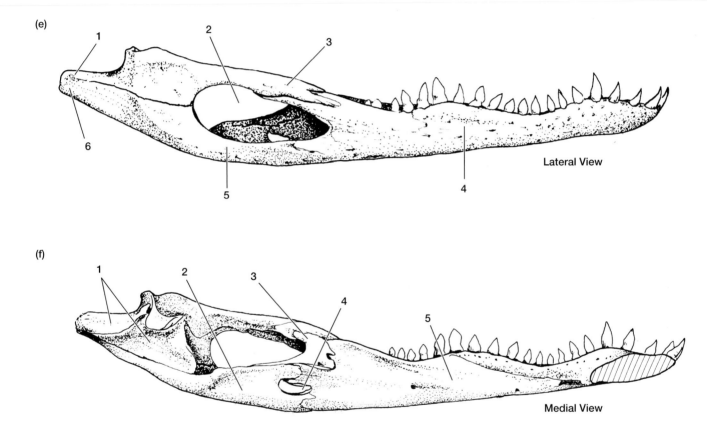

Lateral View

(f)

Medial View

FIGURE 5.32 *Continued.* (d) Posterior view. (e) Lateral view of jaw. (f) Medial view of jaw.

Figure 5.33(a)

1. Temporal emarginations
2. Parietal
3. Postorbital
4. Frontal
5. Prefrontal
6. External nares
7. Postorbital
8. Quadratojugal
9. Squamosal
10. Quadrate
11. Opisthotic
12. Prootic
13. Supraoccipital

Figure 5.33(b)

1. Pterygoid
2. Posterior palatine vacuity
3. Palatine
4. Incisive foramen
5. Internal nares
6. Vomer
7. Quadrate
8. Basisphenoid
9. Occipital condyle
10. Exoccipital
11. Basioccipital
12. Foramina

Figure 5.33(c)

1. Supraoccipital
2. Temporal emarginations
3. Postorbital
4. Jugal
5. Orbit
6. External nares
7. Maxilla (covered by beak)
8. Premaxilla (covered by beak)
9. Dentary
10. Surangular
11. Quadrate
12. Stapes
13. Quadratojugal
14. Squamosal

Figure 5.33(d)

1. Meckelian foramen
2. Meckelian groove
3. Splenial
4. Dentary
5. Angular
6. Prearticular
7. Articular
8. Coronoid

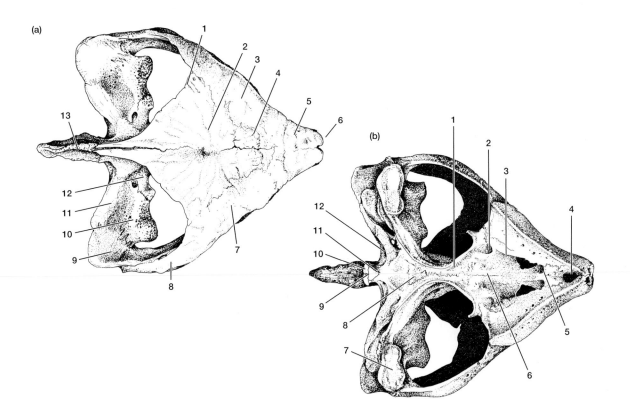

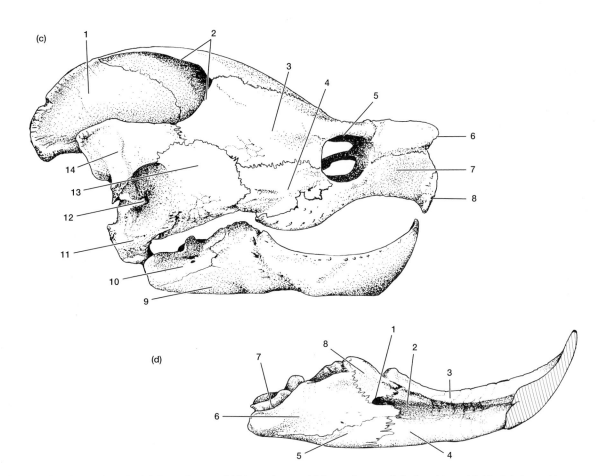

FIGURE 5.33 Skull of turtle. (a) Dorsal view. (b) Ventral view. (c) Lateral view of skull and jaw. (d) Medial view of jaw.

Figure 5.34(a)

1. Articular
2. Dentary
3. Exoccipital
4. Frontal
5. Interorbital septum of the mesethmoid
6. Jugal
7. Jugal bar
8. Lacrimal (prefrontal)
9. Maxilla
10. Maxillary process of the premaxilla
11. Nasal
12. Nasal aperture
13. Nasal process of the premaxilla
14. Orbitosphenoid
15. Palatine
16. Parietal
17. Pterygoid
18. Quadrate
19. Quadratojugal
20. Squamosal
21. Supraoccipital
22. Surangular
23. Premaxilla

Figure 5.34(b)

1. Basioccipital
2. Basisphenoid
3. Exoccipital
4. Jugal
5. Supraoccipital
6. Maxilla
7. Palatine
8. Parasphenoid
9. Premaxilla
10. Pterygoid
11. Quadrate
12. Quadratojugal
13. Foramen magnum

Figure 5.34(c)

1. Frontal
2. Lacrimal (prefrontal)
3. Supraoccipital
4. Jugal
5. Maxillary process of the premaxilla
6. Maxilla
7. Mesethmoid
8. Nasal
9. Nasal process of the premaxilla
10. Parietal
11. Quadrate
12. Quadratojugal
13. Squamosal

Figure 5.34(d)

1. Basioccipital
2. Exoccipital
3. Foramen magnum
4. Frontal
5. Occipital condyle
6. Parietal
7. Supraoccipital

Figure 5.34(e)

1. Basihyal
2. Ceratobranchial
3. Epibranchial
4. Paraglossal
5. Urohyal

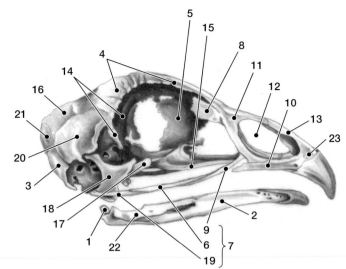

(a)

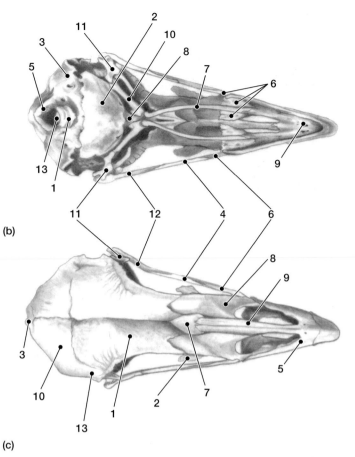

(b)

(c)

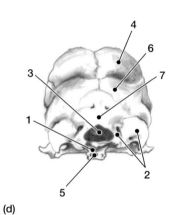

(d)

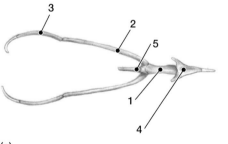

(e)

FIGURE 5.34 Skull of a chicken. (a) Lateral view of skull with jaw. (b) Ventral view of skull with jaw removed. (c) Dorsal view of skull with jaw removed. (d) Posterior view of skull with jaw removed. (e) Hyoid apparatus.

Figure 5.35(a)

1. Occipital bone
2. Parietal
3. Frontal
4. Maxilla
5. Premaxilla
6. External nares
7. Nasal
8. Orbital process of maxilla
9. Orbit
10. Postorbital process
11. Zygomatic arch
12. Temporal fenestra
13. Interparietal
14. Zygomatic process of the temporal
15. Lacrimal
16. Jugal (malar)
17. Nasolacrimal canal
18. Temporal
19. Frontal process of the premaxilla

Figure 5.35(b)

1. Occipital condyle
2. Tympanic bulla
3. Mandibular fossa
4. Vomer
5. Palatine process of maxilla
6. Palatine process of premaxilla
7. Incisive foramen
8. Posterior palatine foramen
9. Palatine process of palatine
10. Presphenoid
11. Pterygoid process
12. Eustachian canal
13. Basisphenoid
14. Foramen magnum
15. Occipital
16. Canine tooth
17. First premolar
18. Third premolar
19. Molar tooth
20. Foramen ovale
21. Foramen rotundum
22. Frontal
23. Stylomastoid foramen
24. Mastoid process
25. External auditory meatus
26. Hamulus
27. Incisors
28. Internal nares
29. Jugal
30. Zygomatic process of the temporal
31. Zygomatic process of the maxilla
32. Infraorbital foramen

Figure 5.35(c)

1. Jugular process
2. Nuchal line
3. Occipital bone
4. Interparietal
5. Squamous portion of temporal
6. Parietal
7. Occipital condyle
8. Zygomatic process of the temporal
9. Alisphenoid
10. Postorbital processes
11. Frontal
12. Lacrimal
13. Nasolacrimal canal
14. Nasal
15. Orbital process of maxilla
16. Frontal process of premaxilla
17. Mastoid process
18. Alveolar process of maxilla
19. Infraorbital canal
20. Jugal (malar)
21. Zygomatic process of maxilla
22. Palatine
23. Zygomatic process of jugal
24. Presphenoid
25. Basisphenoid
26. Hamulus
27. External auditory meatus
28. Tympanic bulla
29. Stylomastoid foramen

Figure 5.35(d)

1. Petrous part of petromastoid
2. Cerebellar fossa
3. Tentorium
4. Cerebral fossa
5. Olfactory fossa
6. Cribiform plate
7. Frontal sinuses
8. Turbinate bone
9. Sphenoid sinus
10. Pterygoid process
11. Sella turcica
12. Internal auditory meatus
13. Frontal
14. Parietal
15. Occipital

Figure 5.35(e)

1. Coronoid process
2. Ramus
3. Body
4. Mental foramina
5. Angular process
6. Condyloid process
7. Mandibular foramen
8. Mandibular symphysis
9. Incisors
10. Canine tooth
11. Premolar teeth
12. Molar tooth

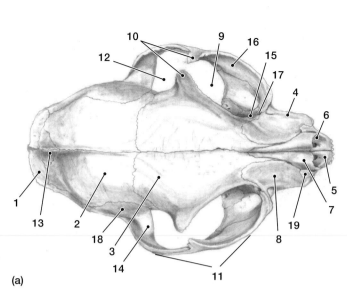

(a)

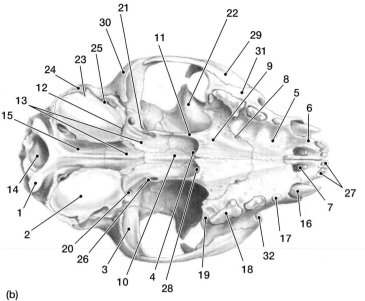

(b)

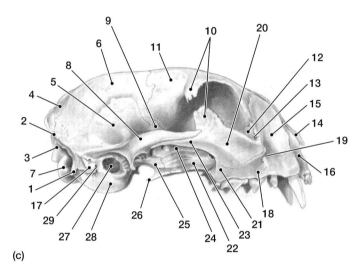

(c)

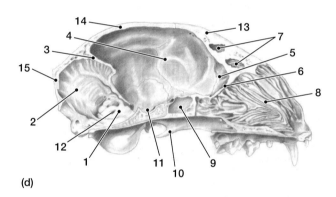

(d)

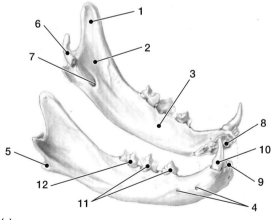

(e)

FIGURE 5.35 Skull of a cat. (a) Dorsal view. (b) Ventral view without the lower jaw. (c) Lateral view without the lower jaw. (d) Sagittal view without the lower jaw. (e) Lower jaw separated at the mandibular symphysis.

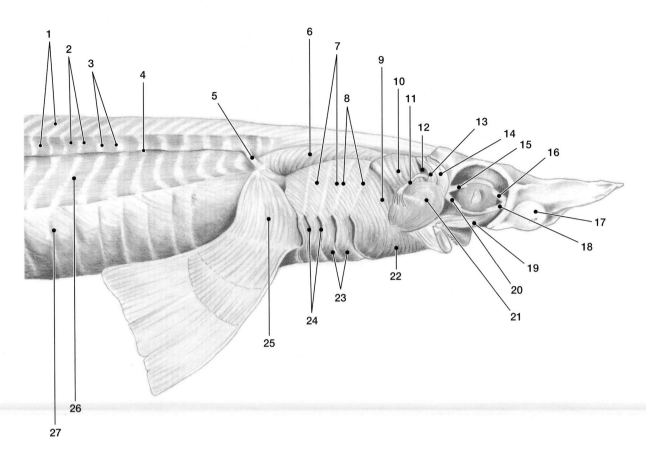

FIGURE 6.6 Lateral view of the trunk, branchial, and cephalic musculature of a shark.

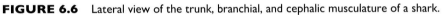

Figure 6.6

1. Dorsal longitudinal bundle
2. Myomeres
3. Myosepta
4. Horizontal septum
5. Scapular process
6. Cucullaris
7. Dorsal constrictors
8. Raphe
9. Second dorsal constrictor
10. Levator hyomandibulae (epihyoideus)
11. Hyomandibular nerve
12. Spiracle
13. Spiracularis
14. Levator palatoquadrati
15. Superior rectus
16. Superior oblique
17. Nasal capsule
18. Inferior oblique
19. Preorbitalis
20. Inferior rectus
21. Adductor mandibulae
22. Intermandibularis
23. Ventral constrictors
24. Gill slits
25. Extensors
26. Lateral longitudinal bundle
27. Ventral longitudinal bundle

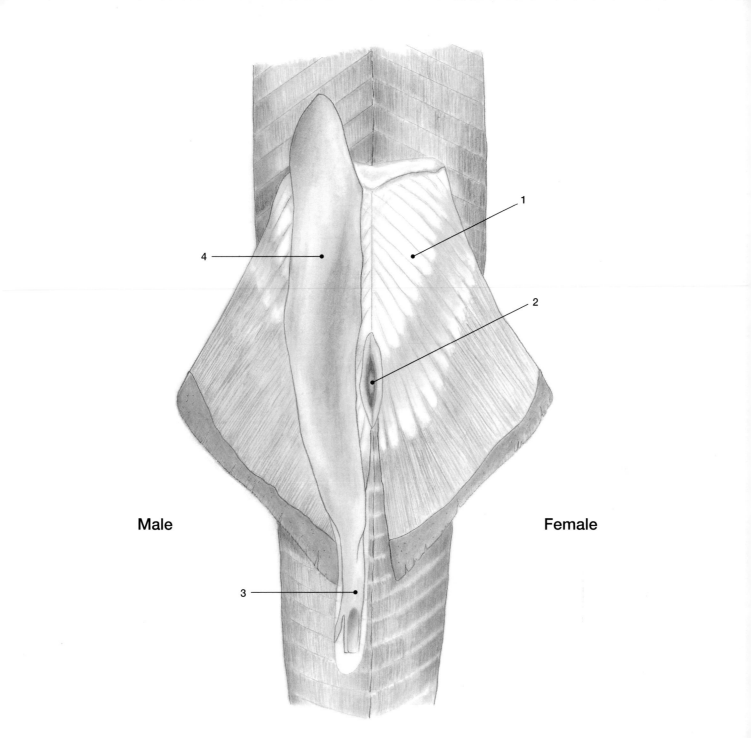

Male   Female

FIGURE 6.7   Ventral view of the pelvic fins of a male (left) and female (right) shark.

Figure 6.7

1. Flexors
2. Cloaca
3. Clasper
4. Siphon

Figure 6.9(a)

1. Thyroid gland
2. Interhyoideus
3. Coracohyoid
4. Common coracoarcual
5. Coracoid bar
6. Flexors (ventral side of fin)
7. Linea alba
8. Ventral longitudinal bundle
9. Ventral constrictors
10. Coracomandibularis
11. Intermandibularis
12. Adductor mandibulae
13. Meckel's cartilage
14. Inferior rectus
15. Inferior oblique
16. Nasal capsule

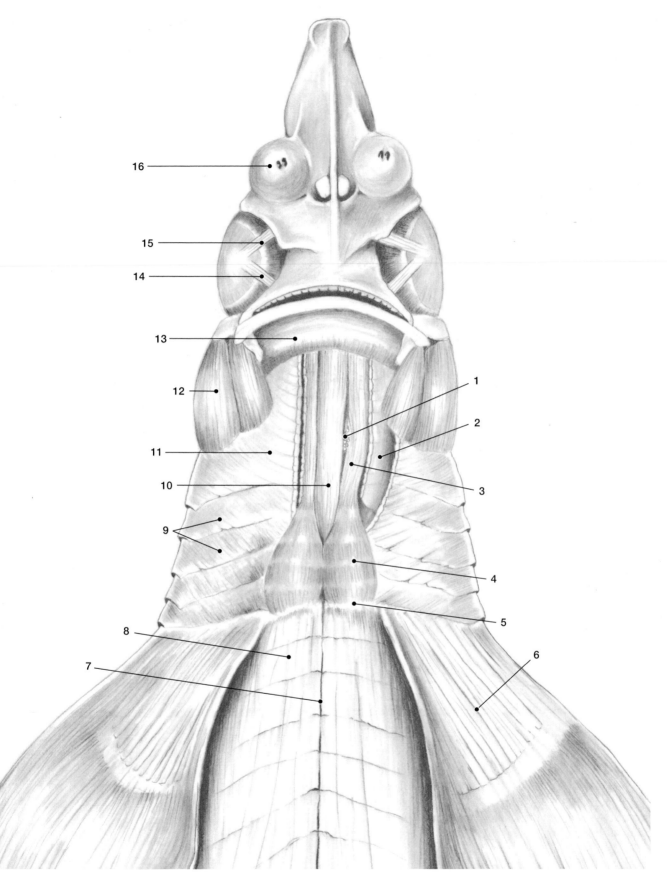

FIGURE 6.9 Trunk, branchial, and cephalic musculature of a shark. (a) Ventral view and (b) dorsal view.

Figure 6.9(b)

1. Levator hyomandibulae (epihyoideus)
2. Gill slits
3. Dorsal constrictors
4. Cucullaris
5. Scapular process
6. Extensors (dorsal side of fin)
7. Ventral longitudinal bundle
8. Lateral longitudinal bundle
9. Horizontal septum
10. Dorsal longitudinal bundles
11. Interarcuals
12. Epibranchial cartilage
13. Pharyngobranchial cartilage
14. Spiracle
15. Adductor mandibulae
16. Spiracularis
17. Levator palatoquadrati
18. Lateral rectus
19. Superior rectus
20. Superior oblique

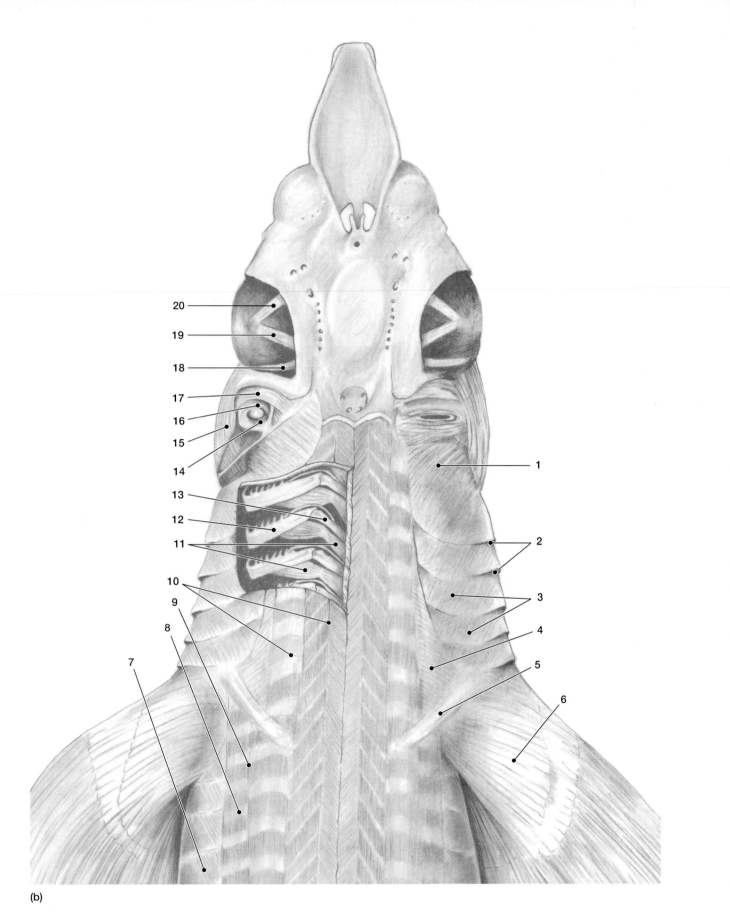

FIGURE 6.9 *Continued.*

Figure 6.10

1. Auditory nerve (VIII)
2. Auricle of cerebellum
3. Cerebral hemisphere
4. Cerebellum
5. Deep ophthalmic branch (V)
6. Diencephalon
7. Dorsal root of spinal nerve
8. Geniculate ganglion
9. Glosspharyngeal nerve (IX)
10. Hyomandibular branch (VII)
11. Hypobranchial nerve
12. Inferior oblique muscle
13. Inferior rectus muscle
14. Infraorbital nerve
15. Lateral rectus muscle
16. Mandibular nerve (V)
17. Medial rectus muscle
18. Medulla oblongata
19. Occipital nerves
20. Oculomotor nerve (III)
21. Olfactory bulb
22. Olfactory sac
23. Olfactory tract
24. Optic lobe
25. Optic nerve
26. Otic capsule
27. Palatine branch (VII)
28. Petrosal ganglion
29. Semicircular canals
30. Spinal cord
31. Spinal ganglion
32. Spiracle
33. Superficial ophthalmic trunk
34. Superior oblique muscle
35. Superior rectus muscle
36. Trochlear nerve (IV)
37. Vagus nerve (X)

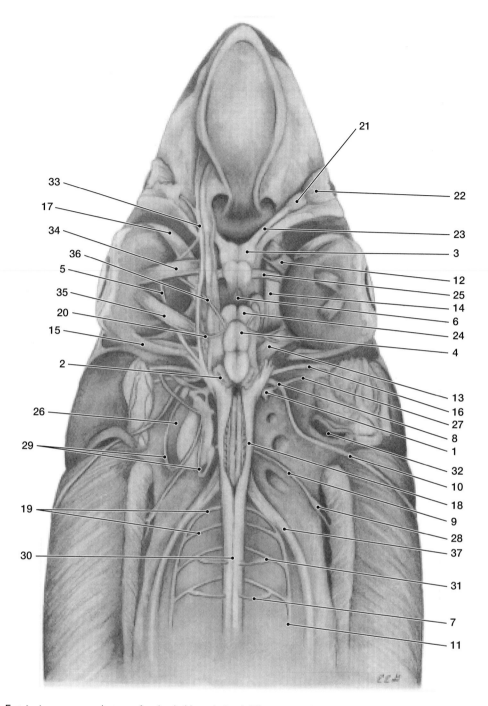

FIGURE 6.10 Extrinsic eye musculature of a shark (dorsal view). The extrinsic eye muscles are derived from somitomeres and rotate the eyeball within the orbit in order to direct the gaze. The roof of the chondrocranium has been removed and the auditory region dissected on the left to reveal the semicircular canals. On the right, the superior oblique, superior rectus, medial rectus, and lateral rectus muscles have been cut to reveal the inferior oblique and inferior rectus muscles.

Figure 6.11

1. Subarcual
2. Rectus cervicis
3. Procoracohumeralis
4. Omoarcual
5. Supracoracoideus
6. Pectoralis
7. Rectus abdominis
8. Linea alba
9. Coracobrachialis
10. Humeroantebrachialis
11. Gular fold
12. Interhyoideus
13. Branchiohyoideus
14. Geniohyoid
15. Intermandibularis
16. Median raphe

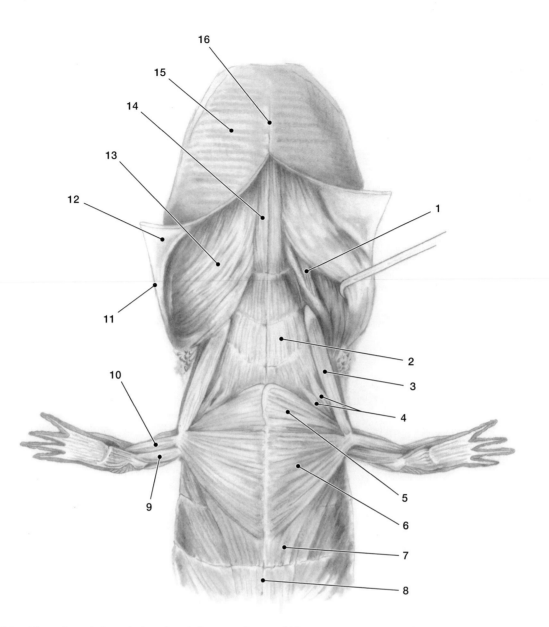

FIGURE 6.11 Ventral trunk, branchial, and cephalic musculature of *Necturus.*

Figure 6.12(a)

1. Horizontal septum
2. Dorsalis trunci
3. Latissimus dorsi
4. Dorsalis scapulae
5. Dilatator laryngis
6. Levatores arcuum
7. Branchiohyoideus
8. Levator mandibulae externus
9. Depressor mandibulae
10. Intermandibularis
11. Interhyoideus
12. Procoracohumeralis
13. Rectus cervicis
14. Pectoriscapularis
15. Cucullaris
16. Humeroantebrachialis
17. Extensors
18. Triceps brachii
19. External oblique

Figure 6.12(b)

1. Dilatator laryngis
2. Levatores arcuum
3. Cut ends of gills
4. Levator mandibulae externus
5. Levator mandibulae anterior
6. Adductor mandibulae
7. Depressor mandibulae
8. Branchiohyoideus
9. Dorsalis trunci

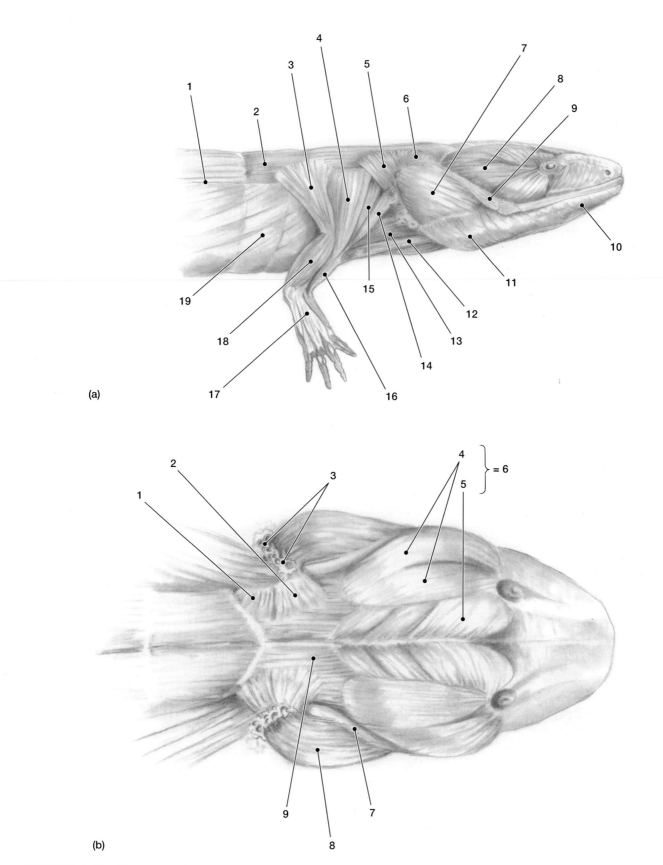

(a)

(b)

FIGURE 6.12 Trunk, branchial, and cephalic musculature of *Necturus*. (a) Lateral and (b) dorsal view.

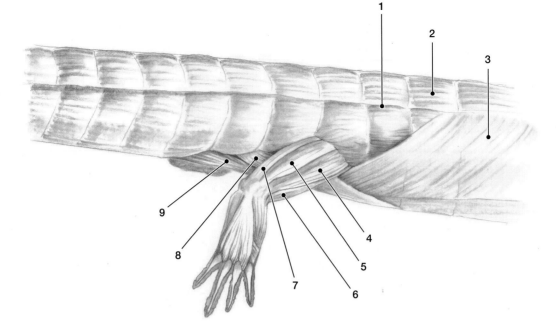

FIGURE 6.13 Lateral trunk and hindlimb musculature of *Necturus*.

Figure 6.13

1. Horizontal septum
2. Dorsalis trunci
3. External oblique
4. Pubo-ischio-femoralis internus
5. Iliotibialis
6. Pubotibialis
7. Ilioextensorius
8. Caudofemoralis
9. Caudocruralis

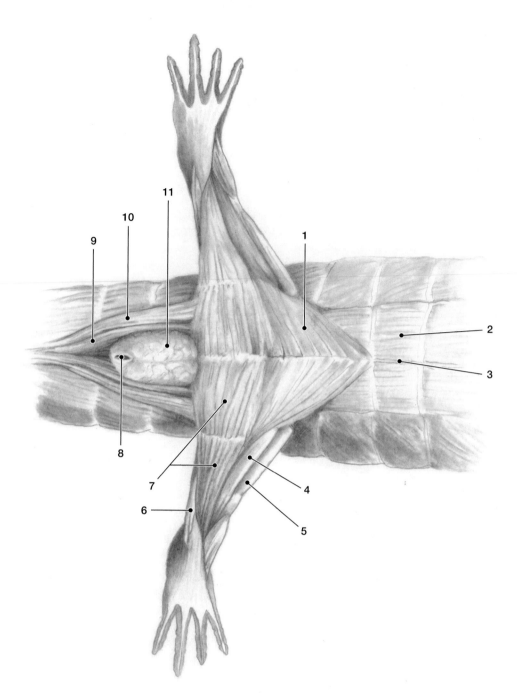

FIGURE 6.14 Ventral trunk and hindlimb musculature of *Necturus*.

Figure 6.14

1. Pubo-ischio-femoralis externus
2. Rectus abdominis
3. Linea alba
4. Pubotibialis
5. Pubo-ischio-femoralis internus
6. Ischioflexorius
7. Pubo-ischio-tibialis
8. Cloacal opening
9. Ischiocaudalis
10. Caudocruralis
11. Cloacal gland

Figure 6.15(a)

1. Digastric
2. Mylohyoid
3. Masseter
4. Geniohyoid
5. Lymph node
6. Salivary glands
7. Sternohyoid
8. Sternomastoid
9. Clavotrapezius
10. Cleidomastoid
11. Clavicle encased in connective tissue
12. Pectoralis major
13. Pectoantebrachialis
14. Clavobrachialis
15. Pectoralis major
16. Epitrochlearis
17. Pectoralis minor
18. Latissimus dorsi
19. Xiphihumeralis

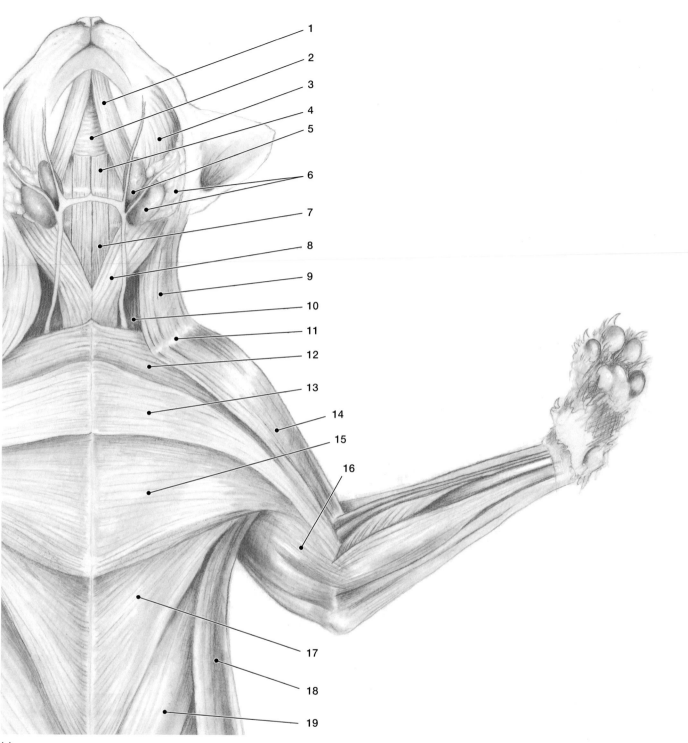

1
2
3
4
5
6
7
8
9
10
11
12
13
14
15
16
17
18
19

(a)

FIGURE 6.15 Superficial ventral trunk, brachial, and cephalic musculature of a cat.

Figure 6.15(b)

1. Cut edge of the mylohyoid
2. Digastric
3. Geniohyoid
4. Hyoglossus
5. Thyrohyoid
6. Levator scapulae ventralis
7. Sternohyoid
8. Sternothyroid
9. Coracobrachialis
10. Subscapularis
11. Transversus costarum
12. Biceps brachii
13. Teres major
14. Medial head of triceps brachii
15. Long head of triceps brachii
16. Scalenus
17. Serratus ventralis
18. Rectus abdominis
19. External intercostal
20. Latissimus dorsi
21. External oblique
22. Transversus
23. Internal oblique
24. Xiphihumeralis
25. Reflected end of pectoantebrachialis
26. Reflected end of pectoralis majo
27. Pectoralis minor
28. Sternomastoid

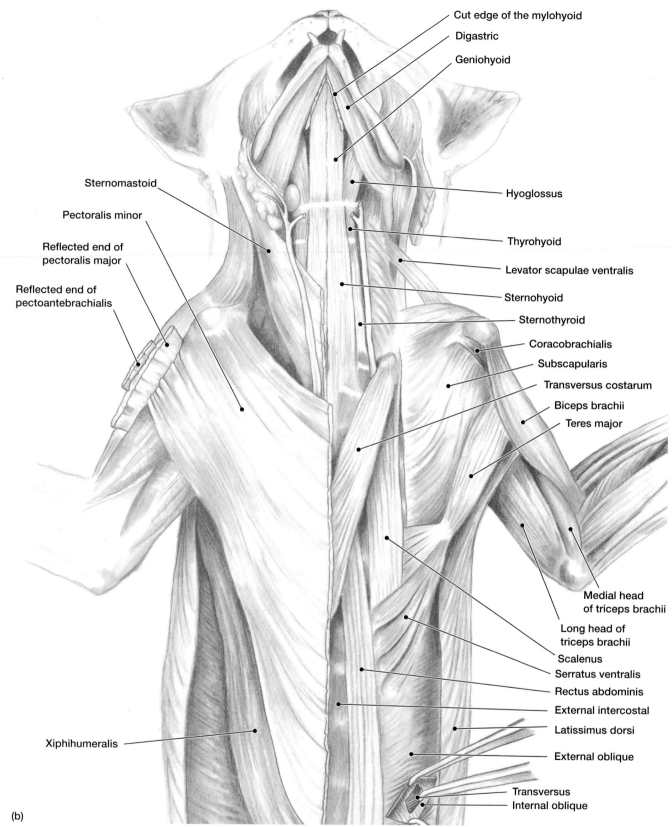

Cut edge of the mylohyoid
Digastric
Geniohyoid
Hyoglossus
Thyrohyoid
Levator scapulae ventralis
Sternohyoid
Sternothyroid
Coracobrachialis
Subscapularis
Transversus costarum
Biceps brachii
Teres major
Medial head of triceps brachii
Long head of triceps brachii
Scalenus
Serratus ventralis
Rectus abdominis
External intercostal
Latissimus dorsi
External oblique
Transversus
Internal oblique

Sternomastoid
Pectoralis minor
Reflected end of pectoralis major
Reflected end of pectoantebrachialis
Xiphihumeralis

(b)

FIGURE 6.15 *Continued,* deep ventral trunk, brachial, and cephalic musculature of a cat.

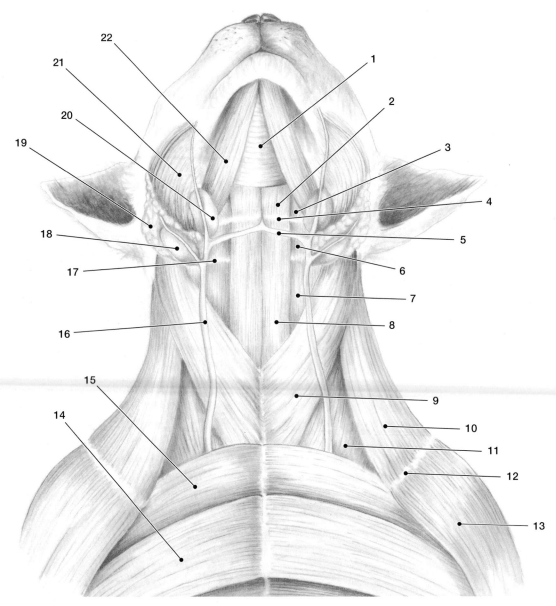

(a)

FIGURE 6.16 Head and neck musculature of a cat. (a) Ventral view. (b) Diagram indicating the origination of the names of some neck muscles based upon their origins and insertions.

Figure 6.16(a)

1. Mylohyoid
2. Geniohyoid
3. Hyoglossus
4. Hyoid
5. Transverse jugular vein
6. Thyrohyoid
7. Sternothyroid
8. Sternohyoid
9. Sternomastoid
10. Clavotrapezius
11. Cleidomastoid
12. Clavicle
13. Clavobrachialis
14. Pectoantebrachialis
15. Pectoralis major
16. External jugular vein
17. Thyroid cartilage
18. Submaxillary gland
19. Parotid gland
20. Lymph node
21. Masseter
22. Digastric

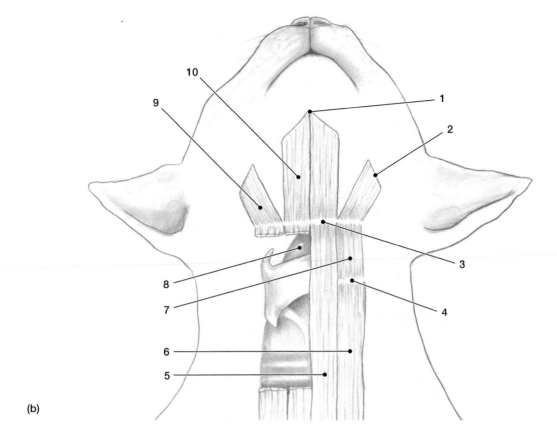

(b)

FIGURE 6.16 *Continued.*

Figure 6.16(b)

1. Chin (genio = chin)
2. Base of tongue (glossus = tongue)
3. Hyoid
4. Thyroid cartilage
5. Sternohyoid
6. Sternothyroid
7. Thyrohyoid
8. Cricothyroid
9. Hyoglossus
10. Geniohyoid

Figure 6.17

1. Spinotrapezius
2. Latissimus dorsi
3. Acromiotrapezius
4. Spinodeltoid
5. Levator scapulae ventralis
6. Clavotrapezius
7. Sternomastoid
8. Submaxillary gland
9. Parotid gland
10. Temporalis
11. Parotid duct
12. Masseter
13. Lymph node
14. Anterior facial vein
15. Posterior facial vein
16. External jugular vein
17. Clavobrachialis
18. Acromiodeltoid
19. Lateral head of triceps brachii
20. Long head of triceps brachii

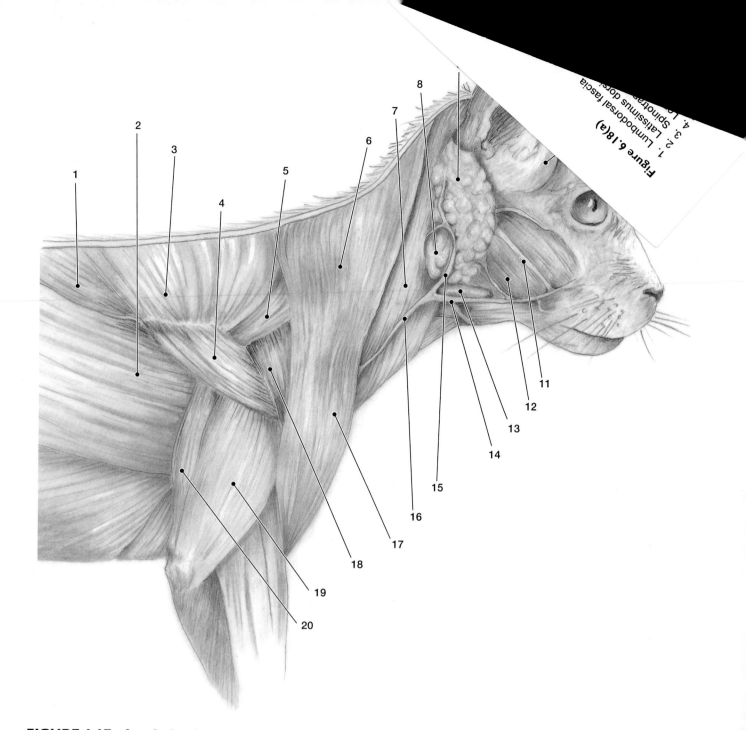

Figure 6.18(a)

1. Lo...
2. Lumbodorsal fascia
3. Spinotrapezius dorsa...
4. Lati...

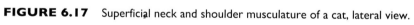

FIGURE 6.17 Superficial neck and shoulder musculature of a cat, lateral view.

                                   r

                          pezius

                  ng head of triceps brachii

5. Lateral head of triceps brachii

6. Extensor digitorum communis

7. Extensor digitorum lateralis

8. Extensor carpi ulnaris

9. Extensor carpi radialis longus

10. Brachioradialis

11. Spinodeltoid

12. Acromiodeltoid

13. Clavobrachialis

14. Levator scapulae ventralis

15. Acromiotrapezius

16. Clavotrapezius

17. Temporalis

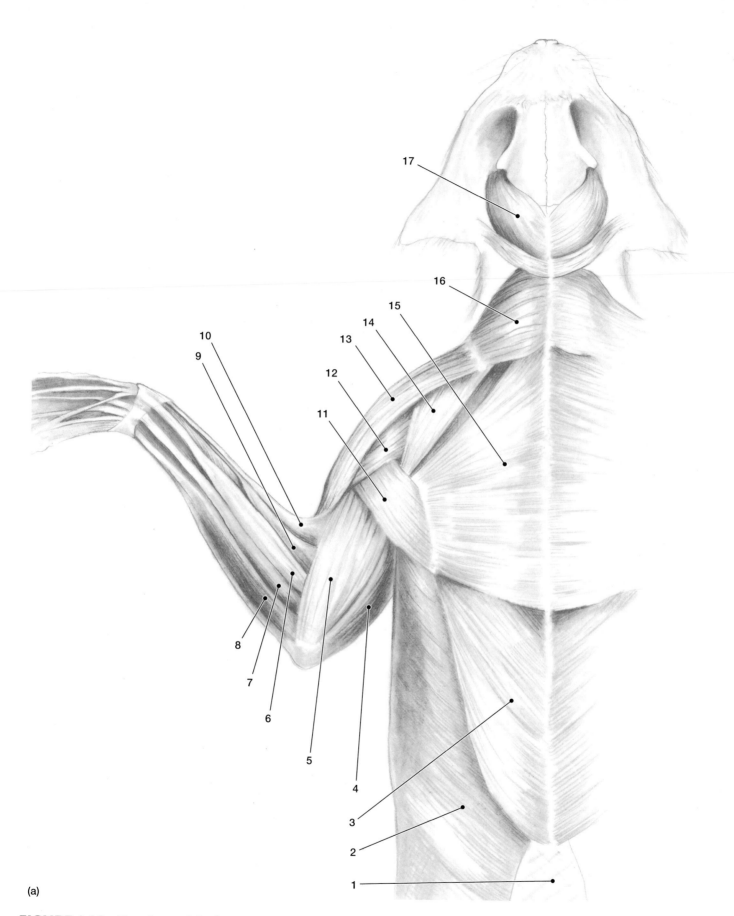

FIGURE 6.18 Musculature of the dorsal left forelimb of a cat. Superficial.

(a)

Figure 6.18(b)

1. Serratus ventralis
2. Teres major
3. Long head of triceps brachii
4. Reflected end of extensor digitorum communis
5. Reflected end of extensor digitorum lateralis
6. Extensor carpi ulnaris
7. Supinator
8. Extensor carpi radialis brevis
9. Abductor pollicis longus
10. Extensor carpi radialis longus
11. Brachialis
12. Medial head of triceps brachii
13. Infraspinatus
14. Spine of scapula
15. Head of humerus
16. Supraspinatus
17. Rhomboideus
18. Rhomboideus capitis
19. Splenius

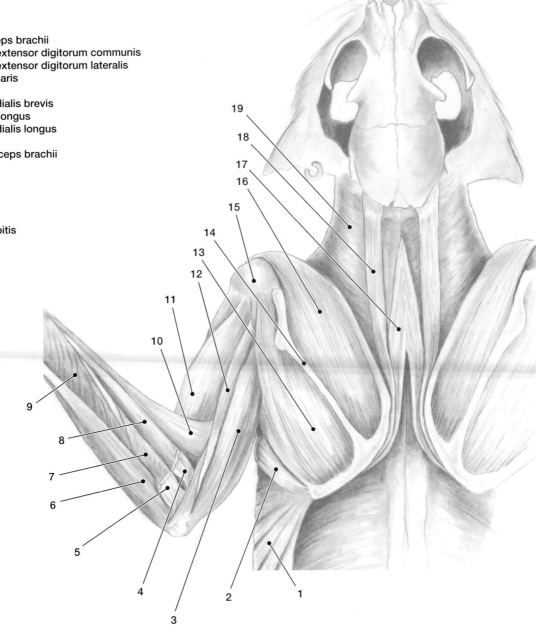

(b)

FIGURE 6.18 *Continued,* Musculature of the dorsal left forelimb of a cat. Deep.

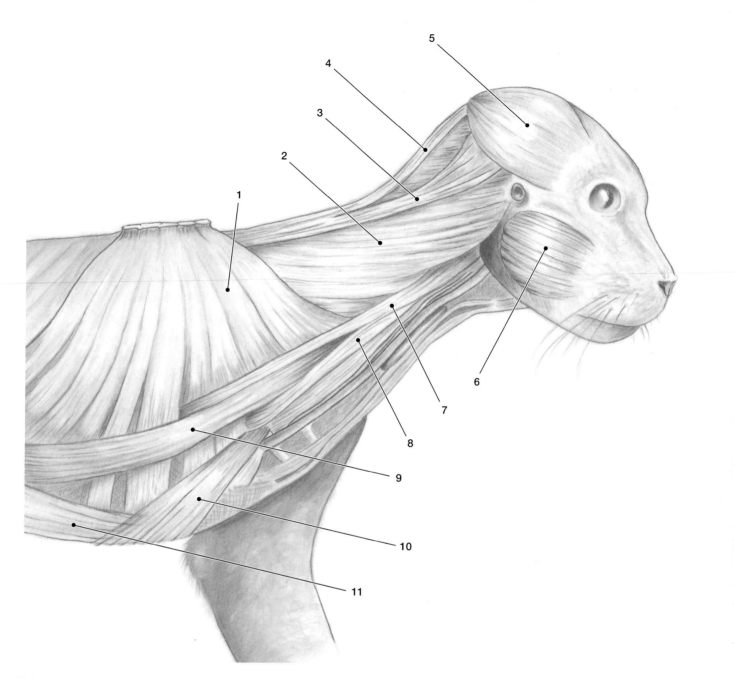

FIGURE 6.19 Lateral thoracic muscles of a cat.

Figure 6.19

1. Serratus ventralis
2. Splenius
3. Rhomboideus capitis
4. Rhomboideus
5. Temporalis
6. Masseter
7. Scalenus posterior
8. Scalenus anterior
9. Scalenus medius
10. Transversus costarum
11. Rectus abdominis

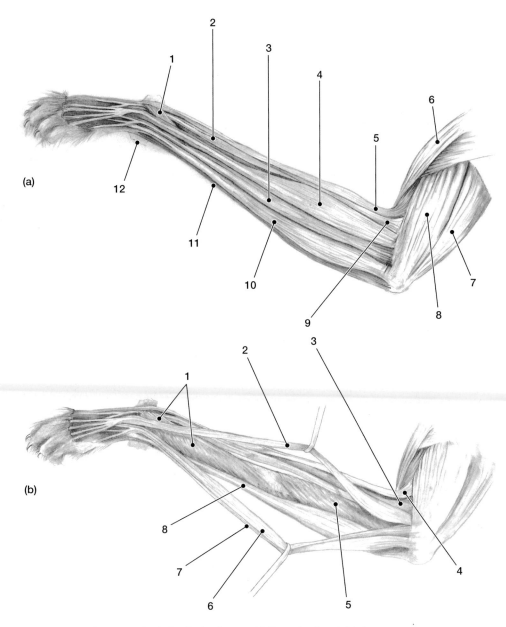

FIGURE 6.20 Musculature of the dorsal left forelimb of a cat. (a) Superficial and (b) deep.

Figure 6.20(a)

1. Abductor pollicis longus
2. Extensor carpi radialis brevis
3. Extensor digitorum lateralis
4. Extensor digitorum communis
5. Brachioradialis
6. Clavobrachialis
7. Triceps brachii (long head)
8. Triceps brachii (lateral head)
9. Extensor carpi radialis (longus)
10. Extensor carpi ulnaris
11. Flexor digitorum profundus
12. Retinaculum

Figure 6.20(b)

1. Abductor pollicis longus
2. Extensor digitorum communis
3. Extensor carpi radialis (longus)
4. Brachioradialis
5. Supinator
6. Extensor digitorum lateralis
7. Extensor carpi ulnaris
8. Flexor digitorum profundus

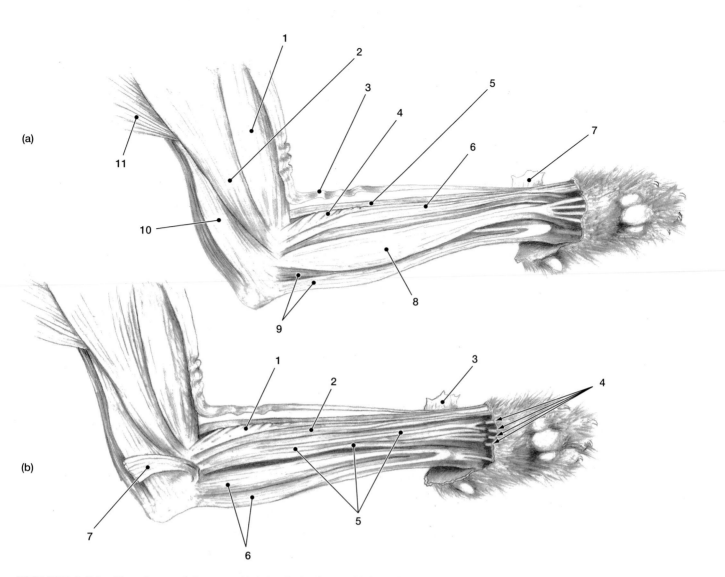

(a)

(b)

FIGURE 6.21 Musculature of the ventral left forelimb of a cat. (a) Superficial and (b) deep.

Figure 6.21(a)

1. Clavobrachialis
2. Pectoantebrachialis
3. Brachioradialis
4. Pronator teres
5. Extensor carpi radialis (longus)
6. Flexor carpi radialis
7. Retinaculum
8. Flexor digitorum superficialis (long head)
9. Flexor carpi ulnaris
10. Epitrochlearis
11. Pectoralis major

Figure 6.21(b)

1. Pronator teres
2. Flexor carpi radialis
3. Retinaculum
4. Cut ends of tendons of the flexor digitorum superficialis (long head)
5. Flexor digitorum profundus
6. Flexor carpi ulnaris
7. Flexor digitorum superficialis (long head)

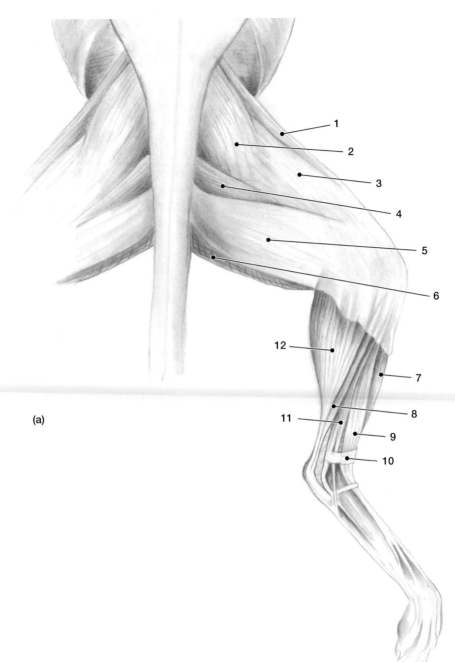

FIGURE 6.22 Lateral musculature of cat hindlimb. Superficial.

(a)

Figure 6.22(a)

1. Sartorius
2. Tensor fasciae latae
3. Fascia lata
4. Caudofemoralis
5. Biceps femoris
6. Semitendinosus
7. Tibialis anterior
8. Soleus
9. Extensor digitorum longus
10. Retinaculum
11. Peroneus
12. Gastrocnemius

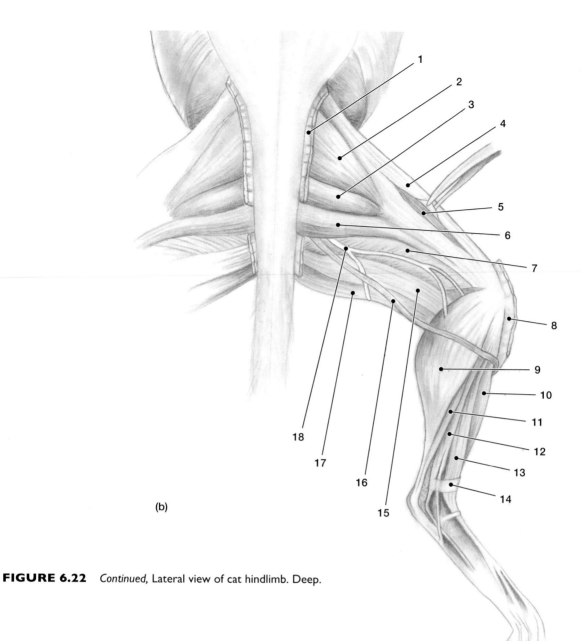

(b)

FIGURE 6.22 *Continued,* Lateral view of cat hindlimb. Deep.

Figure 6.22(b)

1. Reflected edge of tensor fasciae latae
2. Gluteus medius
3. Gluteus maximus
4. Sartorius
5. Vastus lateralis
6. Caudofemoralis
7. Adductor femoris
8. Reflected edge of biceps femoris
9. Gastrocnemius
10. Tibialis anterior
11. Soleus
12. Peroneus
13. Extensor digitorum longus
14. Retinaculum
15. Semimembranosus
16. Tenuissimus
17. Semitendinosus
18. Sciatic nerve

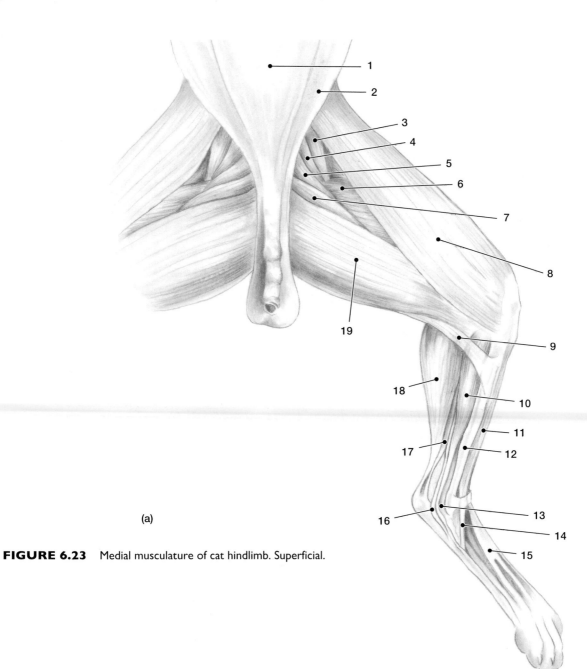

(a)

FIGURE 6.23 Medial musculature of cat hindlimb. Superficial.

Figure 6.23(a)

1. Rectus abdominis
2. External oblique
3. Iliopsoas
4. Pectineus
5. Adductor longus
6. Vastus medialis
7. Adductor femoris
8. Sartorius
9. Semitendinosus
10. Flexor digitorum longus
11. Tibialis anterior
12. Tibia
13. Tendon of tibialis posterior
14. Tendon of tibialis anterior
15. Tendons of extensor digitorum longus
16. Flexor hallucis longus
17. Plantaris
18. Gastrocnemius
19. Gracilis

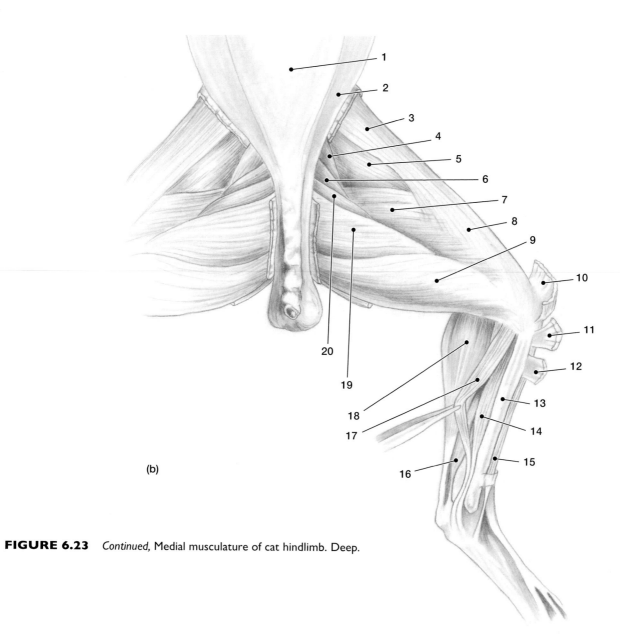

FIGURE 6.23 *Continued,* Medial musculature of cat hindlimb. Deep.

(b)

Figure 6.23(b)

1. Rectus abdominis
2. External oblique
3. Tensor fasciae latae
4. Iliopsoas
5. Rectus femoris
6. Pectineus
7. Vastus medialis
8. Fascia lata
9. Semimembranosus
10. Reflected end of sartorius
11. Reflected end of gracilis
12. Reflected end of semitendinosus
13. Tibia
14. Tibialis posterior
15. Tibialis anterior
16. Plantaris
17. Flexor digitorum longus
18. Gastrocnemius
19. Adductor femoris
20. Adductor longus

Figure 7.3

1. Transverse septum
2. Median lobe of the liver
3. Gallbladder
4. Left lobe of the liver
5. Papillae of the esophagus
6. Gonad (testis or ovary)
7. Rugae of the stomach
8. Gastrohepatic ligament
9. Pylorus of stomach
10. Lesser curvature of the stomach
11. Body of stomach
12. Greater curvature of the stomach
13. Dorsal lobe of the pancreas
14. Spleen
15. Kidney
16. Mesorectum
17. Cloaca
18. Colon
19. Rectal gland
20. Valvular intestine
21. Spiral valve
22. Duodenum
23. Ventral lobe of the pancreas
24. Hepatoduodenal ligament
25. Right lobe of the liver

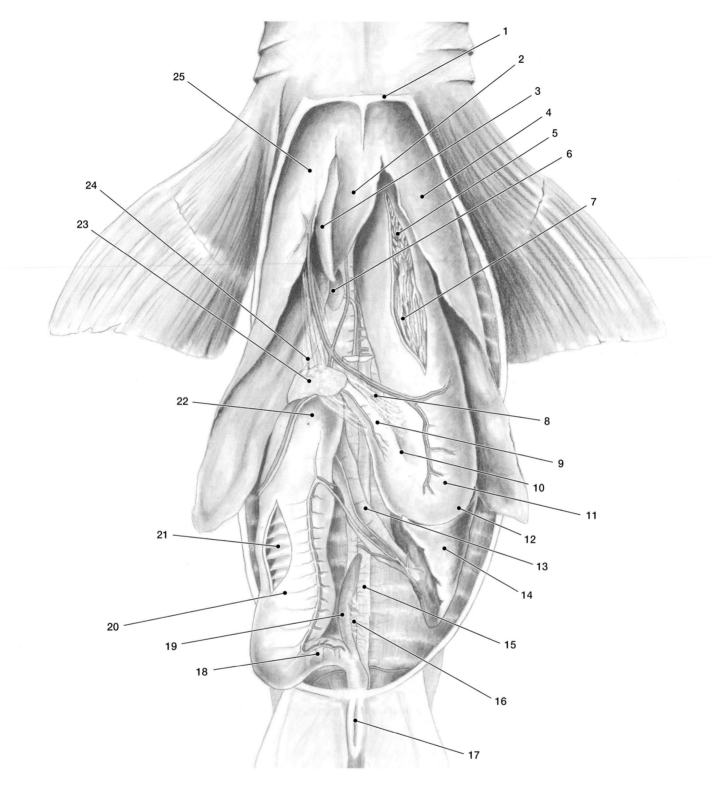

FIGURE 7.3 Ventral view of the pleuroperitoneal cavity of a shark.

Figure 7.4

1. Median lobe of the liver
2. Gallbladder
3. Hepatic artery
4. Celiac artery
5. Dorsal gastric artery and vein
6. Ventral gastric artery and vein
7. Dorsal gastric artery and vein
8. Anterior lienogastric vein
9. Lienomesenteric vein
10. Dorsal lobe of the pancreas
11. Spleen
12. Posterior lienogastric vein
13. Gastrosplenic (lienogastric) artery
14. Kidney
15. Posterior mesenteric artery
16. Rectal gland
17. Colon
18. Valvular intestine
19. Posterior intestinal vein
20. Anterior mesenteric artery
21. Anterior intestinal artery and vein
22. Ventral lobe of the pancreas
23. Pancreaticomesenteric vein
24. Pancreaticomesenteric artery
25. Common bile duct
26. Gastric artery and vein
27. Choledochal vein
28. Right lobe of the liver
29. Hepatic portal vein

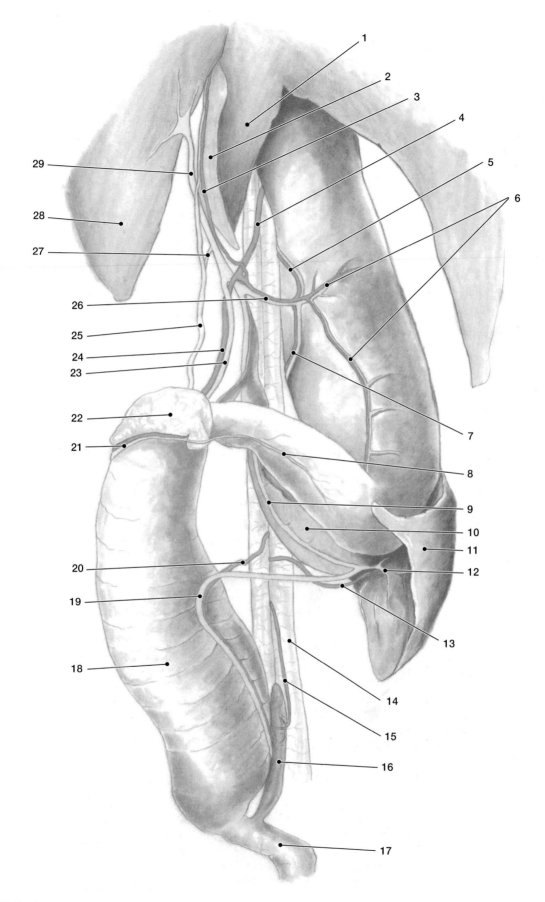

FIGURE 7.4 Blood supply to alimentary canal of a shark.

Figure 7.6

1. Left atrium
2. Ventricle
3. Left lung
4. Pulmonary ligament
5. Oviduct
6. Mesotubarium
7. Dorsal aorta
8. Ovary
9. Mesovarium
10. Postcaval vein (postcava)
11. Kidney
12. Cloacal gland
13. Cloaca
14. Large intestine
15. Urinary bladder
16. Small intestine
17. Mesentery
18. Plicae
19. Gallbladder
20. Liver
21. Duodenum
22. Pancreas
23. Pyloric sphincter
24. Stomach
25. Rugae
26. Gastrosplenic ligament
27. Spleen
28. Gastrohepatic ligament
29. Transverse septum

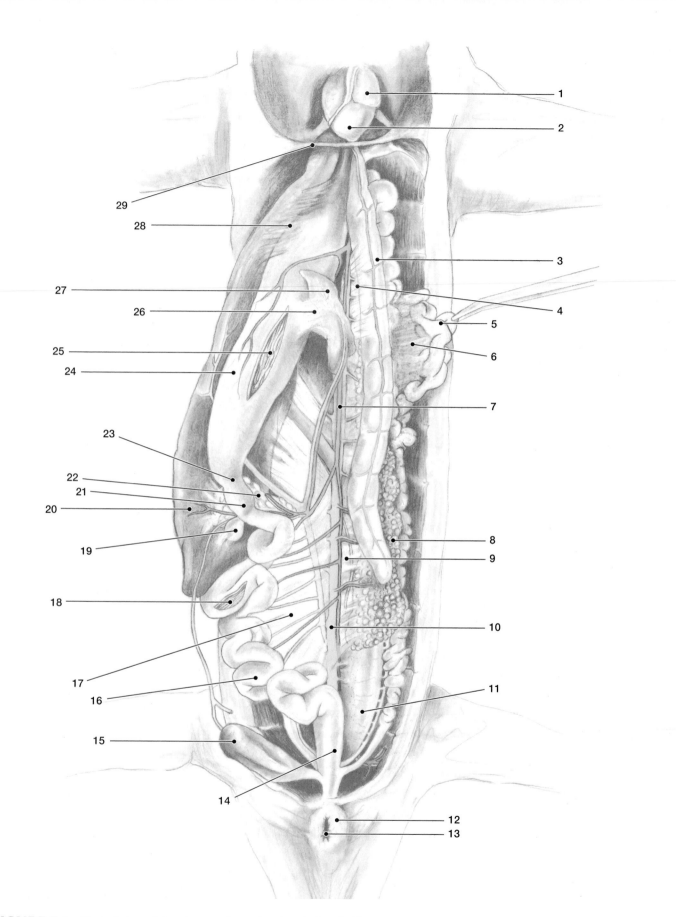

FIGURE 7.6 Ventral view of the pleuroperitoneal cavity of a female *Necturus*.

Figure 7.7

1. Clavotrapezius
2. Posterior facial vein
3. Submaxillary gland
4. Parotid gland
5. Branch of the facial nerve
6. Mandible
7. Molar gland
8. Masseter
9. Parotid duct
10. Digastric
11. Branch of the facial nerve
12. Anterior facial vein
13. Sublingual gland
14. Transverse jugular vein
15. External jugular vein

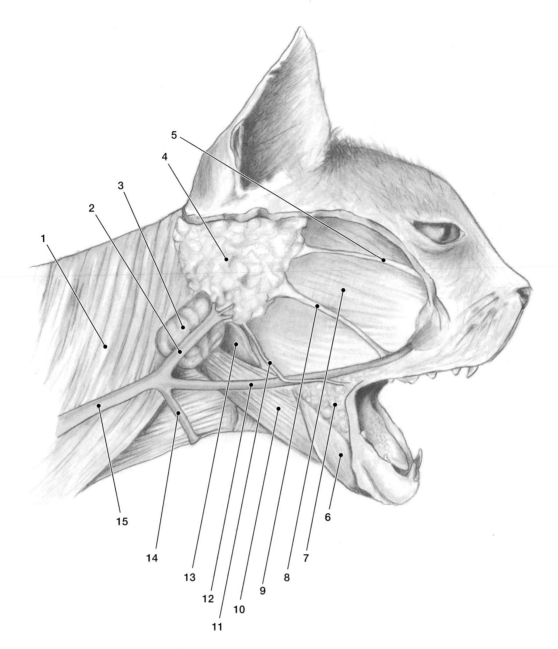

FIGURE 7.7 Lateral view of the neck and face of a cat.

Figure 7.8(a)

1. Spinal cord
2. Esophagus
3. Cerebellum
4. Internal nares
5. Massa intermedia
6. Cerebrum
7. Labia
8. Hard palate
9. Buccal cavity
10. Lingual frenulum
11. Sublingual papilla
12. Soft palate
13. Palatine tonsils
14. Nasopharynx
15. Oropharynx
16. Openings into the auditory tubes
17. Laryngopharynx
18. Hyoid bone
19. Epiglottis
20. Arytenoid cartilage
21. C-shaped cartilages of the trachea

Figure 7.8(b)

1. Filiform papillae
2. Fungiform papillae
3. Vallate papillae
4. Foliate papillae

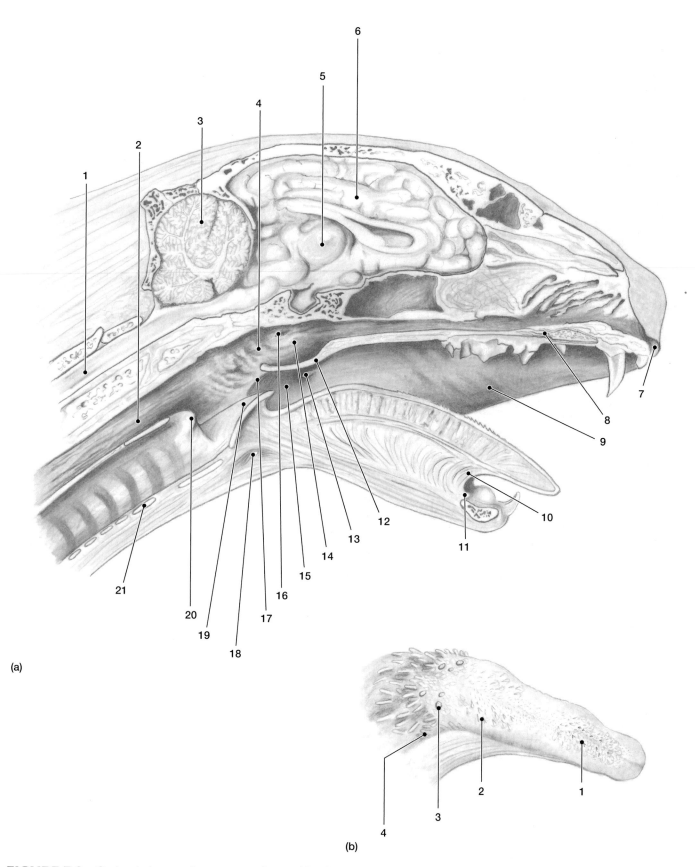

(a)

(b)

FIGURE 7.8 Oral and pharyngeal structures in bisected head and neck of a cat. (a) Midsagittal section. (b) Dorsal three-quarter view of the tongue.

Figure 7.9(a)

1. Thymus
2. Left anterior lobe of the lung
3. Heart
4. Left middle lobe of the lung
5. Left posterior lobe of the lung
6. Diaphragm
7. Left medial lobe of the liver
8. Quadrate lobe of the liver
9. Left lateral lobe of the liver
10. Spleen
11. Urinary bladder
12. Small intestine covered by the greater omentum
13. Right medial lobe of the liver
14. Gallbladder
15. Right posterior lobe of the lung
16. Right middle lobe of the lung
17. Right anterior lobe of the lung

Figure 7.9(b)

1. Epiglottis
2. Thyroid cartilage
3. Cricoid cartilage
4. Trachea

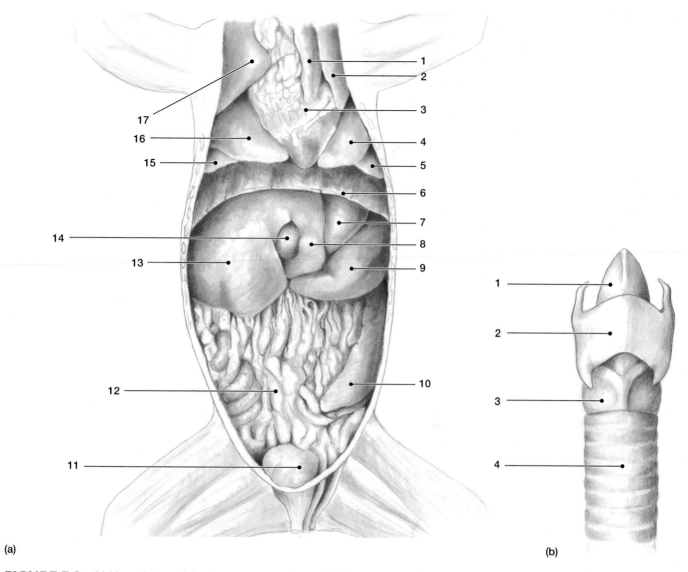

(a)

(b)

FIGURE 7.9   (a) Ventral view of the thoracic cavity of a cat. (b) Ventral view of larynx, cat.

Figure 7.10

1. Left anterior lobe of the lung
2. Left atrium
3. Left middle lobe of the lung
4. Left posterior lobe of the lung
5. Diaphragm
6. Left medial lobe of the liver
7. Quadrate lobe of the liver
8. Left lateral lobe of the liver
9. Body or corpus of the stomach
10. Gastrohepatic ligament
11. Lesser curvature of the stomach
12. Pyloric sphincter
13. Greater curvature of the stomach
14. Gastrosplenic ligament
15. Spleen
16. Small intestine
17. Mesentery
18. Descending colon
19. Urinary bladder
20. Rectum
21. Cecum
22. Ascending colon
23. Transverse colon
24. Pyloric region of the stomach
25. Head of the pancreas
26. Kidney
27. Duodenum
28. Hepatorenal ligament
29. Posterior part of the right lateral lobe of the liver
30. Hepatoduodenal ligament
31. Anterior part of the right lateral lobe of the liver
32. Right medial lobe of the liver
33. Gallbladder
34. Right posterior lobe of the lung
35. Right middle lobe of the lung
36. Coronary artery
37. Right atrium
38. Right anterior lobe of the lung

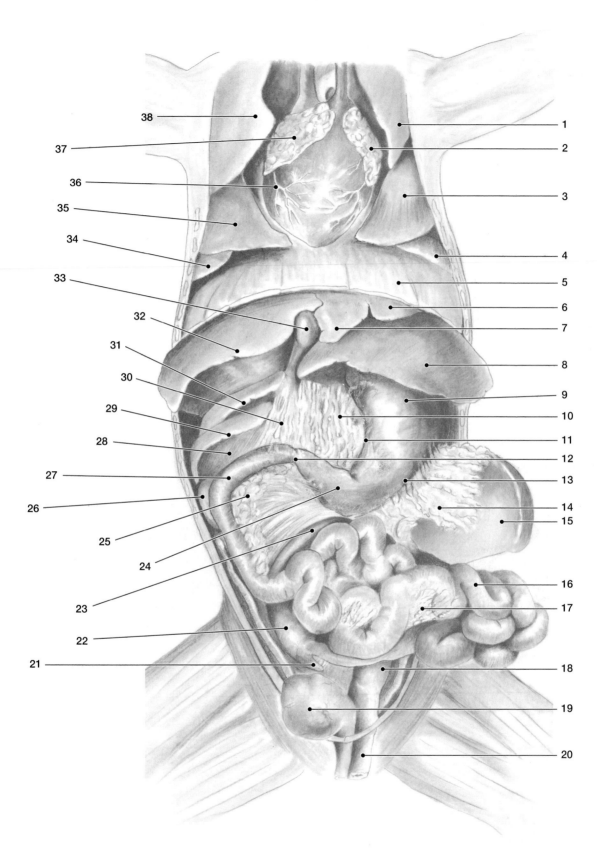

38 —
37 —
36 —
35 —
34 —
33 —
32 —
31 —
30 —
29 —
28 —
27 —
26 —
25 —
24 —
23 —
22 —
21 —

— 1
— 2
— 3
— 4
— 5
— 6
— 7
— 8
— 9
— 10
— 11
— 12
— 13
— 14
— 15
— 16
— 17
— 18
— 19
— 20

FIGURE 7.10 Abdominal structures in a cat, viewed ventrally upon initial inspection.

Figure 8.2

1. Ventral aorta
2. Afferent branchial arteries
3. Conus arteriosus
4. Atrium
5. Ventricle
6. Hepatic sinus
7. Left lobe of liver
8. Stomach
9. Lateral abdominal vein
10. Subscapular vein
11. Brachial vein
12. Common cardinal vein
13. Subclavian vein
14. Inferior jugular vein
15. Sinus venosus
16. Coronary artery
17. Commissural artery

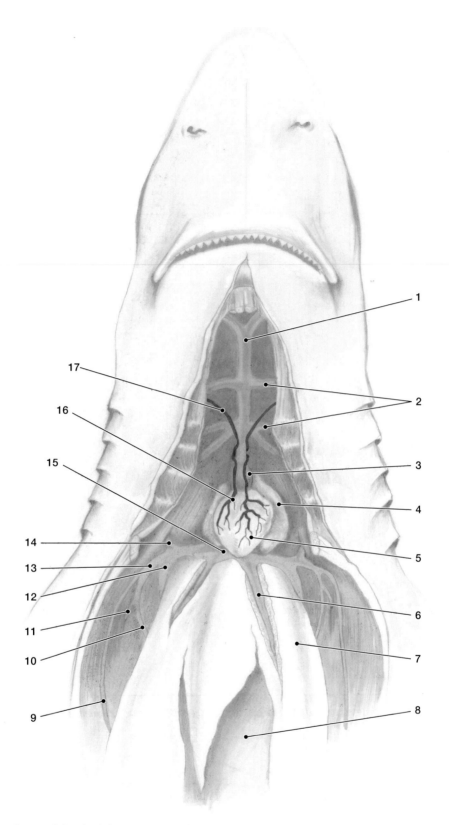

1

17

16

2

15

3

14

4

13

5

12

6

11

7

10

9

8

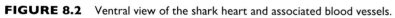

FIGURE 8.2 Ventral view of the shark heart and associated blood vessels.

Figure 8.3

1. Stapedial artery
2. Efferent spiracular artery
3. Spiracle
4. Afferent spiracular artery
5. 1st afferent branchial artery
6. External carotid artery
7. 2nd afferent branchial artery
8. Ventral aorta
9. Commissural artery
10. 3rd afferent branchial artery
11. 4th afferent branchial artery
12. Coracobranchial muscle
13. Coronary artery
14. Conus arteriosus
15. 5th afferent branchial artery
16. Atrium
17. Intertrematic branch
18. Posttrematic branch
19. Pretrematic branch
20. Efferent collector loop
21. Efferent branchial artery
22. Dorsal aorta
23. Subclavian artery
24. Coracoid bar
25. Gill lamellae
26. Interbranchial septum
27. External gill slit
28. Gill pouch
29. Epibranchial cartilage
30. Internal gill slit
31. Adductor muscle
32. Gill raker
33. Spiracle
34. Paired dorsal aortae
35. Hyoidean artery
36. Internal carotid artery

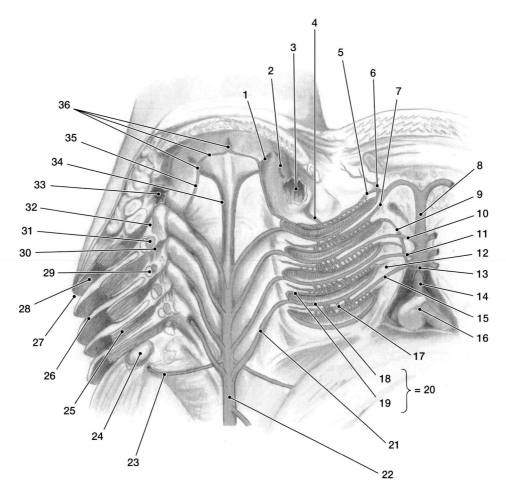

FIGURE 8.3 Ventral view of the branchial structure and vascular supply in a shark. The floor of the pharynx has been cut free caudal to the heart and along the right pharyngeal side.

Figure 8.7(a)

1. Paired dorsal aorta
2. Internal carotid artery
3. Efferent spiracular artery
4. Stapedial artery
5. Spiracle
6. Internal carotid artery
7. Afferent spiracular artery
8. Hyoidean artery
9. External carotid artery
10. Pretrematic branch
11. Posttrematic branch
12. Intertrematic branch
13. Commissural artery
14. Coronary artery
15. Efferent branchial artery
16. Subclavian artery
17. Dorsal aorta
18. Celiac artery
19. Genital arteries
20. Hepatic artery
21. Ventral gastric artery
22. Dorsal gastric artery
23. Pancreaticomesenteric artery
24. Duodenal artery
25. Unnamed branches to pylorus and ventral lobe of the pancreas
26. Anterior intestinal artery
27. Anterior mesenteric artery
28. Gastrosplenic (lienogastric) artery
29. Posterior mesenteric artery
30. Iliac artery
31. Caudal artery

Figure 8.7(b)

1. Sinus venosus
2. Hepatic sinus
3. Inferior jugular vein
4. Common cardinal vein
5. Subclavian vein
6. Lateral abdominal vein
7. Brachial vein
8. Subscapular vein
9. Posterior cardinal sinus
10. Posterior cardinal vein
11. Posterior cardinal vein
12. Efferent renal vein
13. Afferent renal vein
14. Renal portal vein
15. Caudal vein

Figure 8.7(c)

1. Ventral gastric vein
2. Dorsal gastric vein
3. Pyloric vein
4. Anterior lienogastric vein
5. Lienomesenteric vein
6. Posterior intestinal vein
7. Posterior lienogastric vein
8. Anterior intestinal vein
9. Pancreaticomesenteric vein
10. Choledochal vein
11. Hepatic portal vein

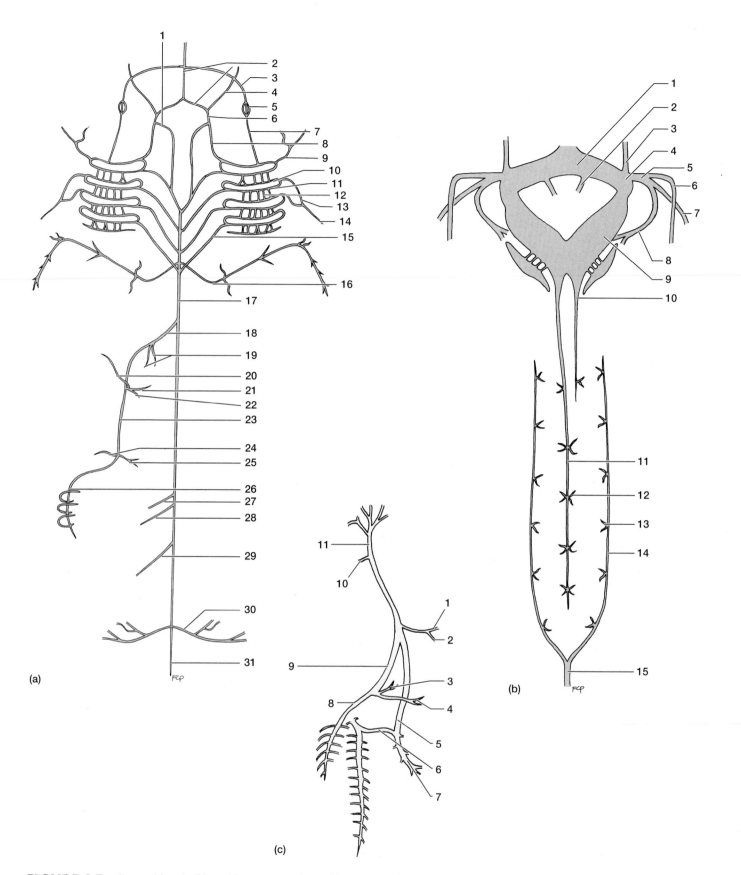

FIGURE 8.7 Artery (a), vein (b), and hepatic portal vein (c) maps in a shark, ventral view. The floor of the pharynx was rotated to the viewer's right.

Figure 8.8(a)

1. Left branch of the ventral aorta
2. Bulbus arteriosus
3. Left atrium
4. Afferent branchial arteries
5. External gills
6. Postcaval vein
7. Common cardinal vein
8. Ventricle
9. Conus arteriosus
10. Right atrium
11. External carotid artery
12. Thyroid gland

Figure 8.8(b)

1. Dorsal aorta
2. Subclavian artery
3. Pulmonary artery
4. Efferent branchial arteries
5. Roots or radices of the aorta
6. Vertebral artery
7. Internal carotid artery
8. Roof of the mouth

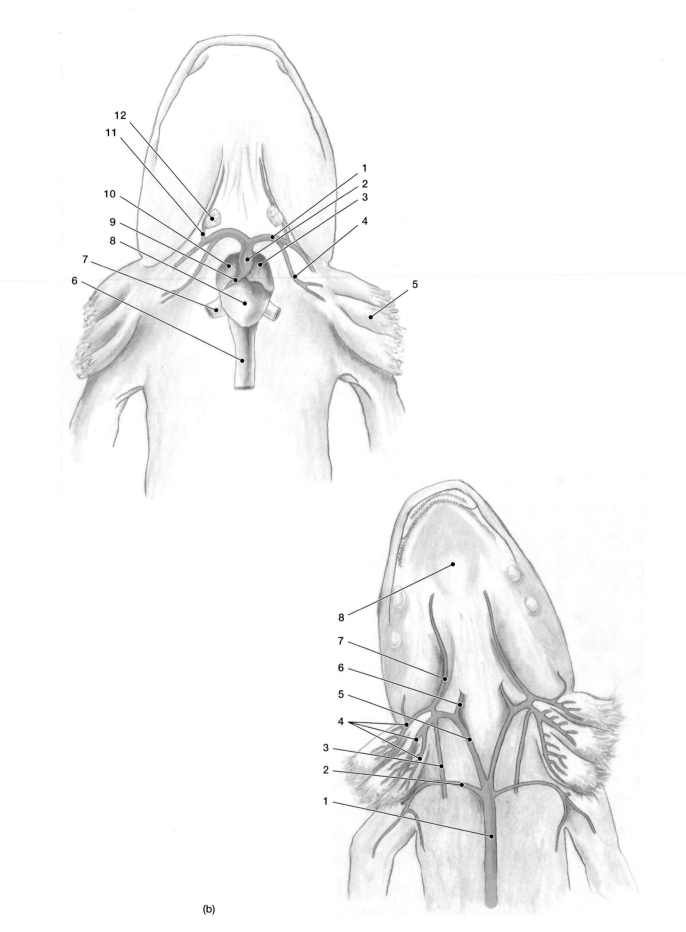

FIGURE 8.8 Heart chambers and associated blood vessels in *Necturus*. (a) Ventral view of heart and associated blood vessels. (b) Ventral view of blood vessels in the roof of the mouth.

Figure 8.12(a)

1. Internal carotid artery
2. Vertebral artery
3. External carotid artery
4. Efferent branchial arteries
5. Pulmonary artery
6. Roots or radices of the aorta
7. Subclavian artery
8. Dorsal gastric artery
9. Ventral gastric artery
10. Splenic artery
11. Hepatic artery
12. Pancreaticoduodenal artery
13. Mesenteric branch of the celiacomesenteric artery
14. Genital arteries
15. Renal arteries
16. Iliac artery
17. Epigastric artery
18. Femoral artery
19. Hypogastric artery
20. Cloacal artery
21. Caudal artery
22. Mesenteric arteries
23. Celiacomesenteric artery
24. Gastric artery
25. Dorsal aorta

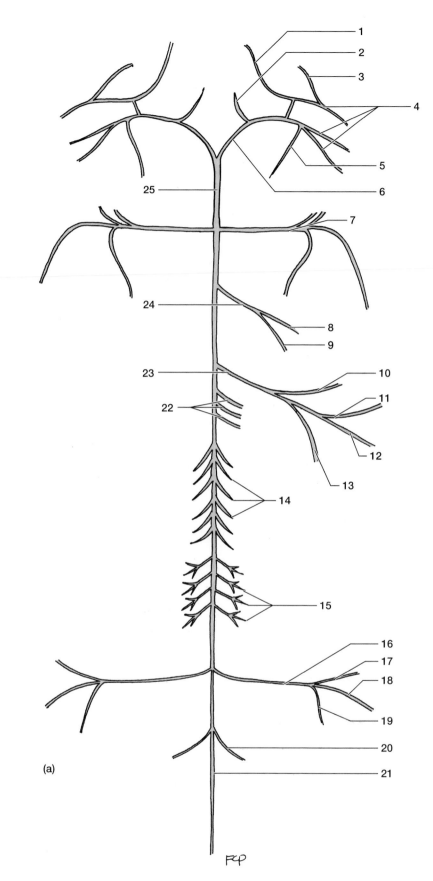

1
2
3
4
5
6
25
7
24
8
9
23
10
11
22
12
13
14
15
16
17
18
19
20
21

(a)

FₒP

FIGURE 8.12 Artery (a), vein (b), and hepatic portal vein (c) maps in *Necturus,* ventral view.

Figure 8.12(b)

1. Common cardinal vein
2. Sinus venosus
3. Hepatic sinus
4. Hepatic veins
5. Postcaval vein
6. Posterior cardinal vein
7. Posterior cardinal vein
8. Postcaval vein
9. Genital vein
10. Renal vein
11. Pelvic vein
12. Femoral vein
13. Caudal vein

Figure 8.12(c)

1. Hepatic portal vein
2. Gastrosplenic vein
3. Mesenteric vein
4. Intestinal veins
5. Pelvic vein
6. Femoral vein
7. Caudal vein
8. Ventral abdominal vein
9. Pancreaticoduodenal vein

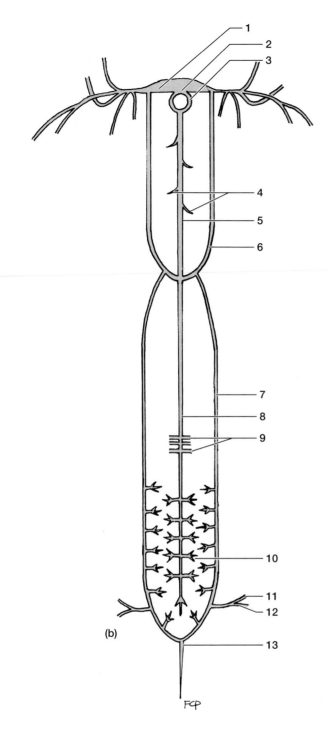

(b)

FCP

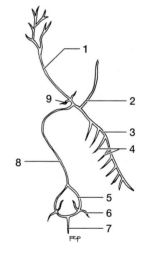

(c)

FCP

FIGURE 8.12 *Continued.*

Figure 8.13

1. External jugular vein
2. Internal jugular vein
3. Right common carotid artery
4. Trachea
5. Esophagus
6. Internal jugular vein
7. External jugular vein
8. Ventral thoracic vein
9. Left vertebral artery and vein
10. Left thyrocervical artery
11. Left costocervical artery and vein
12. Left axillary artery
13. Left common carotid artery
14. Brachiocephalic vein
15. Left internal mammary artery
16. Subclavian artery
17. Phrenic nerve
18. Vagus nerve
19. Internal mammary vein
20. Brachiocephalic artery
21. Systemic arch
22. Pulmonary trunk
23. Right atrium
24. Left atrium
25. Right ventricle
26. Coronary arteries and veins
27. Left ventricle
28. Latissimus dorsi
29. Precava
30. Internal mammary artery
31. Subclavian vein
32. Long thoracic artery and vein
33. Thoracodorsal artery and vein
34. Thoracodorsal shunt
35. Brachial artery and vein
36. Ulnar nerve
37. Median nerve
38. Deep brachial artery and vein
39. Biceps brachii
40. Anterior humeral circumflex artery and vein
41. Subscapular artery and vein
42. Axillary artery and vein
43. Ventral thoracic artery and vein
44. Axillary nerve
45. Subscapular nerve
46. Right transverse scapular artery and vein
47. Thyroid artery and vein

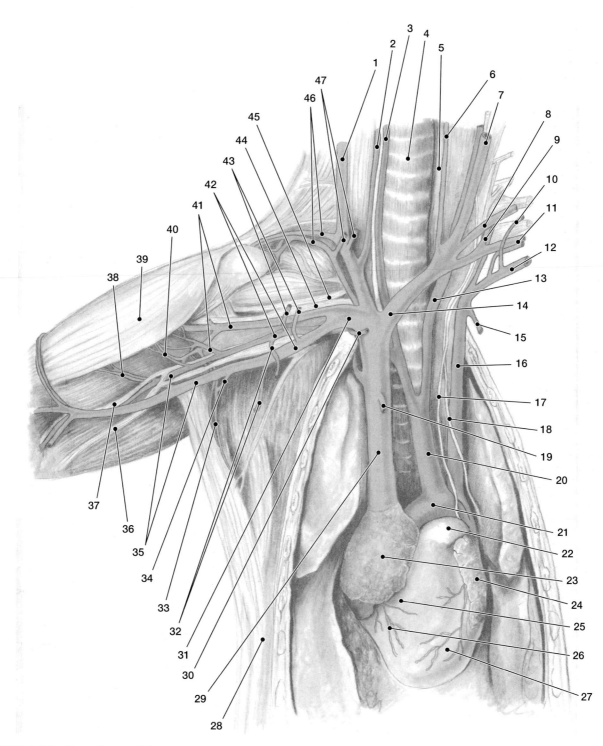

FIGURE 8.13 Ventral view of thoracic components of the circulatory system in a cat.

Figure 8.14

1. Transverse scapular artery
2. Thyroid artery
3. Vagus nerve
4. Right common carotid artery
5. Right thyrocervical artery
6. Left vertebral artery
7. Left thyrocervical artery
8. Left costocervical artery
9. Left axillary artery
10. Trachea
11. Internal mammary arteries
12. Right subclavian artery
13. Left common carotid artery
14. Brachiocephalic artery
15. Left subclavian artery
16. Cut base of the precava
17. Systemic arch
18. Pulmonary trunk
19. Left atrium
20. Right atrium
21. Right ventricle
22. Coronary arteries and veins
23. Left ventricle
24. Latissimus dorsi
25. Phrenic nerve
26. Thoracodorsal artery
27. Ulnar nerve
28. Median nerve
29. Brachial artery
30. Deep brachial artery
31. Thoracodorsal artery
32. Biceps brachii
33. Anterior humeral circumflex artery
34. Subscapular artery
35. Long thoracic artery
36. Ventral thoracic artery
37. Axillary artery

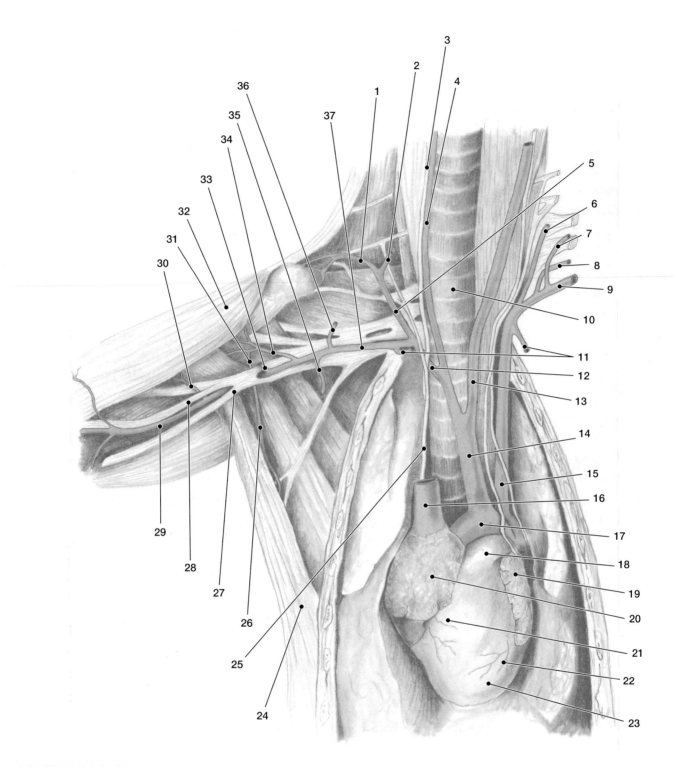

FIGURE 8.14 Ventral view of aortic arch distribution, precava and its tributaries removed, in a cat.

Figure 8.16(a1)

1. Lingual artery
2. External maxillary artery
3. Ascending pharyngeal artery
4. Laryngeal artery
5. Superior thyroid artery
6. Common carotid arteries
7. Vertebral artery
8. Costocervical artery
9. Thyrocervical artery
10. Subscapular artery
11. Thoracodorsal artery
12. Long thoracic artery
13. Ventral thoracic artery
14. Internal mammary artery
15. Subclavian artery
16. Brachiocephalic artery
17. Intercostal arteries
18. Dorsal aorta
19. Esophageal arteries
20. Internal mammary artery
21. Subclavian artery
22. Axillary artery
23. Long thoracic artery
24. Thoracodorsal artery
25. Deep brachial artery
26. Brachial artery
27. Anterior humeral circumflex artery
28. Subscapular artery
29. Ventral thoracic artery
30. Thyrocervical artery
31. Costocervical artery
32. Thyroid artery
33. Transverse scapular artery
34. Muscular artery
35. Occipital artery
36. Posterior auricular artery
37. Superior temporal artery
38. Maxillary artery
39. External carotid artery

Figure 8.16(a2)

1. Dorsal aorta
2. Left gastric artery
3. Celiac artery
4. Splenic artery
5. Anterior splenic artery
6. Superior mesenteric artery
7. Adrenolumbar artery
8. Renal artery
9. Ileocolic artery
10. Internal spermatic or ovarian artery
11. Lumbar arteries
12. Iliolumbar artery
13. External iliac artery
14. Umbilical artery
15. Internal iliac artery
16. Superior gluteal artery
17. Femoral artery
18. Inferior gluteal artery
19. Middle hemorrhoidal artery
20. Caudal artery
21. Deep femoral artery
22. Inferior epigastric artery
23. Inferior mesenteric artery
24. Superior mesenteric artery
25. Intestinal arteries
26. Ileocolic artery
27. Superior mesenteric artery
28. Middle colic artery
29. Posterior pancreaticoduodenal artery
30. Posterior splenic artery
31. Anterior pancreaticoduodenal artery
32. Right gastroepiploic artery
33. Pyloric artery
34. Gastroduodenal artery
35. Cystic artery
36. Hepatic artery

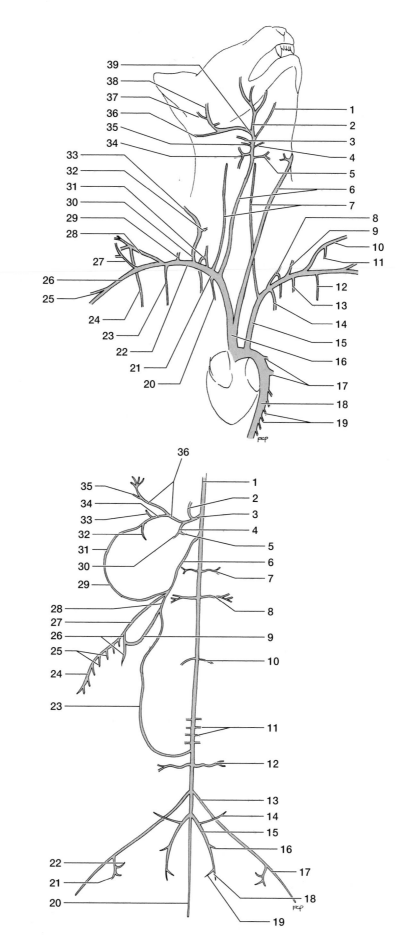

(a1)

(a2)

FIGURE 8.16 Artery (a) and vein (b) maps in a cat, ventral view.

Figure 8.16(b1)

1. Transverse jugular vein
2. Postcava
3. Internal mammary vein
4. Precava
5. Brachiocephalic vein
6. Costocervical vein
7. Long thoracic vein
8. Thoracodorsal vein
9. Thoracodorsal shunt
10. Deep brachial vein
11. Brachial vein
12. Posterior humeral circumflex vein
13. Subscapular vein
14. Axillary vein
15. Ventral thoracic vein
16. Subclavian vein
17. Vertebral vein
18. Thyroid vein
19. Transverse scapular vein
20. Internal jugular vein
21. External jugular vein
22. Posterior facial vein
23. Anterior facial vein

Figure 8.16(b2)

1. Superior phrenic vein
2. Hepatic vein
3. Inferior phrenic vein
4. Left adrenolumbar vein
5. Renal vein
6. Left internal spermatic or ovarian vein
7. Lumbar veins
8. Iliolumbar vein
9. Middle hemorrhoidal vein
10. Inferior gluteal vein
11. Deep femoral vein
12. Femoral vein
13. Internal iliac vein
14. External iliac vein
15. Caudal vein
16. Common iliac vein
17. Right internal spermatic or ovarian vein
18. Right adrenolumbar vein
19. Right phrenic vein
20. Postcava

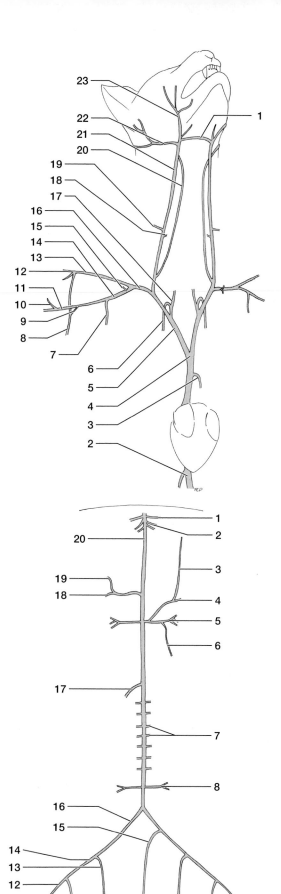

(b1)

(b2)

FIGURE 8.16 *Continued.*

Figure 8.17

1. Hepatic veins
2. Cut end of esophagus
3. Celiac artery
4. Dorsal aorta
5. Superior mesenteric artery
6. Inferior phrenic artery and vein
7. Adrenolumbar artery and vein
8. Renal artery
9. Renal vein
10. Kidney
11. Ovarian vein
12. Ovary
13. Ovarian artery
14. Iliolumbar artery and vein
15. Ureter
16. Urinary bladder
17. Urethra
18. Body of the uterus
19. Uterine horn
20. Cut end of descending colon
21. Inferior mesenteric artery
22. Adrenal gland
23. Postcava

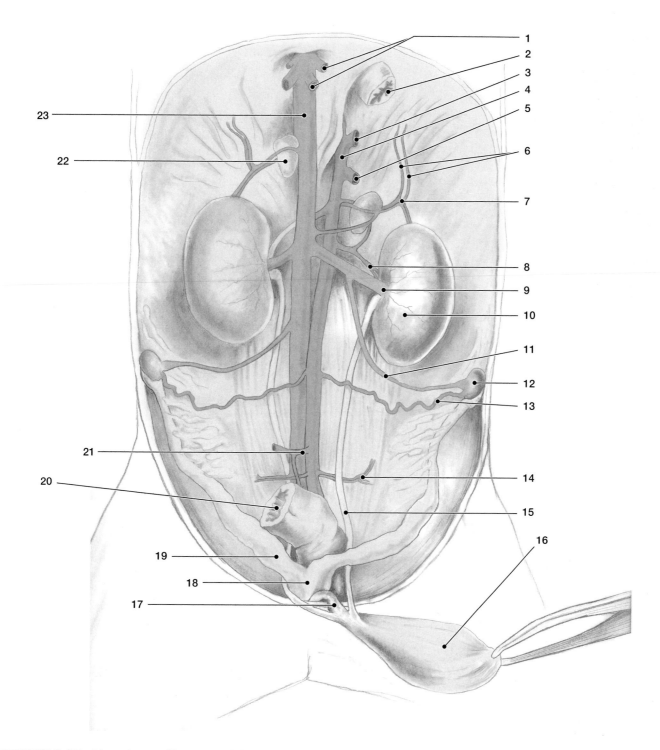

FIGURE 8.17 Ventral view of branches off the dorsal aorta and tributaries to the postcava in the trunk of a female cat.

Figure 8.18

1. Umbilical artery
2. Internal iliac artery and vein
3. External iliac vein
4. Superior gluteal artery
5. Femoral artery and vein
6. Deep femoral artery and vein
7. Inferior epigastric artery
8. Caudal artery and vein
9. Common iliac vein
10. External iliac artery
11. Iliolumbar artery and vein
12. Postcava
13. Dorsal aorta
14. Inferior mesenteric artery

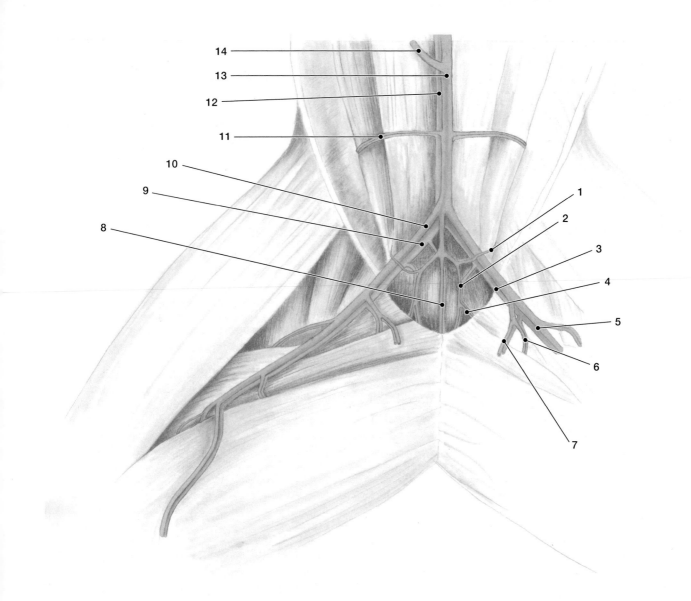

FIGURE 8.18 Blood vessels of the medial hindlimb of a cat, ventral view.

Figure 9.3

1. Testis
2. Mesorchium
3. Ductus deferens
4. Seminal vesicle
5. Sperm sac
6. Cloaca
7. Urogenital papilla
8. Urodeum
9. Coprodeum
10. Colon
11. Accessory urinary duct
12. Kidney
13. Leydig's gland
14. Epididymis

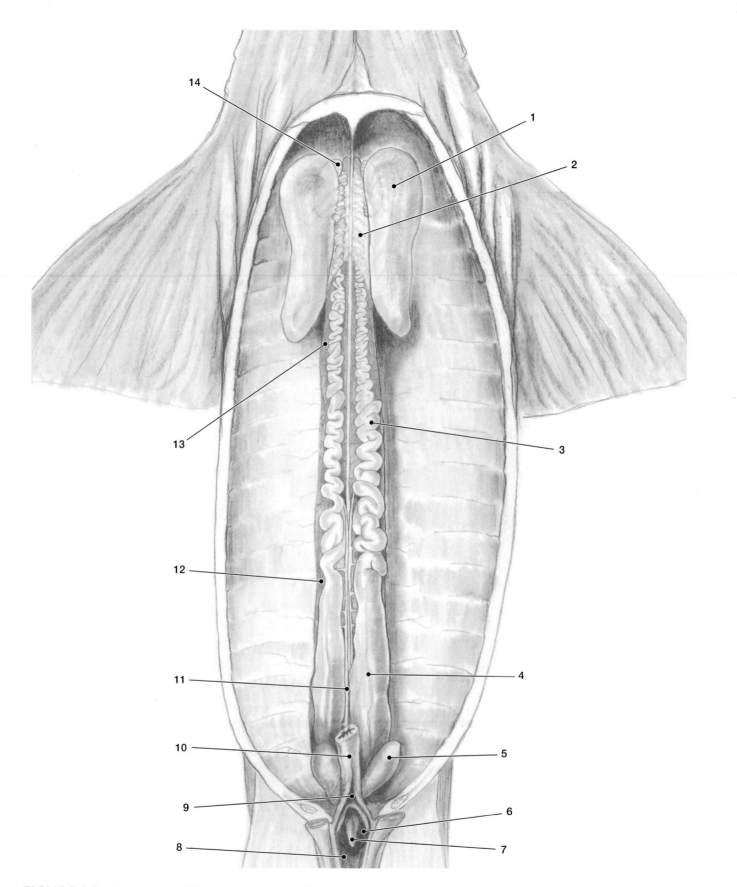

FIGURE 9.3 Ventral view of the urogenital system of a male shark, with the digestive tract removed.

Figure 9.4

1. Ostium
2. Cut end of the esophagus
3. Mesovarium
4. Ovary
5. Oviduct
6. Mesotubarium
7. Uterus
8. Cloaca
9. Urinary papilla
10. Urodeum
11. Coprodeum
12. Cut end of the colon
13. Mesonephric duct
14. Kidney (dorsal to the mesotubarium)
15. Shell gland (dorsal to the ovary)

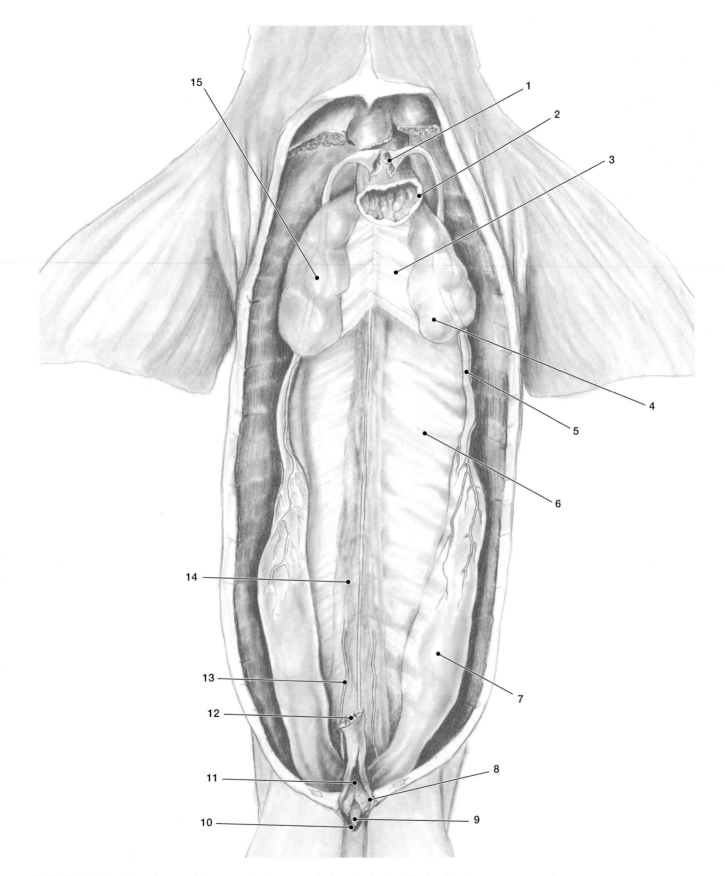

FIGURE 9.4 Ventral view of the urogenital system of a female shark, with the digestive tract removed.

Figure 9.5

1. Müllerian duct (homologue of oviduct)
2. Efferent ductules
3. Mesorchium
4. Testis
5. Urinary collecting tubules
6. Kidney
7. Cut end of intestine
8. Cloaca
9. Cloacal glands
10. Urinary bladder
11. Renal arteries
12. Dorsal aorta
13. Ductus deferens (= mesonephric duct)
14. Epididymis

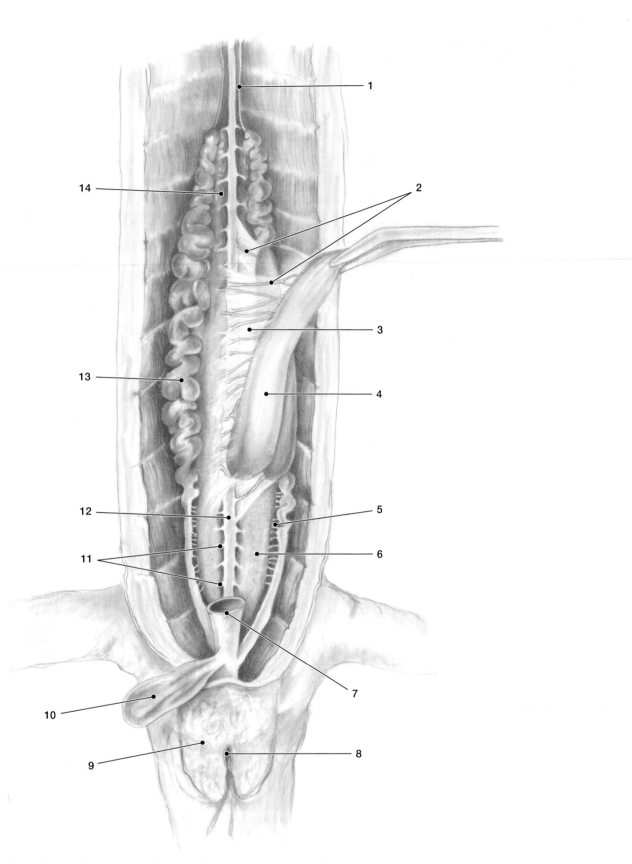

FIGURE 9.5   Ventral view of the urogenital system of a male *Necturus*, with the digestive tract removed.

Figure 9.6

1. Ostium
2. Ovarian arteries and veins
3. Ova
4. Ovary
5. Mesovarium
6. Dorsal aorta
7. Mesonephric duct
8. Cut end of intestine
9. Cloaca
10. Urinary bladder
11. Renal artery
12. Urinary collecting tubules
13. Kidney
14. Oviduct

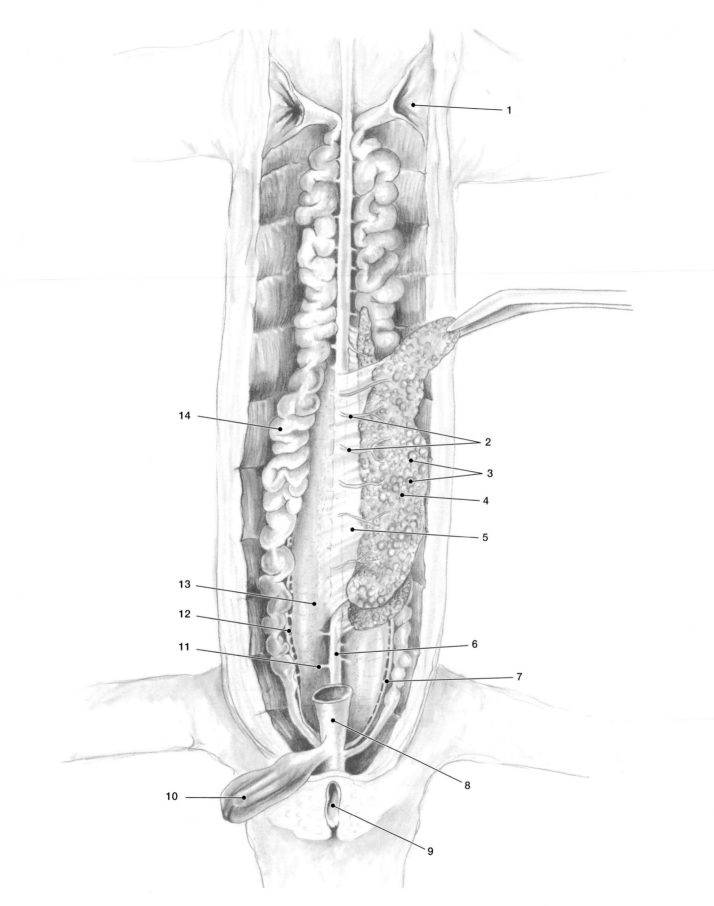

FIGURE 9.6 Ventral view of the urogenital system of a female *Necturus*, with the digestive tract removed.

Figure 9.7

1. Cortex
2. Renal sinus
3. Medulla
4. Ductus deferens
5. Prostate gland
6. Urogenital canal
7. Bulbourethral gland
8. Penis
9. Head of epididymis
10. Body of epididymis
11. Testis
12. Tail of epididymis
13. Cut edges of pubic and ischial symphyses
14. Urethra
15. Fundus of the urinary bladder
16. Vertex of the urinary bladder
17. Cut end of descending colon
18. Ureter
19. Hilus of the kidney

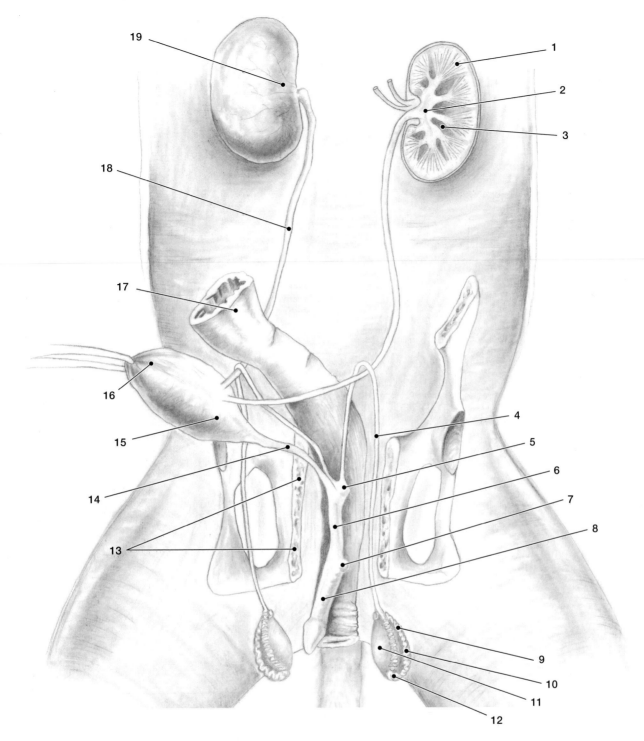

FIGURE 9.7 Ventral view of the urogenital system of a male cat, with the digestive tract removed.

Figure 9.9

1. Ureter
2. Mesosalpinx
3. Ovary
4. Infundibulum
5. Mesometrium
6. Uterine horn
7. Body of the uterus
8. Urethra
9. Vagina
10. Urogenital sinus
11. Anus
12. Urogenital aperture
13. Cut edges of pubic and ischial symphyses
14. Fundus of the urinary bladder
15. Vertex of the urinary bladder
16. Cut end of descending colon
17. Oviduct
18. Hilus of the kidney

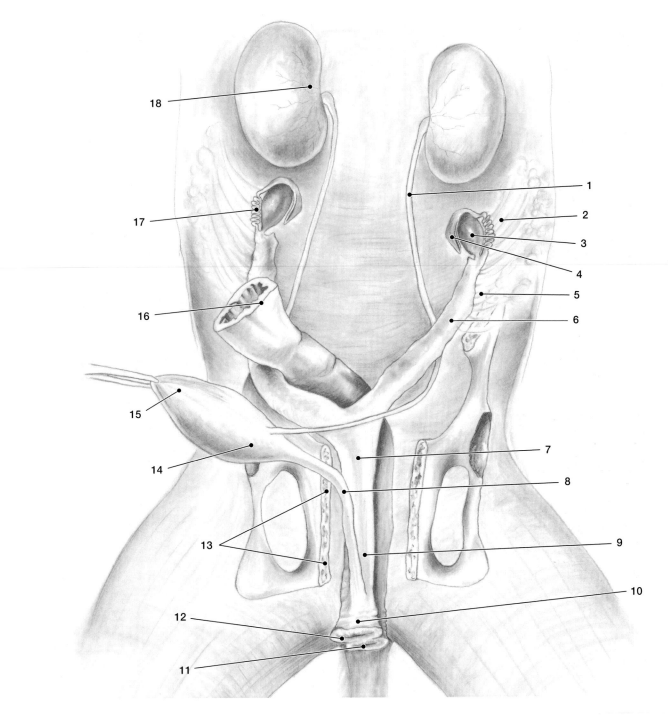

FIGURE 9.9 Ventral view of the urogenital system of a female cat, with the digestive tract removed.

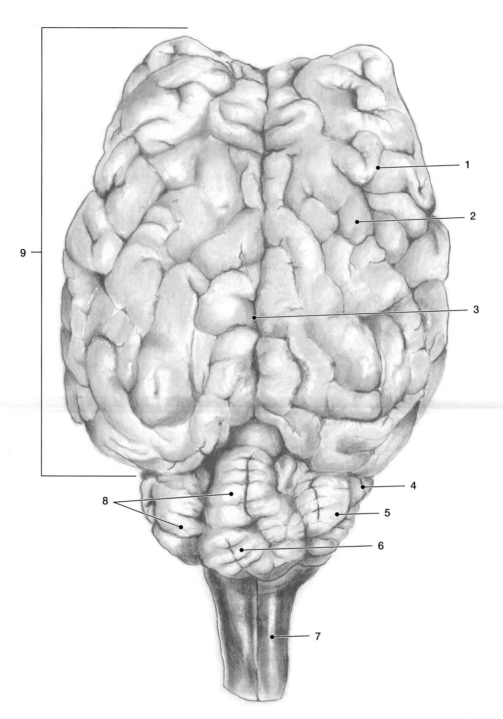

FIGURE 10.10 Dorsal view of the sheep brain.

Figure 10.10

1. Sulcus
2. Gyrus
3. Median sulcus
4. Flocculus of the cerebellum
5. Hemisphere of the cerebellum
6. Vermis of the cerebellum
7. Spinal cord
8. Folia
9. Cerebral hemisphere

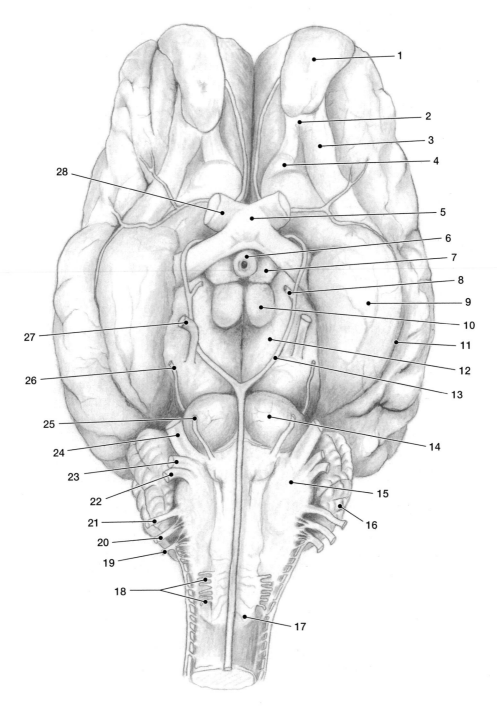

FIGURE 10.11 Ventral view of the sheep brain.

Figure 10.11

1. Olfactory bulb
2. Olfactory tract
3. Lateral olfactory stria
4. Medial olfactory stria
5. Optic chiasm
6. Infundibulum
7. Tuber cinereum
8. Internal carotid artery
9. Pyriform lobe
10. Mammillary body
11. Rhinal fissure
12. Cerebral peduncle
13. Circle of Willis
14. Pons

15. Medulla oblongata
16. Cerebellum
17. Spinal cord
18. Hypoglossal nerve (XII)
19. Spinal accessory nerve (XI)
20. Vagus nerve (X)
21. Glossopharyngeal nerve (IX)
22. Auditory nerve (VIII)
23. Facial nerve (VII)
24. Trigeminal nerve (V)
25. Abducens nerve (VI)
26. Trochlear nerve (IV)
27. Oculomotor nerve (III)
28. Optic nerve (II)

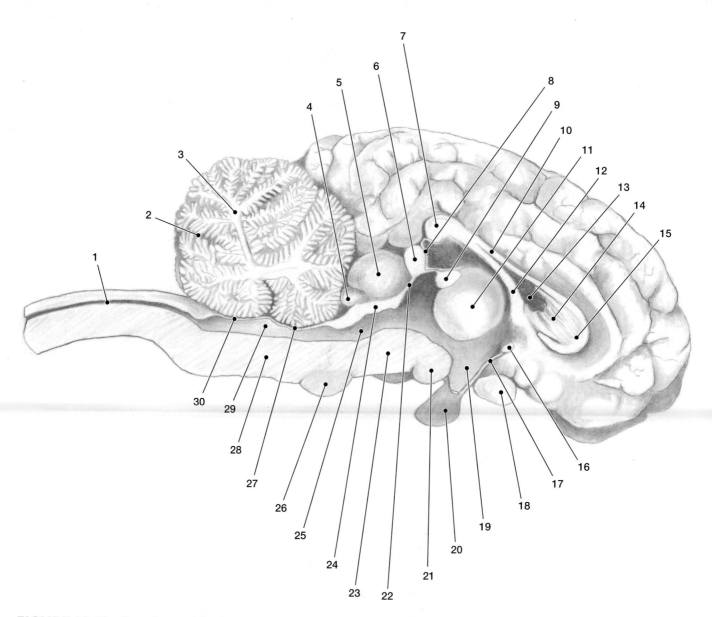

FIGURE 10.12 Sagittal view of the sheep brain.

Figure 10.12

1. Central canal of the spinal cord
2. Cerebellar cortex
3. Arbor vitae
4. Inferior colliculus
5. Superior colliculus
6. Pineal body
7. Splenium
8. Cranial tela choroidea
9. Habenular trigone
10. Corpus callosum
11. Massa intermedia
12. Fornix
13. Lateral ventricle
14. Septum pellucidum
15. Genu

16. Anterior commissure
17. Lamina terminalis
18. Optic chiasm
19. Third ventricle
20. Pituitary gland
21. Mammillary body
22. Posterior commissure
23. Tegmentum (cerebral peduncles)
24. Lamina quadrigemina
25. Aqueduct
26. Pons
27. Medullary vellum
28. Medulla oblongata
29. Fourth ventricle
30. Caudal tela choroidea

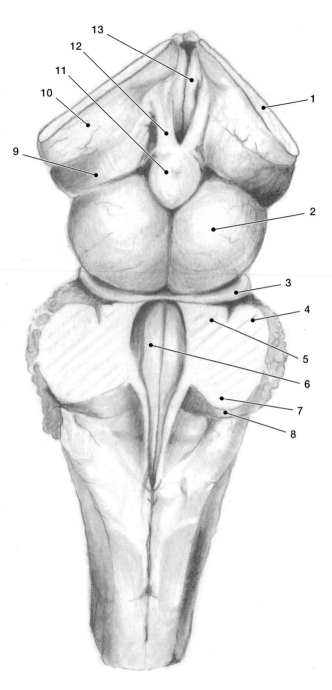

FIGURE 10.13 Dorsal view of the brain stem of a sheep brain, cerebellum removed at cerebellar peduncles.

Figure 10.13

1. Region of cut between the thalamus and the cerebrum
2. Superior colliculus
3. Inferior colliculus
4. Middle cerebellar peduncle
5. Anterior cerebellar peduncle
6. Fourth ventricle
7. Posterior cerebellar peduncle
8. Acoustic stria
9. Medial geniculate body
10. Pulvinar nucleus
11. Pineal body
12. Habenular trigone
13. Massa intermedia